Biotechnology &
Genetic Engineering Reviews

Volume 27

BIOTECHNOLOGY AND GENETIC ENGINEERING REVIEWS

Biotechnology & Genetic Engineering Reviews

Volume 27

Editor:

Stephen E. Harding

Professor of Applied Biochemistry, University of Nottingham

Co-Editor:

Gary G. Adams

Faculty of Medicine and Health Sciences, University of Nottingham

Nottingham University Press
Manor Farm, Main Street, Thrumpton
Nottingham, NG11 0AX, United Kingdom

NOTTINGHAM

First published 2010

British Library Cataloguing in Publication Data
Biotechnology & genetic engineering reviews - Vol. 27
I. Biotechnology – Periodicals

ISBN 978-1-907284-55-7
ISSN 0264-8725

Disclaimer

Every reasonable effort has been made to ensure that the material in this book is true, correct, complete and appropriate at the time of writing. Nevertheless the publishers, the editors and the authors do not accept responsibility for any omission or error, or for any injury, damage, loss or financial consequences arising from the use of the book. Views expressed in the articles are those of the authors and not of the Editors or Publisher.

Cover image: adapted from V Eijsink (see page 343)

Typeset by Nottingham University Press, Nottingham
Printed and bound by Berforts Group, Stevenage Herts SG1 2BH

CONTENTS

Contributors

ADAMS, G.G. *University of Nottingham, Faculty of Medicine and Health Sciences, Insulin Diabetes Experimental Research Group, Clifton Boulevard, Nottingham NG7 2RD UK and University of Nottingham, National Centre for Macromolecular Hydrodynamics, School of Biosciences, University of Nottingham, Sutton Bonington, LE12 5RD, U.K.*

AMPAIRE, A. *Department of Animal Science and Center for Integrated Animal Genomics, Iowa State University, Ames, IA 50011, USA*

BAUMANN, M.K. *Department for Materials Science, ETH Zurich, 8093 Zurich, Swizerland*

BETANCOR, L. *Madrid Institute for Advanced Studies, Campus Universitario de Cantoblanco, c/Einstein, 13 Pabellon C 1ºPlanta, E-28049, Madrid, Spain*

BOVENBERG, R.A.L. *Molecular Cell Biology, Groningen Biomolecular Sciences and Biotechnology Institute (GBB), Kluyver Centre for Genomics of Industrial Fermentation, University of Groningen, P.O. Box 14, Haren, 9750 AA, The Netherlands*

BRANDUARDI, P. *Dipartimento di Biotecnologie e Bioscienze, Università degli Studi di Milano-Bicocca, P.zza della Scienza 2, 20126 Milano, Italy*

BUTTERY, L. *University of Nottingham, School of Pharmacy, Faculty of Science, University Park, Nottingham, NG7 2RD, UK*

CARRILLO, N. *Instituto de Biología Molecular y Celular de Rosario (IBR-CONICET), Universidad Nacional de Rosario, Suipacha 531, S2002LRK Rosario, Argentina*

CECCARELLI, E.A. *Instituto de Biología Molecular y Celular de Rosario (IBR) – CONICET, Universidad Nacional de Rosario, Suipacha 531, (S2002LRK) Rosario, Argentina*

COCHRAN, F.V. *Department of Bioengineering, Cancer Center, Bio-X Program, Stanford University, Stanford CA, USA*

COCHRAN, J.R. *Department of Bioengineering, Cancer Center, Bio-X Program, Stanford University, Stanford CA, USA*

DUSANE, D.H. *Institute of Bioinformatics and Biotechnology, University of Pune, Pune-411 007 India*

EIJSINK, V.G.H. *Department of Chemistry, Biotechnology and Food Science, Norwegian University of Life Sciences, P. O. Box 5003, N-1432 Ås, Norway*

GIDIJALA, L. *Molecular Cell Biology, Groningen Biomolecular Sciences and Biotechnology Institute (GBB), Kluyver Centre for Genomics of Industrial Fermentation, University of Groningen, P.O. Box 14, Haren, 9750 AA, The Netherlands*

GUEBITZ, G.M. *Graz University of Technology, Institute of Environmental Biotechnology, Petersgasse 12/1, A-8010, Graz, Austria*

HAJIREZAEI, M-R. *Leibniz-Institut für Pflanzengenetik und Kulturpflanzenforschung (Leibniz-IPK), Corrensstr. 3, D-06466 Gatersleben, Germany*

HARDING, S.E. *NCMH Laboratory, School of Biosciences, University of Nottingham, Sutton Bonington LE12 5RD, UK*

HOELL, A. *Stord/Haugesund University College, Bjoernsonsgate 45, N-5528 Haugesund, Norway*

JI, H.P. *Stanford Genome Technology Center and Division of Oncology, Department of Medicine, Stanford University School of Medicine, CCSR 3215, 269 Campus Drive, Stanford, California 94305, USA*

KAUFMANN, S. *Department for Materials Science, ETH Zurich, 8093 Zurich, Swizerland*

KIEL, J.A.K.W. *Molecular Cell Biology, Groningen Biomolecular Sciences and Biotechnology Institute (GBB), Kluyver Centre for Genomics of Industrial Fermentation, University of Groningen, P.O. Box 14, Haren, 9750 AA, The Netherlands*

KÖK, M. *Department of Food Engineering, Abant Izzet Baysal University, 14280 Bolu, Turkey*

KUDANGA, T. *Graz University of Technology, Institute of Environmental Biotechnology, Petersgasse 12/1, A-8010, Graz, Austria*

KUMAR, K. *Department for Materials Science, ETH Zurich, 8093 Zurich, Swizerland*

LUCKARIFT, H.R. *Air Force Research Laboratory, AFRL/RXQL, Microbiology and Applied Biochemistry, Tyndall Air Force Base, Florida 32403 and Universal Technology Corporation, 1270 N. Fairfield Road, Dayton, Ohio, 45432, USA*

MARTINDALE, W. *Centre for Food Innovation, Sheffield Hallam University, Sheffield, S1 1WB, UK*

MATTANOVICH, D. *School of Bioengineering, University of Applied Sciences FH Campus Wien, Muthgasse 18, 1190 Vienna, Austria and Department of Biotechnology, BOKU University of Natural Resources and Applied Life Sciences, Muthgasse 18, 1190 Vienna, Austria*

McLEAN, R.J.C. *Department of Biology, Texas State University-San Marcos, 601 University Drive, San Marcos, TX 78666, USA*

MORRIS, G. *University of Nottingham, National Centre for Macromolecular Hydrodynamics, School of Biosciences, University of Nottingham, Sutton Bonington, LE12 5RD, U.K.*

MYLLYKANGAS, S. *Stanford Genome Technology Center and Division of Oncology, Department of Medicine, Stanford University School of Medicine, CCSR 3215, 269 Campus Drive, Stanford, California 94305, USA*

NYANHONGO, G.S. *Graz University of Technology, Institute of Environmental Biotechnology, Petersgasse 12/1, A-8010, Graz, Austria*

ONTERU, S.K. *Department of Animal Science and Center for Integrated Animal Genomics, Iowa State University, Ames, IA 50011, USA*

PORRO, D. *Dipartimento di Biotecnologie e Bioscienze, Università degli Studi di Milano-Bicocca, P.zza della Scienza 2, 20126 Milano, Italy*

PRASETYO, E.N. *Graz University of Technology, Institute of Environmental Biotechnology, Petersgasse 12/1, A-8010, Graz, Austria*

RAHMAN, P.K.S.M. *Chemical and Bioprocess Engineering Group, School of Science and Engineering, Teesside University, Middlesbrough-TS13BA, UK*

REIMHULT, E. *Department for Materials Science, ETH Zurich, 8093 Zurich, Swizerland*

ROTHSCHILD, M.F. *Department of Animal Science and Center for Integrated Animal Genomics, Iowa State University, Ames, IA 50011, USA*

ROVERI, O.A. *Área Biofísica, Facultad de Ciencias Bioquímicas y Farmacéuticas, Universidad Nacional de Rosario, Suipacha 531, (S2002LRK) Rosario, Argentina*

SAUER, M. *School of Bioengineering, University of Applied Sciences FH Campus Wien, Muthgasse 18, 1190 Vienna, Austria and Department of Biotechnology, BOKU University of Natural Resources and Applied Life Sciences, Muthgasse 18, 1190 Vienna, Austria*

SPYCHER, P.R. *Department for Materials Science, ETH Zurich, 8093 Zurich, Swizerland*

STOLNIK, S. *University of Nottingham, School of Pharmacy, Faculty of Science, University Park, Nottingham, NG7 2RD, UK*

VAAJE-KOLSTADA, G. *Department of Chemistry, Biotechnology and Food Science, Norwegian University of Life Sciences, P. O. Box 5003, N-1432 Ås, Norway*

VAN DEN BERG, M.A *DSM Biotechnology Centre (699-0310), P.O. Box 425, 2600 AK, Delft, The Netherlands*

VAN DER KLEI, I.J. *Molecular Cell Biology, Groningen Biomolecular Sciences and Biotechnology Institute (GBB), Kluyver Centre for Genomics of Industrial Fermentation, University of Groningen, P.O. Box 14, Haren, 9750 AA, The Netherlands*

VENUGOPALAN, V.P. *Biofouling and Biofilm Processes Section, Water and Steam Chemistry Division, BARC Facilities, Kalpakkam-603 102 India*

WANG, N. *University of Nottingham, Faculty of Medicine and Health Sciences, Insulin Diabetes Experimental Research Group, Clifton Boulevard, Nottingham NG7 2RD UK*

WEBER, M.M. *Department of Biology, Texas State University-San Marcos, 601 University Drive, San Marcos, TX 78666, USA*

ZINJARDE, S.S. *Institute of Bioinformatics and Biotechnology, University of Pune, Pune-411 007 India*

ZURBRIGGEN, M.D. *Instituto de Biología Molecular y Celular de Rosario (IBR-CONICET), Universidad Nacional de Rosario, Suipacha 531, S2002LRK Rosario, Argentina*

Biotechnology and Genetic Engineering Reviews - Vol. 27, 1-32 (2010)

Biosynthesis of active pharmaceuticals: β-lactam biosynthesis in filamentous fungi

LOKNATH GIDIJALA[1], JAN A.K.W. KIEL[1], ROEL A.L. BOVENBERG[1,2], IDA J. VAN DER KLEI[1], MARCO A VAN DEN BERG [2*]

[1]*Molecular Cell Biology, Groningen Biomolecular Sciences and Biotechnology Institute (GBB), Kluyver Centre for Genomics of Industrial Fermentation, University of Groningen, P.O. Box 14, Haren, 9750 AA, The Netherlands;* [2]*DSM Biotechnology Centre (699-0310), P.O. Box 425, 2600 AK, Delft, The Netherlands*

Abstract

β-lactam antibiotics (e.g. penicillins, cephalosporins) are of major clinical importance and contribute to over 40% of the total antibiotic market. These compounds are produced as secondary metabolites by certain actinomycetes and filamentous fungi (e.g. *Penicillium, Aspergillus* and *Acremonium* species). The industrial producer of penicillin is the fungus *Penicillium chrysogenum*. The enzymes of the penicillin biosynthetic pathway are well characterized and most of them are encoded by genes that are organized in a cluster in the genome. Remarkably, the penicillin biosynthetic pathway is compartmentalized: the initial steps of penicillin biosynthesis are catalyzed by cytosolic enzymes, whereas the two final steps involve peroxisomal enzymes. Here, we describe the biochemical properties of the enzymes of β-lactam biosynthesis in *P. chrysogenum* and the role of peroxisomes in this process. An overview is given

*To whom correspondence may be addressed (Marco.Berg-van-den@DSM.COM)

Abbreviations: A, adenylation; AAA, aminoadipate; 7-ACA, 7-aminocephalosporanic acid; 7-ACCCA, 7-amino-3-carbomoyloxymethyl-3-cephem-4-carboozylic acid; Ad, adipate; 7-ADCA, 7-aminodeacetoxycephalosporanic acid; ACV, L-δ-(α-aminoadipoyl)-L-cysteinyl-D-valine; ACVS, L-δ-(α-aminoadipoyl)-L-cysteinyl-D-valine synthetase; API, active pharmaceutical ingredient; TP, adenosine triphosphate, 6-APA, 6-aminopenicillinic acid; C, condensation; CoA, coenzymeA E, epimerisation, HR, homologous recombination, IAT, isopenicillin N-acyltransferase; IPN, isopenicillinN; IPNS, isopenicillinN synthase; NADPH, nicotinamide adenine dinucleotide phosphate-oxidase; NHEJ, non-homologous end-joining; NRRL, Northern Regional Research Laboratory; NTN, N-terminal nucleophile; PAA, phenyl acetic acid; PenG, penicillinG; PenV, penicillinV; PCL, PAA CoA ligase, PCP, peptidyl carrier protein; PEG, polyethyleneglycol; PEX, peroxin; PKS, polyketide synthase; POA, phenoxy acetic acid; PPP, pentose phosphate pathway; PPTase, phosphopantetheinyl transferase; PTS1, peroxisomal targeting sequence 1; PTS2, peroxisomal targeting sequence 2; T, thiolation; TE, thioesterase;

on strain improvement programs via classical mutagenesis and, more recently, genetic engineering, leading to more productive strains. Also, the potential of using heterologous hosts for the development of novel ß-lactam antibiotics and non-ribosomal peptide synthetase-based peptides is discussed.

Introduction

Microbes produce a large number of secondary metabolites that are of great therapeutic value. Examples are immunosuppressors (e.g cyclosporins), anti-cancer drugs (e.g. doxorubicin) and antibiotics (such as ß-lactam antibiotics and erythromycin) (Clardy and Walsh, 2004; Walsh *et al.*, 2008) (*Figure 1*). Antibiotics are of major clinical importance to combat bacterial infections in man and animals (cattle). Since the late 1930's bacterial infections were treated with synthetic chemical substances derived from sulfanilamide or para-aminobenzenesulfonamide, the so called sulfa drugs. These compounds have been replaced by natural microbial products (*Table 1*), which has revolutionized medicine. The majority of these antibiotics exert their action either by interfering with bacterial cell wall biosynthesis or by inhibiting the protein synthesis machinery (Walsh *et al.*, 2003).

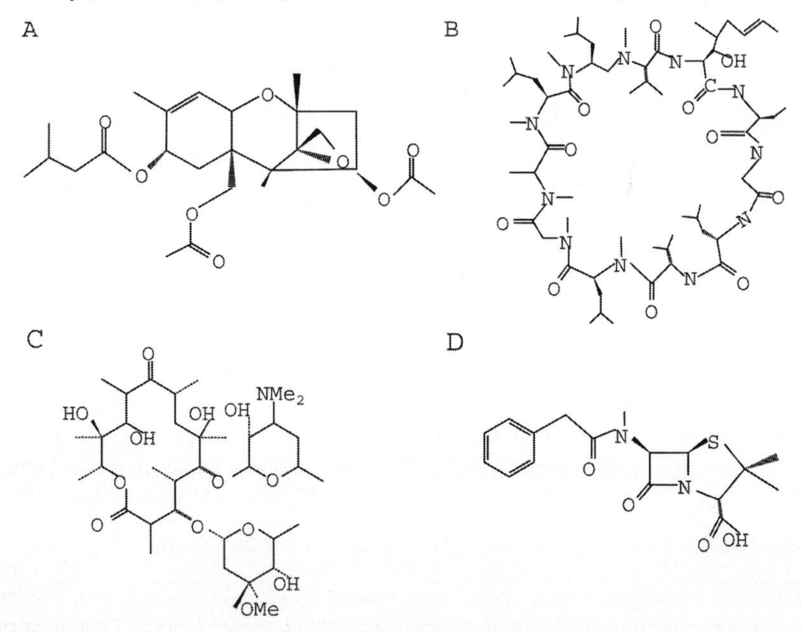

Figure 1. Microbial natural product diversity. Structures of natural products **(A)** T2 mycotoxin, which belongs to the trichothecene mycotoxins produced by filamentous fungi of the genera *Fusarium* and *Trichoderma*; **(B)** Cyclosporin, a cyclic non-ribosomal peptide of 11 amino acids produced by the filamentous fungus *Tolypocladium niveum*; **(C)** Erythromycin, produced by a polyketide synthase, produced in *Streptomyces erythraeus*; **(D)** Penicillin-G, a cyclic non-ribosomal peptide of 3 amino acids produced by the filamentous fungus *Penicillium chrysogenum*.

An important class of antibiotics are the ß-lactam antibiotics (e.g. penems, penicillins and cephaloporins). The antibiotic properties of ß-lactams are governed by the presence of the four-membered ß-lactam ring (*Figure 2*). These compounds can be divided

Table 1. Most commonly used antibiotics produced as secondary metabolites

Antibiotic*	Examples	Source	Ref
Aminoglycosides	streptomycin	*S. griseus,*	(Waksman *et al.*, 1946)
	neomycin	*S. fradiae*	(Dulmage, 1953)
	gentamicin	*Micromonospora sp.*	(Weinstein *et al.*, 1963)
Penicillins	penicillinG	*P. chrysogenum*	(Fleming, 1929)
Cephalosporins	cephalosporinC	*C. acremonium*	(Burton *et al.*, 1951)
Tetracyclines	chlortetracycline	*S. aureofaciens*	(Backus *et al.*, 1954)
	oxytetracycline	*S. rimosus*	(Mindlin *et al.*, 1961)
Macrolides	erythromycin	*S. erythraeus*	(Hung *et al.*, 1965)

* while each class is composed of multiple drugs, each drug is unique in some way e.g. mode of action. Other commonly used antibiotics are the fluoroquinolones, synthetic antibiotics, derivatives of nalidixic acid.

into four major categories based on the second ring structure being either a five- or a six-membered ring fused to the ß-lactam ring. Penicillins and cephalosporins have a sulphur atom in the second ring, which is replaced by an oxygen atom in clavams and by a carbon atom in carbapenams (*Figure 2*). Monobactams have only the ß-lactam ring (Brakhage, 1998).

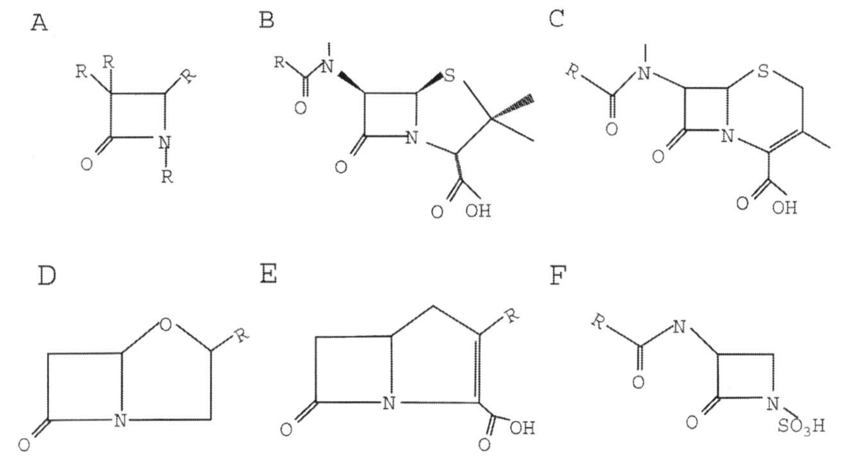

Figure 2. Major classes of ß-lactam antibiotics. (A) represents the characteristic ß-lactam ring, the bio-active part of the various ß-lactams. (B) & (C) show members of the penam class of antibiotics with a sulphur atom in the second ring. Producers of such compounds are found in both filamentous fungi and in certain ascomycete bacteria. (D) represents the clavam class of antibiotics, characterized by the presence of an oxygen atom in the second five-membered ring. These types of ß-lactams are produced by the filamentous bacterium *Streptomyces clavuligerus*. (E) shows the carbapenam class of antibiotics, which are characterized by a second five-membered ring, as produced by *S. clavuligerus* and *Streptomyces olivaceus*. **(F)** represents the monobactam class of antibiotics with only the ß-lactam ring, as produced by *Pseudomonas acidophila* and *Agrobacterium radiobacter*.

Well known producers of ß-lactam antibiotics are the prokaryotes *Streptomyces clavuligerus* (Romero *et al.*, 1997), *Amycolatopsis sp.* and *Nocardia lactamdurans* (Kern *et al.*, 1980) and the eukaryotic filamentous fungi *Aspergillus nidulans*, *Penicillium chrysogenum* and *Acremonium chrysogenum* (Demain and Elander, 1999).

This contribution is focused on the complex biosynthetic pathways and metabolic engineering of secondary metabolites (e.g. penicillins and cephalosporins) in the penicillin producer *P. chrysogenum.*

Principles of penicillin biosynthesis

Commercial ß-lactams used as antibiotics are derivatives of penicillins and cephalosporins (Demain and Elander, 1999). Industrially, penicillin is produced by the filamentous fungus *P. chrysogenum* (Elander, 2003). Genetic analysis of mutants, that produced significantly reduced penicillin levels relative to those of a wild type strain, has allowed identification of the enzymes involved in penicillin production (Queener *et al.*, 1978). The biosynthesis of all naturally occurring penicillins and cephalosporins starts with the condensation of three amino acids: α-aminoadipic acid (AAA), L-cysteine, and L-valine into the tripeptide L-δ-(α-aminoadipoyl)-L-cysteinyl-D-valine (ACV). This tripeptide is synthesized by a single multi-functional enzyme, the non-ribosomal peptide synthetase (NRPS) ACV synthetase (ACVS). In the second stage, ACV is cyclized by the enzyme isopenicillin N (IPN) synthase (IPNS) to form IPN, which contains the characteristic β-lactam ring as well as the five membered thiazolidine ring. Both ACVS and IPNS are cytosolic enzymes. In contrast, the two final steps of penicillin are located in a specialized organelle, the peroxisome (*Figure 3*; see also the section below on the role of peroxisomes during penicillin biosynthesis). Because of this compartmentalization, IPN has to be transported into peroxisomes for further processing. This involves the exchange of the AAA moiety for a novel side chain e.g. phenylacetic acid (PAA) to form penicillinG (PenG) or phenoxyacetic acid (POA) in case of penicillinV (PenV), a reaction catalysed by the enzyme isopenicillin N-acyltransferase (IAT, also known as AatA). In the organelle lumen, IPN is first hydrolyzed into AAA and 6-aminopenicillinic acid (6-APA). To enable the exchange, the side chain (PAA or POA) has to be activated by the enzyme phenylacetyl CoA ligase, PCL (Lamas-Maceiras *et al.*, 2006). As IAT, also PCL is a component of the peroxisome matrix. Currently, it is still unclear how IPN and the final penicillin product(s) are transported across the peroxisomal membrane (see *Figure 3*).

Sequence data of various filamentous fungi revealed the presence of penicillin biosynthetic gene clusters besides *P. chrysogenum* and *A. nidulans* also in *P. griseofulvum, A. oryzae*, *A. flavus*, and *Trichophyton equinum*, suggesting that these organisms can also synthesize β-lactam compounds (Laich *et al.*, 2002). Truncated clusters, which solely contain the genes encoding the first two enzymes of the biosynthesis route (see below), were identified in the marine fungus *Kallichroma tethys* and *Acremonium chrysogenum* (Aharonowitz *et al.*, 1992); Kim *et al.*, 2003). In the latter, these enzymes are part of the cephalosporin biosynthesis pathway.

Non-ribosomal peptide synthetases

EXPLOITING THE POTENTIAL OF NRPSS

Various complex natural peptides are produced by NRPSs, which can accept both proteogenic and non-proteogenic amino acids. Also, they can catalyze multiple amino

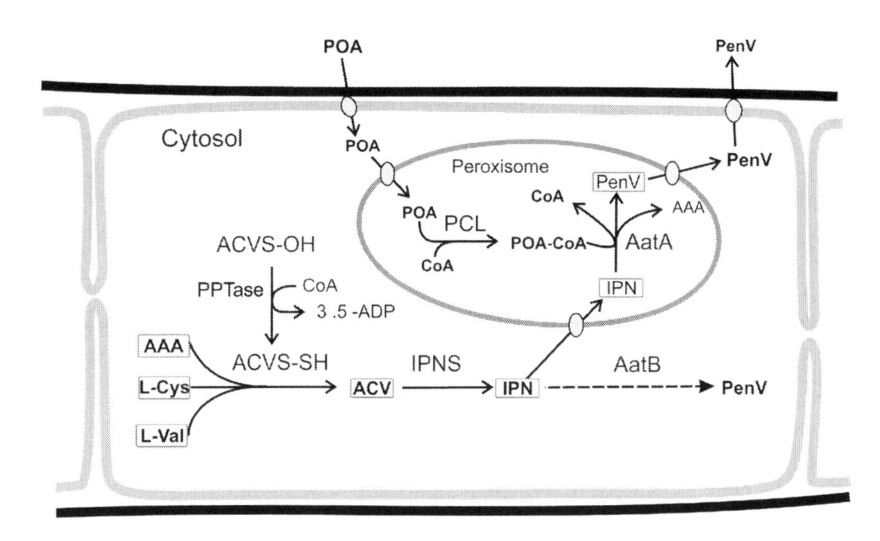

Figure 3. Schematic representation of the enzymes involved in PenV biosynthesis and there subcellular location in filamentous fungi. The penicillin biosynthesis pathway, as present in *P. chrysogenum* and *Aspergillus nidulans*, requires compartmentalization of enzymes: In the cytosol, the non-ribosomal peptide synthetase ACVS (ACVS-OH) is activated by a dedicated PPTase into the active form ACVS-SH, which can produce ACV from the three precursor molecules AAA, L-cysteine and L-valine. The enyzme IPNS subsequently converts ACV into the β-lactam IPN, which is transported into peroxisomes. In this organelle phenoxyacetic acid (POA) is activated by PCL into phenoxyacetyl CoA (POA-CoA), which is used by the NTN hydrolase AatA (IAT) to synthesize PenV. PenV is then exported from the organelle and ultimately secreted into the culture medium.

In addition to this, a secondary route exists in *A. nidulans* (represented by the broken arrow). In this route peroxisomes may not be important for PenV production, because a cytosolic paralog of AatA, designated AatB, may convert IPN into PenV. Currently, it is unknown whether the required POA-CoA is also synthesized in the cytosol.

acid modifications (epimerization, cyclization, etc). The increasing availability of genome sequences of various fungi and bacteria opens the way to identify yet unknown NRPS systems and modify the composition and arrangement of the modules to search for novel peptides of pharmaceutical value. This 'awakening' of sleeping genes has already revealed some new secondary metabolites (Brakhage *et al.*, 2008).

In NRPS proteins, each module exists as a monomer. The total number of modules determines the number of amino acids incorporated into the product. In addition, each module consists of separate domains that are involved in specialized enzymatic reactions during product assembly (Siewers *et al.*, 2009). The now generally accepted multi-domain thiol-template mechanism for NRPSs was initially proposed in 1971 by Lipmann *et al.* (1971). Based on the observation that radio-labeled amino acids were preferentially released from the precipitated synthetases by formic acid, these authors concluded that a thioester linkage must exist between the amino acid substrates and the NRPS. Such a reaction mechanism is analogous to that found for general fatty acid synthases. Much of the information concerning organization and mechanism of NRPS enzymes comes from studies of the enzymes involved in tyrocidine and gramicidin biosynthesis in *Bacillus brevis* (Okuda *et al.*, 1964).

P. CHRYSOGENUM ACVS

The synthesis of penicillins and cephalosporins is initiated by formation of the tri-peptide ACV, catalyzed by the enzyme ACVS, which belongs to a class of amino acid ligases (peptide synthetases). Biochemical characterization of ACVS proteins from different organisms demonstrated that these enzymes have variable molecular masses ranging from 360 to 425 kDa (Jensen *et al.*, 1990; Kallow *et al.*, 1998; van Liempt *et al.*, 1989; Zhang *et al.*, 1992). Despite this variation in molecular mass, a high degree of similarity exists between the individual domains of these orthologs (Kleinkauf *et al.*, 1996). All consist of three distinct modules in which each module contains three distinct concatenated homologous domains each performing a different enzymatic function (cf. *Figure 4*; for details see below).

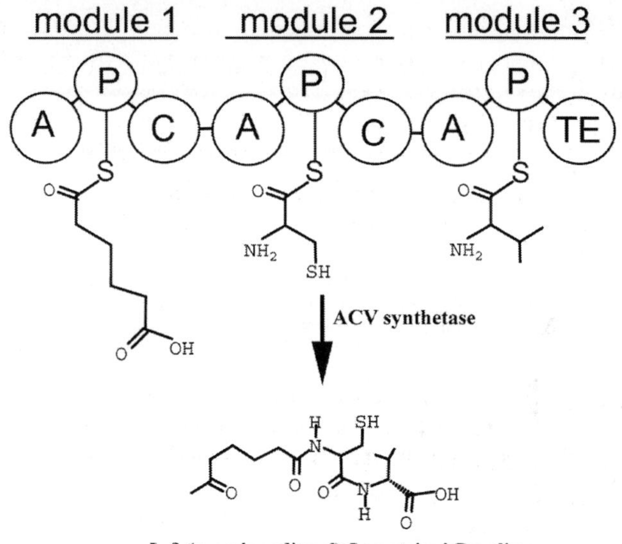

Figure 4. ACV synthetase, a tri-modular non-ribosomal peptide synthetase: The initial activation of the individual amino acids to amino-acyl intermediates is governed by the adenylation domain (A). The activated amino acids are tethered to the acyl carrier domain (P), and peptide bond formation between the amino acids is governed by the condensation domain (C). Epimerization of the last amino acid valine from L to D configuration and release of the peptide product LLD-ACV is governed by C-terminal thioesterase domain (TE).

Synthesis of the tri-peptide ACV occurs via the multi-domain thiol-template mechanism (Byford *et al.*, 1997). Activity is dependent on the availability of ATP and Mg^{+2} as co-factors. Under optimal conditions of saturated substrate concentration it was proposed that for every molecule of the LLD-ACV product formed three molecules of ATP are consumed (Kallow *et al.*, 1998), under sub-optimal conditions this can increase to 20 molecules/molecule tripeptide (Kallow *et al.*, 1998). Recently, using *A. chrysogenum* ACVS it was demonstrated that increasing ATP concentrations exert a substrate inhibition effect, involving the reverse reaction of the aminoacyl adenylate leading to the formation of adenosine (5') tetraphospho(5') adenosine (Kallow *et al.*, 2002). This is consistent with the assumption that in fact ACVS activity is regulated by the prevailing ATP concentrations.

MODULARITY AND PRODUCT ASSEMBLY ALONG NRPSS

THE ADENYLATION DOMAIN

The initial module of each NRPS is the so-called adenylation domain (A domain). The A domain is involved in the activation of specific amino acids into acyl substrates in a reaction that involves ATP hydrolysis. The substrate specificity of the A domain of each of the three modules in ACVS was studied using purified enzyme isolated from *A. nidulans* and various substrate analogues (Etchegaray *et al*., 1997). These experiments indicated that the three A domains of ACVS display a high specificity for respectively AAA, L-cysteine and L-valine. Also, the A domains show sequence homology with acyl CoA ligases (Conti *et al*., 1997). The sequence identity between the A domains of various organisms is 30 to 60%, which implies the presence of a common mechanism involved in substrate activation. Stachelhaus *et al*. (Stachelhaus *et al*., 1995) constructed hybrid proteins by exchanging the A domain of *Bacillus subtilis* SrfA (involved in surfactin biosynthesis) with that of the *P. chrysogenum* cysteine and valine-specifying A domains of ACVS and the phenylalanine, ornithine, and leucine-specifying A domains of the NRPSs involved in gramicidin biosynthesis in *B. brevis*. This domain swap resulted in the synthesis of altered products, although with reduced bioactivity, it allows for further exploration of in search of alternative bioactive peptides.

THE PEPTIDYL CARRIER PROTEIN DOMAIN

The activated acyl substrates produced by the A domains are covalently attached to the NRPS via a thioester bond to the phosphopantetheine arm of the second domain of the module, the carrier domain (also known as thiolation (T) domain or peptidyl carrier protein (PCP) domain). Structural analysis of the PCP domain from *B. brevis* tyrocidine synthetase 3 (*Tyc*C) suggested that this domain exists as a distorted anti-parallel four-helix bundle (Weber *et al*., 2000). This structure is very flexible, due to the presence of a short stretch of amino acids residues (aa 41- 47) that could act as a swinging arm to transport activated, adenylated amino acids from one domain to another. Activation of the PCP domain requires as co-factor a phosphopantetheinyl moiety. This becomes covalently attached to the PCP domain by a phosphopantetheinyl transferase (PPTase) that uses coenzymeA (CoA) as a substrate. CoA is hydrolysed in an ATP dependent manner and the resulting phosphopantetheine moiety then becomes covalently attached to a conserved serine residue in the PCP domain (Schlumbohm *et al*., 1991; Stein *et al*., 1996). Sequence comparisons using various PCP domains showed that only the region surrounding the conserved serine residue (T loop) is highly conserved. Recently the *P. chrysogenum* PPTase was identified (Garcia-Estrada *et al*., 2008a). Deletion of the *P. chrysogenum ppt* gene completely abolished penicillin biosynthesis, and also led to lysine deficiency and absence of green pigmentation in conidiospores, which in *A. nidulans* is formed from a polyketide derivative (Garcia-Estrada *et al*., 2008a; Mayorga *et al*., 1992). This suggests that the *P. chrysogenum* PPTase is not only involved in activation of NRPSs and PKSs, but probably also in activation of the enzyme aminoadipate reductase, a phosphopantetheinyl-requiring enzyme involved in lysine biosynthesis.

Conversion of an inactive apo-PCP domain into its active holo-form was shown *in vitro* using recombinant *Bacillus* PPTase (Sfp) and PCP domains from the *Bacillus* NRPS surfactin synthetase (Quadri *et al.*, 1998). In later studies the very broad range specificity of Sfp for PCP domains was demonstrated (Quadri *et al.*, 1998). Recently, the structure of a portion of the *Bacillus* tyrocidine A synthetase (TycC3) was determined, comprising the PCP N-terminus and the A domain of the third module of the enzyme. Structure analysis suggested that the active (holo) form of the PCP domain exists in two different conformations, the H-state and a transitional A/H state. The transition between the H and A/H states resulted in movement of the phosphopantetheine arm. This suggests that a conformation equilibrium between both states acts in regulating the movement of the activated amino acids (Koglin *et al.*, 2006).

THE CONDENSATION DOMAIN

Peptide bond formation between an amino acyl and thioesterified substrate attached to the individual PCP domains occurs on the condensation domain (C domain). ACVS contains two of these domains. Condensation domains are typically identified by the presence of a signature sequence that contains the motif HHxxxDGWS (the so-called His motif), which is also present in CoA-dependent acyl transferases (Bergendahl *et al.*, 2002; Stachelhaus *et al.*, 1998; Vollenbroich *et al.*, 1993). Recently, the structure of the condensation domain of *Vibrio cholerae* VibH, involved in vibrobactin biosynthesis, was solved. VibH exists as monomer with two modules, the N- and C-terminal module. The His motif is present at the interface between the two domains. The structure of VibH revealed the presence of a solvent channel into which the phosphopantetheine arm enters via its C- terminal end. As a result, the thiol moiety (aminoacyl-SH) will be placed close to the conserved histidine of the His motif, which acts as a general base during the deprotonation reaction (Keating *et al.*, 2002).

In vitro, the substrate specificity of the condensation domain was demonstrated using truncated modules of *Bacillus* gramicidin synthetase, the PheATE (initiation) module having the upstream PCP domain, and the ProCAT (elongation) module having the downstream PCP domain. The authors demonstrated that the condensation domain is specific for the aminoacyl-S-Ppant moiety of the downstream PCP-domain. (Belshaw *et al.*, 1999). This suggests the presence of two binding pockets in the condensation domain namely (i) a broad substrate-specific donor site for accommodating the incoming PCP-bound amino acyl moiety of one module, and (ii) a substrate-specific acceptor site for the aminoacyl-S-Ppant of another module. Condensation results in a peptidyl-S-Ppant moiety becoming bound to the amino acyl-group at the acceptor site thus allowing the growth of the peptide chain in a unidirectional manner

THE THIOESTERASE DOMAIN

Termination of peptide synthesis is governed by the C-terminal module of the NRPS, which comprises the thioesterase (TE) domain. Thioesterases that function in NRPSs can be grouped into three distinct types based on sequence alignments namely (i) non-integrated thioesterases, (ii) integrated thioesterases adjacent to the PCP domain and (iii) integrated thioesterases adjacent to the epimerization domain (see section below).

Structural analysis of the *Bacillus* Srf-TE domain, belonging to the second class of thioesterases, revealed similarity to proteins of the α/β hydroxylase family. The Srf TE domain contains three α-helices that reach over the catalytic site residues serine, histidine and aspartate that form the catalytic triad. Chain release is initiated by the transfer of the peptidyl-acyl chain to the serine residue thus generating an unstable acyl-O-TE intermediate. In the case of release of a linear peptide (e.g. β-lactams or release of the heptapeptide in case of vancomycin biosynthesis), the unstable intermediate is generally hydrolyzed (Shaw-Reid *et al.*, 1999). When the peptide undergoes cyclization, an intra-molecular ester bond is formed (Bruner *et al.*, 2002; Kohli *et al.*, 2003).

ACVS is the only know enzyme belonging to the third class of thioesterases. Mutational analysis of the conserved serine residue in a GxSxG motif of the TE domain of ACVS resulted in only a 50% decrease in LLD-ACV formation, which suggests that this serine residue might not be essential for the release of the tri-peptide. Thus, another, unidentified serine residue involved in the catalytic function of the enzyme is thought to be present in the TE domain of ACVS (Kallow *et al.*, 2000).

THE EPIMERIZATION DOMAIN

Most antimicrobial peptides are characterized by the presence of D-amino acids. This variation in conformation results in an increased potency and stability of the product. Most D-amino acids are generated during peptide assembly by an epimerization domain (E-domain), which epimerizes the PCP-bound aminoacyl/peptidyl intermediates. There is similarity between the conserved signature sequence of the condensation domain and the epimerization domain (HHxxxDxxSW). Along the NRPS's assembly line the E-domain is placed in between the PCP domain and C-domain. Mutational analysis of the conserved residues of the E-domain of the *B. brevis* gramicidin S synthetase PheATE module, the initiation module of the enzyme, has led to the identification of two mutants H753A and Y976A. When tested for chirality of product formation these mutations resulted in formation of 56 % of L,L-Phe-Pro (in wild type only D,L-Phe-Pro), indicating that the preference for D,L-Phe-Pro is still partially retained. This observation has led to the suggestion that the downstream C-domain may also partially determine the chirality of the product (Stachelhaus *et al.*, 2000). When individual domains (E and PCP-E) of the Tyc A module were fused to the A and A-PCP domains of tyrocidine synthetase A, epimerization activity was only observed when the PCP domain came from the same module as the E-domain. Sequence comparison of PCP^C with that of the PCP^E showed a deviation in the core motif (GGHSL versus GGDSI, respectively). The aspartate residue adjacent to the conserved serine in PCP^E was shown to be essential for the epimerization activity (Linne *et al.*, 2001).

Recently, a dual condensation/epimerization domain (C/E-domain) was identified in *Pseudomonas* strains producing arthrofactin, syringomycine and syringopeptin (Balibar *et al.*, 2005). In these hybrids the upstream amino acyl/peptidyl intermediate is epimerized before the condensation reaction takes place but only when the acceptor site of C-domain is loaded with dedicated amino acids. This suggests that the epimerization domain is triggered by conformational changes of the C/E domain.

An alternative way of epimerization was observed during cyclosporine biosynthesis in *Tolypocladium niveum*, where the D-amino acids in the product are generated by a specialized amino acid racemase termed alanine racemase. Such a specialized racemase requires pyridoxal phosphate as co-factor during the enzymatic reaction. Furthermore, these enzymes are generally not part of NRPS proteins (Hoffmann *et al.*, 1994).

Tailoring enzymes involved in penicillin biosynthesis

ISOPENICILLIN N SYNTHASE (IPNS)

IPNS belongs to the class of non-heme iron dependent oxidases, and is required for the cyclization of the tri-peptide LLD-ACV into the β-lactam IPN, which acts as precursor for all β-lactam related compounds. IPNS requires Fe^{2+} as co-factor during the reaction cycle. Unlike most other Fe^{2+} dependent di-oxygenases that incorporate oxygen in the substrate, IPNS reduces molecular oxygen into water during the catalytic conversion of ACV (Baldwin *et al.*, 1988). Sequence analysis of various IPNS isozymes from fungi and bacteria has shown a high level of sequence identity (54-99%), suggesting that this enzyme might have been obtained by filamentous fungi via horizontal gene transfer from lower β-lactam producing prokaryotes (Cohen *et al.*, 1990). Mutational analysis of conserved regions of IPNS has allowed identification of the active site motif HxDxH that is crucial for enzyme activity. Remarkably, changes in other conserved regions have only a minor effect on the function of the enzyme (Borovok *et al.*, 1996). In a separate study it was shown that an RxS motif is involved in substrate binding. Based on analysis of the structure of anaerobically grown IPNS crystals it was proposed that binding of the substrate ACV to Fe^{2+} via its thiol residue leads to uptake of an oxygen molecule thereby oxidizing Fe^{2+} to Fe^{3+}. During the re-conversion of the unstable Fe^{3+} into stable Fe^{2+}, the intermediates thio-aldehyde and a hyperoxide ligand are generated, which actually support closure of the β-lactam ring . Later, Roach *et al.* (1996) succeeded in solving the crystal structure of IPNS containing either substrate ACV, or δ-(L-alpha-aminoadipoyl)-L-cysteinyl-L-S-methylcysteine (ACmC), a modified substrate. This showed that product formation is indeed a two-stage reaction including first the formation of the mono β-lactam ring followed by formation of a highly oxidized iron(IV)-oxo (ferryl) moiety required for thiazolidine ring formation (Burzlaff *et al.*, 1999). Incubation of IPNS in the presence of the substrate analog LLL-ACV considerably decreased the turn-over of the enzyme via stable hydrogen bond formation within the active site. This suggests that the active site of IPNS is well suited to accommodate large, hydrophobic side chains of D-amino acids in the binding pocket close to the iron centre (Howard-Jones *et al.*, 2005).

ISOPENICILLIN N ACYLTRANSFERASE

The final step in the formation of penicillin, the conversion of IPN into penicillin, is mediated by IAT (also known as AatA), a member of the N-terminal nucleophile (NTN) hydrolase family. Members of this family are generally produced as inactive precursors that undergo post-translational activation in an auto-catalytic manner. Cleavage of the precursor leads to the generation of an additional N-terminus, which is a cysteine in IAT. This catalytic group acts as a nucleophile in hydrolysis of the product (Brannigan *et al.*, 1995). Post-translational cleavage of the IAT precursor results the active hetero-dimer, formed by two subunits, the β-subunit corresponding to the C-terminal 29 kDa region and the α-subunit corresponding to an 11 kDa N-terminal region. (Tobin *et al.*, 1990; Whiteman *et al.*, 1990). Mutational analysis of *P. chrysogenum* IAT indicated that cysteine 103 is essential for auto-cleavage of the proenzyme and also for catalytic activity (Tobin *et al.*, 1995). Co-production of wild type IAT with a mutant form unable to be cleaved into the α- and β-subunits in *P. chrysogenum* resulted in decreased penicillin production. The mutant IAT affected auto-cleavage of the WT protein, but had no effect on the activity of the cleaved enzyme (Garcia-Estrada *et al.*, 2008b). This suggests that the post-translational processing of IAT is positively regulated *in trans* by already cleaved IAT molecules via a yet unknown mechanism.

Biochemical characterization of *P. chrysogenum* IAT demonstrated that the enzyme possesses three different enzyme activities (i) IPN amidohydrolase activity, (ii) 6-amidohydrolase activity and (iii) 6-APA acyltransferase activity. The formation of penicillin could be the result of a simple exchange reaction or might require two steps that involve formation of an intermediate, 6-APA. IAT from *P. chrysogenum* displays an enhanced affinity for 6-APA relative to IPN (Alvarez *et al.*, 1993) and requires CoA-activated forms of the PAA or POA side chains (for details see below). Previously, it was suggested that transfer of the side chain PAA inside the active site of the enzyme occurs via the thioesterase motif GxSxG. Sequence comparison of *P. chrysogenum* IAT (PenDE) with its *A. nidulans* ortholog (AatA) indicates the presence of a conserved serine residue that might be involved in cleavage of the CoA moiety from the second substrate PAA-CoA (Alvarez *et al.*, 1993). Mutation of the conserved Ser^{309} of *P. chrysogenum* IAT completely abolished enzyme activity, without inhibiting auto-cleavage, thus suggesting its involvement in substrate acylation (Tobin *et al.*, 1994). Possibly, during benzyl penicillinic acid formation, IPN is first converted into the intermediate 6-APA which remains bound in one of the active site pockets, and consecutive binding of an acyl-CoA moiety at another pocket in the active site leads to a change in enzyme activity. As a result, the new side chain is attached to 6-APA resulting in penicillin formation. In the absence of acyl-CoA, a water molecule binds to the second active site resulting in the release of 6-APA from the enzyme by hydrolysis.

Sequence analysis of the genomes of *Aspergilli* and *P. chrysogenum* indicate that these species contain additional homologs of IAT, that can be classified in three groups, AatA , AatB and AatC, but not all species have all three proteins (Garcia-Estrada *et al.*, 2009) (*Figure 5A*). In all homologs the conserved Gly/Cys motif, which is essential for autocatalytic cleavage of IAT is conserved. This suggests that all proteins represent active NTN hydrolases. Remarkably, *P. chrysogenum* lacks an ortholog of *An*AatB, which was shown to encode an active, cytosolic form of IAT. Instead, the *P. chrysogenum* genome contains a gene encoding an AatC ortholog (designated IAL; Garcia-Estrada *et al.*, 2009), which is absent in *A. nidulans*. The *Pc*AatC protein has no putative peroxisomal targeting sequence, is not expressed and therefore not likely to be involved in penicillin biosynthesis. Syntheny analysis between *P. chrysogenum*, *A. nidulans* and *A. terreus* suggests that in the *A. nidulans* genome the region comprising *aatC* has been deleted (*Figure 5B&C*). These data suggest that the IAT-like proteins in both organisms might have evolved differently and that some of their functions have yet to be deciphered.

A

	Pchr	Afum	Aory	Afla	Acla	Ater	Anid	Anig	Tequ	note
AatA	Yes -ARL*	No	yes -AKL*	yes -AKL*	No	No	yes -ANI*	No	yes -QKL*	4 exons
AatB	No	No	No	No	No	yes	yes	No	yes	4 exons
AatC	Yes	yes	yes	yes	yes	yes	No	yes	No	3 exons

B

Anid	Pchr	Ater
AN6772.2	Pc16g11830	ATEG_07478 (36 %)
AN6773.2	Pc12g10920	ATEG_01070
AN6774.2	Pc18g01290 (54 %)	ATEG_07363
AN6775.2	--	**ATEG_07911**
AN6776.2	--	--
AN6777.2	--	--
AN6778.2	Pc14g00400	ATEG_02189
AN6779.2	Pc13g08080 (38 %)	ATEG_04667
AN6780.2	Pc21g13940 (46%)	ATEG_08673

C

Pchr	Anid	Ater
Pc13g09180	AN3107.2	ATEG_04132
Pc13g09170	AN3106.2	ATEG_04133
Pc13g09160	--	ATEG_04133
Pc13g09150	--	--
Pc13g09140	--	**ATEG_04134**
Pc13g09130	AN3106.2	ATEG_04135
Pc13g09120	AN3106.2	ATEG_04136
Pc13g09110	AN3105.2	ATEG_04137
--------------	AN3104.2	--
Pc13g09100	AN3103.2	ATEG_04138

Figure 5. IAT-like proteins in filamentous fungi. (A) Table showing the occurrence of IAT (AatA)-like proteins in the β-lactam producers *Penicillium chrysogenum* (Pchr; AatA (IAT), Genbank accession number CAP97034; AatC (IAL), CAP91983), *Aspergillus nidulans* (Anid; AatA, CAA37394; AatB, EAA58593) and *Trichophyton equinum* (Tequ; AatA, ABWI01001092; AatB, ABWI01001048) as well as in related *Aspergillus* sp. including *A. fumigatus* (Afum; AatC, EAL92321), *A. oryzae* (Aory; AatA, BAE64316; AatC, BAE55742), *A. flavus* (Afla; AatA, EED50266; AatC, EED57265), *A. clavatus* (Acla; AatC, EAW09828), *A. terreus* (Ater; AatB, EAU32173; AatC, EAU35936) and *A. niger* (Anig; AatC, CAK44265). Protein sequences were taken from Genbank, corrected when required, and placed in three groups (AatA to AatC) based on their sequence similarity. AatA proteins are only observed in species that harbour the penicillin biosynthesis gene cluster, and all of these have a peroxisomal targeting signal type 1 (PTS1; as indicated by the three amino acids in one letter code, the asterisk representing the -COOH group). The very similar AatB proteins are only found in few species including *A. terreus*, which does not produce β-lactams. In addition, the genomes of *A. oryzae* and *A. flavus* contains AatB pseudogenes (*A. oryzae*, BAE54815 and *A. flavus*, AAIH02000032 - not listed) comprising only the first three exons of the gene. (B) & (C). Synteny of alternative AatB and AatC proteins. (B) shows the gene order around the *aatB* gene (AN6775.2) in the *A. nidulans* genome. Comparison with the respecitive orthologs in *P. chrysogenum* and *A. terreus* shows that the region comprising *An*AatB is not conserved in these species. When the percentage identity is indicated, this reflects paralogous proteins rather than true orthologs. (C) shows the synteny around the *P. chrysogenum aatC* gene (Pc13g09140). Despite the high conservation in both *A. nidulans* and *A. terreus*, only in the latter an AatC ortholog is observed (ATEG_04134), while it is lacking in *A. nidulans*. This suggests that during evolution *aatC* has been deleted in *A. nidulans*.

PHENYLACETYL COA LIGASE (PCL)

PAA-CoA ligase (PCL, also known as Phl) belongs to the super-family of adenylate-forming enzymes. PCL activity requires ATP, magnesium and CoA as co-factors. PCL was initially identified in *Pseudomonas putida* as an enzyme involved in degradation of PAA (Olivera *et al.*, 1998). Experiments with the purified enzyme from *P. putida* demonstrated that this enzyme can convert PAA into PAA-CoA, which can be utilized as substrate during *in vitro* penicillin biosynthesis (Fernandez-Valverde *et al.*, 1993). Screening of a *P. chrysogenum* genomic library has led to the identification of the *phl* gene which is involved in the activation of the side chain PAA (or POA) to PAA-CoA (or POA-CoA). Overexpression of *phl* in *P. chrysogenum* led to an 8-fold increase in penicillin production rates. Remarkably, deletion of *phl* resulted in only a 40% reduction in penicillin levels (Lamas-Maceiras *et al.*, 2006). This suggests the existence of other enzymes that can activate PAA. Recently, Koetsier *et al.* (2009b) reported that PCL actually has a very low specificity for PAA and POA (K_{cat}/K_m value of 2.3mM^{-1}.s^{-1}). Surprisingly, PCL showed a high K_{cat}/K_m value (10.8 – 66.3mM^{-1}.s^{-1}) for medium chain fatty acids as trans-cinnamic acid, indicating that the enzyme is actually an acyl-CoA ligase. It is interesting to note that the type strain of *P. chrysogenum* (NRRL1951) produces a mixture of natural penicillins during growth in the presence of PAA (Herschbach *et al.*, 1984), suggesting that also other acyl-CoA moieties are formed and incorporated into β-lactams. Blast analysis using the *P. chrysogenum* genome revealed the presence of multiple putative CoA ligases with 22-36% sequence identity to PCL. Like PCL, most of these contain a C-terminal peroxisomal targeting sequence type 1 (PTS1).

Using subtractive hybridization Wang and co-workers (Wang *et al.*, 2007) have identified PhlB as a close homolog of Phl in *P. chrysogenum*. Biochemical characterization using purified PhlB demonstrated that *in vitro* the enzyme has the capability to convert PAA into PAA-CoA, suggesting that this protein might also play a role in penicillin formation in *P. chrysogenum*. However, recent *in vivo* data show that the absence of this protein (also designated AclA) in *P. chrysogenum* does not affect penicillin formation significantly (Koetsier *et al.*, 2009a). Rather, PhlB/AclA appears to constitute an adipoyl CoA ligase, required for adipate metabolism (Koetsier *et al.*, 2009a).

Role of peroxisomes during penicillin biosynthesis.

Peroxisomes are single membrane-bound organelles characterized by the presence of hydrogen peroxide producing oxidases and catalase to degrade the hydrogen peroxide produced into water and oxygen. Peroxisomes are involved in various metabolic processes (Wanders *et al.*, 2004). Peroxisomal β-oxidation of fatty acids is well conserved from yeast to mammals (Visser *et al.*, 2007). In yeast, β-oxidation is confined to peroxisomes. In contrast, *A. nidulans* harbors components of the β-oxidation pathways both in mitochondria and peroxisomes (Hynes *et al.*, 2008). In mammals, peroxisomes are also involved in α-oxidation of fatty acids. Defects in peroxisome functions lead to severe disorders in man (Wanders *et al.*, 2004).

The formation of peroxisomes depends on the function of the protein products (peroxins) of a set of unique genes, termed *PEX* genes. At present, 32 *PEX* genes have been identified (Kiel *et al.*, 2006). Peroxisome-borne proteins are encoded by nuclear genes and post-translationally sorted into the organellar matrix via either one of the specific peroxisomal targeting signals PTS1 or PTS2. Pex5 or Pex7 act as cargo receptors for PTS1 and PTS2 proteins, respectively (Platta *et al.*, 2007; Schliebs *et al.*, 2006).

In the course of the *P. chrysogenum* strain improvement program it was observed that enhanced penicillin production rates correlated with increasing peroxisome volume fractions in the strains (van den Berg *et al.*, 2008). Consistent with this, Kiel *et al.*

(2005) demonstrated that massive proliferation of peroxisomes by overexpression of the *P. chrysogenum pex11* gene led to a 2-3 fold increase in penicillin production at conditions in which the penicillin biosynthetic enzyme levels were unaltered. This suggested that in fact the increase in peroxisome membrane surface solely is responsible for the increase in production rates by facilitating increased fluxes of substrates (e.g. IPN, ATP, CoA) and/or products (PenG/PenV) over the peroxisomal membrane. The importance of peroxisomes in penicillin biosynthesis was corrobrated in *A. nidulans* where deletion of *pex3* led to a complete loss of peroxisomes that was paralleled by a 60% decrease in penicillin production (Sprote *et al.*, 2009). And, in *P. chrysogenum*, absence of the peroxisomal membrane protein *pex10* resulted in a 77% decrease in β-lactam production (van den Berg *et al.*, 2008). Morover, mislocalization of the protein to the cytosol via a mutation of the C- terminal PTS1 sequence of IAT in *P. chrysogenum* led to a drastic reduction in penicillin production (Muller *et al.*, 1992). All this data together suggests that in *P. chrysogenum* proper localization of IAT to peroxisomes is necessary for normal penicillin biosynthesis. Mislocalization of *A. nidulans* IAT (AatA) to the cytosol resulted in a 35% reduction in penicillin production (Sprote *et al.*, 2009). Recently, in *A. nidulans*, a cytosolic paralog of AatA, designated AatB, was identified (58% sequence identity to AatA). Disruption of An*aatB* also affected penicillin levels but to a much lesser extent than an An*aatA* disruption, suggesting that AatA is the major IAT enzyme of penicillin biosynthesis (Sprote *et al.*, 2008). Hence, these findings imply that penicillin biosynthesis can also take place in the cytosol. It should, however, be noted that *P. chrysogenum* produces (at least) 1000-fold more penicillin than *A. nidulans* and therefore, in *P. chrysogenum* peroxisomes may have become more important than in *A. nidulans*.

It is tempting to speculate on the exact role of the peroxisomes in penicillin biosynthesis. One possibility is that the organelles create a specific bio-chemical environment (e.g. pH, presence of co-factors as ATP and CoA, physical association of IAT and PCL) allowing to increase the rate of the overall biosynthetic process. Notably, generation of PAA-CoA and POA-CoA is reminiscent of the formation of acyl-CoA during β-oxidation. In this respect it is also interesting to note that the formation of PenN from IPN during cephalosporin formation in *Acremonium chrysogenum*, which also requires a CoA-activation step, might also occur inside peroxisomes (Kiel *et al.*, 2009; Ullan *et al.*, 2007; Ullan *et al.*, 2004).

Classical strain improvement

Initiated by the discovery of a *Penicillium* strain with better growth characteristics by the NRRL in Peioria (i.e. the type strain NRRL 1951), several industrial strains with increased penicillin production rates have been isolated via sequential rounds of random mutagenesis and high throughput screening efforts [reviewed by Lein (1986)]. Although this has introduced many mutations in the genome, only a few of these have been identified and characterized (*Figure 6*).

The most well-known mutation that contributed significantly to the increased productivity of industrial *Penicillium* strains is the amplification of the entire penicillin biosynthesis gene cluster (Fierro *et al.*, 1995; Newbert *et al.*, 1997). In the various industrial strains this large genomic region of 50-100 kb is amplified up to 6 times (van den Berg *et al.*, 2008; van den Berg *et al.*, 2007), although even higher numbers are reported by Newbert *et al.* (Newbert *et al.*, 1997), up to an astonishing 50 copies. Analysis of the amplified penicillin gene cluster (*Figure 7*) has shown that besides the three structural biosynthetic genes only a limited number of other ORFs are transcribed significantly during penicillin biosynthesis (van den Berg *et al.*, 2008). One of these ORFs encodes a putative saccharopine dehydrogenase which catalyses the last step of lysine biosynthesis. Indeed, deletion of the entire penicillin biosynthesis gene cluster results in a mild lysine auxotrophy (Fierro *et al.*, 2006). For most other genes no biochemical data confirming the actual function are available.

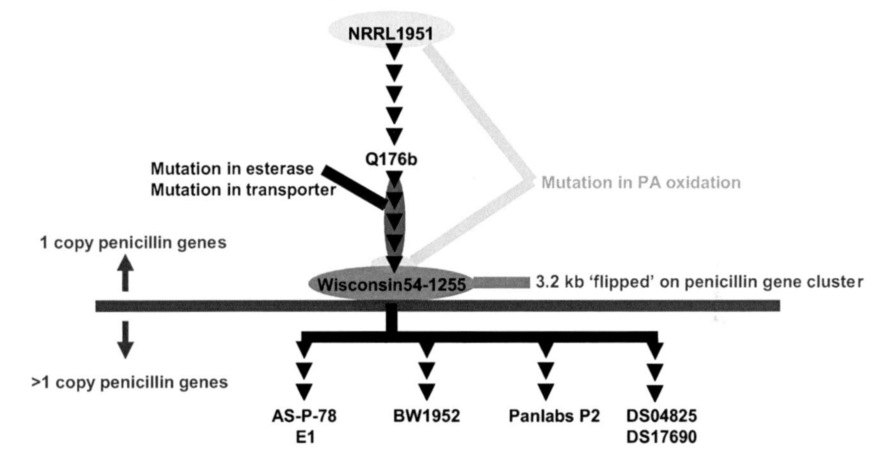

Figure 6. ***P. chrysogenum* industrial strain lineages and identified mutations.** All strains are derived from the type strain NRRL1951. AS-P-78 and E1 are strains from Antibioticos; BW1952 is a strain from Beecham; P2 is a strain from Panlabs; DS04825 and DS17690 are strains from DSM. The black triangles indicate mutagenesis and screening rounds (for the industrial strains the exact number of screening rounds is not known). The occurrence of the identified mutations are indicated and described in details in the text.

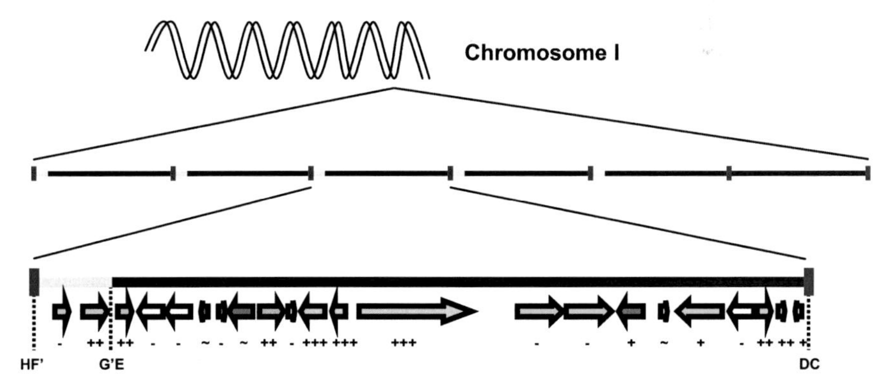

Figure 7. Organization and transcriptional activity of the penicillin biosynthetic genes in *P. chrysogenum* DS17690. The amplified gene cluster is located on chromosome I (Fierro *et al.*, 1993; Xu *et al.*, 2005) in 6 repeated copies. One cluster contains 22 genes (see for details van den Berg *et al.*, 2008). The colored arrows indicate the significance of the Blastp results: hypothetical (white), weak similarity (pink), similar (purple), strong similarity (light green), known protein (light blue). The relative transcription rate is indicated (-, 0; ~, just above detection level; +; detected; ++, good; +++, strong). The hexameric repeats as identified by Fierro *et al.* (1993) are indicated using the same letter coding as in the original publication.

One of the differences between the original Fleming strain (*Penicillium notatum*) and current industrial strains (high penicillin producers) is the catabolism of PAA. The modern production strains have a point mutation in the *phaA* gene encoding PAA hydroxylase, an enzyme involved in PAA catabolism, leading to an inability to utilize PAA as sole carbon source. Expression of the *phaA* gene from *P. notatum* in *P. chrysogenum* has led to a decrease in the penicillin productivity (Rodríguez-Sáiz *et al.*, 2005). Surprisingly, while the function of this PAA catabolism pathway is nearly redundant in high producing strains the transcription of the encoding genes is amongst the highest throughout the genome (Harris *et al.*, 2009a).

Three other mutations have been identified, but unfortunately for none of these a functional relation with productivity has been established. A small region of the 56 kb region amplified in most industrial strains is flipped in the Wisconsin54-1255 strain as compared to the NRRL1951 strain. This region contains the *erg25* gene, encoding C-4 methyl sterol oxidase, one of the later steps in ergosterol biosynthesis. Ergosterol, as a component of the cell wall, can influence the influx of penicillin (Hillenga, 1999) and although this mutation might suggest a relation, this was not confirmed (van den Berg *et al.*, 2007). Recently, a second functional *erg25* gene was identified, *Pcerg25B* (Wang *et al.*, 2008), implying redundancy.

The other two mutations were identified via AFLP analyses, a technology marketed by KeyGene (Wageningen, The Netherlands). One represents a 14-bp duplication in a putative esterase (Pc12g04030) causing a frame-shift, the other is a point mutation in a member of the ABC superfamily (Pc12g00440), but so far no further characterization has been performed (M.A. van den Berg, unpublished data).

Classical mutagenesis is still the common approach applied in *Penicillium* industrial strain improvement programs but suffers from specific limitations as it is random, unpredictable, time consuming and laborious. It is likely that during strain improvement also other possible limiting factors, such as precursor generation or transport capacity of intermediates across the peroxisome and plasma membrane, have been improved. However, the actual mutations causing these changes have not yet been identified. As generally the mutation(s) underlying the increased productivity of the new production strains remain unresolved, this might also lead to the accumulation of unwanted mutations. Initial comparative studies between different strains have identified large genomic regions deleted from the genome of higher producing strains. These regions contain up to 30 genes, most of which are not or lowly expressed in the low producing strains (van den Berg *et al.*, 2008). As the functions of these gene products are not known (most are annotated as hypothetical proteins) the relation between these deletions and increased productivity can not be established.

Metabolic engineering approaches

Genetic modification may adapt specific target genes in order to change the metabolic fluxes that are favorable or unfavorable for a particular pathway (Koffas *et al.*, 1999). Comparisons with optimization of a factory line have been made in the past (van den Berg *et al.*, 1999). Various analytical methods are applied to identify the bottlenecks involved in a particular pathway. For example, metabolic flux analysis using high penicillin-producing strains of *P. chrysogenum*, demonstrated that a step-wise increase in the metabolic demand for NADPH resulted in decreased penicillin production rates suggesting co-factor availability as potential bottleneck during penicillin biosynthesis (van Gulik *et al.*, 2000). The pentose phosphate pathway (PPP) is involved in maintaining *in vivo* reducing equivalents (NADPH). Recently, it was demonstrated that during penicillin producing conditions the PPP was more active towards the oxidative branch point, which is the major source for NADPH production (Kleijn *et al.*, 2007). Jorgensen *et al.* (1995) observed that the supply of the three precursor amino acids AAA, valine and cysteine to a fed-batch culture increased PenV production by about 20%, indicating that precursor availability may limit penicillin production. This was confirmed by MicroArray analysis of high producing strains, which showed increased transcription rates of genes involved in the biosynthesis of these amino acid precursors (van den Berg *et al.*, 2008).

Genetic engineering tools are needed to allow for a targeted approach in order to relieve the various bottle-necks in contrast to the lengthy and undirected classical strain improvement programs using mutagenesis and screening. The most relevant features of an efficient toolset are the availability of gene expression cassettes (i.e. strong, constitutive promoters and efficient terminators), selectable markers, a replicating plasmid, a transformation method and efficient gene targeting (to construct gene deletions). Several promoters are available for use in *P. chrysogenum*, but most authors use either the homologous *pcbC* promoter (from the gene encoding IPNS; see for example Hillenga, 1999) or the heterologous *gpdA* promoter (from the *A. nidulans* gene encoding glyceeraldehyde-3-phosphate dehydrogenase), which e.g. has been used to equip *P. chrysogenum* for the direct biosynthesis of cephalosporins (see below). Only few plasmids able to replicate in *P. chrysogenum* have been described (Fierro *et al.*, 1996), containing selectable markers like the *ble* gene (Kolar *et al.*, 1988) encoding for resistance against phleomycin, or the *amdS* gene (Beri and Turner, 1987), encoding acetamidase supporting growth on acetamide as the sole nitrogen source. The transformation system commonly applied is the PEG method using protoplasts to transform DNA. The most recent improvement of the *P. chrysogenum* toolbox is the optimalization of gene targeting. Filamentous fungi (as well as higher eukaryotes) have two pathways to integrate DNA: a very active Non-Homologous End-Joining (NHEJ) pathway, resulting in random integration, and a relatively silent Homologous Recombination (HR) pathway, resulting in typically less than 2% of the transformants containing the DNA at the anticipated genomic locus [see for example (Casqueiro *et al.*, 1999)]. To facilitate the generation of targeted integrations (i.e. to create knock-out mutants), the NHEJ pathway was deleted and the resulting strains demonstrated a huge increase in correct gene targeting with more than 50% of the transformants obtained having the correct gene deletion (Hoff *et al.*, 2009; Snoek *et al.*, 2009).

Several authors reported successful application of these tools to enhance the PenG productivity. One approach is to increase the amino acid pools. Filamentous fungi including yeast have a characteristic lysine biosynthesis pathway involving AAA as intermediate. It was shown in *P. chrysogenum* that deletion of the *lys2* gene encoding aminoadipate reductase led to an accumulation of the precursor AAA and a subsequent increase in PenG production (Casqueiro *et al.*, 1998). Another approach is to overexpress the biosynthetic genes, encoding the crucial steps in penicillin biosynthesis. By doing this Theilgaard *et al.* (Theilgaard *et al.*, 2001) showed that there is a delicate balance mainly between the first two enzymes, ACVS and IPNS, controlling the flux through the pathway. Optimizing this can lead to improved productivities. The most recent example is the proliferation of peroxisomes as reported by Kiel *et al.* (2005).

However, this is not straightforward, as the industrial strains are already optimized in productivity and yield for more than 6 decades. With the availability of the complete genome sequence of the laboratory strain Wisconsin54-1255 (van den Berg *et al.*, 2008) and Affymetrics MicroArrays, *P. chrysogenum* strain improvement has entered a new era. This allows for a genome wide comparison of different strains, but also to study individual strains in time under producing conditions. The first publications reported differential studies between the laboratory strain and a high producing derivative containing an estimated 6 copies of the penicillin biosynthetic gene cluster (Harris *et al.*, 2009a; van den Berg *et al.*, 2008). Only the single comparative study of two strains under two different conditions lead already to the identification 2470 genes which were differentially transcribed in at least one of the four comparisons, which is 18% of all genes in the genome!! Besides the penicillin biosynthetic genes, also the transcription of genes encoding enzymes involved in the biosynthesis of the amino acids cysteine, valine and aminoadipic acid correlate with PenG productivity. This clearly suggests that also in primary metabolism there might be room for further improvement, but with these large datasets it is difficult to select which gene(s) to modify. For example, cysteine supply has been reported to be the rate limiting step in ACV biosynthesis (van den Berg *et al.*, 1999), which nicely corroborates with the mRNA data. And whilst two cysteine biosynthetic pathways are known, there is some evidence suggesting that the transsulfuration pathway is actually used to produce the cysteine to be incorporated in the antibiotic (Evers *et al.*, 2004). This is confirmed by the mRNA studies, which show that besides early stages of serine (and cysteine) biosynthesis, also a homolog of O-acetylhomoserine (thiol)-lyase (Pc12g05420), a key enzyme in the transsulfuration pathway, correlates with PenG production. However, stoichiometric modeling suggests that a more significant increase in yield can be obtained if cysteine would be produced via the direct sulfhydrylation pathway (Jorgensen *et al.*, 1995). It must be noted that cysteine is a toxic molecule and cells will limit its accumulation, implying that its overproduction will not be an easy task.

Cephalosporin biosynthesis

ADIPOYL-7-AMINODEACETOXYCEPHALOSPORANIC ACID (AD-7-ADCA)

The natural producers of cephalosporins are *Streptomycetes* and the fungus *Acremonium chrysogenum* (also known as *Cephalosporium acremonium*). The biosynthesis of this group of β-lactam antibiotics starts similar as for penicillins (*Figure 8*). However, it deviates at the level of the first β-lactam compound, IPN. Here, the cephalosporin-producing organisms epimerize IPN into penicillin N (PenN), which is the substrate for expandase (Baldwin *et al.*, 1987; Kovacevic *et al.*, 1989), catalyzing the ring-expansion of the 5-membered penicillin ring into the 6-membered cephalosporin ring (*Figure 8*).

The high PenG productivities of *P. chrysogenum* and the availability of the expandase genes inspired the Eli Lilly R&D team in 1990 for the heterologous expression of the *S. clavuligerus cefE* gene in a *P. chrysogenum* strain resulting in the direct biosynthesis of cephalosporin (Cantwell *et al.*, 1990). This was the start of 10 years

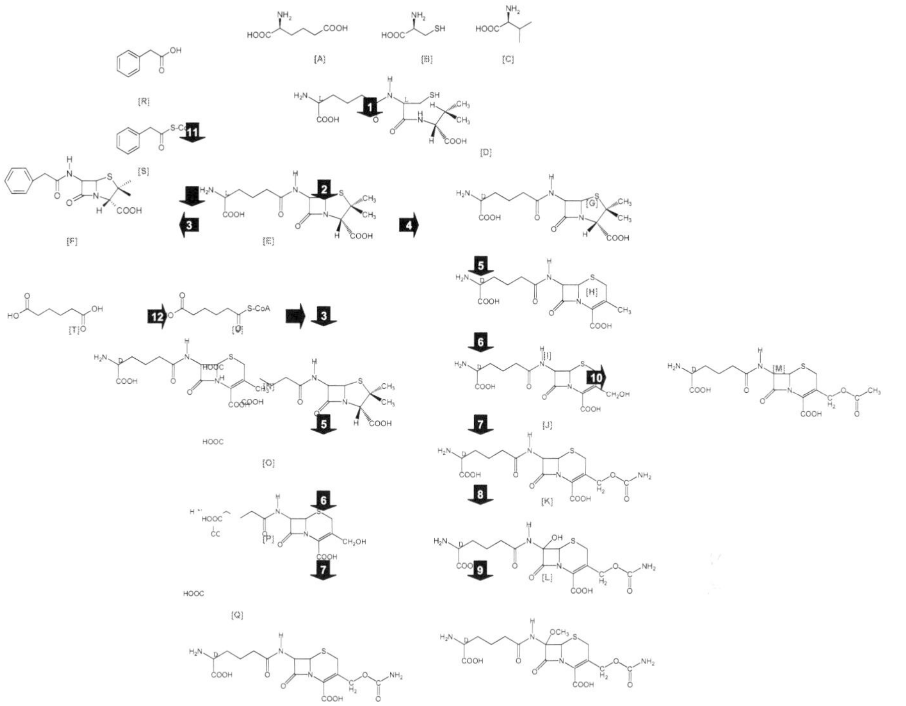

Figure 8. Overview of the various enzymes and intermediates in β-lactam biosynthesis routes (i.e. penicillins, cephalosporins and cephamycins). The arrows represent the following enzymes: 1, L-α-aminoadipyl-L-cysteinyl-D-valine synthetase; 2, isopenicillin N synthase; 3, acyl-CoA:6-aminopenicillanic acid/isopenicillin N acyltransferase; 4, penicillin N epimerase; 5, deacetoxycephalosporinC synthase (expandase); 6, deacetylcephalosporinC hydroxylase; 7, 3'-hydroxymethylcephem-O-carbamoyltransferase; 8, O-carbamoyl-deacetylcephalosporinC hydroxylase; 9, methyltransferase; 10, acetyltransferase; 11, phenylacetyl-CoA ligase; 12, adipoyl-CoA-ligase. The capitals in brackets represent the following intermediates: [A], L-α-aminoadipic acid; [B], L-cysteine; [C], L-valine; [D], L-α-aminoadipoyl-L-cysteinyl-D-valine (ACV); [E], isopenicillin N (IPN); [F], penicillinG (PenG); [G], penicillin N (PenN); [H], deacetoxycephalosporinC (DAOC); [I], deacetylcephalosporinC (DAC); [J], O-carbamoyl-DAC (OCDAC); [K], 7-α-hydroxy-OCDAC; [L], cephamycin C; [M], cephalosporinC (CPC); [N], adipoyl-6-aminopenicillinic acid (Ad-6-APA); [O], adipoyl-7-aminodeacetoxycephalosporanic acid (Ad-7-ADCA); [P], adipoyl-7-aminohydroxycephalosporanic acid (Ad-7-AHCA); [Q], adipoyl-7-amino-3-carbomoyloxymethyl-3-cephem-4-carboozylic acid (Ad-7-ACCCA); [R], phenylacetic acid; [S]. phenylacetyl-CoA; [T], adipic acid; [U], adipoyl-CoA.

of strain and process improvements (see for example Cantwell *et al.*, 1992; Xie *et al.*, 2001) leading to the development of an industrial production process of adipoyl-7-amino-3-deacetoxycephalosporanic acid (ad-7-ADCA), a cephalosporin precursor for 7-ADCA, at DSM in Delft. In this process adipic acid is fed as the side chain precursor to an industrial *P. chrysogenum* strain expressing the expandase, which was shown to be active on adipoyl-6-APA.

OTHER CEPHALOSPORINS PRODUCED BY *P. CHRYSOGENUM*

Introduction of other genes from the cephalosporin biosynthetic pathways from *Streptomycetes* and/or *A. chrysogenum* leads to the accumulation of several interesting cephalosporins in *P. chrysogenum*. For example, the introduction of the *A. chrysogenum* expandase/hydroxylase (*cefEF*) and *Streptomyces clavuligerus* carbamoyltransferase (*cmcH*) genes and the presence of adipate in the feed led to the accumulation of a novel carbamoylated cephalosporin, adipoyl-7-amino-3-carbamoyloxymethyl-3-cephem-4-carboxylic acid or ad-7-ACCCA (Harris *et al.*, 2009b). This compound can be used as building block for semi-synthetic cephalosporins, like cefuroxime. The productivity of ad-7-ACCCA was further increased by the overexpression the *A. chrysogenum cefT* gene, encoding a cephalosporinC transporter of the Major Facilitator Superfamily, resulting in a 2-fold increase with a concomitant decrease in penicillin by-product formation (Nijland *et al.*, 2008).

Moreover, *P. chrysogenum* was also engineered to produce an adipoyl-version of cephalosporinC by introducing the expandase, hydroxylase and acetyltransferase activities (Nieboer and Bovenberg, 1999).

TAILORING ENZYMES

All β-lactams produced by fermentation are converted into Active Pharmaceutical Ingredients (APIs) as ampicillin, amoxicillin, cefaclor and cefuroxime. Nowadays, most of these semi-synthetic routes start with an enzymatic cleavage of the side chain from the β-lactam nucleus. In the case of cephalosporinC, this is a two-step conversion catalyzed by D-amino acid oxidase and glutaryl-7-ACA-acylase, leading to keto-adipyl-7-aminocephalosporanic acid or 7-ACA (Conlon *et al.*, 1995). For compounds like PenG and ad-7-ADCA this starts with specialized acylases, of which the catalytic properties are improved via directed evolution (Otten *et al.*, 2007; Sio *et al.*, 2004; Sio *et al.*, 2002) leading to 6-APA and 7-ADCA, respectively.

Depending on the API to be produced, either chemical and/or enzymatic steps can be used for the synthesis starting from the available intermediates (6-APA for penicillins; 7-ACA and 7-ADCA for cephalosporins). Enzymatic synthesis of cephalexin using 7-ACA and phenylglycine methylester was already reported in the early 1980's (Rhee *et al.*, 1980). This concept is applicable towards other APIs too (Yang *et al.*, 2003; Youshko *et al.*, 2004).

Future developments will lead to further short-cut routes and the application of enzymatic conversions, like using different precursors (for example 7-ACCCA in the case of cefuroxime synthesis) or different concepts like expanding of APIs like the recently reported direct conversion of ampicillin into cephalexin (Thakur *et al.*, 2009). All these developments will lead to further improved versions of the production process for active β-lactam antibiotics, resulting in a decreased carbon footprint.

Perspectives

Exploiting the full potential of the microorganisms producing natural products offers a challenging task. Often biosynthesis of these products is influenced by various environmental factors, such as carbon/nitrogen/phosphorus ratios in medium, pH, temperature, growth rate (see for example Brown and Peterson, 1950). Characterization of the biosynthetic pathways has also revealed a diverse array of enzymes not commonly associated with other cellular functions, including primary metabolism. Many of the novel enzymes can be potentially used as biocatalysts in the synthesis and derivation of pharmaceutical compounds and drug intermediates. These biocatalysts can either replace existing multi-step semi-synthetic routes or introduce structural diversities that are otherwise inaccessible by chemical synthesis.

Genome sequencing of filamentous fungi as *P. chrysogenum* and *A. nidulans* has revealed various surprises. Amongst them are a number of genes encoding enzymes with putative functions in β-lactam biosynthesis, like hypothetical beta-lactamases, cephalosporin esterases and isopenicillin N epimerase, which were not anticipated on beforehand and the actual function of the encoded enzymes needs to be deciphered. Also, both genomes contain around 10 gene clusters comprising NRPS encoding genes. For some the putative functions was hypothesized (Harris *et al.*, 2009a; van den Berg *et al.*, 2008), but for most it remains a challenging task to identify the compounds produced by such unknown clusters, either because they are not expressed or the end-products are below the detection limit. Transfer of these biosynthesis routes from one organism to another organism might help in solving the complex biosynthesis routes, improving existing products and may also lead to development of new products.

Optimizing growth and production conditions for the qualitative and quantitative production of natural products is a complex process. Combinatorial synthesis by combined expression of natural product biosynthesis genes originating from diverse organisms may result in the development of new products. A similar approach was recently demonstrated in yeast by Evolva (Naesby *et al.*, 2009). Yeast model systems provide a potential attractive platform for production of natural products. Some of the advantages of yeast are the similarity of the primary metabolism with that of multi-cellular eukaryotes (filamentous fungi, plants), the availability of a very efficient genetic toolbox and mutant libraries. Yeast species are well known for their established fermentation procedures and powerful genetic tools (Gellissen and Hollenberg, 1997). Furthermore, *Saccharomyces cerevisiae* and methylotrophic yeast species are considered safe organisms and are already in use for industrial production of various biocatalysts/human vaccines of importance (Gellissen, 2000; Hollenberg and Gellissen, 1997). Recently, the first steps towards the production of β-lactams in yeast were reported (Gidijala *et al.*, 2008; Gidijala *et al.*, 2007; Lutz *et al.*, 2005; Siewers *et al.*, 2009) and will stimulate the further use of yeast as a powerful platform for improving natural product diversity, and is also an excellent starting point for developing semi-synthetic products.

Acknowledgements

L.G. and J.A.K.W.K. were financially supported by the Netherlands Ministry of Economic Affairs and the B-Basic partner organizations (www.b-basic.nl) through B-Basic, a public-private NWO-ACTS programme (ACTS = Advanced Chemical Technologies for Sustainability). This project was carried out within the research programme of the Kluyver Centre for Genomics of Industrial Fermentation, which is part of the Netherlands Genomics Initiative / Netherlands Organization for Scientific Research.

References

AHARONOWITZ, Y., COHEN, G. AND MARTIN, J. F. (1992). Penicillin and cephalosporin biosynthetic genes: structure, organization, regulation, and evolution. *Annu Rev Microbiol* **46,** 461-95.

ALVAREZ, E., MEESSCHAERT, B., MONTENEGRO, E., GUTIERREZ, S., DIEZ, B., BARREDO, J. L. AND MARTIN, J. F. (1993). The isopenicillin-N acyltransferase of *Penicillium chrysogenum* has isopenicillin-N amidohydrolase, 6-aminopenicillanic acid acyltransferase and penicillin amidase activities, all of which are encoded by the single penDE gene. *Eur J Biochem* **215,** 323-32.

BACKUS, E. J., DUGGAR, B. M. AND CAMPBELL, T. H. (1954). Variation in *Streptomyces aureofaciens*. *Ann N Y Acad Sci* **60,** 86-101.

BALDWIN, J. E., AND ABRAHAM, E. (1988). The biosynthesis of penicillins and cephalosporins. *Nat Prod Rep* **5,** 129-45.

BALDWIN, J. E., ADLINGTON, R. M., COATES, J. B., CRABBE, M. J., CROUCH, N. P., KEEPING, J. W., KNIGHT, G. C., SCHOFIELD, C. J., TING, H. H., VALLEJO, C. A. AND *ET AL.* (1987). Purification and initial characterization of an enzyme with deacetoxycephalosporinC synthetase and hydroxylase activities. *Biochem J* **245,** 831-41.

BALIBAR, C. J., VAILLANCOURT, F. H. AND WALSH, C. T. (2005). Generation of D amino acid residues in assembly of arthrofactin by dual condensation/epimerization domains. *Chem Biol* **12,** 1189-200.

BELSHAW, P. J., WALSH, C. T. AND STACHELHAUS, T. (1999) Aminoacyl-CoAs as probes of condensation domain selectivity in nonribosomal peptide synthesis. *Science* **284,** 486-9.

BERGENDAHL, V., LINNE, U., AND MARAHIEL, M. A. (2002) Mutational analysis of the C-domain in nonribosomal peptide synthesis. *Eur J Biochem* **269,** 620-9.

BERI, R. K., AND TURNER, G. (1987) Transformation of *Penicillium chrysogenum* using the Aspergillus nidulans amdS gene as a dominant selective marker. *Curr Genet* **11,** 639-41.

BOROVOK, I., LANDMAN, O., KREISBERG-ZAKARIN, R., AHARONOWITZ, Y., AND COHEN, G. (1996) Ferrous active site of isopenicillin N synthase: genetic and sequence analysis of the endogenous ligands. *Biochemistry* **35,** 1981-7.

BRAKHAGE, A. A. (1998) Molecular regulation of beta-lactam biosynthesis in filamentous fungi. *Microbiol Mol Biol Rev* **62,** 547-85.

BRAKHAGE, A. A., SCHUEMANN, J., BERGMANN, S., SCHERLACH, K., SCHROECKH, V., AND HERTWECK, C. (2008) Activation of fungal silent gene clusters: a new avenue to drug discovery. *Prog Drug Res* **66,** 1, 3-12.

BRANNIGAN, J. A., DODSON, G., DUGGLEBY, H. J., MOODY, P. C., SMITH, J. L., TOMCHICK, D. R., AND MURZIN, A. G. (1995) A protein catalytic framework with an N-terminal nucleophile is capable of self-activation. *Nature* **378,** 416-9.

BROWN, W. E., AND PETERSON, W. H. (1950) Factors Affecting Production of Penicillin in Semi-Pilot Plant Equipment. *Industrial & Engineering Chemistry* **42,** 1769-1774.

BRUNER, S. D., WEBER, T., KOHLI, R. M., SCHWARZER, D., MARAHIEL, M. A., WALSH, C. T., AND STUBBS, M. T. (2002) Structural basis for the cyclization of the lipopeptide antibiotic surfactin by the thioesterase domain SrfTE. *Structure* **10,** 301-10.

BURTON, H. S., AND ABRAHAM, E. P. (1951) Isolation of antibiotics from a species of Cephalosporium; cephalosporins P1, P2, P3, P4, and P5) *Biochem J* **50,** 168-74.

Burzlaff, N. I., Rutledge, P. J., Clifton, I. J., Hensgens, C. M., Pickford, M., Adlington, R. M., Roach, P. L., and Baldwin, J. E. (1999) The reaction cycle of isopenicillin N synthase observed by X-ray diffraction. *Nature* **401,** 721-4.

Byford, M. F., Baldwin, J. E., Shiau, C. Y., and Schofield, C. J. (1997) The Mechanism of ACV Synthetase. *Chem Rev* **97,** 2631-2650.

Cantwell, C., Beckmann, R., Whiteman, P., Queener, S. W., and Abraham, E. P. (1992) Isolation of deacetoxycephalosporinC from fermentation broths of *Penicillium chrysogenum* transformants: construction of a new fungal biosynthetic pathway. *Proc Biol Sci* **248,** 283-9.

Cantwell, C. A., Beckmann, R. J., Dotzlaf, J. E., Fisher, D. L., Skatrud, P. L., Yeh, W. K., and Queener, S. W. (1990) Cloning and expression of a hybrid *Streptomyces clavuligerus cefE* gene in *Penicillium chrysogenum. Curr Genet* **17,** 213-21.

Casqueiro, J., Gutierrez, S., Banuelos, O., Fierro, F., Velasco, J., and Martin, J. F. (1998) Characterization of the *lys2* gene of *Penicillium chrysogenum* encoding alpha-aminoadipic acid reductase. *Mol Gen Genet* **259,** 549-56.

Casqueiro, J., Gutierrez, S., Banuelos, O., Hijarrubia, M. J., and Martin, J. F. (1999) Gene targeting in *Penicillium chrysogenum*: disruption of the *lys2* gene leads to penicillin overproduction. *J Bacteriol* **181,** 1181-8.

Clardy, J., and Walsh, C. (2004) Lessons from natural molecules. *Nature* **432,** 829-37.

Cohen, G., Shiffman, D., Mevarech, M., and Aharonowitz, Y. (1990) Microbial isopenicillin N synthase genes: structure, function, diversity and evolution. *Trends Biotechnol* **8,** 105-11.

Conlon, H. D., Baqai, J., Baker, K., Shen, Y. Q., Wong, B. L., Noiles, R., and Rausch, C. W. (1995) Two-step immobilized enzyme conversion of cephalosporinC to 7-aminocephalosporanic acid. *Biotechnol Bioeng* **46,** 510-3.

Conti, E., Stachelhaus, T., Marahiel, M. A., and Brick, P. (1997) Structural basis for the activation of phenylalanine in the non-ribosomal biosynthesis of gramicidin S. *Embo J* **16,** 4174-83.

Demain, A. L., and Elander, R. P. (1999) The beta-lactam antibiotics: past, present, and future. *Antonie Van Leeuwenhoek* **75,** 5-19.

Dulmage, H. T. (1953) The production of neomycin by Streptomyces fradiae in synthetic media. *Appl Microbiol* **1,** 103-6.

Elander, R. P. (2003) Industrial production of beta-lactam antibiotics. Appl *Microbiol Biotechnol* **61,** 385-92.

Etchegaray, A., Dieckmann, R., Kennedy, J., Turner, G., and von Dohren, H. (1997) ACV synthetase: expression of amino acid activating domains of the *Penicillium chrysogenum* enzyme in *Aspergillus nidulans. Biochem Biophys Res Commun* **237,** 166-9.

Evers, M. E., Trip, H., van den Berg, M. A., Bovenberg, R. A., and Driessen, A. J. (2004) Compartmentalization and transport in beta-lactam antibiotics biosynthesis. *Adv Biochem Eng Biotechnol* **88,** 111-35.

Fernandez-Valverde, M., Reglero, A., Martinez-Blanco, H., and Luengo, J. M. (1993) Purification of *Pseudomonas putida* acyl coenzyme A ligase active with a range of aliphatic and aromatic substrates. *Appl Environ Microbiol* **59,** 1149-54.

Fierro, F., Barredo, J. L., Diez, B., Gutierrez, S., Fernandez, F. J., and Martin, J. F. (1995) The penicillin gene cluster is amplified in tandem repeats linked by conserved hexanucleotide sequences. *Proc Natl Acad Sci U S A* **92,** 6200-4.

Fierro, F., Garcia-Estrada, C., Castillo, N. I., Rodriguez, R., Velasco-Conde, T., and Martin, J. F. (2006) Transcriptional and bioinformatic analysis of the 56.8 kb DNA region amplified in tandem repeats containing the penicillin gene cluster in *Penicillium chrysogenum. Fungal Genet Biol* **43,** 618-29.

Fierro, F., Gutierrez, S., Diez, B., and Martin, J. F. (1993) Resolution of four large chromosomes in penicillin-producing filamentous fungi: the penicillin gene cluster is located on chromosome II (9.6 Mb) in *Penicillium notatum* and chromosome I (10.4 Mb) in *Penicillium chrysogenum. Mol Gen Genet* **241,** 573-8.

Fierro, F., Kosalkova, K., Gutierrez, S., and Martin, J. F. (1996) Autonomously replicating plasmids carrying the AMA1 region in *Penicillium chrysogenum. Curr Genet* **29,** 482-9.

Fleming, A. (1929) On the antibacterial action of cultures of a penicillium, with special reference to their use in the isolation of *B. influenzae*. *Bull World Health Organ* **79,** 780-90.

Garcia-Estrada, C., Ullan, R. V., Velasco-Conde, T., Godio, R. P., Teijeira, F., Vaca, I., Feltrer, R., Kosalkova, K., Mauriz, E., and Martin, J. F. (2008a) Post-translational enzyme modification by the phosphopantetheinyl transferase is required for lysine and penicillin biosynthesis but not for roquefortine or fatty acid formation in *Penicillium chrysogenum. Biochem J* **415,** 317-24.

Garcia-Estrada, C., Vaca, I., Fierro, F., Sjollema, K., Veenhuis, M., and Martin, J. F. (2008b) The unprocessed preprotein form IATC103S of the isopenicillin N acyltransferase is transported inside peroxisomes and regulates its self-processing. *Fungal Genet Biol* **45,** 1043-52.

Garcia-Estrada, C., Vaca, I., Ullan, R. V., van den Berg, M. A., Bovenberg, R. A., and Martin, J. F. (2009) Molecular characterization of a fungal gene paralogue of the penicillin penDE gene of *Penicillium chrysogenum. BMC Microbiol* **9,** 104.

Gellissen, G. (2000) Heterologous protein production in methylotrophic yeasts. *Appl Microbiol Biotechnol* **54,** 741-50.

Gellissen, G., and Hollenberg, C. P. (1997) Application of yeasts in gene expression studies: a comparison of *Saccharomyces cerevisiae*, *Hansenula polymorpha* and *Kluyveromyces lactis* -- a review. *Gene* **190,** 87-97.

Gidijala, L., Bovenberg, R. A., Klaassen, P., van der Klei, I. J., Veenhuis, M., and Kiel, J. A. (2008) Production of functionally active *Penicillium chrysogenum* isopenicillin N synthase in the yeast *Hansenula polymorpha. BMC Biotechnol* **8,** 29.

Gidijala, L., van der Klei, I. J., Veenhuis, M., and Kiel, J. A. (2007) Reprogramming Hansenula polymorpha for penicillin production: expression of the *Penicillium chrysogenum pcl* gene. *FEMS Yeast Res* **7,** 1160-7.

Harris, D. M., van der Krogt, Z. A., Klaassen, P., Raamsdonk, L. M., Hage, S., van den Berg, M. A., Bovenberg, R. A., Pronk, J. T., and Daran, J. M. (2009a) Exploring and dissecting genome-wide gene expression responses of *Penicillium chrysogenum* to phenylacetic acid consumption and penicillinG production. *BMC Genomics* **10,** 75.

Harris, D. M., Westerlaken, I., Schipper, D., van der Krogt, Z. A., Gombert, A. K., Sutherland, J., Raamsdonk, L. M., van den Berg, M. A., Bovenberg, R. A., Pronk, J. T., and Daran, J. M. (2009b) Engineering of *Penicillium chrysogenum* for fermentative production of a novel carbamoylated cephem antibiotic precursor. *Metab Eng* **11,** 125-37.

Hersbach, G.J.M., van der Beek, C.P., and Van Dijck, P.W.M. (1984) The penicillins: properties, biosynthesis and fermentation. In: E.J. Vandamme, Editor, Biotechnology of industrial antibiotics vol. 3, Marcel Dekker, NY (1984), pp. 45–140.

Hillenga, D. (1999) Transport processes in penicillin biosynthesis. University of groningen, Groningen.

Hoff, B., Kamerewerd, J., Sigl, C., Zadra, I., and Kuck, U. (2009) Homologous recombination in the antibiotic producer *Penicillium chrysogenum*: strain DeltaPcku70 shows up-regulation of genes from the HOG pathway. *Appl Microbiol Biotechnol* in press.

Hoffmann, K., Schneider-Scherzer, E., Kleinkauf, H., and Zocher, R. (1994) Purification and characterization of eucaryotic alanine racemase acting as key enzyme in cyclosporin biosynthesis. *J Biol Chem* **269,** 12710-4.

Hollenberg, C. P., and Gellissen, G. (1997) Production of recombinant proteins by methylotrophic yeasts. *Curr Opin Biotechnol* **8,** 554-60.

Howard-Jones, A. R., Rutledge, P. J., Clifton, I. J., Adlington, R. M., and Baldwin, J. E. (2005) Unique binding of a non-natural L,L,L-substrate by isopenicillin N synthase. *Biochem Biophys Res Commun* **336,** 702-8.

Hung, P. P., Marks, C. L., and Tardrew, P. L. (1965) The Biosynthesis and Metabolism of Erythromycins by *Streptomyces erythreus. J Biol Chem* **240,** 1322-6.

Hynes, M. J., Murray, S. L., Khew, G. S., and Davis, M. A. (2008) Genetic analysis of the role of peroxisomes in the utilization of acetate and fatty acids in *Aspergillus nidulans. Genetics* **178,** 1355-69.

Jensen, S. E., Wong, A., Rollins, M. J., and Westlake, D. W. (1990) Purification and partial characterization of delta-(L-alpha-aminoadipyl)-L-cysteinyl-D-valine synthetase from *Streptomyces clavuligerus. J Bacteriol* **172,** 7269-71.

Jorgensen, H., Nielsen, J., Villadsen, J., and Mollgaard, H. (1995) Metabolic flux distributions in Penicillium chrysogenum during fed-batch cultivations. *Biotechnol Bioeng* **46,** 117-31.

Kallow, W., Kennedy, J., Arezi, B., Turner, G., and von Dohren, H. (2000) Thioesterase domain of delta-(l-alpha-Aminoadipyl)-l-cysteinyl-d-valine synthetase: alteration of stereospecificity by site-directed mutagenesis. *J Mol Biol* **297,** 395-408.

Kallow, W., Pavela-Vrancic, M., Dieckmann, R., and von Dohren, H. (2002) Nonribosomal peptide synthetases-evidence for a second ATP-binding site. *Biochim Biophys Acta* **1601,** 93-9.

Kallow, W., von Dohren, H., and Kleinkauf, H. (1998) Penicillin biosynthesis: energy requirement for tripeptide precursor formation by delta-(L-alpha-aminoadipyl)-L-cysteinyl-D-valine synthetase from *Acremonium chrysogenum. Biochemistry* **37,** 5947-52.

Keating, T. A., Marshall, C. G., Walsh, C. T., and Keating, A. E. (2002) The structure of VibH represents nonribosomal peptide synthetase condensation, cyclization and epimerization domains. *Nat Struct Biol* **9,** 522-6.

Kern, B. A., Hendlin, D., and Inamine, E. (1980) L-lysine epsilon-aminotransferase involved in cephamycin C synthesis in *Streptomyces lactamdurans. Antimicrob Agents Chemother* **17,** 679-85.

Kiel, J. A., van den Berg, M. A., Fusetti, F., Poolman, B., Bovenberg, R. A., Veenhuis, M., and van der Klei, I. J. (2009) Matching the proteome to the genome: the microbody of penicillin-producing *Penicillium chrysogenum* cells. *Funct Integr Genomics* **9,** 167-84.

KIEL, J. A., VAN DER KLEI, I. J., VAN DEN BERG, M. A., BOVENBERG, R. A., AND VEENHUIS, M. (2005) Overproduction of a single protein, Pc-Pex11p, results in 2-fold enhanced penicillin production by *Penicillium chrysogenum. Fungal Genet Biol* **42,** 154-64.

KIEL, J. A., VEENHUIS, M., AND VAN DER KLEI, I. J. (2006) PEX genes in fungal genomes: common, rare or redundant. *Traffic* **7,** 1291-303.

KIM, C. F., LEE, S. K., PRICE, J., JACK, R. W., TURNER, G., AND KONG, R. Y. (2003) Cloning and expression analysis of the pcbAB-pcbC beta-lactam genes in the marine fungus *Kallichroma tethys. Appl Environ Microbiol* **69,** 1308-14.

KLEIJN, R. J., LIU, F., VAN WINDEN, W. A., VAN GULIK, W. M., RAS, C., AND HEIJNEN, J. J. (2007) Cytosolic NADPH metabolism in penicillin-G producing and non-producing chemostat cultures of *Penicillium chrysogenum. Metab Eng* **9,** 112-23.

KLEINKAUF, H., AND VON DOHREN, H. (1996) A nonribosomal system of peptide biosynthesis. *Eur J Biochem* **236,** 335-51.

KOETSIER, M. J., GOMBERT, A.K., FEKKEN, S., BOVENBERG, R.A.L., VAN DEN BERG, M.A., KIEL, J.A.K.W., JEKEL, P.A., JANSSEN, D.B., PRONK, J.T., VAN DER KLEI, I.J., AND DARAN, J.M. (2009a) The *Penicillium chrysogenum aclA* gene encodes a broad-substrate-specificity acyl-coenzyme A ligase involved in activation of adipic acid, a side-chain precursor for cephem antibiotics. *Fungal Genet Biol* in press.

KOETSIER, M. J., JEKEL, P. A., VAN DEN BERG, M. A., BOVENBERG, R. A., AND JANSSEN, D. B. (2009b) Characterization of a phenylacetate-CoA ligase from *Penicillium chrysogenum. Biochem J* **417,** 467-76.

KOFFAS, M., ROBERGE, C., LEE, K., AND STEPHANOPOULOS, G. (1999) Metabolic engineering. *Annu Rev Biomed Eng* **1,** 535-57.

KOGLIN, A., MOFID, M. R., LOHR, F., SCHAFER, B., ROGOV, V. V., BLUM, M. M., MITTAG, T., MARAHIEL, M. A., BERNHARD, F., AND DOTSCH, V. (2006) Conformational switches modulate protein interactions in peptide antibiotic synthetases. *Science* **312,** 273-6.

KOHLI, R. M., AND WALSH, C. T. (2003) Enzymology of acyl chain macrocyclization in natural product biosynthesis. *Chem Commun (Camb)* **7,** 297-307.

KOLAR, M., PUNT, P. J., VAN DEN HONDEL, C. A., AND SCHWAB, H. (1988) transformation of *Penicillium chrysogenum* using dominant selection markers and expression of an *Escherichia coli lacZ* fusion gene. *Gene* **62,** 127-34.

KOVACEVIC, S., WEIGEL, B. J., TOBIN, M. B., INGOLIA, T. D., AND MILLER, J. R. (1989) Cloning, characterization, and expression in *Escherichia coli* of the *Streptomyces clavuligerus* gene encoding deacetoxycephalosporinC synthetase. *J Bacteriol* **171,** 754-60.

LAICH, F., FIERRO, F., AND MARTIN, J. F. (2002) Production of penicillin by fungi growing on food products: identification of a complete penicillin gene cluster in Penicillium griseofulvum and a truncated cluster in *Penicillium verrucosum. Appl Environ Microbiol* **68,** 1211-9.

LAMAS-MACEIRAS, M., VACA, I., RODRIGUEZ, E., CASQUEIRO, J., AND MARTIN, J. F. (2006) Amplification and disruption of the phenylacetyl-CoA ligase gene of *Penicillium chrysogenum* encoding an aryl-capping enzyme that supplies phenylacetic acid to the isopenicillin N-acyltransferase. *Biochem J* **395,** 147-55.

LEIN, J. (1986) The Panlabs penicillin strain improvement program. Butterworth Publishers, Boston.

LINNE, U., DOEKEL, S., AND MARAHIEL, M. A. (2001) Portability of epimerization domain and role of peptidyl carrier protein on epimerization activity in nonribosomal peptide synthetases. *Biochemistry* **40,** 15824-34.

LIPMANN, F. (1971) Attempts to map a process evolution of peptide biosynthesis. *Science* **173**, 875–84.

LUTZ, M. V., BOVENBERG, R. A., VAN DER KLEI, I. J., AND VEENHUIS, M. (2005) Synthesis of *Penicillium chrysogenum* acetyl-CoA:isopenicillin N acyltransferase in *Hansenula polymorpha:* first step towards the introduction of a new metabolic pathway. *FEMS Yeast Res* **5,** 1063-7.

MAYORGA, M. E., AND TIMBERLAKE, W. E. (1992) The developmentally regulated *Aspergillus nidulans wA* gene encodes a polypeptide homologous to polyketide and fatty acid synthases. *Mol Gen Genet* **235,** 205-12.

MINDLIN, S. Z., ALIKHANIAN, S. I., VLADIMIROV, A. V., AND MIKHAILOVA, G. R. (1961) A new hybrid strain of an oxytetracycline-producing organism, *Streptomyces rimosus. Appl Microbiol* **9,** 349-53.

MULLER, W. H., BOVENBERG, R. A., GROOTHUIS, M. H., KATTEVILDER, F., SMAAL, E. B., VAN DER VOORT, L. H., AND VERKLEIJ, A. J. (1992) Involvement of microbodies in penicillin biosynthesis. *Biochim Biophys Acta* **1116,** 210-3.

NAESBY. M., NIELSEN, S.V., NIELSEN, C.A., GREEN, T., TANGE, T.O., SIMÓN, E., KNECHTLE, P., HANSSON, A., SCHWAB, M.S., TITIZ, O., FOLLY, C., ARCHILA, R.E., MAVER, M., VAN SINT FIET, S., BOUSSEMGHOUNE, T., JANES, M., KUMAR, A.S., SONKAR, S.P., MITRA, P.P., BENJAMIN, V.A., KORRAPATI, N., SUMAN, I., HANSEN, E.H., THYBO, T., GOLDSMITH, N., AND SORENSEN, A.S. (2009) Yeast artificial chromosomes employed for random assembly of biosynthetic pathways and production of diverse compounds in Saccharomyces cerevisiae. *Microb Cell Fact* **13**, 45.

NEWBERT, R. W., BARTON, B., GREAVES, P., HARPER, J., AND TURNER, G. (1997) Analysis of a commercially improved *Penicillium chrysogenum* strain series: involvement of recombinogenic regions in amplification and deletion of the penicillin biosynthesis gene cluster. *J Ind Microbiol Biotechnol* **19,** 18-27.

NIEBOER, M., AND BOVENBERG, R.A. (1999) Improved *in vivo* production of cehpalosporins.

NIJLAND, J. G., KOVALCHUK, A., VAN DEN BERG, M. A., BOVENBERG, R. A., AND DRIESSEN, A. J. (2008) Expression of the transporter encoded by the *cefT* gene of *Acremonium chrysogenum* increases cephalosporin production in *Penicillium chrysogenum. Fungal Genet Biol* **45,** 1415-21.

OKUDA, K., UEMURA, I., BODLEY, J. W., AND WINNICK, T. (1964) Further Aspects of Gramicidin and Tyrocidine Biosynthesis in the Cell-Free System of *Bacillus Brevis. Biochemistry* **3,** 108-13.

OLIVERA, E. R., MINAMBRES, B., GARCIA, B., MUNIZ, C., MORENO, M. A., FERRANDEZ, A., DIAZ, E., GARCIA, J. L., AND LUENGO, J. M. (1998) Molecular characterization of the phenylacetic acid catabolic pathway in *Pseudomonas putida U*: the phenylacetyl-CoA catabolon. *Proc Natl Acad Sci U S A* **95,** 6419-24.

OTTEN, L. G., SIO, C. F., REIS, C. R., KOCH, G., COOL, R. H., AND QUAX, W. J. (2007) A highly active adipyl-cephalosporin acylase obtained via rational randomization. *FEBS J* **274,** 5600-10.

PLATTA, H. W., AND ERDMANN, R. (2007) The peroxisomal protein import machinery. *FEBS Lett* **581,** 2811-9.

QUADRI, L. E., WEINREB, P. H., LEI, M., NAKANO, M. M., ZUBER, P., AND WALSH, C. T.

(1998) Characterization of Sfp, a *Bacillus subtilis* phosphopantetheinyl transferase for peptidyl carrier protein domains in peptide synthetases. *Biochemistry* **37,** 1585-95.

QUEENER, S. W., SEBEK, O. K., AND VEZINA, C. (1978) Mutants blocked in antibiotic synthesis. *Annu Rev Microbiol* **32,** 593-636.

RHEE, D. K., LEE, S. B., RHEE, J. S., RYU, D. D., AND HOSPODKA, J. (1980) Enzymatic biosynthesis of cephalexin. *Biotechnol Bioeng* **22,** 1237-47.

ROACH, P. L., CLIFTON, I. J., HENSGENS, C. M., SHIBATA, N., LONG, A. J., STRANGE, R. W., HASNAIN, S. S., SCHOFIELD, C. J., BALDWIN, J. E., AND HAJDU, J. (1996) Anaerobic crystallisation of an isopenicillin N synthase.Fe(II).substrate complex demonstrated by X-ray studies. *Eur J Biochem* **242,** 736-40.

RODRÍGUEZ-SÁIZ, M., DÍEZ, B., AND BARREDO, J. L. (2005) Why did the Fleming strain fail in penicillin industry? *Fungal Genet Biol* **42,** 464-470.

ROMERO, J., MARTIN, J. F., LIRAS, P., DEMAIN, A. L., AND RIUS, N. (1997) Partial purification, characterization and nitrogen regulation of the lysine epsilon-aminotransferase of *Streptomyces clavuligerus. J Ind Microbiol Biotechnol* **18,** 241-6.

SCHLIEBS, W., AND KUNAU, W. H. (2006) PTS2 co-receptors: diverse proteins with common features. *Biochim Biophys Acta* **1763,** 1605-12.

SCHLUMBOHM, W., STEIN, T., ULLRICH, C., VATER, J., KRAUSE, M., MARAHIEL, M. A., KRUFT, V., AND WITTMANN-LIEBOLD, B. (1991) An active serine is involved in covalent substrate amino acid binding at each reaction center of gramicidin S synthetase. *J Biol Chem* **266,** 23135-41.

SHAW-REID, C. A., KELLEHER, N. L., LOSEY, H. C., GEHRING, A. M., BERG, C., AND WALSH, C. T. (1999) Assembly line enzymology by multimodular nonribosomal peptide synthetases: the thioesterase domain of *E. coli EntF* catalyzes both elongation and cyclolactonization. *Chem Biol* **6,** 385-400.

SIEBER, S. A., LINNE, U., HILLSON, N. J., ROCHE, E., WALSH, C. T., AND MARAHIEL, M. A. (2002) Evidence for a monomeric structure of nonribosomal Peptide synthetases. *Chem Biol* **9,** 997-1008.

SIEWERS, V., CHEN, X., HUANG, L., ZHANG, J., AND NIELSEN, J. (2009) Heterologous production of non-ribosomal peptide LLD-ACV in *Saccharomyces cerevisiae. Metab Eng* in press.

SIO, C. F., AND QUAX, W. J. (2004) Improved beta-lactam acylases and their use as industrial biocatalysts. *Curr Opin Biotechnol* **15,** 349-55.

SIO, C. F., RIEMENS, A. M., VAN DER LAAN, J. M., VERHAERT, R. M., AND QUAX, W. J. (2002) Directed evolution of a glutaryl acylase into an adipyl acylase. *Eur J Biochem* **269,** 4495-504.

SNOEK, I. S., VAN DER KROGT, Z. A., TOUW, H., KERKMAN, R., PRONK, J. T., BOVENBERG, R. A., VAN DEN BERG, M. A., AND DARAN, J. M. (2009) Construction of an *hdfA Penicillium chrysogenum* strain impaired in non-homologous end-joining and analysis of its potential for functional analysis studies. *Fungal Genet Biol* **46,** 418-26.

SPROTE, P., BRAKHAGE, A. A., AND HYNES, M. J. (2009) Contribution of peroxisomes to penicillin biosynthesis in A*spergillus nidulans. Eukaryot Cell* **8,** 421-3.

SPROTE, P., HYNES, M. J., HORTSCHANSKY, P., SHELESTY, E., SCHARF, D. H., WOLKE, S. M., AND BRAKHAGE, A. A. (2008) Identification of the novel penicillin biosynthesis gene *aatB* of *Aspergillus nidulans* and its putative evolutionary relationship to this fungal secondary metabolism gene cluster. *Mol Microbiol* **70,** 445-61.

Stachelhaus, T., Mootz, H. D., Bergendahl, V., and Marahiel, M. A. (1998) Peptide bond formation in nonribosomal peptide biosynthesis. Catalytic role of the condensation domain. *J Biol Chem* **273,** 22773-81.

Stachelhaus, T., Schneider, A., and Marahiel, M. A. (1995) Rational design of peptide antibiotics by targeted replacement of bacterial and fungal domains. *Science* **269,** 69-72.

Stachelhaus, T., and Walsh, C. T. (2000) Mutational analysis of the epimerization domain in the initiation module PheATE of gramicidin S synthetase. *Biochemistry* **39,** 5775-87.

Stein, T., Vater, J., Kruft, V., Otto, A., Wittmann-Liebold, B., Franke, P., Panico, M., McDowell, R., and Morris, H. R. (1996) The multiple carrier model of nonribosomal peptide biosynthesis at modular multienzymatic templates. *J Biol Chem* **271,** 15428-35.

Thakur, D., Roy, M. K., and Bora, T. C. (2009) Expandase-like activity mediated cell-free conversion of ampicillin to cephalexin by *Streptomyces* sp. DRS I. *Biotechnol Lett* **31,** 1059-64.

Theilgaard, H., van den Berg, M.A., Mulder, C., Bovenberg, R.A., and Nielsen, J. (2001) Quantitative analysis of Penicillium chrysogenum Wis54-1255 transformants overexpressing the penicillin biosynthetic genes. Biotechnol Bioeng **72,** 379-88.

Tobin, M. B., Cole, S. C., Kovacevic, S., Miller, J. R., Baldwin, J. E., and Sutherland, J. D. (1994) Acyl-coenzyme A: isopenicillin N acyltransferase from *Penicillium chrysogenum*: effect of amino acid substitutions at Ser227, Ser230 and Ser309 on proenzyme cleavage and activity. *FEMS Microbiol Lett* **121,** 39-46.

Tobin, M.B., Cole, S.C., Miller, J. R., Baldwin, J. E., and Sutherland, J. D. (1995) Amino-acid substitutions in the cleavage site of acyl-coenzyme A:isopenicillin N acyltransferase from *Penicillium chrysogenum*: effect on proenzyme cleavage and activity. *Gene* **162,** 29-35.

Tobin, M. B., Fleming, M. D., Skatrud, P. L., and Miller, J. R. (1990) Molecular characterization of the acyl-coenzyme A:isopenicillin N acyltransferase gene (*penDE*) from *Penicillium chrysogenum* and *Aspergillus nidulans* and activity of recombinant enzyme in *Escherichia coli. J Bacteriol* **172,** 5908-14.

Ullan, R. V., Campoy, S., Casqueiro, J., Fernandez, F. J., and Martin, J. F. (2007) DeacetylcephalosporinC production in *Penicillium chrysogenum* by expression of the isopenicillin N epimerization, ring expansion, and acetylation genes. *Chem Biol* **14,** 329-39.

Ullan, R. V., Casqueiro, J., Naranjo, L., Vaca, I., and Martin, J. F. (2004) Expression of *cefD2* and the conversion of isopenicillin N into penicillin N by the two-component epimerase system are rate-limiting steps in cephalosporin biosynthesis. *Mol Genet Genomics* **272,** 562-70.

van den Berg, M. A., Albang, R., Albermann, K., Badger, J. H., Daran, J. M., Driessen, A. J., Garcia-Estrada, C., Fedorova, N. D., Harris, D. M., Heijne, W. H., Joardar, V., Kiel, J. A., Kovalchuk, A., Martin, J. F., Nierman, W. C., Nijland, J. G., Pronk, J. T., Roubos, J. A., van der Klei, I. J., van Peij, N. N., Veenhuis, M., von Dohren, H., Wagner, C., Wortman, J., and Bovenberg, R. A. (2008) Genome sequencing and analysis of the filamentous fungus Penicillium chrysogenum. *Nat Biotechnol* **26,** 1161-8.

van den Berg, M. A., Bovenberg, R. A., de Laat, W. T., and van Velzen, A. G. (1999) Engineering aspects of beta-lactam biosynthesis. *Antonie Van Leeuwenhoek* **75,** 155-61.

van den Berg, M. A., Westerlaken, I., Leeflang, C., Kerkman, R., and Bovenberg, R. A. (2007) Functional characterization of the penicillin biosynthetic gene cluster of *Penicillium chrysogenum* Wisconsin54-1255) *Fungal Genet Biol* **44,** 830-44.

van Gulik, W. M., de Laat, W. T., Vinke, J. L., and Heijnen, J. J. (2000) Application of metabolic flux analysis for the identification of metabolic bottlenecks in the biosynthesis of penicillin-G. *Biotechnol Bioeng* **68,** 602-18.

van Liempt, H., von Dohren, H., and Kleinkauf, H. (1989) delta-(L-alpha-aminoadipyl)-L-cysteinyl-D-valine synthetase from *Aspergillus nidulans*. The first enzyme in penicillin biosynthesis is a multifunctional peptide synthetase. *J Biol Chem* **264,** 3680-4.

Visser, W. F., van Roermund, C. W., Ijlst, L., Waterham, H. R., and Wanders, R. J. (2007) Metabolite transport across the peroxisomal membrane. *Biochem J* **401,** 365-75.

Vollenbroich, D., Kluge, B., D'Souza, C., Zuber, P., and Vater, J. (1993) Analysis of a mutant amino acid-activating domain of surfactin synthetase bearing a serine-to-alanine substitution at the site of carboxylthioester formation. *FEBS Lett* **325,** 220-4.

Waksman, S. A., Schatz, A. and Reilly, H. C. (1946) Metabolism and the Chemical Nature of *Streptomyces griseus. J Bacteriol* **51,** 753-9.

Walsh, C. (2003) Where will new antibiotics come from? *Nat Rev Microbiol* **1,** 65-70.

Walsh, C. T. (2008) The chemical versatility of natural-product assembly lines. *Acc Chem Res* **41,** 4-10.

Wanders, R. J. (2004) Peroxisomes, lipid metabolism, and peroxisomal disorders. *Mol Genet Metab* **83,** 16-27.

Wang, F. Q., Liu, J., Dai, M., Ren, Z. H., Su, C. Y., and He, J. G. (2007) Molecular cloning and functional identification of a novel phenylacetyl-CoA ligase gene from *Penicillium chrysogenum. Biochem Biophys Res Commun* **360,** 453-8.

Wang, F. Q., Zhao, Y., Dai, M., Liu, J., Zheng, G. Z., Ren, Z. H., and He, J. G. (2008) Cloning and functional identification of C-4 methyl sterol oxidase genes from the penicillin-producing fungus *Penicillium chrysogenum. FEMS Microbiol* Lett **287,** 91-9.

Weber, T., Baumgartner, R., Renner, C., Marahiel, M. A., and Holak, T. A. (2000) Solution structure of PCP, a prototype for the peptidyl carrier domains of modular peptide synthetases. *Structure* **8,** 407-18.

Weinstein, M. J., Luedemann, G. M., Oden, E. M., Wagman, G. H., Rosselet, J. P., Marquez, J. A., Coniglio, C. T., Charney, W., Herzog, H. L., and Black, J. (1963) Gentamicin, a New Antibiotic Complex from *Micromonospora. J Med Chem* **6,** 463-4.

Whiteman, P. A., Abraham, E. P., Baldwin, J. E., Fleming, M. D., Schofield, C. J., Sutherland, J. D., and Willis, A. C. (1990) Acyl coenzyme A: 6-aminopenicillanic acid acyltransferase from *Penicillium chrysogenum* and *Aspergillus nidulans. FEBS Lett* **262,** 342-4.

Xie, Y., Van de Sandt, E., de Weerd, T., and Wang, N. H. (2001) Purification of adipoyl-7-amino-3-deacetoxycephalosporanic acid from fermentation broth using stepwise elution with a synergistically adsorbed modulator. *J Chromatogr A* **908,** 273-91.

XU, Z., VAN DEN BERG, M. A., SCHEURING, C., COVALEDA, L., LU, H., SANTOS, F. A., UHM, T., LEE, M. K., WU, C., LIU, S., AND ZHANG, H. B. (2005) Genome physical mapping from large-insert clones by fingerprint analysis with capillary electrophoresis: a robust physical map of *Penicillium chrysogenum. Nucleic Acids Res* **33,** e50.

YANG, L., AND WEI, D. Z. (2003) Enhanced enzymatic synthesis of a semi-synthetic cephalosprin, cefaclor, with in situ product removal. *Biotechnol Lett* **25,** 1195-8.

YOUSHKO, M. I., MOODY, H. M., BUKHANOV, A. L., BOOSTEN, W. H., AND SVEDAS, V. K. (2004) Penicillin acylase-catalyzed synthesis of beta-lactam antibiotics in highly condensed aqueous systems: beneficial impact of kinetic substrate supersaturation. *Biotechnol Bioeng* **85,** 323-9.

ZHANG, J., WOLFE, S., AND DEMAIN, A. L. (1992) Biochemical studies on the activity of delta-(L-alpha-aminoadipyl)-L-cysteinyl-D-valine synthetase from *Streptomyces clavuligerus. Biochem J* **283,** 691-8.

Biotechnology and Genetic Engineering Reviews - Vol. 27, 33-56 (2010)

Engineering the future. Development of transgenic plants with enhanced tolerance to adverse environments

MATIAS D. ZURBRIGGEN[1], MOHAMMAD-REZA HAJIREZAEI[2] AND NESTOR CARRILLO[1]*

[1]*Instituto de Biología Molecular y Celular de Rosario (IBR-CONICET), Universidad Nacional de Rosario, Suipacha 531, S2002LRK Rosario, Argentina;*
[2]*Leibniz-Institut für Pflanzengenetik und Kulturpflanzenforschung (Leibniz-IPK), Corrensstr. 3, D-06466 Gatersleben, Germany*

Abstract

Environmental stresses - especially drought and salinity - and iron limitation are the primary causes of crop yield losses. Therefore, improvement of plant stress tolerance has paramount relevance for agriculture, and vigorous efforts are underway to design stress-tolerant crops. Three aspects of this ongoing research are reviewed here. First, attempts have been made to strengthen endogenous plant defences, which are characterised by intertwined, hierarchical gene networks involved in stress perception, signalling, regulation and expression of effector proteins, enzymes and metabolites. The multigenic nature of this response requires detailed knowledge of the many actors

*To whom correspondence should be addressed (carrillo@ibr.gov.ar)

Abbreviations: ABA, abscisic acid; ABRE, abscisic acid-responsive element; ALR, NADPH-dependent aldose/aldehyde reductase; AREB1/ABF2, bZIP transcription factor; APX, ascorbate peroxidase; BADH, betaine aldehyde dehydrogenase; DRE, dehydration-responsive element; FAD7, ω-3 fatty acid desaturase; Fd, ferredoxin; Fld, flavodoxin; FNR, ferredoxin-NADP$^+$ reductase; FTR, ferredoxin-thioredoxin reductase; GS, glutathione synthase; HR, hypersensitive response; HSP, heat-shock protein; JERF3, tomato jasmonate- and ethylene-responsive factor 3; LCD, localised cell death; LEA, late embryogenesis abundant; MAP, mitogen-activated protein; MATE, multidrug and toxin efflux; NHX1, Na^+/H^+ antiporter; PETC, photosynthetic electron transport chain; ROS, reactive oxygen species; SOD, superoxide dismutase; SOS, salt overly sensitive; TERF1, tomato ethylene-responsive factor 1; Trx, thioredoxin; UV, ultraviolet; WT, wild-type; *Xcv*, *Xanthomonas campestris* pv. *vesicatoria*.

and interactions involved in order to identify proper intervention points, followed by significant engineering of the prospective genes to prevent undesired side-effects. A second important aspect refers to the effect of concurrent stresses as plants normally meet in the field (*e.g.*, heat and drought). Recent findings indicate that plant responses to combined environmental hardships are somehow unique and cannot be predicted from the addition of the individual stresses, underscoring the importance of programming research within this conceptual framework. Finally, the photosynthetic microorganisms from which plants evolved (*i.e.*, algae and cyanobacteria) deploy a totally different strategy to acquire stress tolerance, based on the substitution of stress-vulnerable targets by resistant isofunctional proteins that could take over the lost functions under adverse conditions. Reintroduction of these ancient traits in model and crop plants has resulted in increased tolerance to environmental hardships and iron starvation, opening a new field of opportunities to increase the endurance of crops growing under suboptimal conditions.

Introduction

Adverse environmental situations, including inappropriate soils, nutrient deficit, abiotic stresses, pathogens, industrial pollutants and other consequences of anthropocentric activities, constitute the most important negative factor affecting the yield of agricultural production worldwide (Boyer, 1982; Umezawa *et al.*, 2006; Vij and Tyagi, 2007; Vinocur and Altman, 2005). When growing in their natural habitats, wild plants encounter similar environmental challenges in the course of their lifetimes and accordingly, during millennia of adaptive evolution, they have developed numerous strategies to survive and set seeds under these unfavourable conditions. Selection of crops by humans, instead, has been conducted with a bias towards their productivity in agriculture, and along the process of domestication, many traits associated with stress tolerance that were present in the wild ancestors of modern cultivars have been lost. An estimation made on eight major crops in which average yields were compared with record yields obtained under supposedly optimal conditions indicates that the combination of biotic and abiotic stresses resulted in losses in the range of 60-90 % (*Figure 1*). These observations suggest that there is plenty of room for yield improvement and therefore, engineering stress tolerance in plants has paramount economic relevance.

The design of novel strategies to accomplish this purpose requires a thorough understanding of the molecular mechanisms underlying plant tolerance to adverse environments, and of the built-in and inducible systems that plants set in motion to overcome the adverse situation. Given the diversity of stresses that a plant may face in the field and the many different effects these conditions might exert on physiology, development and reproduction, this might seem as a hopeless endeavour. However, recent findings have shown that plant responses to different environmental insults, although displaying idiosyncratic features, proceed through formally analogous pathways involving stress perception, signal transduction and regulatory networks. They affect the expression of downstream stress-related genes and metabolites which, in turn, attempt to protect and repair biomolecules and membranes, and re-establish homeostasis (Vinocur and Altman, 2005). These protective pathways operate in the

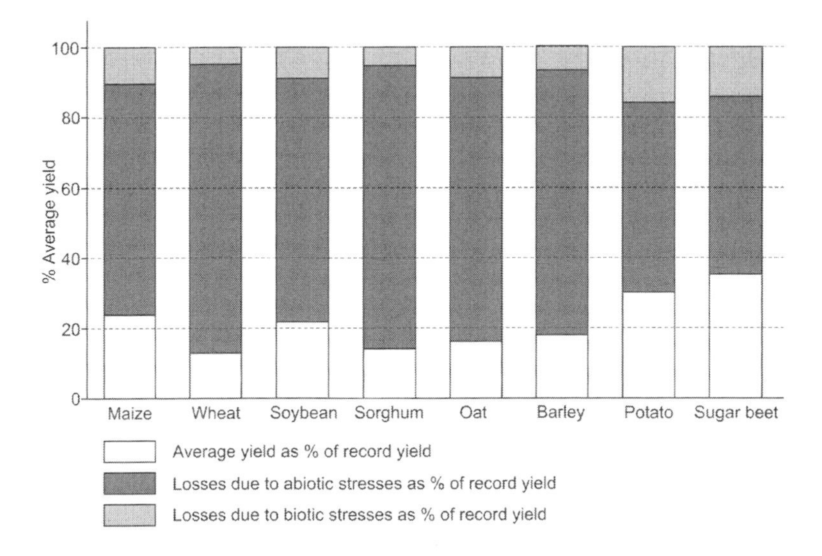

Figure 1. Yield losses of eight representative crops due to environmental stress. Average yields and losses caused by abiotic and biotic stresses are illustrated as a percentage of the record yield obtained for each of the crops: maize, wheat, soybean, sorghum, oat, barley, potato and sugar beet. Data taken from Buchanan *et al.* (2000).

form of hierarchically ordered cascades and display a considerable degree of cross-talk, among them and with other metabolic and developmental processes of the cell. The knowledge gained through these studies allowed the identification of potential intervention points amenable to be manipulated in order to strengthen endogenous defences. Accordingly, several comprehensive studies on the molecular mechanisms of nutritional and environmental stress tolerance, many of them profiling a large number of stress-related genes and signal transduction systems, have been published in recent years (reviewed in Kim and Guerinot, 2007; Seki *et al.*, 2003; Vinocur and Altman, 2005; Zhang *et al.*, 2004). Attempts to achieve increased tolerance based on boosting of endogenous systems and through the genetic manipulation of specific plant genes have been pursued actively, with variable degree of success (Flowers, 2004; Ito *et al.*, 2006; Mittler, 2006; Vij and Tyagi, 2007; Wang *et al.*, 2003).

On the other hand, comparison of the stress responses of plants with those deployed by photosynthetic microorganisms which share common ancestors with them (*i.e.*, algae and cyanobacteria) against similar environmental challenges allowed identification of defence mechanisms that have been lost in plants, but could be reintroduced with advantage, opening a completely new field of opportunities (Zurbriggen *et al.*, 2008). Finally, it is worth noting that the simultaneous occurrence of several stresses, rather than a single one, is the common situation in the field and is most lethal to crops. This agronomically relevant situation has been, however, rarely addressed by molecular biologists studying stress tolerance in plants (Mittler, 2006). Acknowledgment of the limitations of a one-sided approach led to stimulating research on concurrent stresses, which in turn revealed that plant responses to a combination of environmental insults is somehow unique and cannot be directly predicted from those invoked by the individual challenges.

In this article we review these recent advances, their strengths and possibilities, and the potential of a strategy based on the use of cyanobacterial genes and mechanisms. Emphasis will be given on the possibility of achieving wide range tolerance to a combination of stresses as plants normally face in the field.

Engineering plant endogenous defences for improved abiotic stress tolerance

As indicated, plant perception of an adverse environment triggers a complex response governed by networks of signalling factors, transcriptional regulators and downstream responsive genes that ameliorate the damage undergone by the organism (*Figure 2*). Different sources of stress (drought, chilling, salinity) display both common and unique features with respect to other responses and to metabolic and developmental pathways of the plant. Significant cross-talk and overlap exist among them, which could be synergistic or antagonistic (Mittler, 2006). Although this observation opens the possibility that tolerance to multiple sources of stress could be gained through a single transgenic intervention, it also limits the number of useful intervention points, and often requires sophisticated regulation of the introduced gene(s) to prevent undesirable impacts on plant growth and development (Gutterson and Zhang, 2004).

Engineering strategies for abiotic stress tolerance (*Figure 2*, *Table 1*) have relied on expression of either *i*) genes involved in signalling and regulatory pathways (Seki *et al.*, 2003; Shinozaki *et al.*, 2003); *ii*) genes that encode proteins conferring tolerance, such as heat-shock proteins (HSPs), late embryogenesis abundant (LEA) proteins and antioxidant enzymes (Vinocur and Altman, 2005; Wang *et al.*, 2004); or *iii*) enzymes involved in detoxification pathways or in the synthesis of protective metabolites (Apse and Blumwald, 2002; Chen and Murata, 2002; Park *et al.*, 2004). We will briefly discuss the results obtained with these various approaches.

Stress-responsive genes involved in signalling cascades and in expression control, including those encoding mitogen-activated protein (MAP) and salt overly sensitive (SOS) kinases, phospholipases and transcription factors such as the dehydration-responsive element (DRE) binding protein, have been extensively characterised (Qiu *et al.*, 2002; Shou *et al.*, 2004; Umezawa *et al.*, 2006; Zhang *et al.*, 2004). Since they operate high in the hierarchy of stress responses, they are good candidates to obtain broad-range tolerance. Several of these signal transduction and transcription factors have been introduced into plants and found to increase stress tolerance (Vij and Tyagi, 2007, see also Table 1). For instance, constitutive expression of the tobacco MAP kinase kinase kinase 1 in maize activates an oxidative signal cascade that leads to chilling, heat and salinity tolerance in the transformants, providing significant protection to photosynthesis under these conditions (Shou *et al.*, 2004). Other examples are the overexpression of the rice *OsDREB1a* gene (encoding a drought-responsive transcriptional regulator) and sunflower *Hahb-4* coding sequence (a transcription factor of the homeodomain-leucine zipper family, regulated by water availability and abscisic acid) in transgenic *Arabidopsis*, which resulted in increased freezing and high-salt or drought tolerance, respectively (Dezar *et al.*, 2005; Ito *et al.*, 2006; Manavella *et al.*, 2006). This approach is not without problems, however, since constitutive expression of regulators also affects stress-unrelated pathways and is often associated with growth handicaps and alterations in basic metabolism (Dubouzet *et al.*, 2003; Ito *et al.*, 2006). These drawbacks could be bypassed, in principle, by the use of stress-inducible instead of constitutive promoters, but the situation highlights the difficulties of manipulating key regulators in the absence of general rules to identify proper intervention points (Gutterson and Zhang, 2004).

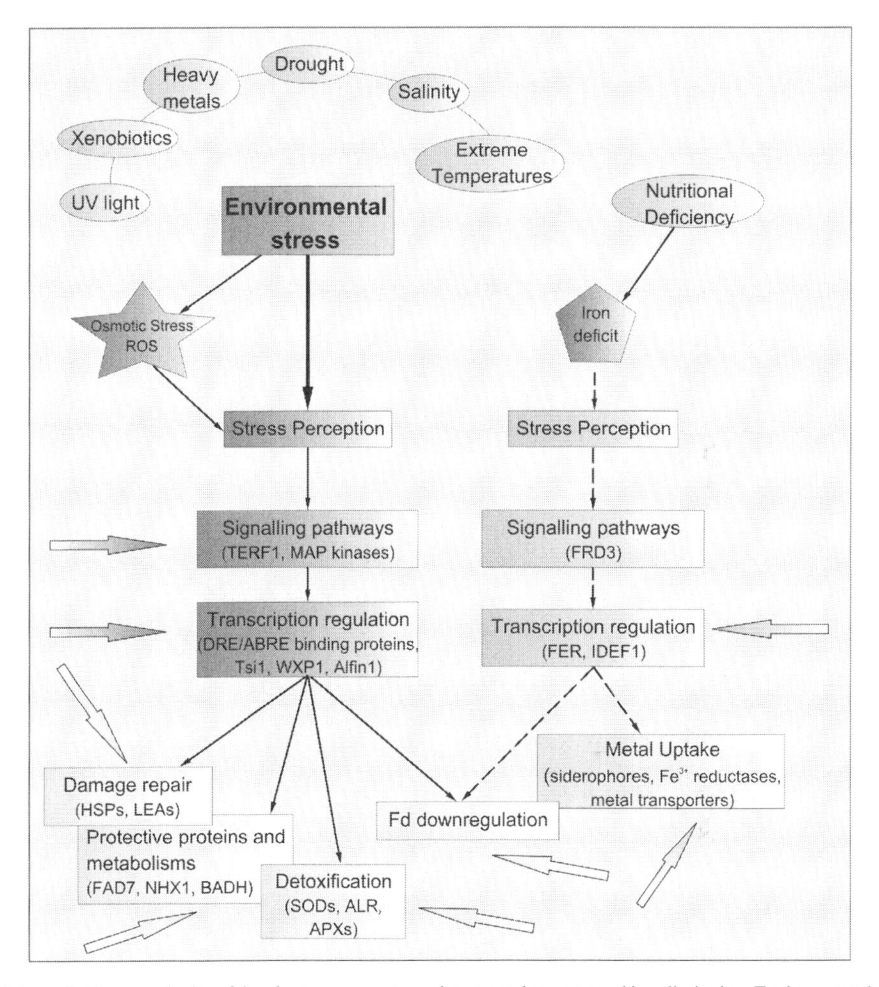

Figure 2. The complexity of the plant responses to environmental stresses and iron limitation. Environmental adversities (drought, extreme temperatures, salinity, UV light, xenobiotics) could cause secondary stresses such as oxidative and osmotic stress. Initial stress signals (*e.g.*, iron status, changes in membrane fluidity or light quality) trigger downstream signalling and transcriptional control cascades, which activate stress-responsive mechanisms to re-establish homeostasis, eliminate toxic compounds and protect and repair damaged proteins and membranes. Some components of the iron restriction and environmental stress cascades are included as examples: tomato ethylene-response factor (TERF1) conferring osmotic stress tolerance; drought- and abscisic acid-responsive element (DRE/ABRE) binding proteins; mitogen-activated protein (MAP) kinases; tobacco transcription factor (Tsi1) conferring tolerance to osmotic stress; WXP1 and Alfin1, alfalfa transcription factors conferring drought and salinity tolerance; respectively; FRD3, ferric reductase defective, a multidrug and toxin efflux (MATE)-type protein acting as sensor of shoot iron status; FER, tomato basic helix-loop-helix transcription factor involved in iron deficit response; IDEF1, rice transcription factor binding the iron deficiency-responsive *cis*-acting element; Fe-chelate reductases; broad-range metal transporters; enzymes for the synthesis of phytosiderophores; FAD7, ω-3 fatty acid desaturase, NHX1, Na^+/H^+ antiporter; BADH, betaine aldehyde dehydrogenase; SOD, superoxide dismutase; APX, ascorbate peroxidase; ALR, NADPH-dependent aldose/aldehyde reductase; HSP, heat-shock proteins; LEA, late embryogenesis abundant proteins. Intervention points that have been engineered in model or crop plants are indicated by arrows.

Table 1. Molecular networks and intervention points in plant stress responses. More comprehensive list of genes engineered in model and crop species can be found in Guerinot (2007), Umezawa *et al.* (2006), Vij and Tyagi (2007).

Strategies to increase abiotic stress tolerance in plants	Selected examples	Function
Engineering of signalling components and transcriptional regulators	Signalling components: members of MAP kinase cascades, Ca^{++}-dependent kinases, protein phosphatases Transcriptional regulators: Tsi1, WXP1 and Alfin1, DRE/ABRE-binding proteins, AREB1/ABF2 Hormones, signalling molecules: ethylene- and jasmonate-responsive factors (TERF1, JERF3), ABA biosynthesis and catabolism	Drought, osmotic and salt tolerance
Overexpression of enzymes and proteins conferring tolerance	Heat-shock proteins: HSP101, HSP17.7, mitochondrial small HSP	Protein folding-stabilisation, translational modulation
	LEA proteins	Protein and membrane stabilisation
	Ferritin	Iron homeostasis/scavenging
	ATPases and high and low affinity ion transporters (Na^+/H^+ antiporters), aquaporins	Regulation of osmolarity, stomatal behavior, sequestration of toxic ions
	Enzymatic and non-enzymatic antioxidant systems: peroxiredoxin, catalase, Mn-, Fe- and Cu/Zn-SOD, ascorbate peroxidase, glutathione reductase, ALR, glyoxilases	Detoxification of ROS
Manipulation of enzymes involved in detoxification pathways and synthesis of protective metabolties	Fatty acid desaturase and glycerol-3-phosphate acyltransferases	Chilling stress tolerance
	Betaine aldehyde dehydrogenase, choline mono-oxygenase, trehalose-6-phosphate synthase, mannitol-1-phosphate dehydrogenase, glutathione synthase/reductase	Osmotic adjustment, protein and membrane protection, ROS scavenging
Manipulation of regulators of the iron deficiency response	Transcriptional regulators: FER, IDEF1 Ferric-chelate reductase Nicotianamine synthase	Improvement of iron uptake

Abbreviations: ABA, abscisic acid; ABRE, abscisic acid-responsive element; Alfin1, alfalfa transcription factor; ALR, NADPH-dependent aldose/aldehyde reductase; AREB1/ABF2, bZIP transcription factor; DRE, dehydration-responsive element; FER, ferric-chelate reductase; HSP, heat-shock protein; IDEF1, rice transcription factor; JERF3, tomato jasmonate- and ethylene-responsive factor 3; LEA, late embryogenesis abundant; MAP, mitogen-activated protein; SOD, superoxide dismutase; TERF1, tomato ethylene-responsive factor 1; Tsi1, alfalfa transcription factor, WXP1, alfalfa transcription factor.

Many environmental insults, including high temperature, salinity and drought can cause denaturation and inactivation of biomolecules. LEA proteins, HSPs, and molecular chaperones in general, are expected to provide protection by controlling the proper folding and assembly of proteins. Indeed, a positive correlation between the levels of several HSPs and stress tolerance has been reported (Wang *et al.*, 2004). Introduction of an *Arabidopsis* HSP101 in rice resulted in increased thermotolerance (Katiyar-Agarwal *et al.*, 2003), and overexpression of LEA proteins has been shown to confer better drought endurance in a number of cases (Villalobos *et al.*, 2004).

Increased generation of reactive oxygen species (ROS), such as hydrogen peroxide and the superoxide radical, is also commonplace during many different stress situations (Mittler *et al.*, 2004). ROS play a dual role in plants, participating in signal transduction but at the same time imposing an additional oxidative stress (*Figure 2*), which is responsible for a fair share of the damage undergone by the stressed plant. Then, manipulation of ROS metabolism offers yet another opportunity to prevent stress-dependent damage. However, the approach is far from straightforward, since different members of the ROS gene network of plants have been shown to respond differently to different stress treatments (Mittler, 2006; Mittler *et al.*, 2004). Accordingly, variable levels of tolerance have been obtained by expressing ROS-scavenging enzymes in transgenic plants. Enhanced yield and survival could be obtained under some environmental stresses, which may be exemplified by superoxide dismutase-overexpressing alfalfa, pea and tobacco plants, which displayed increased tolerance to freezing and drought (reviewed in Vij and Tyagi, 2007). In other cases, however, expression of the same enzyme failed to result in augmented tolerance to oxidative stress (McKersie *et al.*, 2000).

Oxidative and osmotic stresses usually accompany many different adverse situations and contribute to the damage by negatively affecting cellular components and functions (*Figure 2*). A wide range of metabolites that can prevent these detrimental effects have been identified, including amino acids (*e.g.*, proline), quaternary and other amines (glycine-betaine, polyamines), sugars and sugar alcohols (mannitol, trehalose) and antioxidants (glutathione, ascorbate, tocopherols). Strategies for the metabolic engineering of stress physiology involve both enhanced production of desired compounds and elimination of toxic by-products (Capell and Christou, 2004). However, modulation of a single enzymatic reaction, even if rate-limiting, is generally regulated by the tendency of cells to restore metabolic homeostasis, limiting the potential of this approach. For instance, transgenic poplar trees expressing a bacterial gene encoding glutathione synthase (GS) failed to show significant changes in foliar glutathione contents despite increases in GS activities of up to 100-fold relative to control plants, indicating that glutathione synthesis is tightly controlled and not easily amenable to manipulation (Foyer *et al.*, 1995). Attempts to overproduce trehalose and mannitol in rice and wheat, respectively, led to only modest increases, although the transgenic plants were indeed more tolerant to abiotic stress (Abebe *et al.*, 2003; Penna, 2003). Targeting multiple steps in the same route could help to control metabolic fluxes in a more predictable manner. For instance, co-targeting of various steps in the corresponding pathways was found to be a successful strategy to overproduce glycine-betaine and trehalose in plants (Jang *et al.*, 2003; Yilmaz and Bulow, 2002). Overall metabolic profiling of plants under stress has therefore become a most important tool to understand stress-induced changes in protective metabolites (Rizhsky *et al.*, 2004).

Plant responses to iron deprivation

Deficiency of essential nutrients in poor lands is also a major agricultural concern and among them, iron deficit ranks top. Paradoxically, iron is the fourth most abundant element on Earth, but its bioavailabity is compromised in the presence of an oxygen-rich atmosphere due to precipitation as insoluble oxides and salts (Guerinot, 2007). The situation is particularly critical in alkaline, calcareous soils, which cover about one-third of the planet surface and constitute a formidable deterrent for agriculture (Kim and Guerinot, 2007).

In general, plants react to Fe deficit by optimising metal uptake from scarcely available sources. In response to iron limitation, dicotyledoneous species and nongraminaceous monocots display a reduction-based strategy, characterized by increased proton extrusion to the rizosphere to improve Fe^{+3} solubility, and enhanced accumulation of ferric-chelate reductases (to reduce Fe^{+3} to Fe^{+2}) and broad-range metal transporters in the root cell plasma membrane to favour Fe^{2+} intake (Kim and Guerinot, 2007). Grasses, on the other hand, have evolved a chelation-based response, releasing phytosiderophores of the mugineic acid family into the surrounding soil. These compounds bind Fe^{+3} and the complexes are then taken up by the roots (Mori, 1999).

The signalling pathways governing systemic responses to iron deficit are also illustrated in *Figure 2*. They involve shoot sensing of the nutrient status, and root devices to perceive the long-distance signals coming from the shoot (Schmidt, 2003). A likely candidate for sensing shoot information is the product of the *Arabidopsis FRD3* gene, a protein belonging to the multidrug and toxin efflux (MATE) family which is expressed in roots (Rogers and Guerinot, 2002). *FRD3* mutants fail to detect nutrient status and over-accumulate iron in all tissues. A number of downstream regulatory genes, including members of the MAP kinase cascades, 14-3-3 proteins and leucine zipper transcription factors, are strongly up-regulated upon iron restriction in *Arabidopsis* (Thimm *et al.*, 2001). Among them, the tomato *fer* gene (orthologue of the *Arabidopsis FIT* genes), which encodes a basic helix-loop-helix transcription factor, appears to be particularly important. Tomato *fer* mutants are completely unable to deploy an iron-deficit response and can only survive under iron-sufficient conditions (Brumbarova and Bauer, 2005), suggesting that FER acts at a high level of hierarchy in the iron-dependent signal cascade. Although FER has no direct function in iron status sensing (Schmidt, 2003), it could likely integrate the information of signalling molecules such as nitric oxide (Graziano and Lamattina, 2007). These regulatory genes are attractive targets for genetic engineering strategies aimed at increasing tolerance to iron deficiency, but an in-depth understanding of the molecular mechanisms involved is mandatory.

So far, attempts to improve growth and yield under iron starvation have included overexpression in rice of a mutated ferric-chelate reductase (Ishimaru *et al.*, 2007), and of nicotianamine aminotransferase, an enzyme involved in mugineic acid synthesis (Takahashi *et al.*, 2001). In both cases, increase in grain yield but not in grain iron contents was observed. Indeed, soybean plants constitutively expressing ferric-chelate reductase displayed growth penalties (Vasconcelos *et al.*, 2006), which were avoided when an iron deficit-dependent promoter was used (Ishimaru *et al.*, 2007).

Stress combinations and the "Stress Matrix"

Individually, adverse environmental situations such as water deficit, chilling or salinity have been the subject of intense research (reviewed in Umezawa *et al.*, 2006; Vij and Tyagi, 2007; Vinocur and Altman, 2005). However, plants growing in the field are regularly exposed to a combination of stresses as is frequent in drought- and heat-stricken regions or in areas where drought is combined with high salt concentrations in the soil. Although in a few cases sequential plant exposure to different sources of stress leads to the phenomenon of cross-hardening, as illustrated by the observation that ozone treatment can induce the ultraviolet (UV)-B and pathogen responses (Sandermann, 2004), the opposite situation is the rule. The combination of stresses usually results in higher damage and yield losses than the sum of the individual hardships, as exemplified in the case study of drought and heat wave (Mittler, 2006, and references therein).

The relationships and cross-talk among different stress responses provide a rationale to understand this behaviour. When plants are exposed to concurrent stresses they trigger defence responses that might be in some cases conflicting or antagonistic. In the case of heat, for example, plants open their stomata to favour transpiration and cooling of their leaves. Drought, instead, leads to stomatal closure to prevent water losses. When heat and water deficit are combined, some aspects of each individual stress prevail over the other. Under this situation, plants are no longer able to open their stomata and leaf temperature increases (Rizhsky *et al.*, 2002). By contrast, synthesis of proline, regarded as important for drought protection, is strongly suppressed during simultaneous water deficit and heat stress (Rizhsky *et al.*, 2004). High temperatures lead to significant increases of respiration, with little effect on photosynthetic rates. Water limitation, on the other hand, results in strong inhibition of photosynthesis without affecting respiratory activity (Rizhsky *et al.*, 2002; Rizhsky *et al.*, 2004). Plants subjected to a combination of both stresses display the unique phenotype of high respiration, typical of high temperatures, and low photosynthesis as in water deficit conditions (Mittler, 2006; Rizhsky *et al.*, 2002; Rizhsky *et al.*, 2004), which represents a drastic reprogramming of central metabolism.

Transcriptome profiling of *Arabidopsis* plants subjected to heat, drought or a combination of both revealed unanticipated features of the response to concurrent stresses. Only about half of the genes whose expression was modified by either high temperatures or water deficit were also affected during the combined situation; the remainder were not (Mittler, 2006; Rizhsky *et al.*, 2002; Rizhsky *et al.*, 2004). Even more striking, of the 1833 genes induced or repressed when drought was combined with high temperatures, 772 (42 %) were specific of the combination (Mittler, 2006; Rizhsky *et al.*, 2004). Metabolite profiling led to similar observations (Rizhsky *et al.*, 2004). Although not so extensively documented, other examples of synergistic or antagonistic relationships between different stress responses have been reported (Bowler and Fluhr, 2000; Sandermann, 2004; Walter, 1989). Mittler (2006) has proposed a "stress matrix" which illustrates the reported interactions (negative or positive) between various types of environmental hardships. The conclusion is somehow surprising: development of tolerance to combined stresses might require a response which is largely unique, and dedicated pathways specific for the particular stress combination might be activated (Mittler, 2006).

Cross-talk between co-activated cascades can be mediated at the various levels depicted in *Figure 2*, and several reports indicate integration between different networks involving kinases, hormones, receptors and transcriptional regulators (Anderson *et al.*, 2004; Bowler and Fluhr, 2000; Cardinale *et al.*, 2002; Casal, 2002; Mittler *et al.*, 2004; Suzuki *et al.*, 2005; Xiong and Yang, 2003). By comparison with the many studies performed with single stress sources, our knowledge on the molecular mechanisms underlying plant tolerance to a combination of environmental insults lags way behind. This limited understanding might in part explain why some transgenic plants developed in the laboratory with enhanced tolerance to a particular stress situation failed to show a better endurance when tested under field conditions (Mittler, 2006).

On the other hand, genome-wide analyses of transcript abundance in *Arabidopsis* indicate that there is little or no overlap between iron deficit and other environmental stress responses. A common feature, however, did emerge from inspection of transcriptome profiling studies: universal down-regulation of the chloroplast iron-sulphur protein ferredoxin (Fd) (Thimm *et al.*, 2001; Zimmermann *et al.*, 2004), a result confirmed by biochemical experiments (Tognetti *et al.*, 2006; Tognetti *et al.*, 2007b). Since Fd plays a pivotal role in electron distribution to many oxido-reductive routes, its decline could be extremely harmful to the stressed organism and significantly contribute to damage and growth arrest, making it a promising intervention point for transformation. Given its importance, the functions of chloroplast Fd will be discussed in some detail in the following chapter.

A biochemical digression: role and expression of ferredoxin in photosynthetic organisms

Ferredoxins are small, soluble [2Fe-2S] proteins which participate in many different oxido-reductive processes in prokaryotes, plants and animals. In chloroplasts, Fd is rapidly reduced (~ 800 s^{-1}) at the level of photosystem I (Cassan *et al.*, 2005) and delivers low-potential reducing equivalents to various electron-consuming reactions of the plastid (*Figure 3*). Enzyme partners of Fd include ferredoxin-$NADP^+$ reductase (FNR) for $NADP^+$ reduction, nitrite reductase and glutamate-oxoglutarate amino transferase for nitrogen assimilation and amino acid synthesis, sulphite reductase and fatty acid desaturase (Hase *et al.*, 2006). In addition, Fd acts as electron donor for ascorbate regeneration (Miyake and Asada, 1994), for thioredoxin (Trx) reduction via Fd-Trx reductase (Schürmann and Buchanan, 2008); and for the synthesis of phytochromobilin, the chromophore of the light sensor phytochrome and one of the initial intermediates in the synthesis of chlorophyll (Muramoto *et al.*, 1999). Fd also helps to relieve the electron pressure on the photosynthetic electron transport chain (PETC) when photochemical efficiency is low due to CO_2 shortage and/or excess illumination. Under such conditions, the PETC becomes overreduced and electrons or light energy might be passed straight to oxygen, generating ROS that could damage all types of biomolecules (Apel and Hirt, 2004). Fd alleviates this situation, either by returning the surplus of reducing equivalents to the PETC via cyclic electron flow (Yamamoto *et al.*, 2006), or by delivering them to the cell cytosol through Trx-dependent activation of the malate valve (Scheibe, 2004).

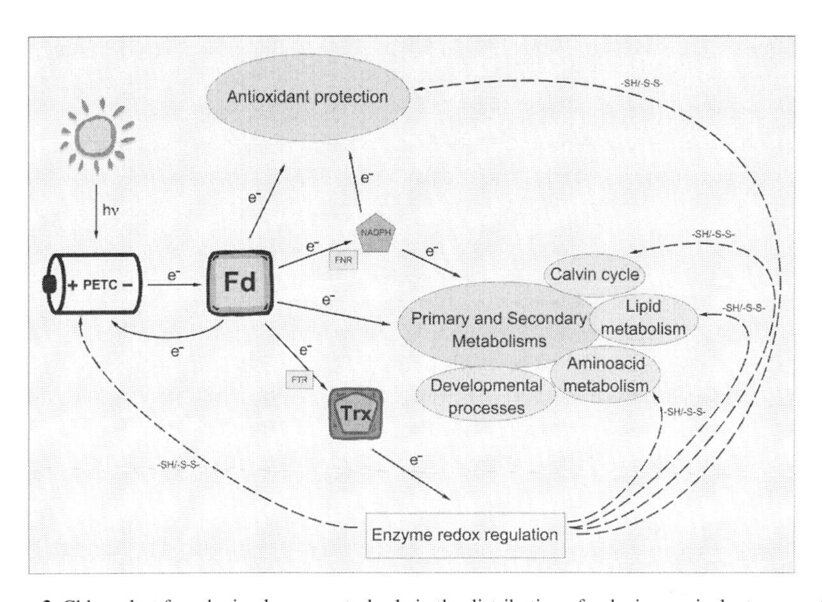

Figure 3. Chloroplast ferredoxin plays a central role in the distribution of reducing equivalents generated during photosynthesis. Electrons originating in the photosynthetic electron transport chain (PETC) may be transferred via Fd to FNR for $NADP^+$ photoreduction, generating the NADPH necessary for the Calvin cycle and other biosynthetic and protective pathways including glutathione reduction. Reduced Fd is also the electron donor for nitrogen and sulphur assimilation via nitrite and sulphite reductases, for fatty acid desaturation by fatty acid desaturase and for glutamate synthesis mediated by glutamate-oxoglutarate aminotransferase. Still other Fd molecules will participate in thioredoxin (Trx) reduction via Fd-thioredoxin reductase (FTR). Reduced Trx will then activate key target enzymes through reduction of their critical cysteines (-SH/-S-S- exchange), resulting in the maintenance and/or stimulation of the Calvin cycle, the malate valve, biosynthetic pathways and many other metabolic routes. Dissipative systems requiring Fd include regeneration of active peroxiredoxins, the most abundant peroxidase of chloroplasts, and of ascorbate, a step essential for the function of the Mehler-Asada cycle. Fd also regulates the distribution of reducing equivalents between lineal and cyclic electron flow via Fd-ubiquinol reductase. Finally, it participates in developmental processes through the synthesis of phytochromobilin, the chromophore of the light sensor phytochrome, by donating electrons to two key enzymes of the pathway: heme oxygenase and phytochromobilin synthase.

Most of these Fd activities are also found in cyanobacteria, the closest living relatives of the ancient endosymbiont that gave origin to modern-day chloroplasts (Zurbriggen *et al.*, 2007). The amounts of Fd in these photosynthetic microorganisms fluctuate in response to different environmental stimuli, including down-regulation by iron deficit, environmental adversities and oxidative stress (Mazouni *et al.*, 2003; Singh *et al.*, 2004). In higher plants, Fd is encoded by a small gene family whose members display tissue-specific expression and perform specialised tasks in different organs (Hanke *et al.*, 2004). Chloroplast Fd is induced by light through a redox-based signalling pathway (Petracek *et al.*, 1998), whereas steady-state levels of Fd transcripts (Thimm *et al.*, 2001; Zimmermann *et al.*, 2004) and protein (Tognetti *et al.*, 2006; Tognetti *et al.*, 2007b) are decreased under iron deficit and a plethora of environmental adversities, including drought, chilling, salinity and UV exposure. The mechanisms underlying this effect are not completely understood, although post-transcriptional regulation could be implicated (Petracek *et al.*, 1998). Impaired assembly of Fe-S clusters due to iron shortage and their destruction by ROS probably contribute to the decline.

Ferredoxins are very ancient proteins that evolved during anaerobic times when ROS were absent and iron, plentiful (Tognetti *et al.*, 2008), and are not very well suited for the present oxygen-rich atmosphere. Since Fd participates in all aspects of plant life, including central metabolism, protective devices, enzyme regulation and morphogenesis, and at the same time is a critical target of virtually all adverse conditions, it has become an Achilles' heel for plant welfare under suboptimal environments (Tognetti *et al.*, 2008). Several lines of evidence indicate that Fd decrease compromises cell survival: *i*) Fd is essential for the viability of cyanobacteria as demonstrated by targeted disruption of the gene (Poncelet *et al.*, 1998); *ii*) Fd down-regulation using antisense RNA technology in potato indicated that its levels could not be decreased below 50 % of wild-type (WT) contents without fatally affecting plant fitness (Holtgrefe et al., 2003); *iii*) knockout (Voss *et al.*, 2008) and knock-down (Hanke and Hase, 2008) *Arabidopsis* lines in which either the major leaf Fd or both isoforms were depleted displayed growth arrest, reproductive handicaps and partial inactivation of photosynthesis.

Unfortunately, increase of plant Fd amounts in transgenic plants proved to be a difficult task, because its expression is redox-regulated at the post-transcriptional level from sequences that lie within the region encoding the structural protein, so this control cannot be overcome by promoter engineering (Petracek *et al.*, 1998). One possible approach to circumvent this problem comes from studies on photosynthetic microorganisms. They are exposed to similar hardships as plants, especially for those species living in the open oceans, where environmental and nutritional conditions are usually extreme. Indeed, analysis of the protective responses elicited by algae and cyanobacteria to iron starvation and environmental stresses provided the clue and the genetic resources to develop multiple stress tolerance in plants.

Stress responses of algae and cyanobacteria

Photosynthetic microorganisms respond to adverse environments by a strategy which is completely different from those elicited by plants, and that involves substitution of stress-sensitive enzymes and proteins by resistant isofunctional versions (Erdner *et al.*, 1999; Palenik *et al.*, 2006). It is interesting to note that these microorganisms could, in response to iron deficiency, substitute key iron-containing proteins with others induced under these circumstances, and which are able to perform the same physiological roles but utilising alternative prosthetic groups (Fillat *et al.*, 1995; McKay *et al.*, 1997). For instance, upon iron deprivation, some oceanic diatoms synthesise the electron transporter plastocyanin that uses Cu to functionally replace the hemoprotein cytochrome c_6, whereas cyanobacteria of the genus *Synechoccocus* substitute their iron-dependent ribonucleotide reductase and superoxide dismutase with Co- and Ni-containing isoenzymes (Palenik *et al.*, 2003; Peers and Price, 2006). A most conspicuous example of this strategy is induction of flavodoxin (Fld) expression to take over the functions of stress-vulnerable Fd. Flavodoxins are soluble electron carriers containing flavin mononucleotide as prosthetic group, whose properties as redox shuttles largely match those of Fd (Sancho, 2006). While Fds are present in virtually all types of organisms, the Fld gene displays a rather limited distribution. Within photosynthetic species, Fld is common, but not universal, among cyanobacteria. The gene entered the algal world during the endosymbiotic event that gave origin to photosynthetic eukaryotes and spread into all major algal lineages, with the corresponding product being targeted to chloroplasts (Zurbriggen *et al.*, 2007). However, Fld presence is less frequent than in cyanobacteria and generally confined to algae living in the open oceans (Erdner *et al.*, 1999; Zurbriggen *et al.*, 2007). The Fld gene is not found in the plant genome, and the considerable adaptive advantages derived from its expression and stress-dependent induction were irreversibly lost somewhere during the evolution from green algae to vascular plants. Noteworthy, cyanobacterial Fld is still able to productively interact *in vitro* with plant enzymes whose prokaryotic ancestors used this flavoprotein as usual or occasional substrate (Nogués *et al.*, 2004; Scheller, 1996; Tognetti *et al.*, 2006), raising the possibility that introduction of Fld in plants could improve stress tolerance as it occurs in microorganisms. Results obtained by following this lead are discussed in the next chapters.

Broad-range tolerance to environmental adversities and iron starvation by expression of a cyanobacterial flavodoxin in plants

Transgenic tobacco plants expressing a cyanobacterial Fld were generated to probe the possibilities of the strategy mentioned above, *i.e.*, Fld substitution of decaying endogenous Fd in plants upon stress. By introduction or not of a chloroplast-targeting transit peptide, lines were designed in which Fld was localised in chloroplasts (*pfld*) or in the cytosol (*cfld*), respectively. In both cases the genes were placed under control of a constitutive promoter. Independent lines were selected with Fld accumulation in chloroplasts ranging from 20 μM up to 80 μM, the approximate concentration of endogenous Fd in leaf plastids. Transformed lines did not exhibit phenotypic differences and/or growth penalties in relation to WT plants when grown under controlled conditions (Tognetti *et al.*, 2006).

Fld-expressing plants were able to survive and reproduce in iron-deficient substrates and soil, a condition most favourable to Fd decay (*Figure 4A*). Fld-expressing plants did not improve iron accumulation relative to their WT counterparts and developed a normal response to iron limitation, including induction of several genes involved in metal uptake and mobilisation (Tognetti *et al.*, 2007b). The results indicate that the presence of the bacterial protein does not interfere with sensing or signalling pathways related to the iron deficit response, but allowed the transformants to grow with iron shares lower than WT plants. Iron-starved WT specimens underwent a general decrease in CO_2 fixation capacity together with a strong down-regulation of metabolic activities. On the contrary, metabolite profile analysis of *pfld* plants indicated that the contents of many central metabolites of the Calvin cycle, energy storage and anabolic routes were preserved upon stress (Tognetti *et al.*, 2007b). Fld was indeed able to engage in electron delivery to Trx-dependent routes that normally rely on Fd. For instance, the activation states of enzymes of the Calvin cycle such as phosphoribulokinase and fructose-1,6-bisphosphatase were hardly affected in Fld-expressing lines but declined sharply in non-transformed siblings by the nutritional deficiency (Tognetti *et al.*, 2007b).

Transformants from *pfld* lines also excelled when assayed for their tolerance towards abiotic stress. For instance, while exposure of WT and *cfld* leaves to the redox-cycling herbicide paraquat that generates superoxide radicals in chloroplasts led to necrotic lesions and wilting, the *pfld* counterparts suffered little tissue damage (Tognetti *et al.*,

Figure 4. Increased tolerance of transgenic plants expressing a cyanobacterial flavodoxin in chloroplasts to iron deficit and environmental stresses. Prior to treatments, all species were grown for the times indicated at 400 μmol quanta m^{-2} s^{-1} and 25°C to provide a 14-h photoperiod. (A) Four-week old WT, and transgenic *pfld* (for **p**lastid-targeted ***fl***avo***d***oxin) and *cfld* (for **c**ytosolic-targeted ***fl***avo***d***oxin) tobacco plants grown in Murashige-Skoog-agar medium were transferred to Fe-free hydroponic Hoagland solution containing 10 mM $CaCO_3$, pH 8.0, and photographed 29 days after transfer. (B) Leaves from 2-month old *pfld* and WT tobacco plants grown in soil were exposed for 18 h to a focused light beam of 2,000 μmol quanta m^{-2} s^{-1} through a blueprint representing the logo of the University of Rosario ("The undying flame"). The treatment led to imprinting of the logo on the surface of a WT leaf, whereas *pfld* tissue showed only a blur. (C) Fully expanded undetached leaves of 4 week-old potato plants were dipped for 30 s in a solution containing 50 μM paraquat and 0,1% v/v Tween 20, illuminated at 400 μmol quanta m^{-2} s^{-1}, and photographed after 20 h of treatment. Lines: WT (*Solanum tuberosum* cv Solara); transgenic *Stpfld*21, *Stpfld*22, *Stpfld*34 and *Stpfld*36 (expressing Fld in chloroplasts); transgenic *Stcfld*27 and *Stcfld*30 (expressing Fld in the cytosol). (D) Six week-old potato plants were subjected to drought by withholding water for 10 days. Lines: WT (*Solanum tuberosum* cv Solara); *Stpfld*33 and *Stpfld*36. (E) Exposure of 6 week-old tomato plants to paraquat was carried out as described in Panel C, and leaves were photographed after 13 h (top) and 37 h (bottom) of treatment. Lines: WT (*Solanum lycopersicum* cv Moneymaker); transgenic *Slpfld*8, *Slpfld*60 and *Slpfld*66 (expressing Fld in chloroplasts). (F) Six week-old *Brassica* plants were subjected to drought by withholding water for 10 days (above). Lines: WT (*Brassica oleracea* DH1012); *Bopfld*25-29, *Bopfld*47-5 and *Bopfld*43-2 (transgenic plants expressing Fld in chloroplasts). Below are shown immunoblots performed with antiserum raised against Fld. There is a positive correlation between Fld expression levels and the tolerance of the different lines to drought stress. For instance, the high-expressing lines *Bopfld*25-29 and *Bopfld*47-5 performed better under water deficit than *Bopfld*43-2 or WT plants. *Brassica* transformation was carried out by Dr. Penny Sparrow, John Innes Centre (JIC), Norwich, UK (http://www.bract.org).

2006). Fld-expressing plants were also tolerant to oxidative stress conditions imposed upon exposure of plants to high light intensities (*Figure 4B*), extreme temperatures, water deficiency, UV-AB radiation and nitrotoluene derivatives (Tognetti *et al.*, 2007a; Tognetti *et al.*, 2006). Chloroplast-targeted Fld displayed a general protective effect evidenced in the form of intact cell structures and membranes, lower ROS accumulation, decreased lipid peroxidation and preservation of the photosynthetic activity in contrast to WT siblings (Tognetti *et al.*, 2006). Fld also played a decisive role in ROS detoxification under oxidative stress conditions by mediating the Trx-dependent reductive regeneration of peroxiredoxin, the most abundant chloroplast peroxidase, and modulation of the activity of NADPH-malate dehydrogenase (Tognetti *et al.*, 2006).

Attempted infection of plants by pathogens elicits a complex defensive response. In many non-host and incompatible host interactions it includes the induction of defence-associated genes and a form of localised cell death (LCD), purportedly designed to restrict pathogen advance, collectively known as the hypersensitive response (HR). It is preceded by an oxidative burst, generating ROS that are proposed to cue subsequent deployment of the HR, although neither the origin nor the precise role played by active oxygen species in the execution of this response are completely understood. We used *pfld* plants as a probe to address these questions. Infiltration of WT leaves with high titers of *Xanthomonas campestris* pv. *vesicatoria* (*Xcv*), a non-host pathogen, resulted in ROS accumulation in chloroplasts, followed by the appearance of localised lesions typical of the HR. In contrast, chloroplast ROS build-up and LCD were significantly reduced in *Xcv*-inoculated plants expressing plastid-targeted Fld (Zurbriggen *et al.*, 2009). Metabolic routes normally inhibited by the pathogen were protected in the transformants, whereas other aspects of the HR, including induction of defence-associated genes and synthesis of salicylic and jasmonic acid, proceeded as in inoculated WT plants (Zurbriggen *et al.*, 2009). Therefore, experiments performed with Fld-expressing lines allowed us to propose that ROS generated in chloroplasts during this non-host interaction are essential for the progress of LCD, but do not contribute to the induction of pathogenesis-related genes or other signalling components of the response.

In all cases studied, the protective effect conferred by Fld to stressed plants was unavoidably subordinated to its plastid location and dosage. Accordingly, transformants expressing low plastidic or high cytosolic Fld contents displayed WT levels of stress tolerance (Tognetti *et al.*, 2007a; Tognetti *et al.*, 2006; Tognetti *et al.*, 2008; Tognetti *et al.*, 2007b).

Experiments carried out with the transgenic tobacco plants revealed that Fld is able to functionally substitute plant Fd in several of its electron-donation roles when the levels of the iron-sulphur protein decrease upon stress. Fld manages to sustain a normal electron supply to various key pathways and routes involved in redox homeostasis, photosynthesis, carbon and nitrogen metabolisms (*Figure 3*). It is interesting to note that Fld can successfully engage in electron exchange reactions with the usual enzyme partners of plant Fd, even though these proteins have significantly evolved and diverged from their cyanobacterial homologues. The answer to this conundrum is probably related to the nature of the interaction between Fd and its partners, which relies on general electrostatic contacts more than on specific interactions between active site and substrate residues. This feature allows Fd to functionally interact with several different redox partners which bear no structural resemblance among them (Hase *et al.*, 2006). It is this promiscuous behaviour that allows Fld expression in chloroplasts to exert such pleiotropic effects.

ROS accumulation is responsible for a great part of cell and tissue damage in plants exposed to hostile environments, and this effect is largely caused by deficient electron distribution due to Fd shortage. In a similar way, Fd down-regulation is one of the hallmarks of the iron deficiency response in plants and photosynthetic microorganisms. Therefore, by restoring proper electron distribution to productive routes and preventing ROS build-up in chloroplasts and cells, Fld confers wide-range tolerance to many different sources of stress.

Stress tolerance of flavodoxin-expressing transgenic crops

Introduction of Fld into tobacco, a model system, enabled us not only to verify the mechanistic model but also to substantiate the feasibility of the substitutive strategy as a general biotechnological tool aimed to obtain increased plant tolerance to environmental hardships. It followed then as part of the conceptual framework to transfer the technology to other plant species, particularly crops. Transformation of two species related to tobacco (tomato and potato) with a gene encoding a plastid-targeted Fld yielded essentially the same tolerant phenotypes when the transgenic plants were exposed to paraquat (*Figure 4C,E*) or water deprivation (*Figure 4D*). Fld-expressing *Brassica* plants were also more tolerant to drought conditions than their non-transformed siblings, in an extent strictly dependent on Fld dose (*Figure 4F*). Likewise, preliminary results indicate that this strategy conveys tolerance towards oxidative stress conditions in monocots (Zurbriggen *et al.*, unpublished results).

Mono- and dicotyledoneous plants, despite having the same basic mechanisms of tolerance to abiotic stresses, may utilise different molecular players (transcription factors, signalling relay components, defence metabolites), so transferring knowledge from model species such as *Arabidopsis* or tobacco to monocot crops is not always possible (Tester and Bacic, 2005). In this respect, Fld expression equally prevented abiotic stress damage in both major plant groups, highlighting its potential as a general protectant. The results suggest that the Fld effects observed in model plants could be thus used as a guide for the development of stress-tolerant crops. However, challenging with pathogens and exposure to mixed stresses, situations normally occurring in the field, and yield analyses are still missing to be pursued in order to complete the assessment of this transgenic approach.

Taking into account the broad-range beneficial effects of Fld expression on stress physiology, it is not clear why a genetic trait that confers so obvious advantages was not selected during the evolution of vascular plants. Our hypothesis is that tolerance to iron deficit was the most important adaptive value of Fld, and that the gene was lost because during the evolutionary pathway from cyanobacteria to plants, from the open oceans to the firm land, photosynthetic microorganisms passed through a stage in the coastal regions in which the services of Fld as an inducible backup for Fd were no longer required because iron was both abundant and readily accessible (Zurbriggen *et al.*, 2007). Unlike the open oceans where iron tends to be chronically deficient, coastal waters typically contain high concentrations of this transition metal due to inputs from land and sediments (Palenik *et al.*, 2006). Under such conditions, Fld expression would not be induced, and selection presumably eliminated the Fld gene from the algal precursor genome.

Conclusions and perspectives

Abiotic stresses and iron limitation are the primary causes of crop yield loss worldwide. Therefore, plant biotechnologies aimed at overcoming environmental hardships need to be quickly designed, with genetic engineering being at the forefront. Three major issues pertaining to this goal have been reviewed in this article, and the state-of-the-art situation discussed in some detail. Boosting of endogenous plant defence systems likely represents the mainstream in stress tolerance research (Vij and Tyagi, 2007; Vinocur and Altman, 2005). When increasing the dosage of endogenous plant genes, it is important to keep in mind that any tolerance mechanism must always be assessed with respect to its potential cross-talk with other stress-related, developmental or metabolic genes and mechanisms, especially if the relevant gene product operates high in the cascade of the plant stress response (*Figure 2*). On the other hand, the higher damage caused to agriculture by combinations of concurrent stresses underscores the need to develop crops with tolerance to multiple stresses (Mittler, 2006). Information obtained from the limited number of studies performed with combined stresses indicates that to draw sound conclusions, it is not enough to investigate each of the individual challenges separately. The stress combination should be regarded as a new state of abiotic stress and studied accordingly (Mittler, 2006).

Research on the mechanisms of stress tolerance deployed by photosynthetic microorganisms should be pursued with vigour in order to identify genes and systems that are no longer present in plants but could be reintroduced with advantage. The example of Fld indicates that this approach has some unique characteristics that make it an attractive alternative to be applied alone or in combination with other strategies. Tolerance to multiple sources of stress and to iron deficit could be gained by transformation with a single gene. Its product operates downstream of the plant response cascade to stress and therefore, it does not interfere with developmental programmes even when expressed constitutively. Indeed, Fld expression remains unnoticed in phenotypic terms when the transformants are grown under normal conditions (Tognetti *et al.*, 2006; Tognetti *et al.*, 2007b). Finally, the prokaryotic origin of the flavodoxin gene precludes undesired effects such as silencing, co-suppression or endogenous regulatory networks, allowing a more customised manipulation of this trait.

Several important aspects related with stress tolerance in plants have not been addressed here. They should be taken into consideration, however, when developing research programmes that aim to have an impact in agriculture. For instance, most current studies use short-term treatments rather than to follow the effects of stress over longer periods that more closely reflect the field situation. Responses during short and long periods might differ significantly. The consequences of adverse environments are generally assessed at the level of survival, physiology or biochemistry, mostly on vegetative tissues, while other factors such as yield data and fruit development are of great practical importance. Finally, cycles of stress and recovery are prevalent under natural conditions and agricultural practice. The degree of recovery from stress is as relevant as the response to stress in agricultural terms.

Acknowledgements

We are indebted to Dr. Penny Sparrow, JIC, Norwich, UK (http://www.bract.org) for preparation of the *Brassica* transgenic plants. This research was supported by grants from the National Agency for the Promotion of Science and Technology (ANPCyT, Argentina), the National Research Council (CONICET, Argentina), the German Academic Exchange Service (DAAD, Germany) and the European Molecular Biology Organization (EMBO, European Union).

References

ABEBE, T., GUENZI, A.C., MARTIN, B. and CUSHMAN, J.C. (2003). Tolerance of mannitol-accumulating transgenic wheat to water stress and salinity. *Plant Physiology* **131**, 1748-55.

ANDERSON, J.P., BADRUZSAUFARI, E., SCHENK, P.M., MANNERS, J.M., DESMOND, O.J., EHLERT, C., MACLEAN, D.J., EBERT, P.R. and KAZAN, K. (2004). Antagonistic interaction between abscisic acid and jasmonate-ethylene signaling pathways modulates defense gene expression and disease resistance in *Arabidopsis*. *Plant Cell* **16**, 3460-79.

APEL, K. and HIRT, H. (2004). Reactive oxygen species: metabolism, oxidative stress, and signal transduction. *Annual Review of Plant Biology* **55**, 373-99.

APSE, M.P. and BLUMWALD, E. (2002). Engineering salt tolerance in plants. *Current Opinion in Biotechnology* **13**, 146-50.

BOWLER, C. and FLUHR, R. (2000). The role of calcium and activated oxygens as signals for controlling cross-tolerance. *Trends in Plant Science* **5**, 241-6.

BOYER, J.S. (1982). Plant productivity and environment. *Science* **218**, 443-48.

BRUMBAROVA, T. and BAUER, P. (2005). Iron-mediated control of the basic helix-loop-helix protein FER, a regulator of iron uptake in tomato. *Plant Physiology* **137**, 1018-26.

BUCHANAN, B.B., GRUISSEM, W. and JONES, R.L. (2000). *Biochemistry and Molecular Biology of Plants*. Rockville, MD, USA: American Society of Plant Phsyiologists.

CAPELL, T. and CHRISTOU, P. (2004). Progress in plant metabolic engineering. *Current Opinion in Biotechnology* **15**, 148-54.

CARDINALE, F., MESKIENE, I., OUAKED, F. and HIRT, H. (2002). Convergence and divergence of stress-induced mitogen-activated protein kinase signaling pathways at the level of two distinct mitogen-activated protein kinase kinases. *Plant Cell* **14**, 703-11.

CASAL, J.J. (2002). Environmental cues affecting development. *Current Opinion in Plant Biology* **5**, 37-42.

CASSAN, N., LAGOUTTE, B. and SÉTIF, P. (2005). Ferredoxin-NADP$^+$ reductase. Kinetics of electron transfer, transient intermediates, and catalytic activities studied by flash-absorption spectroscopy with isolated photosystem I and ferredoxin. *Journal of Biological Chemistry* **280**, 25960-72.

CHEN, T.H. and MURATA, N. (2002). Enhancement of tolerance of abiotic stress by metabolic engineering of betaines and other compatible solutes. *Current Opinion in Plant Biology* **5**, 250-7.

DEZAR, C.A., GAGO, G.M., GONZÁLEZ, D.H. and CHAN, R.L. (2005). *Hahb-4*, a sunflower homeobox-leucine zipper gene, is a developmental regulator and confers drought tolerance to *Arabidopsis thaliana* plants. *Transgenic Research* **14**, 429-40.

Dubouzet, J.G., Sakuma, Y., Ito, Y., Kasuga, M., Dubouzet, E.G., Miura, S., Seki, M., Shinozaki, K. and Yamaguchi-Shinozaki, K. (2003). *OsDREB* genes in rice, *Oryza sativa* L., encode transcription activators that function in drought-, high-salt- and cold-responsive gene expression. *Plant Journal* **33**, 751-63.

Erdner, D.L., Price, N.M., Doucette, D.G., Peleato, M.L. and Anderson, D.M. (1999). Characterization of ferredoxin and flavodoxin as markers of iron limitation in marine phytoplankton. *Marine Ecology Progress Series* **184**, 43-53.

Fillat, M.E., Peleato, M.L., Razquin, P. and Gómez-Moreno, C. (1995). Effects of iron deficiency in photosynthetic electron transport and nitrogen fixation in the cyanobacterium *Anabaena*: flavodoxin induction adaptative response. In *Iron Nutrition in Soils and Plants* ed. J. Abadía, pp 315-21. Dordrecht, The Netherlands: Kluwer Academic Publishers.

Flowers, T.J. (2004). Improving crop salt tolerance. *Journal of Experimental Botany* **55**, 307-19.

Foyer, C.H., Souriau, N., Perret, S., Lelandais, M., Kunert, K.J., Pruvost, C. and Jouanin, L. (1995). Overexpression of glutathione reductase but not glutathione synthetase leads to increases in antioxidant capacity and resistance to photoinhibition in poplar trees. *Plant Physiology* **109**, 1047-57.

Graziano, M. and Lamattina, L. (2007). Nitric oxide accumulation is required for molecular and physiological responses to iron deficiency in tomato roots. *Plant Journal* **52**, 949-60.

Guerinot, M.L. (2007). It's elementary: enhancing Fe^{3+} reduction improves rice yields. *Proceedings of the National Academy of Sciences of the United States of America* **104**, 7311-2.

Gutterson, N. and Zhang, J.Z. (2004). Genomics applications to biotech traits: a revolution in progress? *Current Opinion in Plant Biology* **7**, 226-30.

Hanke, G.T. and Hase, T. (2008). Variable photosynthetic roles of two leaf-type ferredoxins in *Arabidopsis*, as revealed by RNA interference. *Photochemistry and Photobiology* **84**, 1302-9.

Hanke, G.T., Kimata-Ariga, Y., Taniguchi, I. and Hase, T. (2004). A post genomic characterization of *Arabidopsis* ferredoxins. *Plant Physiology* **134**, 255-64.

Hase, T., Schürmann, P. and Knaff, D.B. (2006). The interaction of ferredoxin with ferredoxin-dependent enzymes. In *Photosystem I: The Light-Driven Plastocyanin-Ferredoxin Oxidoreductase* ed. J.H. Golbeck, pp 477-98. Dordrecht, The Netherlands: Springer.

Holtgrefe, S., Bader, K.P., Horton, P., Scheibe, R., von Schaewen, A. and Backhausen, J.E. (2003). Decreased content of leaf ferredoxin changes electron distribution and limits photosynthesis in transgenic potato plants. *Plant Physiology* **133**, 1768-78.

Ishimaru, Y., Kim, S., Tsukamoto, T. *et al.* (2007). Mutational reconstructed ferric chelate reductase confers enhanced tolerance in rice to iron deficiency in calcareous soil. *Proceedings of the National Academy of Sciences of the United States of America* **104**, 7373-8.

Ito, Y., Katsura, K., Maruyama, K., Taji, T., Kobayashi, M., Seki, M., Shinozaki, K. and Yamaguchi-Shinozaki, K. (2006). Functional analysis of rice DREB1/CBF-type transcription factors involved in cold-responsive gene expression in transgenic rice. *Plant and Cell Physiology* **47**, 141-53.

Jang, I.C., Oh, S.J., Seo, J.S. *et al.* (2003). Expression of a bifunctional fusion of the

Escherichia coli genes for trehalose-6-phosphate synthase and trehalose-6-phosphate phosphatase in transgenic rice plants increases trehalose accumulation and abiotic stress tolerance without stunting growth. *Plant Physiology* **131**, 516-24.

Katiyar-Agarwal, S., Agarwal, M. and Grover, A. (2003). Heat-tolerant basmati rice engineered by over-expression of hsp101. *Plant Molecular Biology* **51**, 677-86.

Kim, S.A. and Guerinot, M.L. (2007). Mining iron: iron uptake and transport in plants. *FEBS Letters* **581**, 2273-80.

Manavella, P.A., Arce, A.L., Dezar, C.A., Bitton, F., Renou, J.P., Crespi, M. and Chan, R.L. (2006). Cross-talk between ethylene and drought signalling pathways is mediated by the sunflower Hahb-4 transcription factor. *Plant Journal* **48**, 125-37.

Mazouni, K., Domain, F., Chauvat, F. and Cassier-Chauvat, C. (2003). Expression and regulation of the crucial plant-like ferredoxin of cyanobacteria. *Molecular Microbiology* **49**, 1019-29.

McKay, R.M., Geider, R.J. and LaRoche, J. (1997). Physiological and biochemical response of the photosynthetic apparatus of two marine diatoms to Fe stress. *Plant Physiology* **114**, 615-22.

McKersie, B.D., Murnaghan, J., Jones, K.S. and Bowley, S.R. (2000). Iron-superoxide dismutase expression in transgenic alfalfa increases winter survival without a detectable increase in photosynthetic oxidative stress tolerance. *Plant Physiology* **122**, 1427-37.

Mittler, R. (2006). Abiotic stress, the field environment and stress combination. *Trends in Plant Science* **11**, 15-9.

Mittler, R., Vanderauwera, S., Gollery, M. and van Breusegem, F. (2004). Reactive oxygen gene network of plants. *Trends in Plant Science* **9**, 490-8.

Miyake, C. and Asada, K. (1994). Ferredoxin-dependent photoreduction of the monodehydroascorbate radical in spinach thylakoids. *Plant and Cell Physiology* **34**, 539-49.

Mori, S. (1999). Iron acquisition by plants. *Current Opinion in Plant Biology* **2**, 250-3.

Muramoto, T., Kohchi, T., Yokota, A., Hwang, I. and Goodman, H.M. (1999). The *Arabidopsis* photomorphogenic mutant hy1 is deficient in phytochrome chromophore biosynthesis as a result of a mutation in a plastid heme oxygenase. *Plant Cell* **11**, 335-48.

Nogués, I., Tejero, J., Hurley, J.K. *et al.* (2004). Role of the C-terminal tyrosine of ferredoxin-nicotinamide adenine dinucleotide phosphate reductase in the electron transfer processes with its protein partners ferredoxin and flavodoxin. *Biochemistry* **43**, 6127-37.

Palenik, B., Brahamsha, B., Larimer, F.W. *et al.* (2003). The genome of a motile marine *Synechococcus*. *Nature* **424**, 1037-42.

Palenik, B., Ren, Q., Dupont, C.L. *et al.* (2006). Genome sequence of *Synechococcus* CC9311: insights into adaptation to a coastal environment. *Proceedings of the National Academy of Sciences of the United States of America* **103**, 13555-9.

Park, E.J., Jeknic, Z., Sakamoto, A., DeNoma, J., Yuwansiri, R., Murata, N. and Chen, T.H. (2004). Genetic engineering of glycinebetaine synthesis in tomato protects seeds, plants, and flowers from chilling damage. *Plant Journal* **40**, 474-87.

Peers, G. and Price, N.M. (2006). Copper-containing plastocyanin used for electron transport by an oceanic diatom. *Nature* **441**, 341-4.

PENNA, S. (2003). Building stress tolerance through over-producing trehalose in transgenic plants. *Trends in Plant Science* **8**, 355-7.

PETRACEK, M.E., DICKEY, L.F., NGUYEN, T.T., GATZ, C., SOWINSKI, D.A., ALLEN, G.C. and THOMPSON, W.F. (1998). Ferredoxin-1 mRNA is destabilized by changes in photosynthetic electron transport. *Proceedings of the National Academy of Sciences of the United States of America* **95**, 9009-13.

PONCELET, M., CASSIER-CHAUVAT, C., LESCHELLE, X., BOTTIN, H. and CHAUVAT, F. (1998). Targeted deletion and mutational analysis of the essential (2Fe-2S) plant-like ferredoxin in *Synechocystis* PCC6803 by plasmid shuffling. *Molecular Microbiology* **28**, 813-21.

QIU, Q.S., GUO, Y., DIETRICH, M.A., SCHUMAKER, K.S. and ZHU, J.K. (2002). Regulation of SOS1, a plasma membrane Na+/H+ exchanger in *Arabidopsis thaliana*, by SOS2 and SOS3. *Proceedings of the National Academy of Sciences of the United States of America* **99**, 8436-41.

RIZHSKY, L., LIANG, H. and MITTLER, R. (2002). The combined effect of drought stress and heat shock on gene expression in tobacco. *Plant Physiology* **130**, 1143-51.

RIZHSKY, L., LIANG, H., SHUMAN, J., SHULAEV, V., DAVLETOVA, S. and MITTLER, R. (2004). When defense pathways collide. The response of *Arabidopsis* to a combination of drought and heat stress. *Plant Physiology* **134**, 1683-96.

ROGERS, E.E. and GUERINOT, M.L. (2002). FRD3, a member of the multidrug and toxin efflux family, controls iron deficiency responses in *Arabidopsis*. *Plant Cell* **14**, 1787-99.

SANCHO, J. (2006). Flavodoxins: sequence, folding, binding, function and beyond. *Cellular and Molecular Life Sciences* **63**, 855-64.

SANDERMANN, H. (2004). Molecular ecotoxicology: from man-made pollutants to multiple environmental stresses. In *Molecular Ecotoxicology of Plants (Ecological Studies)* ed. H. Sandermann, pp 1-16. Dordrecht, The Netherlands: Springer Verlag.

SCHEIBE, R. (2004). Malate valves to balance cellular energy supply. *Physiologia Plantarum* **120**, 21-26.

SCHELLER, H.V. (1996). *In vitro* cyclic electron transport in barley thylakoids follows two independent pathways. *Plant Physiology* **110**, 187-94.

SCHMIDT, W. (2003). Iron solutions: acquisition strategies and signaling pathways in plants. *Trends in Plant Science* **8**, 188-93.

SCHÜRMANN, P. and BUCHANAN, B.B. (2008). The ferredoxin/thioredoxin system of oxygenic photosynthesis. *Antioxidants and Redox Signaling* **10**, 1235-74.

SEKI, M., KAMEI, A., YAMAGUCHI-SHINOZAKI, K. and SHINOZAKI, K. (2003). Molecular responses to drought, salinity and frost: common and different paths for plant protection. *Current Opinion in Biotechnology* **14**, 194-9.

SHINOZAKI, K., YAMAGUCHI-SHINOZAKI, K. and SEKI, M. (2003). Regulatory network of gene expression in the drought and cold stress responses. *Current Opinion in Plant Biology* **6**, 410-7.

SHOU, H., BORDALLO, P., FAN, J.B., YEAKLEY, J.M., BIBIKOVA, M., SHEEN, J. and WANG, K. (2004). Expression of an active tobacco mitogen-activated protein kinase kinase kinase enhances freezing tolerance in transgenic maize. *Proceedings of the National Academy of Sciences of the United States of America* **101**, 3298-303.

SINGH, A.K., LI, H. and SHERMAN, L.A. (2004). Microarray analysis and redox control of gene expression in the cyanobacterium *Synechocystis* sp. PCC 6803. *Physiologia*

Plantarum **120**, 27-35.

SUZUKI, N., RIZHSKY, L., LIANG, H., SHUMAN, J., SHULAEV, V. and MITTLER, R. (2005). Enhanced tolerance to environmental stress in transgenic plants expressing the transcriptional coactivator multiprotein bridging factor 1c. *Plant Physiology* **139**, 1313-22.

TAKAHASHI, M., NAKANISHI, H., KAWASAKI, S., NISHIZAWA, N.K. and MORI, S. (2001). Enhanced tolerance of rice to low iron availability in alkaline soils using barley nicotianamine aminotransferase genes. *Nature Biotechnology* **19**, 466-9.

TESTER, M. and BACIC, A. (2005). Abiotic stress tolerance in grasses. From model plants to crop plants. *Plant Physiology* **137**, 791-3.

THIMM, O., ESSIGMANN, B., KLOSKA, S., ALTMANN, T. and BUCKHOUT, T.J. (2001). Response of *Arabidopsis* to iron deficiency stress as revealed by microarray analysis. *Plant Physiology* **127**, 1030-43.

TOGNETTI, V.B., MONTI, M.R., VALLE, E.M., CARRILLO, N. and SMANIA, A.M. (2007a). Detoxification of 2,4-dinitrotoluene by transgenic tobacco plants expressing a bacterial flavodoxin. *Environmental Science and Technology* **41**, 4071-6.

TOGNETTI, V.B., PALATNIK, J.F., FILLAT, M.F., MELZER, M., HAJIREZAEI, M.R., VALLE, E.M. and CARRILLO, N. (2006). Functional replacement of ferredoxin by a cyanobacterial flavodoxin in tobacco confers broad-range stress tolerance. *Plant Cell* **18**, 2035-50.

TOGNETTI, V.B., ZURBRIGGEN, M., VALLE, E.M., CARRILLO, N., MORANDI, E.N., MELZER, M., HAJIREZAEI, M.R. and FILLAT, M.F. (2008). Recovering the cyanobacterial heritage in land plants: the case of flavodoxin. In *Flavins and Flavoproteins* eds. S. Frago, C. Gómez-Moreno and M. Medina, pp 527-36. Zaragoza, Spain: Prensas Universitarias de Zaragoza.

TOGNETTI, V.B., ZURBRIGGEN, M.D., MORANDI, E.N., FILLAT, M.F., VALLE, E.M., HAJIREZAEI, M.R. and CARRILLO, N. (2007b). Enhanced plant tolerance to iron starvation by functional substitution of chloroplast ferredoxin with a bacterial flavodoxin. *Proceedings of the National Academy of Sciences of the United States of America* **104**, 11495-500.

UMEZAWA, T., FUJITA, M., FUJITA, Y., YAMAGUCHI-SHINOZAKI, K. and SHINOZAKI, K. (2006). Engineering drought tolerance in plants: discovering and tailoring genes to unlock the future. *Current Opinion in Biotechnology* **17**, 113-22.

VASCONCELOS, M., ECKERT, H., ARAHANA, V., GRAEF, G., GRUSAK, M.A. and CLEMENTE, T. (2006). Molecular and phenotypic characterization of transgenic soybean expressing the *Arabidopsis* ferric chelate reductase gene, FRO2. *Planta* **224**, 1116-28.

VIJ, S. and TYAGI, A.K. (2007). Emerging trends in the functional genomics of the abiotic stress response in crop plants. *Plant Biotechnology Journal* **5**, 361-80.

VILLALOBOS, M.A., BARTELS, D. and ITURRIAGA, G. (2004). Stress tolerance and glucose insensitive phenotypes in *Arabidopsis* overexpressing the CpMYB10 transcription factor gene. *Plant Physiology* **135**, 309-24.

VINOCUR, B. and ALTMAN, A. (2005). Recent advances in engineering plant tolerance to abiotic stress: achievements and limitations. *Current Opinion in Biotechnology* **16**, 123-32.

VOSS, I., KOELMANN, M., WOJTERA, J., HOLTGREFE, S., KITZMANN, C., BACKHAUSEN, J.E. and SCHEIBE, R. (2008). Knockout of major leaf ferredoxin reveals new redox-regulatory adaptations in *Arabidopsis thaliana*. *Physiologia Plantarum* **133**, 584-98.

WALTER, M.H. (1989). The induction of phenylpropanoid biosynthetic enzymes by ultraviolet light or fungal elicitor in cultured parsley cells is overridden by a heat-shock treatment. *Planta* **177**, 1-8.

WANG, W., VINOCUR, B. and ALTMAN, A. (2003). Plant responses to drought, salinity and extreme temperatures: towards genetic engineering for stress tolerance. *Planta* **218**, 1-14.

WANG, W., VINOCUR, B., SHOSEYOV, O. and ALTMAN, A. (2004). Role of plant heat-shock proteins and molecular chaperones in the abiotic stress response. *Trends in Plant Science* **9**, 244-52.

XIONG, L. and YANG, Y. (2003). Disease resistance and abiotic stress tolerance in rice are inversely modulated by an abscisic acid-inducible mitogen-activated protein kinase. *Plant Cell* **15**, 745-59.

YAMAMOTO, H., KATO, H., SHINZAKI, Y. *ET AL.* (2006). Ferredoxin limits cyclic electron flow around PSI (CEF-PSI) in higher plants--stimulation of CEF-PSI enhances non-photochemical quenching of Chl fluorescence in transplastomic tobacco. *Plant and Cell Physiology* **47**, 1355-71.

YILMAZ, J.L. and BULOW, L. (2002). Enhanced stress tolerance in *Escherichia coli* and *Nicotiana tabacum* expressing a betaine aldehyde dehydrogenase/choline dehydrogenase fusion protein. *Biotechnology Progress* **18**, 1176-82.

ZHANG, J.Z., CREELMAN, R.A. and ZHU, J.K. (2004). From laboratory to field. Using information from *Arabidopsis* to engineer salt, cold, and drought tolerance in crops. *Plant Physiology* **135**, 615-21.

ZIMMERMANN, P., HIRSCH-HOFFMANN, M., HENNIG, L. and GRUISSEM, W. (2004). GENEVESTIGATOR. *Arabidopsis* microarray database and analysis toolbox. *Plant Physiology* **136**, 2621-32.

ZURBRIGGEN, M.D., CARRILLO, N., TOGNETTI, V.B., MELZER, M., PEISKER, M., HAUSE, B. and HAJIREZAEI, M.R. (2009). Chloroplast-generated reactive oxygen species play a major role in localized cell death during the non-host interaction between tobacco and *Xanthomonas campestris* pv. *vesicatoria*. *Plant Journal* **60**, 962-73.

ZURBRIGGEN, M.D., TOGNETTI, V.B. and CARRILLO, N. (2007). Stress-inducible flavodoxin from photosynthetic microorganisms. The mystery of flavodoxin loss from the plant genome. *IUBMB Life* **59**, 355-60.

ZURBRIGGEN, M.D., TOGNETTI, V.B., FILLAT, M.F., HAJIREZAEI, M.R., VALLE, E.M. and CARRILLO, N. (2008). Combating stress with flavodoxin: a promising route for crop improvement. *Trends in Biotechnology* **26**, 531-7.

Biotechnology and Genetic Engineering Reviews - Vol. 27, 57-94 (2010)

Phage display and molecular imaging: expanding fields of vision in living subjects

FRANK V. COCHRAN AND JENNIFER R. COCHRAN*

Department of Bioengineering, Cancer Center, Bio-X Program, Stanford University, Stanford CA, USA

Abstract

In vivo molecular imaging enables non-invasive visualization of biological processes within living subjects, and holds great promise for diagnosis and monitoring of disease. The ability to create new agents that bind to molecular targets and deliver imaging probes to desired locations in the body is critically important to further advance this field. To address this need, phage display, an established technology for the discovery and development of novel binding agents, is increasingly becoming a key component of many molecular imaging research programs. This review discusses the expanding role played by phage display in the field of molecular imaging with a focus on *in vivo* applications. Furthermore, new methodological advances in phage display that can be directly applied to the discovery and development of molecular imaging agents are described. Various phage library selection strategies are summarized and compared, including selections against purified target, intact cells, and *ex vivo* tissue, plus *in vivo* homing strategies. An outline of the process for converting polypeptides obtained from phage display library selections into successful *in vivo* imaging agents is provided, including strategies to optimize *in vivo* performance. Additionally, the use

To whom correspondence may be addressed (cochran1@stanford.edu)

Abbreviations: scFv, single chain variable fragment; Fab, fragment antigen binding; ELISA, Enzyme-Linked Immunosorbent Assay; NEB, New England Biolabs; PET, positron emission tomography; SPECT, single photon emission computed tomography; NIR, near infrared; PEG, polyethylene glycol; SPARC, secreted protein acidic and rich in cysteine; VCAM-1, vascular cell adhesion molecule; MRI, magnetic resonance imaging; DOTA, 1,4,7,10-tetraazacyclododecane-1,4,7,10-tetraacetic acid; IC_{50}, half maximal inhibitory concentration; CT, computed tomography; K_D, equilibrium dissociation constant; EGFR, epidermal growth factor receptor; EGFR-ECD, EGFR extracellular domain; $A\beta_{42}$, amyloid-beta; MMP, matrix metalloprotease; uPAR, urokinase-type plasminogen activator receptor; VEGF, vascular endothelial growth factor; IL-11, Interleukin-11; IL-11αR, Interleukin-11 α-receptor.

of phage particles as imaging agents is also described. In the latter part of the review, a survey of phage-derived *in vivo* imaging agents is presented, and important recent examples are highlighted. Other imaging applications are also discussed, such as the development of peptide tags for site-specific protein labeling and the use of phage as delivery agents for reporter genes. The review concludes with a discussion of how phage display technology will continue to impact both basic science and clinical applications in the field of molecular imaging.

Introduction

In vivo molecular imaging combines the principles of molecular and cellular biology with physics and chemistry to enable the non-invasive visualization of biological processes within living subjects (Mather 2009). Given this capability, molecular imaging holds great potential for both basic science and the diagnosis and monitoring of disease in clinical settings. Many fundamental breakthroughs in recent years have led to molecular imaging agents for various diseases, including cancer (Weissleder *et al.* 2008), Alzheimer's disease (Nordberg 2009), and atherosclerotic cardiovascular disease (Desai *et al.* 2009). Furthermore, a variety of molecular imaging modalities exist where an imaging probe is delivered to a specific molecular target. Examples include the delivery of radioisotopes for single photon emission computed tomography (SPECT) (Franc *et al.* 2008) and positron emission tomography (PET) (Gambhir 2002), photon-emitting fluorophores for fluorescence imaging (Rao *et al.* 2007), and paramagnetic contrast agents for magnetic resonance imaging (MRI) (Sosnovik *et al.* 2007). Multimodal approaches that combine the strengths of individual imaging modalities are also being developed (Lee *et al.* 2009).

Molecular targets, such as cell surface receptors, are often identified as key players in a disease through an understanding of the underlying biochemistry, or are uncovered by genomics and proteomics investigations. Many natural ligands for cell surface receptor targets have been explored for use as imaging agents; however, modifications are often required to improve target selectivity and *in vivo* stability, and to optimize biodistribution and pharmacokinetic profiles. Important clinical examples of ligand-based imaging agents are somatostatin analogs that target tumors expressing the somatostatin receptor (Reubi *et al.* 2008). While reports continuously appear where natural ligands are labeled with probes for various *in vivo* imaging studies, the number of natural ligands is finite, and many attractive targets may not have a known ligand that can be developed into a molecular imaging agent. Therefore, the ability to discover and engineer entirely new agents that bind to specific molecular targets is becoming increasingly important to further advance the field of molecular imaging.

Over the last few decades, the technology to discover polypeptides that bind to cellular targets or to engineer molecular recognition has been extensively developed. Hybridoma technology for the production of monoclonal antibodies was an early and important means of discovering novel target binders. Yet, since most of these antibodies are of mouse origin, they require "humanization" or development in transgenic animals to reduce immunogenicity in the body (Jakobovits *et al.* 2007). Furthermore, conversion into antibody fragment formats (scFv, Fab, diabody, etc.) is required to improve pharmacokinetics and biodistribution for imaging in living subjects (Presta 2008). The antibody engineering approaches described above, with a focus on molecular imaging applications, have recently been reviewed (Wu 2009). In parallel, display technologies have been increasingly used as an alternative approach of antibody discovery (Hoogenboom 2005; Lonberg 2008). In this regard, phage display has made enormous strides in the discovery and engineering of antibodies along with other classes of targeting polypeptides. Appreciating the value and potential of this technology, many molecular imaging groups have incorporated phage display as a core component of their research program (Landon *et al.* 2003; Sergeeva *et al.* 2006; Newton *et al.* 2008; Tavitian *et al.* 2009).

Many classes of polypeptides have been engineered using phage display; however, to date, most efforts aimed at *in vivo* molecular imaging applications have focused on peptides less than 20 residues in length, which is likely due to the commercial availability of phage peptide libraries. Compared to larger proteins, such as intact antibodies, peptides are attractive as *in vivo* imaging agents for several reasons, including efficient production by highly-optimized synthetic methods, improved organ and tumor penetration, and rapid clearance from non-target areas (Landon *et al.* 2004b; Mori 2004; Krumpe *et al.* 2006a). While many peptides have been obtained from phage display library selections, binding affinity to the desired target is often weak, requiring additional mutagenesis and selections. However, a high target binding affinity often cannot be obtained with small peptides, which lack a large surface area for molecular recognition and possess conformational flexibility that result in entropic penalties upon binding. To address these shortcomings, structural motifs have been used to stabilize and present a recognition site that is randomized and selected for binding interactions. The concept of using a structural motif or tertiary fold to support a randomized binding surface is referred to as an alternative scaffold. This name is used to distinguish these structural folds from that of antibodies. Many alternative scaffolds have been engineered by phage display to present constrained epitopes with larger surface areas necessary for high affinity binding interactions against a desired target (Friedman *et al.* 2009a). Affibody molecules are one such example that has been extensively developed for molecular imaging application, as discussed in detail below (Tolmachev *et al.* 2007).

This review will discuss the expanding role of phage display in the field of *in vivo* molecular imaging. We will also cover new methodological advances in phage display technology that can be directly applied to the discovery and development of novel molecular imaging agents. By tying this information together, we will summarize common trends, discuss current limitations and challenges, and highlight important examples of the application of phage display to *in vivo* molecular imaging. We will end with a discussion of how new advances in phage display technology will contribute to the future of molecular imaging.

Phage display background

Phage display is a robust and validated technique that has been widely used for the discovery of new binding agents and affinity maturation of existing ones (Smith *et al.* 1997; Kehoe *et al.* 2005). Since its inception, the methodology has been greatly expanded in scope and highly optimized; extensive descriptions of phage display technology and established protocols have been described in detail (Clackson *et al.* 2004; Sidhu 2005). A significant benefit of phage display is that basic laboratory bench skills along with common molecular biology laboratory equipment and services are all that is required.

In phage display, polypeptides are expressed as genetic fusions to coat proteins that compose the phage particle, where they are accessible to engage in binding interactions with target molecules. Each phage particle displays from one to many copies of the same polypeptide, depending on the specific format used. A library of DNA encoding random polypeptide sequences is produced and inserted into the appropriate location within the phage genome, resulting in each phage particle containing a single mutant. Phage library sizes of ~10^9 members are common, and several rounds of selection can be performed in a high-throughput manner to isolate clones that bind a variety of targets (*Figure 1*). DNA recovered from individual clones is sequenced to determine the identity of the corresponding amino acid sequences. This genotype-phenotype link is one of the most important components of phage display technology, as without it there would be no easy way to identify polypeptide variants isolated from library selections. Phage display systems can be divided into two main classifications based on strain type: filamentous phage and lytic phage. Further subdivisions are based on the type of format, such as how the polypeptide of interest is displayed and the degree of display levels. The differences that exist between various phage display libraries should encourage researchers to utilize several library types when searching for binders against a desired target (Kuzmicheva *et al.* 2009).

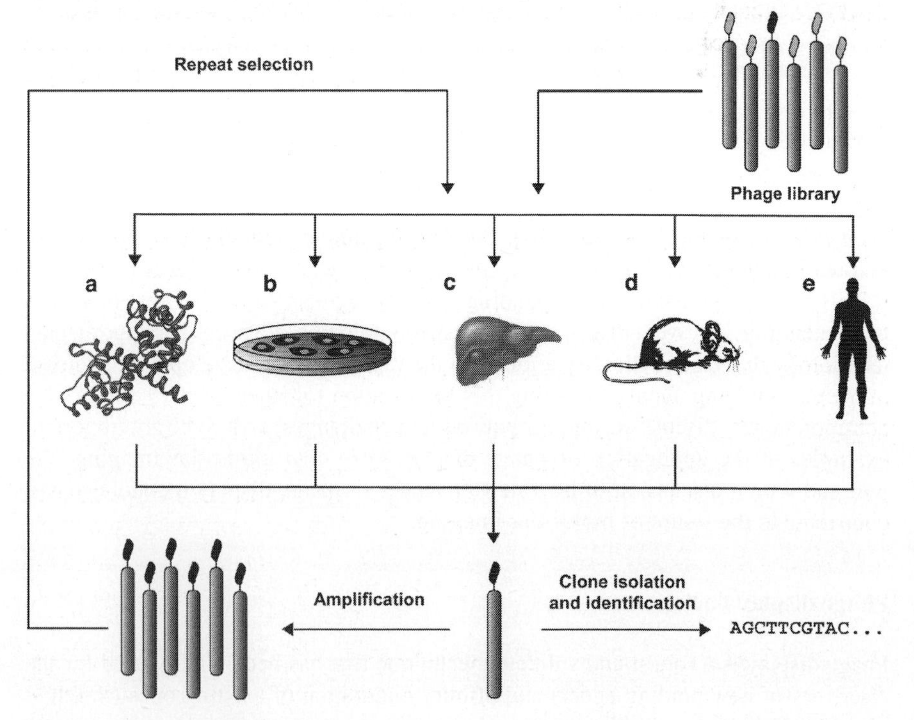

Figure 1. Phage display library selection strategies using purified target (a), cells in culture (b), *ex vivo* tissue (c), *in vivo* homing in small animals (d), and *in vivo* homing in humans (e).

FILAMENTOUS PHAGE DISPLAY

Filamentous phage strains, such as M13, infect *E. coli* and possess a circular single-stranded DNA genome. Polypeptide display is done by genetic fusion to phage coat proteins, typically pIII and pVIII. After almost two decades of development, researchers now have many options for the design and creation of filamentous phage libraries (Pini *et al.* 2004). In addition, M13 phage peptide libraries are commercially available from New England Biolabs (NEB). These prefabricated libraries, in which peptides are displayed by fusion to the pIII coat protein through a flexible linker sequence, are currently available in three formats: random linear peptides of 7 or 12 amino acids in length, and a cyclic format where a random 7 amino acid sequence is constrained by a putative intramolecular disulfide bond formed by flanking cysteine residues. Despite an abundance of choices, new M13 phage display formats continually appear. For example, a library displaying random peptides on the pVIII coat protein was recently developed for regulated control of peptide density on the phage surface through use of an arabinose-inducible promoter (Fagerlund *et al.* 2008). This regulation of display levels provides a convenient means of modulating stringency and avidity effects during the course of library selections. A potential limitation of filamentous phage is that the degree of display is often sequence-dependant. For example, sequences with high positive charge were reported to have relatively low levels of display (Imai *et al.* 2008). In addition, proteins that fold fast and are very stable in the cytoplasm suffer from reduced export efficiency to the periplasm (Steiner *et al.* 2006). To circumvent poor display levels in these instances, an alternative translocation pathway was accessed by use of an appropriate signal sequence, allowing efficient display and library selections of proteins refractory to conventional filamentous phage systems (Steiner *et al.* 2008).

LYTIC PHAGE DISPLAY

Polypeptide display on the capsid shell of lytic T7 phage is also common, and commercial kits for preparing libraries with this format are available from Novagen (Rosenberg *et al.* 1996). Peptides of up to 50 residues in length can be displayed in high copy number (415 per phage), and larger peptides and proteins of up to 1200 residues can be displayed in low copy number (≤ 1 per phage). T7 phage are released by cell lysis as part of their life cycle, making this system better suited than standard M13 phage systems for displaying certain cytoplasmic proteins, which may not be amenable for export through the periplasm and cell membrane. Another advantage is that T7 phage replicate faster than filamentous phage, which shortens experimental time. A recent study analyzed and compared the sequence diversity of the pre-made M13 phage linear 12-mer and cyclic 7-mer peptide libraries from NEB to the same type of peptide libraries constructed for display on T7 phage using the kit sold by Novagen (Krumpe *et al.* 2006b). Through a bioinformatics-assisted computational approach, the T7 phage libraries were found to have fewer amino acid biases, increased peptide diversity, and more normal distributions of peptide net charge and hydropathy in comparison to the M13 libraries.

OTHER COMBINATORIAL TECHNOLOGIES

Several other display technologies have been developed for peptide and protein engineering alongside of phage display in the last two decades. Bacterial (Daugherty 2007) and yeast surface display (Gai *et al.* 2007; Pepper *et al.* 2008) have emerged as powerful platforms, and also hold great promise for developing *in vivo* molecular imaging agents. The larger size of bacteria and yeast cells compared to phage particles allows library screening by fluorescence-activated cell sorting (FACS) using a fluorescently labeled target (Moore *et al.* 2009). Furthermore, as eukaryotic organisms, yeast are capable of processing large proteins with complex folds that may contain post-translational modifications; yet potential hyperglycosylation of proteins by yeast may be deleterious to target binding. Alternatively, *in vitro* display systems, such as ribosome and mRNA display have also been developed (Lipovsek *et al.* 2004). In these systems, polypeptide libraries are tethered to ribosomes or mRNA and screened in a manner similar to phage display. Since cell transformation is not required, larger library sizes can be generated with these platforms (~10^{13}-10^{14}) compared to other display technologies (10^{8}-10^{11}). In addition, mutation of library DNA in between each round of selection is convenient because the genetic material is recovered and amplified by PCR, allowing a broader search of sequence space. Aptamer technology (Perkins *et al.* 2007) and combinatorial chemistry (Aina *et al.* 2007) are alternative platforms for the discovery and development of molecular imaging agents based on nucleic acids and small molecules, respectively.

Phage display library selection strategies

Phage display library selections have been reviewed extensively (Clackson *et al.* 2004; Sidhu 2005). *Figure 1* provides an overview of selection strategies commonly used to develop molecular imaging agents.

SELECTIONS AGAINST PURIFIED TARGETS, INTACT CELLS, AND EX VIVO TISSUE

Early phage library selections relied upon targets that were obtained from a natural source or through recombinant DNA technology (*Figure 1a*). While effort is required to produce and purify a target of interest, this approach allows one to address hypotheses about the role of a specifically defined target within a living organism. In these selections, a phage library is incubated under controlled conditions for a specified length of time with a target that is either immobilized or present in solution. Alternatively, intact cells, typically as adherent monolayers, are used for presenting a target of interest (*Figure 1b*). This approach removes the significant bottleneck of producing and purifying a desired target in sufficient quantities for library selections. Phage library selections against intact cell lines without a pre-defined target in mind are also common. This selection strategy presents an opportunity to isolate phage clones that bind to interesting new targets, which can be identified using biochemical methods such as affinity chromatography, immunoprecipitation, photo- and chemical crosslinking, and mass spectrometry coupled with bioinformatics analysis. In addition, selection strategies based on cell internalization are important because they can be used to localize and trap an imaging agent inside cells (Kelly *et al.* 2005). A phage selection strategy with intact cells in suspension followed by separation using differential centrifugation has also identified many potentially useful peptides (Giordano *et al.* 2001). Phage libraries can be selected against sections of tissue as well (*Figure 1c*). In most of the above cases, sufficient washing is used to discard non-bound clones, followed by recovery of library members that bind using a variety of means, such as pH changes, or competition with natural ligand. Negative selection steps are often used to deplete the library of members that bind undesired molecules in order to increase the binding specificity of recovered clones. The recovered phage are amplified through infection of an *E. coli* host, and used in additional rounds of selection to obtain an enriched pool of binders. Progress is monitored during iterative rounds of selection by measuring the enrichment of the recovered pool of phage clones. After the final round of selection, individual clones are isolated and sequenced. A strong consensus among recovered sequences indicates that the best clones were selected from the library. However, false positives may arise from retention of clones that bind non-specifically or from clones with genetic advantages for propagation that are not related to target binding (Brammer *et al.* 2008). A recent report described an approach to address issues with non-specific charge interactions by genetically modifying phage coat proteins or using chemical agents to suppress non-specific interactions that occur between the negative charge on the phage coat surface with positively charged targets (Lamboy *et al.* 2008). Phage clones recovered from library selections are then individually validated for target binding; yet, this is usually a time-consuming step that can strongly benefit from new methods to rapidly prioritize hits. Some examples of increasing throughput include a fluorescence-based method for rapid ranking of recovered clones (Landon *et al.* 2004a), and a computational approach for the graphical display of ELISA data to depict affinity and specificity rankings (Kelly *et al.* 2006a). If desired, additional mutagenesis can be introduced into the recovered phage DNA to further improve properties such as binding affinity (Orlova *et al.* 2006; Friedman *et al.* 2008).

SELECTIONS IN LIVING ORGANISMS

It is not uncommon for polypeptides identified by the *in vitro* selection methods described above to perform poorly as imaging agents in a living organism. These disappointments likely result from selection conditions that do not accurately reproduce the relevant physiological environment or from undesired binding interactions. In addition, poor performance of phage-derived imaging agents *in vivo* can result from minimal signals due to low target expression levels, poor serum stability, or sub-optimal pharmacokinetic profiles. Target accessibility is also an important factor to consider (Ozawa *et al.* 2008). To address many of these problems, phage display library selections can be performed within the context of a living organism. This technique is referred to as *in vivo* homing (*Figure 1d,e*). This strategy allows the identification of library members that possess the desired phenotype (i.e. binding to target) under potentially more realistic conditions.

In vivo homing with phage libraries was first reported over a decade ago (Pasqualini *et al.* 1996). In brief, a phage display library is intravenously injected, allowed to circulate for a specified length of time, the target tissue, organ or tumor is removed, and phage that localize to a site of interest are recovered and amplified. Subsequent rounds of selection are performed with amplified phage until an enriched pool of binders is obtained. Individual clones are then isolated and sequenced in a manner similar to that for clones obtained through *in vitro* selection methods. This methodology is highly developed and comprehensively described in recent reviews (Trepel *et al.* 2008; Newton *et al.* 2009). Moreover, these efforts have led to the discovery of phage clones that localize in organ and tumor vasculature, and have shown that expression patterns within the vasculature give rise to unique molecular addresses that can be exploited to deliver therapeutics and imaging agents with high specificity (Kolonin *et al.* 2001). *In vivo* homing methodology has been expanded to select for phage clones that can extravasate from the vasculature and penetrate into tumors (Newton *et al.* 2006), and for phage clones specific for atherosclerotic plaques (Kelly *et al.* 2006b), among other tissue. Furthermore, *in vivo* homing has also been applied to human subjects (Arap *et al.* 2002; Krag *et al.* 2006). To refine *in vivo* homing protocols, a thorough biodistribution study with immunodeficient mice was performed, and circulation time was found to be a critical variable, depending on the organ or tissue being targeted (Zou *et al.* 2004).

The process of developing phage-derived *in vivo* molecular imaging agents

Converting peptides and proteins identified from phage library selections into imaging agents that perform as intended in the context of a living organism is not often straightforward. A series of validation and optimization steps are required to ensure that the peptide or protein retains activity when produced in soluble form and when conjugated with an imaging moiety. Furthermore, the *in vivo* performance of the candidate imaging agent requires sufficient optimization and validation for the desired application.

A common risk in any display technology project is that a peptide or protein obtained from a library screen may not retain the selected property when independent of the display format. For example, the apparent binding affinity of a selected peptide may significantly decrease when it is produced in soluble form. Possible explanations for this phenomenon include avidity artifacts due to multivalent binding which can occur under certain phage selection conditions, and structural influences imposed by genetic fusion to the phage surface that contribute to high affinity binding. Attachment of a radiolabel or dye molecule for imaging could be especially problematic due to modification of a chemical group that participates in binding, or distortion of the conformation required for binding. Diminished or abolished target binding may also occur due to unanticipated steric hindrance within the binding interface. In one case, attachment of a radiolabel to a disulfide-constrained peptide was shown to significantly reduce its ability to bind to tumor-specific integrin receptors (Haubner 2006). To minimize potential interference with binding, a linker region is commonly employed to separate the imaging moiety from the binding site. In addition, molecular modeling can be used to map the conformational profile of a flexible peptide to help predict the influence of conjugation on the bioactive conformation (Zanuy *et al.* 2009). It should also be noted that conjugation to hydrophobic groups could cause the peptide or protein to have reduced water solubility. An interesting example where chemical conjugation had beneficial effects involved a 12-residue peptide that was selected from the NEB linear 12-mer library for binding to a human thyroid cancer cell line (Zitzmann *et al.* 2007). C-terminal conjugation of a DOTA chelating group for ^{111}In SPECT imaging led to improved cell binding affinity and increased serum stability (Mier *et al.* 2007). Circular dichroism spectroscopy revealed that the conjugated peptide contains more α-helical secondary structure in comparison to the unconjugated peptide. In this instance, the chelating group was proposed to act as a surrogate for structural constraints imposed by coat protein fusion in the display format.

When *in vivo* applications are intended, properties such as serum and metabolic stability, as well as pharmacokinetics and tissue biodistribution must be considered. First, peptides and proteins can rapidly degrade upon exposure to proteases present in serum and tissues. Incorporation of D-amino acids (Doronina *et al.* 2008), disulfide bonds (Kimura *et al.* 2009b), and backbone cyclization (Li *et al.* 2002) have produced peptides with increased stability towards proteolysis. N-terminal acetylation and C-terminal amidation are also commonly used to prevent proteolytic degradation (Brinckerhoff *et al.* 1999). Second, rapid clearance is desired to obtain minimal background levels for molecular imaging applications; however, this must be balanced with the ability to elicit specific uptake in desired tissue. Chemical modifications with polyethylene glycol (PEG) or albumin binding peptides have been used to increase circulation time and enhance pharmacokinetic profiles (Chen *et al.* 2004b). In addition, peptides with higher target binding affinity (Kimura *et al.* 2009a) and multivalent target binding (Liu 2006) have been shown to elicit increased imaging signals compared to weaker binding and monomeric peptides; yet, in some cases monovalent proteins have been shown to exhibit better tumor uptake compared to higher affinity dimeric versions (Cheng *et al.* 2008). The latter observation is interpreted to result from the effects of molecular size on tissue diffusion. Third, tissue biodistribution is also important, as many molecular imaging agents elicit undesired uptake in organs such as the kidneys and liver. Chemical modifications, such as PEG, have been shown to reduce non-specific liver uptake (Chen *et al.* 2004b). In addition, modifying the chemical properties of peptide sequences (Ekblad *et al.* 2008), or increasing specificity by altering the binding affinity to other target family members (Wang *et al.* 2007a) have been used to improve the tissue biodistibution of *in vivo* molecular imaging agents.

Phage-derived peptides are often conjugated to nanoparticles, which have attractive properties for *in vivo* imaging, such as fluorescence and magnetic resonance (McCarthy *et al.* 2008). In addition, since many phage-derived peptides bind targets with relatively weak affinity (K_D ~ μM), nanoparticle conjugation is used to improve overall binding affinity through avidity effects resulting from polyvalency and multimerization. This approach is often used to avoid additional affinity maturation, or when high affinity binders cannot be identified through subsequent phage library selections. However, a recent study indicated that surface density as well as intrinsic affinity are important factors to consider when conjugating peptides or proteins to nanoparticles (Zhou *et al.* 2007). In this study, nanoparticles that contained high surface densities of antibody fragments with affinities ranging from 0.9 nM to 264 nM showed no difference in overall binding to tumor cells. An additional factor to consider is that the tissue-penetrating properties of targeted nanoparticles are dependant upon their size and shape (Decuzzi *et al.* 2009). Recent computational modeling studies are providing insight into the complex interplay among factors responsible for optimal tumor uptake and retention (Schmidt *et al.* 2009).

Intact phage particles as imaging agents

Phage particles themselves have been examined as imaging agents to avoid the loss of binding that can occur when selected polypeptides are decoupled from the phage surface. This approach also takes advantage of the rapid amplification of phage by *E. coli*, and removes the need for chemical synthesis or recombinant expression, followed by purification and characterization. Imaging probes can be attached to the phage particle surface coat proteins using standard amine-reactive isothiocyanate or succinimidyl ester conjugation chemistry. In one initial report, M13 phage were conjugated to mercaptoacetyltriglycine and radiolabeled with ^{99m}Tc for imaging bacteria in a mouse model of infection (Rusckowski *et al.* 2004). Conjugation chemistry has also been used to attach fluorophores to phage coat proteins (Jaye *et al.* 2004), including fluorophores with emission maxima greater than 800 nm, which provide improved tissue penetration of emitted light with minimal competition from background autofluorescence (Hilderbrand *et al.* 2005). Fluorescently-labeled phage clones have been tested in mice against multiple targets, such as SPARC (secreted protein acidic and rich in cysteine) and VCAM-1 (vascular cell adhesion molecule) (Kelly *et al.* 2006c). In addition, M13 phage have been labeled with probes that exhibit protonation-dependent fluorescence, with the goal of monitoring pH in disease microenvironments (Hilderbrand *et al.* 2008). In these examples, conjugation of imaging moieties to phage coat proteins did not appear to affect target binding, nor did it significantly suppress *E. coli* infectivity required for further phage propagation. Furthermore, since phage particles contain many copies of coat proteins, chemical conjugation allows for attachment of a high number of imaging moieties to generate strong imaging signals. Innovative methods for producing fluorescent T7 phage have also been reported, including use of a translational frameshift site for the regulated expression of enhanced yellow fluorescent protein on the phage capsid (Slootweg *et al.* 2006), and incorporation of a europium complex within the phage capsid (Liu *et al.* 2005). In addition to radiolabels and fluorescent probes, magnetic resonance imaging (MRI) contrast agents have also been conjugated to phage particles. For example, an M13 phage clone selected for binding to phosphatidylserine (Laumonier *et al.* 2006) was chemically coupled to iron oxide nanoparticles and used to detect apoptosis in the livers of mice (Segers *et al.* 2007). As an alternative to direct chemical conjugation, the phage surface can be modified to contain recognition elements that bind imaging probes noncovalently. In one example, an M13 phage clone was created to display an Arg-Gly-Asp integrin-binding peptide sequence on pIII and a streptavidin-binding peptide sequence on the pVIII coat protein that allowed attachment to streptavidin-coated fluorescent nanoparticles (Chen *et al.* 2004a). These noncovalent phage-nanoparticle assemblies bound to integrin-expressing cancer cells as demonstrated by flow cytometry and fluorescence microscopy.

One drawback to using phage particles as imaging agents in a living organism is sub-optimal clearance time and accumulation in the reticuloendothelial system due to their high molecular weight. This slow clearance reduces signal contrast between target and non-target areas and increases toxicity concerns when radiolabels are employed. To overcome these obstacles, a pre-targeting strategy was developed in mice (Newton *et al.* 2007). In an example of this approach, a phage clone was created that displayed α-melanocyte–stimulating hormone on pIII for targeting the melanocortin-1 receptor, and was chemically biotinylated on pVIII. After systemic administration, sufficient time was allowed for targeting and clearance of phage from non-target areas before a radiolabeled streptavidin imaging probe was injected. While this approach addresses the issue of high background signal *in vivo*, its clinical translation is limited due to the immunogenicity of streptavidin (Breitz *et al.* 2000).

Examples of phage-derived *in vivo* molecular imaging agents

PHAGE-DERIVED IN VIVO MOLECULAR IMAGING AGENTS FROM LIBRARY SELECTIONS USING PRE-DEFINED TARGETS

MOLECULAR IMAGING AGENTS OBTAINED USING PURIFIED TARGETS

An example of a phage-derived antibody used for molecular imaging is the scFv L19 (Pini *et al.* 1998; Viti *et al.* 1999). This antibody fragment was discovered and developed by selecting a phage antibody library of >300 million clones based on human antibody repertoires against the ED-B domain of fibronectin, an important target for tumor angiogenesis. Affinity maturation of a recovered clone by combinatorial mutation of six strategically selected residues in the heavy chain variable domain resulted in a 27-fold increase in binding affinity. A further 28-fold affinity improvement was achieved by mutating two residues in the light chain. The resulting scFv bound to the ED-B domain of fibronectin with a K_D = 54 pM, as determined by surface plasmon resonance. An ^{123}I-labeled dimeric form of L19 was shown to localize to tumor lesions in human patients with aggressive types of lung and colorectal cancer (Santimaria *et al.* 2003). In another study, L19 was recombinantly and chemically modified for radiolabeling with ^{99m}Tc to image F9 teratocarcinoma tumors in mice (Berndorff *et al.* 2006). In addition, PET imaging of the tumor neovasculature in mice was reported with ^{76}Br and ^{124}I-labeled L19 (Rossin *et al.* 2007; Tijink *et al.* 2009). These longer-lived radiohalogen isotopes (half life = 16.2 hours for ^{76}Br, 4.18 days for ^{124}I) were chosen over ^{18}F (half life = 110 minutes) due to the relatively long clearance time of the L19 antibody fragment from the blood pool.

Galectin-3 is a carbohydrate binding protein that plays an important role in cancer-related processes, such as cell adhesion, tumor invasion, and metastasis (Perillo *et al.* 1998). Filamentous phage libraries consisting of random 6-mer peptides flanked by cysteine residues or linear 15-mer peptides were selected against purified, recombinant galectin-3 to identify target binders (Zou *et al.* 2005). These peptides bound the galetin-3 carbohydrate recognition domain and were further examined as potential tumor imaging agents. One candidate peptide (ANTPCGPYTHDCPVKR) was conjugated to a DOTA chelating group through a Gly-Ser-Gly linker at the N-terminus and radiolabeled with ^{111}In. Using a competition binding assay, radiolabeled peptide was shown to bind to galectin-3 expressed on the surface of cancer cells with a half maximal inhibitory concentration (IC_{50}) value of approximately 200 nM. Using SPECT imaging coupled with computed tomography (CT), the radiolabeled peptide was used to image PC3-M prostate carcinoma (Deutscher *et al.* 2009) and MDA-MB-435 breast tumor xenografts (Kumar *et al.* 2008), the latter of which has recently been proposed to be of melanoma origin instead of breast cancer (Rae *et al.* 2007).

Imaging apoptosis is important for applications such as assessing tumor response to therapy, monitoring the course of myocardial infarction, and diagnosing early stages of neurodegenerative disease. Phosphatidylserine, an anionic phospholipid, is translocated from the cytosolic side of the plasma membrane to the extracellular surface in apoptotic cells, and can be used as an imaging marker for apoptosis. The NEB cyclic 7-mer library was selected against surface immobilized phosphatidylserine to yield a phage clone displaying the sequence CLSYYPSYC (Thapa *et al.* 2008). The peptide was synthesized and conjugated to fluorescein through its N-terminal amino group. Peptide binding to apoptotic cells was measured using fluorescence microscopy and flow cytometry, and binding specificity to phosphatidylserine was validated by competition with annexin V, a protein commonly used to detect apoptosis through phosphatidylserine binding. Tumor apoptosis was imaged in H460 mouse tumor xenografts that were treated with a single dose of the anti-cancer drug camptothecin 24 hours prior to intravenous injection of the fluorescent peptide.

ErbB2 (HER2) is an important diagnostic and therapeutic target as it is expressed in many cancers, including 20-30% of breast tumors (Tolmachev 2008). No natural ligand is currently known to directly bind ErbB2, which presents an opportunity for phage display to identify novel target binders. A linear 6 amino acid peptide library displayed using the M13 fUSE5 phage vector (Smith *et al.* 1993) was selected against recombinantly expressed ErbB2-extracellular domain (ErbB2-ECD) that was biotinylated and surface immobilized to streptavidin-coated plates (Karasseva *et al.* 2002). A phage clone that displayed a peptide with the sequence KCCYSL was identified. After production by solid phase synthesis, electrospray ionization mass spectrometry analysis found that the peptide exists as a mixture of oxidized and reduced monomer in solution without measurable amounts of a dimeric form. The binding affinity to recombinant ErbB2-ECD, as determined by fluorescence titration experiments in solution, was reported to be K_D ~300 nM (Kumar *et al.* 2007). A DOTA chelating group was added to the N-terminus of the peptide through a Gly-Ser-Gly linker and labeled with ^{111}In. Using a competition binding assay, radiolabeled peptide was shown to bind to MDA-MB-435 cancer cells expressing ErbB2 with an IC_{50} value of 42 nM. The ^{111}In-DOTA-GSG-KCCYSL peptide conjugate was also used for SPECT/CT imaging of MDA-MB-435 human cancer xenografts in mice.

A high profile account of *in vivo* molecular imaging of ErbB2 has been reported with engineered affibody molecules, a class of alternative scaffold proteins that has been developed for molecular recognition (Orlova *et al.* 2007a; Tolmachev *et al.* 2007; Nygren 2008). These 3-helix bundle proteins are derived from one of the IgG-binding domains of staphylococcal protein A (Nord *et al.* 1997) and are approximately 58 residues in length (*Figure* 2). The helical structure serves as a framework for presenting a binding surface of 13 randomized amino acid positions on helix 1 and helix 2. This recognition site, which was structurally characterized by nuclear magnetic resonance spectroscopy (Lendel *et al.* 2006), provides a large surface area for high affinity binding interactions through a greater number of molecular contacts than is possible with a small peptide. Affibody molecules attracted interest as potential imaging agents as their relatively small size (~7 kDa) was expected to provide good tumor penetration and fast clearance from non-target tissue. Furthermore, their small size allows efficient preparation by solid phase peptide synthesis and site-specific incorporation of chelating groups or other chemical functionalities for attachment of imaging moieties (Orlova *et al.* 2007b). In addition, affibody molecules maintain stability upon exposure to harsh conditions during chemical conjugation, such as pH extremes and elevated temperatures up to 50 °C.

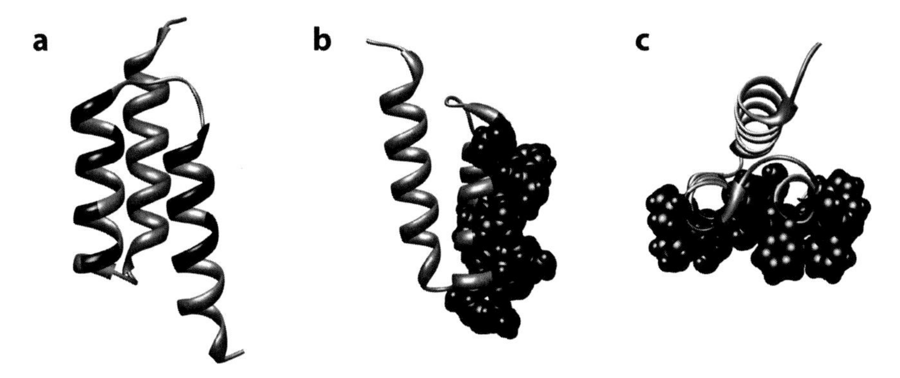

Figure 2. Affibody molecule structure. Ribbon representation of the three-helix bundle scaffold structure (gray) with the 13 randomized positions depicted in black (a). Side (b) and top (c) views of an affibody molecule with space-filling depictions to illustrate the molecular recognition surface. PDB ID: 2B87. Molecular graphics images were produced using the UCSF Chimera package (Pettersen *et al.* 2004).

A filamentous phage affibody molecule library was selected against biotinylated recombinant ErbB2-ECD that was immobilized on streptavidin-coated magnetic beads (Wikman *et al.* 2004). After four rounds of selection, one variant was produced in *E. coli* and found to bind to ErbB2-ECD with a K_D = 50 nM as determined by surface plasmon resonance. Dimerization of this affibody molecule by genetic fusion of two monomeric units led to an increase in binding affinity to ErbB2 by an order of magnitude, resulting in a K_D of ~ 3 nM (Steffen *et al.* 2005; Steffen *et al.* 2006). Next, a second phage library was made by randomizing positions found to vary in a sequence alignment of clones recovered from the first-generation library (Orlova *et al.* 2006). Further selection with this second phage library against ErbB2-ECD led to an affinity-maturated variant (K_D = 22 pM). Using a variety of different isotopes (^{125}I, ^{99m}Tc, ^{111}In, and ^{18}F), many imaging studies were performed in mice with these ErbB2-targeted affibody molecules (Steffen *et al.* 2006; Orlova *et al.* 2007b; Ekblad *et al.* 2008; Kiesewetter *et al.* 2008). Several studies also investigated the effects of labeling chemistry and chelator composition on the influence of biodistribution and tumor targeting in mice (Ahlgren *et al.* 2008; Tran *et al.* 2008; Tolmachev *et al.* 2009b; Ekblad *et al.* 2009). Initial clinical studies in human patients with these ErbB2-targeting affibody molecules, labeled with ^{111}In for SPECT imaging or ^{68}Ga for PET imaging, have also been reported (Baum *et al.* 2006). In another study, oxime conjugation chemistry was used to prepare ErbB2-specific monomeric and dimeric affibody molecules labeled with ^{18}F for PET imaging (Cheng *et al.* 2008). Unexpectedly, the monomer showed better *in vivo* performance despite the higher ErbB2 binding affinity of the dimer (*Figure 3*). This result was interpreted to arise from increased tumor penetration of the monomer due to its smaller size. Further engineering led to a two-helix version of an affibody molecule by removal of helix 3, which is not involved in the binding interface, followed by stabilization with chemical crosslinking (Webster *et al.* 2009). These truncated affibody molecules are expected to retain high affinity binding

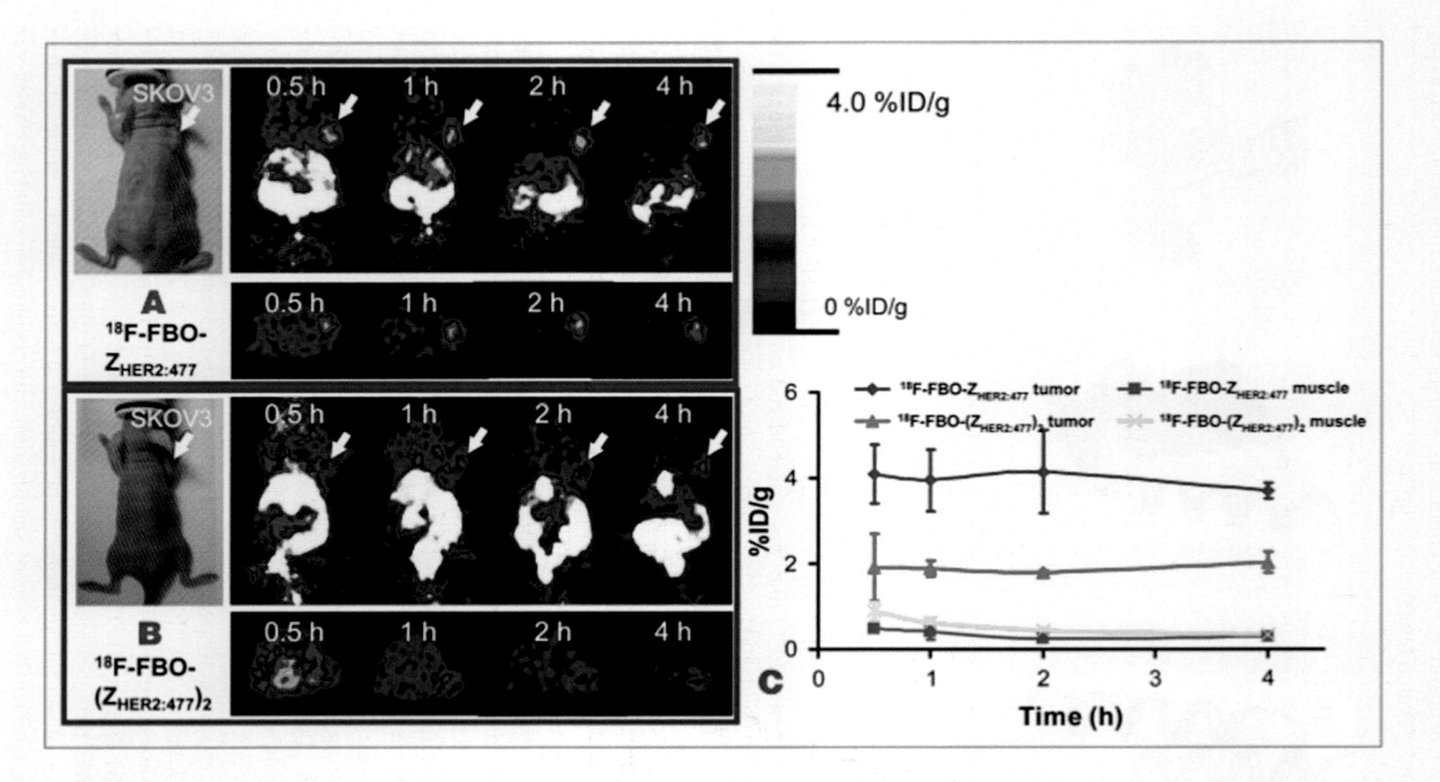

Figure 3. Coronal (top) and transaxial (bottom) PET images of nude mice bearing SKOV3 tumors on right shoulder at 0.5, 1, 2, and 4 hours after tail vein injection of monomer (^{18}F-FBO-$Z_{HER2:477}$) (A) and dimer (^{18}F-FBO-$(Z_{HER2:477})_2$) (B). Arrows indicate location of tumors. (C) Tumor and muscle time-activity curves derived from multiple-time-point PET images of mice bearing SKOV3 tumors. Data are shown as mean ± SD %ID/g (%ID/g = injected dose per gram of tissue) (n = 3). Reprinted by permission from the Society of Nuclear Medicine (Cheng *et al.* 2008).

to target along with increased tumor penetration. The two-helix affibody molecule was validated to bind to cell-surface expressed ErbB2, yet preliminary *in vivo* studies with a monomeric ^{68}Ga-labeled version showed lower absolute tumor uptake compared to other monomeric 3-helix affibodies used to image ErbB2 (Ren *et al.* 2009). This result was attributed to sub-optimal specific activity of the probe and a drop in binding affinity for the ErbB2 target resulting from removal of helix 3. Additional protein engineering and improved methods for probe attachment will likely correct these shortcomings. Recently, affibody molecules have been conjugated to AlexaFluor maleimide dyes (Invitrogen) through C-terminal cysteine thiols for NIR fluorescence imaging of ErbB2-expressing xenograft tumors in mice (Lee *et al.* 2008).

Affibody molecules have also been engineered to bind epidermal growth factor receptor (EGFR), another important target for tumor imaging (Mishani *et al.* 2008). A filamentous phage affibody molecule library was incubated with biotinylated recombinant EGFR extracellular domain (EGFR-ECD) and bound phage clones were captured using streptavidin magnetic beads. Four rounds of selection identified phage clones that displayed affibody molecules with binding affinities of 130-185 nM as measured by surface plasmon resonance (Friedman *et al.* 2007). These affibody molecules were expressed in *E. coli* with a free cysteine residue at the C-terminus for site-specific conjugation of fluorescent and radionuclide labels, and were shown to bind to native EGFR expressed on A431 human epidermoid carcinoma cells. A second-generation phage library was created by randomization of positions that were found to vary in a sequence alignment of recovered clones from the first-generation library. Selections with this second-generation phage library led to the identification of clones with low nanomolar affinity (~5-10 nM), as determined by surface plasmon resonance to EGFR-ECD and binding to A431 cells by flow cytometry (Friedman *et al.* 2008). Imaging studies with A431 xenograft tumors in mice using an ^{111}In-labeled monomer or dimer found that while the dimer had better tumor retention of radioactivity, the absolute uptake values were higher for the monomer, despite the monomer having a lower binding affinity compared to the dimer, similar to the case of the ErbB2-binding affibody molecules discussed above (Tolmachev *et al.* 2009a).

An affibody with bispecific targeting for two different receptors was recently reported (Friedman *et al.* 2009b). A gene encoding the dimeric form of an ErbB2-targeted affibody was fused to a gene encoding an EGFR-targeted affibody dimer with separation by a flexible $(Gly_4\text{-}Ser)_3$ linker. This construct was expressed in *E coli*, labeled with a fluorophore through a C-terminal cysteine, and demonstrated to bind ErbB2 and EGFR simultaneously by surface plasmon resonance, fluorescence microscopy, and a radioactivity-based cell-cell interaction assay. *In vivo* imaging studies with these bispecific affibody molecules are expected to be forthcoming.

Elevated expression levels of gelatinases, which are matrix metalloproteases (MMPs) involved in tumor growth, angiogenesis, and metastasis, are associated with poor patient prognosis, making this class of enzymes important for targeted cancer therapy and imaging (Van De Wiele *et al.* 2006). Selecting filamentous phage libraries against surface-immobilized MMP-9 led to a cyclic peptide with the sequence CTTHWGFTLC (cCTT), which bound and inhibited the activity of both MMP-2 and MMP-9 (Koivunen *et al.* 1999). Low accumulation in MMP-expressing tumors in mice was observed with cCTT peptides radiolabeled with ^{125}I, ^{64}Cu, or ^{111}In (Kuhnast *et al.* 2004; Sprague *et al.* 2006; Hanaoka *et al.* 2007). Furthermore, a ^{99m}Tc-labeled version of this peptide required incorporation into liposomes for KS1767 xenograft tumor imaging in mice (Medina *et al.* 2005). While cCTT was only active at concentrations in the micromolar range, serum degradation was also implicated as a main reason for sub-optimal *in vivo* performance. Successful strategies to improve serum stability were incorporation of non-natural amino acids through intein-directed production (Bjorklund *et al.* 2003), and conversion of cCTT into a peptidomimetic composed entirely of β-amino acids (Mukai *et al.* 2008). In another study, a structure-function analysis of cCTT led to the modified sequence KAHWGFTLD. Here, the disulfide bond of cCTT was replaced with an amide bond between the side chain amino group of lysine and the side chain carboxylate of aspartate resulting in increased resistance to proteolysis and a four-fold increase in gelatinase inhibition (Wang *et al.* 2009). In the same study, conjugation of the NIR dye Cy5.5 to the lysine α amine of this improved analog allowed optical imaging of intratibial prostate PC3 tumors and intracranially implanted U87MG brain tumors in mice.

In addition to selecting phage libraries against cancer targets, amyloid-beta ($A\beta_{42}$), the main component of amyloid plaques, was used as a target for developing phage-derived *in vivo* imaging agents for Alzheimer's disease (Larbanoix *et al.* 2008). In this study, the NEB cyclic 7-mer peptide library was selected against purified mouse $A\beta_{42}$. Two isolated phage clones bound $A\beta_{42}$ with affinities in the picomolar range as confirmed by ELISA; however, the binding affinity dropped to the micromolar range when the peptides were decoupled from the phage surface and produced synthetically. The peptides were conjugated to an iron oxide particle for imaging and to improve binding through avidity effects. A preliminary *in vivo* MRI imaging study with these peptide-functionalized contrast agents showed promising results in an Alzheimer's disease transgenic mouse model.

Phage display has also been used to identify peptides for imaging blood clots that can cause life-threatening conditions, such as heart attack, stroke, and pulmonary embolism. Phage peptide library selections against human fibrin, a insoluble polymer formed during the clotting cascade, led to the discovery and development of an 11 amino acid peptide MRI imaging agent, EP-2104R, that demonstrated 100-1000-fold higher selectivity for fibrin compared to other blood plasma proteins (Overoye-Chan *et al.* 2008). Nuclear magnetic resonance spectroscopy measurements, obtained on a diamagnetic lanthanum analog that enabled spectral interpretation and assignment, revealed that the conjugated peptide adopts one dominant and well-defined conformation in solution. EP-2104R was found to bind equally to two sites on human fibrin with a $K_D = 1.7$ μM. *In vivo* signal enhancement was obtained by incorporating four chelating groups on the peptide (two on each terminus) for the complexation of four gadolinium ions. Transmetallation competition experiments demonstrated that EP-2104R was highly inert toward gadolinium loss, which is an important property for patient safety. Remarkably, EP-2104R is the first MRI molecular imaging agent to advance to human trials (Spuentrup *et al.* 2008).

MOLECULAR IMAGING AGENTS OBTAINED USING CELL LINES EXPRESSING A TARGET OF INTEREST

The urokinase-type plasminogen activator receptor (uPAR) is an important target related to cancer due to its involvement in early tumor development and metastasis (Romer *et al.* 2004). A 15-mer peptide that binds uPAR was discovered in one of the earliest reports of phage library selections against targets presented in the context of intact cells (Goodson *et al.* 1994). This peptide was further optimized by combinatorial chemistry, resulting in a more stable, affinity-matured 9-mer peptidomimetic analog that bound uPAR with a K_D of approximately 0.4 nM (Ploug *et al.* 2001). In a recent study, this peptidomimetic was conjugated to DOTA, radiolabeled with ^{64}Cu, and used to image U87MG human glioblastoma xenografts in mice by PET (Li *et al.* 2008).

In another example, selections with the NEB linear 12-mer library identified a peptide that bound to Met, a receptor tyrosine kinase that is overexpressed in many types of human cancers (Zhao *et al.* 2007). A negative selection step was performed where the phage library was first incubated with a cell line that was devoid of Met, followed by selecting the reduced pool of phage clones against two tumor cell lines that expressed Met. A phage clone displaying a peptide with the sequence YLFSVHWPPLKA was recovered and validated for Met binding *in vitro*. Preliminary studies with ^{125}I-labeled peptide showed promise for imaging Met expression in a human SK-LMS-1/HGF leiomyosarcoma xenograft mouse model.

Hepsin, a transmembrane serine protease, is expressed at high levels in prostate cancer but at significantly lower levels in normal tissue, making it an important biomarker for early detection of the disease by *in vivo* molecular imaging (Magee *et al.* 2001). The NEB linear 7-mer phage library was selected against PC3 prostate cancer cells that were transfected to express hepsin on their surface (Kelly *et al.* 2008b). Negative selections against untransfected PC3 cells were included to remove library members that bound to unwanted cell surface molecules for improved target specificity. A recovered phage clone displaying the sequence IPLVVPL was labeled with fluorescein and shown to bind to cell surface-expressed hepsin by flow cytometry and fluorescence microscopy. The peptide was then synthesized, conjugated to fluorescein through a C-terminal linker, and coupled to an iron oxide nanoparticle that was fluorescently labeled with the NIR dye Cy5.5. A typical preparation resulted in approximately 11 peptides per nanopartcle. These hepsin-targeted nanoparticles bound specifically to hepsin-expressing LNCaP prostate tumor xenografts, but not to hepsin-negative PC3 prostate tumor xenografts.

PHAGE-DERIVED *IN VIVO* MOLECULAR IMAGING AGENTS FROM LIBRARY SELECTIONS WITHOUT A PRE-DEFINED TARGET

MOLECULAR IMAGING AGENTS OBTAINED USING INTACT CELLS AND EX VIVO TISSUE

Molecular imaging agents can greatly aid endoscopic examinations for early detection of colon cancer and in rapid assessment of response to therapy. Selection of the NEB cyclic 7-mer library against HT-29 colon cancer cells identified a peptide with the sequence CPIEDRPMC (Kelly *et al.* 2003). This peptide was synthesized with a C-terminal linker sequence for labeling with a NIR fluorophore and used to image orthotopic colonic tumors in mice with a custom-built microendoscope (Kelly *et al.* 2004). Preliminary results suggested that the $\alpha_5\beta_1$ integrin receptor is the putative target of the peptide. A more recent report demonstrated the success of phage-derived imaging agents in a clinical setting for early detection of premalignant tissue in the colon (Hsiung *et al.* 2008). In this study, the NEB linear 7-mer library was selected against freshly resected colon biopsy specimens, and negative selections against nonmalignant epithelial cells and healthy tissue were included to isolate peptides with increased binding specificity. This strategy led to the discovery of a phage clone displaying a peptide with the sequence VRPMPLQ that bound 20-fold greater to HT-29 human adenocarcinoma-derived cells than nonmalignant human intestinal cells. This peptide was synthesized and a fluorescein group was attached to the N-terminus via a hexanoic acid linker. Topical administration of the fluorescently-labeled heptapeptide in patients undergoing screening colonoscopy led to the detection of dysplastic polyps with up to 50 times more fluorescence than normal mucosa using a confocal microendoscope (*Figure 4 a,b*). Histology of biopsy specimens confirmed the presence of both dysplastic and normal crypts (*Figure 4 c,d*). The binding target of this peptide is currently unknown; however, this selection strategy using excised tissue from patients can potentially be extended to other malignancies amenable to endoscopic detection, such as cancers of the mouth, stomach, and esophagus.

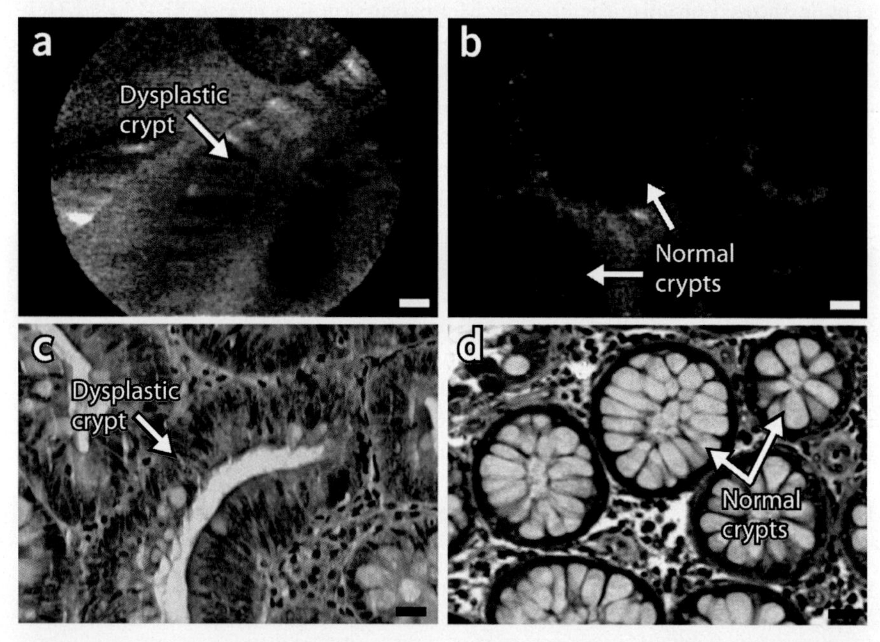

Figure 4. *In vivo* confocal endoscopic fluorescence images of fluorescein-VRPMPLQ peptide binding to dysplastic colon polyp (a) and adjacent normal mucosa (b). Histology of dysplastic colon polyp (c) and normal mucosa (d) stained with hematoxylin and eosin (H&E) stain. Reprinted by permission from Macmillan Publishers Ltd (Hsiung *et al.* 2008).

If pancreatic cancer is found early, surgical removal can provide a cure. Unfortunately, non-specific symptoms often result in late detection after the cancer has spread to an advanced stage when therapeutic treatment becomes much less effective. To develop a molecular imaging agent for early diagnosis of pancreatic cancer, the NEB linear 7-mer peptide library was selected against cell lines derived from a mouse model of human pancreatic cancer that represent early stages of the disease (Kelly *et al.* 2008a). These cell lines have been genetically engineered to reproduce many of the molecular features of pancreatic ductal adenocarcinoma, a common form of pancreatic cancer. One discovered peptide candidate (KTLLPTP) was synthesized and used for multimodal imaging by conjugation to fluorescent iron oxide MRI-active nanoparticles. This new imaging agent allowed detection of small pancreatic ductal adenocarcinoma and precursor lesions with both MRI and intravital confocal microscopy in mice. By using photo-crosslinking chemistry and mass spectrometry, the target of the selected phage clone was identified as plectin-1, a cytoplasmic protein not previously linked to pancreatic cancer. The relatively low expression levels of plectin-1 in normal pancreatic ductal cells, along with its localization in the cytoplasm, suggests that membrane-localized plectin-1 may be used as a biomarker for the early detection of pancreatic cancer in high risk patients. This finding also inspires additional studies to further examine the role of membrane-localized plectin-1, which may provide additional insight into cancer biology.

In another example, a strategy for selecting T7 phage library members that bound to plasma clots, but not to liquid plasma components, identified two cyclic peptides: CGLIIQKNEC (CLT1) and CNAGESSKNC (CLT2) (Pilch *et al.* 2006). Studies in mice strongly indicated that these peptides recognize fibrin-fibrinogen complexes formed by clotting plasma proteins that leaked into the extravascular space in tumors and other lesions. In a preliminary MRI imaging study, the CLT1 peptide was conjugated at the N-terminus with a chelating group for labeling with gadolinium and used to image fibronectin-fibrin complexes within mice bearing HT-29 human colon carcinoma xenografts (Ye *et al.* 2008).

MOLECULAR IMAGING AGENTS OBTAINED THROUGH PHAGE SELECTIONS IN LIVING ORGANISMS

Early *in vivo* homing studies in mice involved phage-displayed peptides that bound to integrins (Ruoslahti 2000). These cell surface receptors mediate cell adhesion through interactions with Arg-Gly-Asp (RGD)-containing proteins, such as fibronectin and vitronectin. Without question, the discovery and development of RGD-containing peptides that bind to α_v integrins overexpressed on endothelial cells of the tumor vasculature have resulted in a large class of molecular imaging agents with many variations and iterations that span a wide range of modalities. This large body of work will not be discussed here, since it has been extensively reviewed elsewhere (Haubner 2006; Liu 2006; Cai *et al.* 2008).

In vivo homing selections are being used to develop molecular imaging agents that monitor tumor response to therapy as an alternative to current methods that rely on tumor volume measurement or repeated biopsy. Significant changes in tumor volume usually happen only after a patient is treated for a prolonged length of time, and certain disease sites are not readily accessible to repeated biopsies. To address these issues, a remarkable study applied *in vivo* homing with a T7 phage peptide library to identify clones that only recognized tumors that were responsive to therapy (Han *et al.* 2008). In these studies, lung carcinoma or glioblastoma xenograft models were treated with vascular endothelial growth factor (VEGF) receptor tyrosine kinase inhibitors in combination with radiation prior to *in vivo* homing selections. Sequencing of the recovered phage clones revealed GSXV as a common binding motif, where X is a variable amino acid. Further analysis showed that the clone with the highest binding specificity displayed a peptide with the sequence HVGGSSV. This peptide was synthesized, labeled with a NIR fluorophore and used to image the therapeutic response of various tumor types in mice (*Figure 5*). This peptide is purported to bind a marker in the vasculature that becomes accessible during the course of endothelial response to therapy. Non-invasive imaging of pharmacodynamic responses of tumors to therapy can provide rapid assessment of treatment efficacy, shortening time spent on ineffective treatment regimens and greatly accelerating drug development.

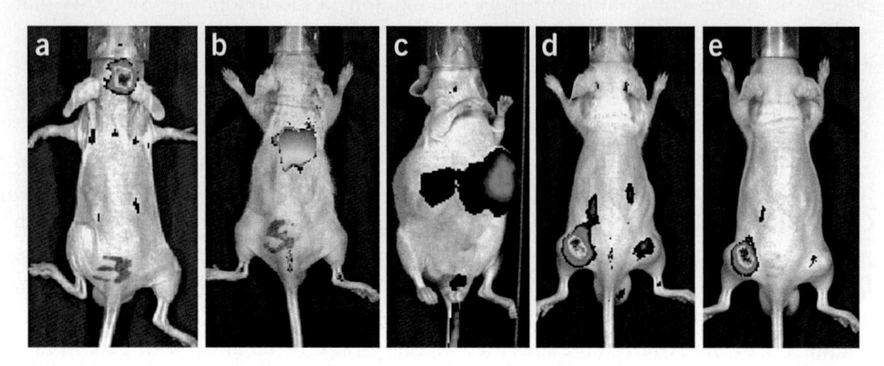

Figure 5. Peptide binding detects response to therapy in many tumor models. All tumor types had a significant increase in signal compared to untreated controls. Tumor development was induced by the following methods: D54 human glioblastoma cells were injected into the cerebra (a), H460 lung cancer cells were injected through the tail vein to develop pulmonary metastases (b), HT29 human colon cancer cells were injected into the spleen to develop liver metastases (c), and PC3 prostate cancer cells (d) and MDA-MB-231 breast cancer cells (e) were injected subcutaneously into the hind limbs of nude mice. The tumor-bearing mice were treated with a VEGF receptor tyrosine kinase inhibitor for 1 hour before irradiation. Cy7-labeled HVGGSSV peptide was injected intravenously 4 hours after treatment. Shown are NIR images obtained 48 hours after peptide injection. Reprinted by permission from Macmillan Publishers Ltd (Han *et al.* 2008).

In vivo homing methodology has also been extended for selecting phage clones that extravasate from the vasculature (Newton *et al.* 2006). This strategy incorporated a pre-clearing step in non-tumor-bearing mice to reduce the likelihood of obtaining phage clones that target the normal vasculature or other organs. The resulting pool of unbound phage clones was then injected into human PC3 prostate tumor-bearing mice, the tumors were resected, and phage clones that bound and internalized into tumor cells were recovered. This approach led to selection of a clone from a filamentous phage 15-mer peptide library that displayed the sequence IAGLATPGWSHWLAL. Optical imaging of human PC-3 carcinoma tumor xenografts was performed with this phage clone after it was labeled with the commercial NIR fluorophore Alexa Fluor 680. This phage clone was also used for ^{111}In-based SPECT/CT imaging of PC-3 carcinoma tumor xenografts in mice with pre-targeting strategies incorporating biotin-streptavidin/avidin interactions similar to the melanocortin-1 receptor imaging example described above (Newton-Northup *et al.* 2009).

In another example, *in vivo* homing has been used to develop a molecular imaging agent for atherosclerosis by selecting the NEB linear 7-mer phage peptide library for clones that were internalized in plaque-associated endothelial cells within atherosclerotic lesions present in mice (Kelly *et al.* 2006b). A peptide was identified (VHPKQHR) with sequence homology to a known ligand for VCAM-1. This peptide was synthesized with a cysteine at the C-terminus, and was conjugated to an iron oxide magnetic nanoparticle that contained the NIR fluorophore Cy5.5. These multi-modal targeted nanoparticles were used for MRI and optical imaging of plaques in a mouse model of atherosclerosis (Nahrendorf *et al.* 2006).

Encouraged by results obtained in mice, *in vivo* homing was also performed in humans (Arap *et al.* 2002). An M13 phage peptide library was injected into a human subject and allowed to circulate for 15 minutes followed by tissue biopsies to recover phage from various organs. A phage clone was identified that targeted the prostate vasculature. Interestingly, the sequence of this peptide (CGRRAGGSC) was found to have homology to Interleukin-11 (IL-11), an FDA-approved drug for treating low platelet levels caused by chemotherapy. Furthermore, this peptide was also shown to bind to Interleukin-11 α-receptor (IL-11αR). Additional characterization of the binding interaction between the peptide and IL-11αR was reported in a structure-function study using binding assays, site-directed mutagenesis, and nuclear magnetic resonance spectroscopy (Cardo-Vila *et al.* 2008). This study identified the key residues within the peptide that mediated interaction with the target and found that receptor binding does not occur within previously characterized interacting sites in the IL-11/IL-11αR complex. The peptide was further developed into a dual-modality imaging agent by conjugating it to both a NIR fluorophore and a chelating group for labeling with ^{111}In through an added N-terminal lysine residue (Wang *et al.* 2007b). This agent was used to image subcutaneous MDA-MB-231 xenograft tumors in mice by both NIR fluorescence and SPECT/CT.

Phage-derived peptide tags for imaging applications

Phage display has also been applied to molecular imaging for uses other than the discovery and development of high affinity target binders. One of these applications is the production of peptide tags for attaching imaging moieties to proteins in a site selective manner. These novel tools can be used for a variety of imaging experiments in both cells and living subjects.

In one case, phage display was used to evolve a new peptide substrate for yeast biotin ligase, which upon genetic fusion to a protein of interest enables site-specific labeling with imaging probes conjugated to streptavidin (Chen *et al.* 2007). Selections were performed by incubating phage peptide libraries with biotin and yeast biotin ligase, followed by recovery of biotinylated phage clones with streptavidin-coated beads. A negative selection step was used to remove peptide sequences that were biotinylated by endogenous *E. coli* biotin ligase during production of phage in bacteria. Attachment of biotin to a lysine residue within this novel 15-residue sequence, TTNWVAQAFKMTFDP, was shown to occur with high specificity among a background of other cellular proteins. Furthermore, this peptide tag was reactive when placed at the N or C terminus, or within an internal position of a fusion protein. Importantly, this yeast biotin ligase-peptide substrate pair is orthogonal to *E. coli* biotin ligase and its corresponding peptide substrate, which allows both of these tags to be used in the same experiment. Cell culture studies successfully demonstrated peptide substrate specificity by two-color quantum dot labeling of cell surface proteins. This new biotinylation tag is expected to find use in applications such as simultaneous single molecule imaging of trafficking and localization of two different cellular proteins.

Phage display has also been used to identify peptide sequences that non-covalently bind to fluorophores. While this approach was first investigated over a decade ago (Rozinov *et al.* 1998; Marks *et al.* 2004), a more recent report described a peptide sequence termed "IQ-tag" that was identified by selecting the NEB linear 7-mer library against a NIR fluorophore (Kelly *et al.* 2007). Prior to library selections, the NIR fluorophore was covalently coupled to bovine serum albumin and the resulting conjugate was surface immobilized. A peptide with the sequence IQSPHFF was found to bind this fluorescent dye with 100 nM affinity as demonstrated by surface plasmon resonance. *In vivo* performance was validated in mice by optical imaging of implanted HEK-293T cells expressing the IQ-tag fused to the platelet-derived growth factor receptor after injection of a NIR fluorophore (*Figure 6*). The relatively small size of this peptide tag also permits site-specific fluorophore attachment to targeting proteins under mild conditions and with minimal disruption to binding or activity.

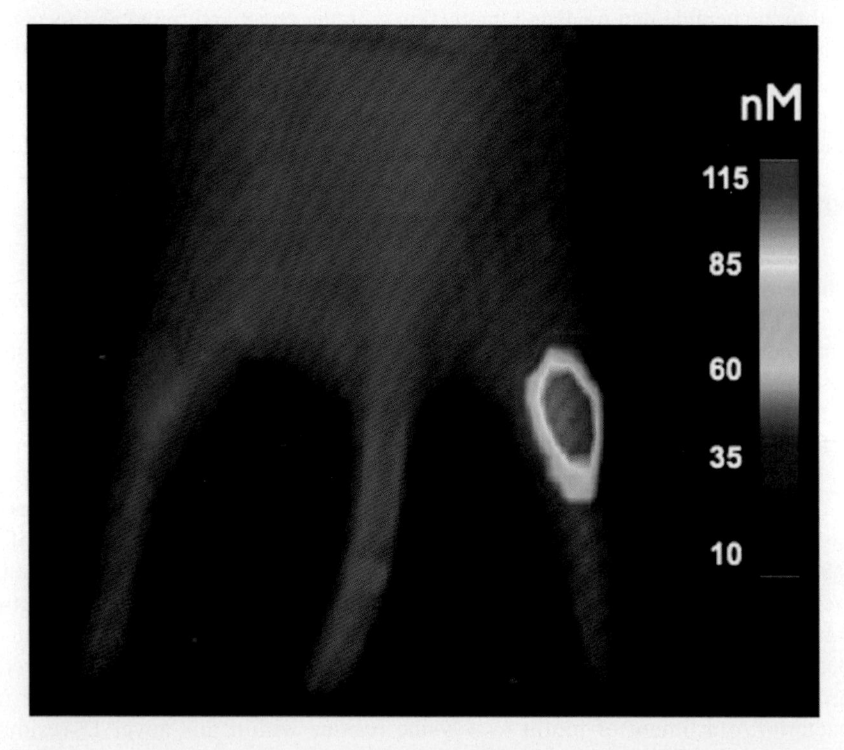

Figure 6. A mouse implanted with HEK-293T cells expressing the IQ-tag peptide fused to platelet-derived growth factor receptor (right flank) or without expression (left flank) was injected with a NIR fluorophore and imaged with fluorescence-mediated tomography. Reprinted from (Kelly *et al.* 2007).

Phage as delivery agents for reporter gene imaging

A new class of viral particles that combine the targeting ability of phage and the transgene delivery capabilities of eurokyotic viruses has recently been reported

(Hajitou *et al.* 2006). This system is based on a genetic chimera of the adeno-associated virus with the M13-derived fd-tet phage. Gene delivery is mediated through target binding and cell internalization after systemic administration (*Figure 7*). Moreover, these hybrid prokaryotic-eukaryotic vectors can be produced cheaply and in high yield in bacteria. The prototype vector displayed an RGD-containing peptide for α_v integrin binding and transgene delivery to tumors in mice. Gene delivery and expression of herpes simplex virus thymidine kinase resulted in the intracellular trapping of an ^{18}F-labeled substrate for PET imaging. Detailed protocols for the design, construction, and production of these novel vectors, along with mammalian cell experiments *in vitro* and tumor targeting in living subjects have been described in detail (Hajitou *et al.* 2007). In addition, this strategy was further applied to monitor and predict drug response in a nude rat model of soft tissue sarcoma (Hajitou *et al.* 2008). The ability to specifically deliver reporter genes to cells via targeted phage particles is expected to be a valuable tool for studying many biological phenomena.

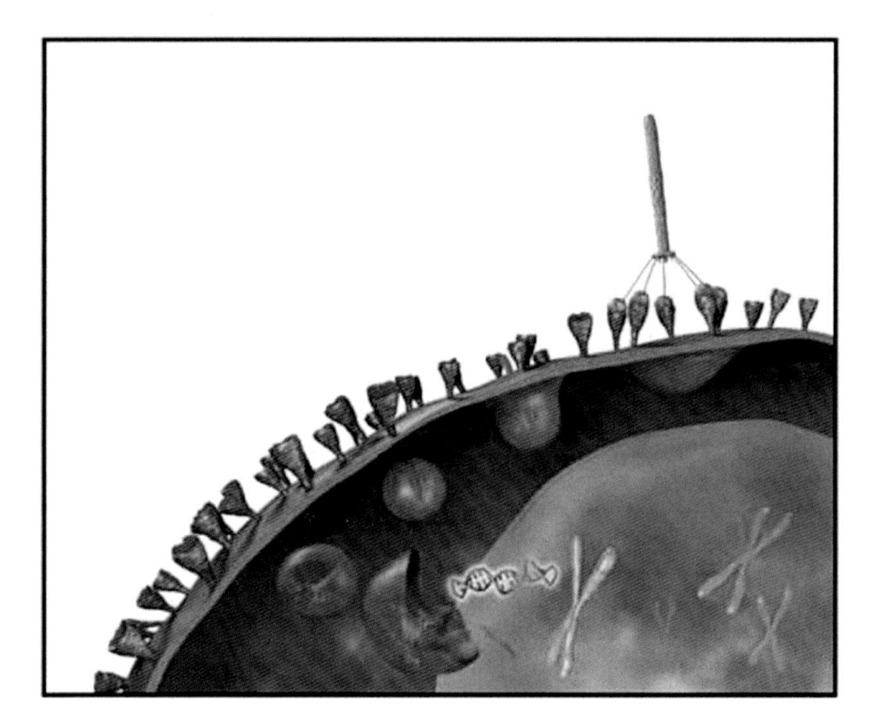

Figure 7. Binding of the targeted adeno-associated virus/phage hybrid particle to a specific cell surface receptor in the target tissue and internalization after systemic administration. Reprinted by permission from Macmillan Publishers Ltd (Hajitou *et al.* 2007).

Horizons

Phage display methodology is continually expanding in scope, and these advances hold great promise for *in vivo* molecular imaging applications. One example is the encoding of non-natural amino acids into phage polypeptide libraries to increase functional diversity available for molecular recognition beyond the 20 common amino acids (Liu *et al.* 2008). While this achievement is ground breaking, other pioneering studies have demonstrated that molecular recognition can also be achieved with protein libraries containing reduced amino acid diversity, with tyrosine playing a dominant role (Koide *et al.* 2009). In addition, integrating different display technologies allows one to take advantage of the particular strong points of each platform to increase the chances of success. For example, in the first report of engineering an affibody molecule for high affinity using cell surface display, a large library was first selected using a phage format, followed by transferring the enriched DNA pool to a bacterial display format, which allowed sorting by flow cytometry for quantitative discrimination between clones (Kronqvist *et al.* 2008). Besides affibodies, other alternative scaffolds are increasingly being developed for molecular imaging applications. One recent example is cystine knot miniproteins (also known as knottins), which are ~ 3-5 kDa peptides that contain disulfide-constrained loops with high tolerance to sequence diversity. Yeast surface display was used to engineer knottins that bound to tumor-related integrin receptors with low nanomolar affinity (Kimura *et al.* 2009b), and radiolabeled versions of these engineered miniproteins had favorable biodistribution and pharmacokinetic profiles in integrin-expressing human tumor xenograft models, including low kidney and liver uptake and retention (Kimura *et al.* 2009a). Additionally, bacterial surface display has been used to discover a small peptide employed for targeted ultrasound (Brown *et al.* 2000; Weller *et al.* 2005). In another emerging area, phage engineered to interact with nanomaterials offer great promise for *in vivo* imaging applications (Petrenko 2008; Mao *et al.* 2009). For example, self-assembling hydrogels composed of phage and gold nanoparticles have recently been reported (Souza *et al.* 2006; Souza *et al.* 2008). These novel materials retain the receptor binding and internalization attributes of the peptide displayed on the phage, and are capable of producing signals based on enhanced fluorescence and dark-field microscopy, surface-enhanced Raman scattering, and NIR photon-to-heat conversion. Another example describes T7 phage displaying a 15-residue peptide that is biotinylated during replication by bacterial biotin ligase, which in turn are captured by streptavidin-coated fluorescent quantum dot nanocrystals for rapid and highly sensitive detection of bacteria (Edgar *et al.* 2006). A noteworthy strategy was reported where a phage display-derived peptide was used to target iron oxide nanoparticles to tumor stroma by binding to clotted plasma protein, which induces further local clotting that creates new binding sites for additional accumulation of nanoparticles and amplification of imaging signal (Simberg *et al.* 2007). Furthermore, advances in computational design methods and biophysical analysis are demonstrating success in improving the binding affinity of phage-derived peptides for their target (Hao *et al.* 2008). Lastly, phage library selections are also being performed with innovative devices, such as a cuvette type quartz-crystal microbalance (Nishiyama *et al.* 2009) and a microfluidic magnetic separator (Liu *et al.* 2009).

Conclusions

Phage display is increasingly becoming a key component of many molecular imaging research programs for the discovery and engineering of new targeting molecules. The availability of commercial libraries and ease of use have made this technology accessible to researchers across several disciplines. While many phage library selections are performed against pre-defined targets, the use of cell lines, tissue, and *in vivo* homing selections have also been valuable. These latter strategies avoid the need to identify a specific target at the outset, allow library selections under conditions potentially more relevant to a clinical setting, and can lead to the discovery of unanticipated and interesting targets. The full potential of phage display is far from being completely explored; many library formats and selection strategies have not been fully exploited for the production of molecular imaging agents. The successful and rapid translation of phage-derived molecular imaging agents into the clinic remains a challenge, but new methods and tools are becoming available for optimizing *in vivo* performance. In conclusion, phage display will continue to be a significant driving force and a key player in enabling *in vivo* molecular imaging to deliver on its promise for both basic science and clinical applications.

Acknowledgements

The authors thank Bertrand Lui, Douglas Jones, Sarah Moore, and Niv Papo for helpful comments and discussions. We are grateful for funding from the Sidney Kimmel Foundation for Cancer Research.

References

AHLGREN, S., ORLOVA, A., ROSIK, D., SANDSTROM, M., SJOBERG, A., BAASTRUP, B., WIDMARK, O., FANT, G., FELDWISCH, J. and TOLMACHEV, V. (2008) Evaluation of maleimide derivative of DOTA for site-specific labeling of recombinant affibody molecules. *Bioconjugate Chemistry* **19**, 235-243.

AINA, O.H., LIU, R., SUTCLIFFE, J.L., MARIK, J., PAN, C.X. and LAM, K.S. (2007) From combinatorial chemistry to cancer-targeting peptides. *Molecular Pharmaceutics* **4**, 631-651.

ARAP, W., KOLONIN, M., TREPEL, M., LAHDENRANTA, J., CARDO-VILA, M., GIORDANO, R., MINTZ, P.J., ARDELT, P.U., YAO, V.J., VIDAL, C.I., CHEN, L., FLAMM, A., VALTANEN, H., WEAVIND, L.M., HICKS, M.E., POLLOCK, R.E., BOTZ, G.H., BUCANA, C.D., KOIVUNEN, E., CAHILL, D., TRONCOSO, P., BAGGERLY, K.A., PENTZ, R.D., DO, K.A., LOGOTHETIS, C.J. and PASQUALINI, R. (2002) Steps toward mapping the human vasculature by phage display. *Nature Medicine* **8**, 121-127.

BAUM, R.P., ORLOVA, A., TOLMACHEV, V. and FELDWISCH, J. (2006) A novel molecular imaging agent for diagnosis of recurrent HER2 positive breast cancer. First time in human study using an Indium-111- or Gallium-68-labeled Affibody molecule. *European Journal of Nuclear Medicine and Molecular Imaging* **33 Suppl 14**, S91.

BERNDORFF, D., BORKOWSKI, S., MOOSMAYER, D., VITI, F., MULLER-TIEMANN, B., SIEGER, S., FRIEBE, M., HILGER, C.S., ZARDI, L., NERI, D. and DINKELBORG, L.M. (2006) Imaging of tumor angiogenesis using ^{99m}Tc-labeled human recombinant anti-ED-B fibronectin antibody fragments. *Journal of Nuclear Medicine* **47**, 1707-1716.

BJORKLUND, M., VALTANEN, H., SAVILAHTI, H. and KOIVUNEN, E. (2003) Use of intein-directed peptide biosynthesis to improve serum stability and bioactivity of a gelatinase inhibitory peptide. *Combinatorial Chemistry and High Throughput Screening* **6**, 29-35.

BRAMMER, L.A., BOLDUC, B., KASS, J.L., FELICE, K.M., NOREN, C.J. and HALL, M.F. (2008) A target-unrelated peptide in an M13 phage display library traced to an advantageous mutation in the gene II ribosome-binding site. *Analytical Biochemistry* **373**, 88-98.

BREITZ, H.B., WEIDEN, P.L., BEAUMIER, P.L., AXWORTHY, D.B., SEILER, C., SU, F.-M., GRAVES, S., BRYAN, K. and RENO, J.M. (2000) Clinical optimization of pretargeted radioimmunotherapy with antibody-streptavidin conjugate and ^{90}Y-DOTA-Biotin. *Journal of Nuclear Medicine* **41**, 131-140.

BRINCKERHOFF, L.H., KALASHNIKOV, V.V., THOMPSON, L.W., YAMSHCHIKOV, G.V., PIERCE, R.A., GALAVOTTI, H.S., ENGELHARD, V.H. and SLINGLUFF, C.L., JR. (1999) Terminal modifications inhibit proteolytic degradation of an immunogenic MART-1(27-35) peptide: implications for peptide vaccines. *International Journal of Cancer* **83**, 326-334.

BROWN, C., MODZELEWSKI, R., JOHNSON, C. and WONG, M. (2000) A novel approach for the identification of unique tumor vasculature binding peptides using an *E. coli* peptide display library. *Annals of Surgical Oncology* **7**, 743-749.

CAI, W., NIU, G. and CHEN, X. (2008) Imaging of integrins as biomarkers for tumor angiogenesis. *Current Pharmaceutical Design* **14**, 2943-2973.

CARDO-VILA, M., ZURITA, A., GIORDANO, R., SUN, J., RANGEL, R., GUZMAN-ROJAS, L., ANOBOM, C., VALENTE, A., ALMEIDA, F.C., LAHDENRANTA, J., KOLONIN, M., ARAP, W. and PASQUALINI, R. (2008) A ligand peptide motif selected from a cancer patient is a receptor-interacting site within human interleukin-11. *PLoS ONE* **3**, e3452.

CHEN, I., CHOI, Y. and TING, A. (2007) Phage display evolution of a peptide substrate for yeast biotin ligase and application to two-color quantum dot labeling of cell surface proteins. *Journal of the American Chemical Society* **129**, 6619-6625.

CHEN, L., ZURITA, A., ARDELT, P.U., GIORDANO, R., ARAP, W. and PASQUALINI, R. (2004a) Design and validation of a bifunctional ligand display system for receptor targeting. *Chemistry & Biology* **11**, 1081-1091.

CHEN, X., HOU, Y., TOHME, M., PARK, R., KHANKALDYYAN, V., GONZALES-GOMEZ, I., BADING, J.R., LAUG, W.E. and CONTI, P.S. (2004b) Pegylated Arg-Gly-Asp peptide: ^{64}Cu labeling and PET imaging of brain tumor $\alpha_v\beta_3$-integrin expression. *Journal of Nuclear Medicine* **45**, 1776-1783.

CHENG, Z., DE JESUS, O.P., NAMAVARI, M., DE, A., LEVI, J., WEBSTER, J.M., ZHANG, R., LEE, B., SYUD, F.A. and GAMBHIR, S.S. (2008) Small-animal PET imaging of human epidermal growth factor receptor type 2 expression with site-specific ^{18}F-labeled protein scaffold molecules. *Journal of Nuclear Medicine* **49**, 804-813.

CLACKSON, T. and LOWMAN, H.B., Eds (2004). *Phage Display: A Practical Approach.* Oxford University Press, New York.

DAUGHERTY, P.S. (2007) Protein engineering with bacterial display. *Current Opinion in Structural Biology* **17**, 474-480.

DECUZZI, P., PASQUALINI, R., ARAP, W. and FERRARI, M. (2009) Intravascular delivery of particulate systems: does geometry really matter? *Pharmaceutical Research* **26**, 235-243.

DESAI, M.Y. and SCHOENHAGEN, P. (2009) Emergence of targeted molecular imaging in atherosclerotic cardiovascular disease. *Expert review of cardiovascular therapy* **7**,

197-203.

DEUTSCHER, S., FIGUEROA, S.D. and KUMAR, S.R. (2009) Tumor targeting and SPECT imaging properties of an ^{111}In-labeled galectin-3 binding peptide in prostate carcinoma. *Nuclear Medicine and Biology* **36**, 137-146.

DORONINA, S.O., BOVEE, T.D., MEYER, D.W., MIYAMOTO, J.B., ANDERSON, M.E., MORRIS-TILDEN, C.A. and SENTER, P.D. (2008) Novel peptide linkers for highly potent antibody-auristatin conjugate. *Bioconjugate Chemistry* **19**, 1960-1963.

EDGAR, R., MCKINSTRY, M., HWANG, J., OPPENHEIM, A.B., FEKETE, R.A., GIULIAN, G., MERRIL, C., NAGASHIMA, K. and ADHYA, S. (2006) High-sensitivity bacterial detection using biotin-tagged phage and quantum-dot nanocomplexes. *Proceedings of the National Academy of Sciences of the United States of America* **103**, 4841-4845.

EKBLAD, T., TRAN, T., ORLOVA, A., WIDSTROM, C., FELDWISCH, J., ABRAHMSEN, L., WENNBORG, A., KARLSTROM, A. and TOLMACHEV, V. (2008) Development and preclinical characterisation of ^{99m}Tc-labelled Affibody molecules with reduced renal uptake. *European Journal of Nuclear Medicine and Molecular Imaging* **35**, 2245-2255.

EKBLAD, T., ORLOVA, A., FELDWISCH, J., WENNBORG, A., KARLSTROM, A.E. and TOLMACHEV, V. (2009) Positioning of ^{99m}Tc-chelators influences radiolabeling, stability and biodistribution of Affibody molecules. *Bioorganic and Medicinal Chemistry Letters* **19**, 3912-3914.

FAGERLUND, A., MYRSET, A. and KULSETH, M. (2008) Construction and characterization of a 9-mer phage display pVIII-library with regulated peptide density. *Applied Microbiology and Biotechnology* **80**, 925-936.

FRANC, B.L., ACTON, P.D., MARI, C. and HASEGAWA, B.H. (2008) Small-animal SPECT and SPECT/CT: Important tools for preclinical investigation. *Journal of Nuclear Medicine* **49**, 1651-1663.

FRIEDMAN, M., NORDBERG, E., HOIDEN-GUTHENBERG, I., BRISMAR, H., ADAMS, G.P., NILSSON, F.Y., CARLSSON, J. and STAHL, S. (2007) Phage display selection of Affibody molecules with specific binding to the extracellular domain of the epidermal growth factor receptor. *Protein Engineering, Design and Selection* **20**, 189-199.

FRIEDMAN, M., ORLOVA, A., JOHANSSON, E., ERIKSSON, T.L., HOIDEN-GUTHENBERG, I., TOLMACHEV, V., NILSSON, F.Y. and STAHL, S. (2008) Directed evolution to low nanomolar affinity of a tumor-targeting epidermal growth factor receptor-binding affibody molecule. *Journal of Molecular Biology* **376**, 1388-1402.

FRIEDMAN, M. and STAHL, S. (2009a) Engineered affinity proteins for tumour-targeting applications. *Biotechnology and Applied Biochemistry* **53**, 1-29.

FRIEDMAN, M., LINDSTROM, S., EKERLJUNG, L., ANDERSSON-SVAHN, H., CARLSSON, J., BRISMAR, H., GEDDA, L., FREJD, F.Y. and STAHL, S. (2009b) Engineering and characterization of a bispecific HER2 x EGFR-binding affibody molecule. *Biotechnology and Applied Biochemistry* **54**, 121-131.

GAI, S.A. and WITTRUP, K.D. (2007) Yeast surface display for protein engineering and characterization. *Current Opinion in Structural Biology* **17**, 467-473.

GAMBHIR, S.S. (2002) Molecular imaging of cancer with positron emission tomography. *Nature Reviews Cancer* **2**, 683-693.

GIORDANO, R.J., CARDO-VILA, M., LAHDENRANTA, J., PASQUALINI, R. and ARAP, W. (2001) Biopanning and rapid analysis of selective interactive ligands. *Nature Medicine* **7**, 1249-1253.

GOODSON, R.J., DOYLE, M.V., KAUFMAN, S.E. and ROSENBERG, S. (1994) High-affinity urokinase receptor antagonists identified with bacteriophage peptide display. *Proceedings of the National Academy of Sciences of the United States of America* **91**, 7129-7133.

HAJITOU, A., LEV, D.C., HANNAY, J.A., KORCHIN, B., STAQUICINI, F.I., SOGHOMONYAN, S., ALAUDDIN, M.M., BENJAMIN, R.S., POLLOCK, R.E., GELOVANI, J.G., PASQUALINI, R. and ARAP, W. (2008) A preclinical model for predicting drug response in soft-tissue sarcoma with targeted AAVP molecular imaging. *Proceedings of the National Academy of Sciences of the United States of America* **105**, 4471-4476.

HAJITOU, A., RANGEL, R., TREPEL, M., SOGHOMONYAN, S., GELOVANI, J.G., ALAUDDIN, M.M., PASQUALINI, R. and ARAP, W. (2007) Design and construction of targeted AAVP vectors for mammalian cell transduction. *Nature Protocols* **2**, 523-531.

HAJITOU, A., TREPEL, M., LILLEY, C.E., SOGHOMONYAN, S., ALAUDDIN, M.M., MARINI, F.C., RESTEL, B.H., OZAWA, M.G., MOYA, C.A., RANGEL, R., SUN, Y., ZAOUI, K., SCHMIDT, M., VON KALLE, C., WEITZMAN, M.D., GELOVANI, J.G., PASQUALINI, R. and ARAP, W. (2006) A hybrid vector for ligand-directed tumor targeting and molecular imaging. *Cell* **125**, 385-398.

HAN, Z., FU, A., WANG, H., DIAZ, R., GENG, L., ONISHKO, H. and HALLAHAN, D. (2008) Noninvasive assessment of cancer response to therapy. *Nature Medicine* **14**, 343-349.

HANAOKA, H., MUKAI, T., HABASHITA, S., ASANO, D., OGAWA, K., KURODA, Y., AKIZAWA, H., IIDA, Y., ENDO, K., SAGA, T. and SAJI, H. (2007) Chemical design of a radiolabeled gelatinase inhibitor peptide for the imaging of gelatinase activity in tumors. *Nuclear Medicine and Biology* **34**, 503-510.

HAO, J., SEROHIJOS, A.W., NEWTON, G., TASSONE, G., WANG, Z., SGROI, D., DOKHOLYAN, N. and BASILION, J. (2008) Identification and rational redesign of peptide ligands to CRIP1, a novel biomarker for cancers. *PLoS Computational Biology* **4**, e1000138.

HAUBNER, R. (2006) $\alpha_v\beta_3$-integrin imaging: a new approach to characterise angiogenesis? *European Journal of Nuclear Medicine and Molecular Imaging* **33 Suppl 1**, S54-S63.

HILDERBRAND, S.A., KELLY, K.A., NIEDRE, M. and WEISSLEDER, R. (2008) Near infrared fluorescence-based bacteriophage particles for ratiometric pH imaging. *Bioconjugate Chemistry* **19**, 1635-1639.

HILDERBRAND, S.A., KELLY, K.A., WEISSLEDER, R. and TUNG, C.H. (2005) Monofunctional near-infrared fluorochromes for imaging applications. *Bioconjugate Chemistry* **16**, 1275-1281.

HOOGENBOOM, H. (2005) Selecting and screening recombinant antibody libraries. *Nature Biotechnology* **23**, 1105-1116.

HSIUNG, P.L., HARDY, J., FRIEDLAND, S., SOETIKNO, R., DU, C., WU, A., SAHBAIE, P., CRAWFORD, J., LOWE, A., CONTAG, C. and WANG, T. (2008) Detection of colonic dysplasia in vivo using a targeted heptapeptide and confocal microendoscopy. *Nature Medicine* **14**, 454-458.

IMAI, S., MUKAI, Y., TAKEDA, T., ABE, Y., NAGANO, K., KAMADA, H., NAKAGAWA, S., TSUNODA, S. and TSUTSUMI, Y. (2008) Effect of protein properties on display efficiency using the M13 phage display system. *Pharmazie* **63**, 760-764.

JAKOBOVITS, A., AMADO, R.G., YANG, X., ROSKOS, L. and SCHWAB, G. (2007) From

XenoMouse technology to panitumumab, the first fully human antibody product from transgenic mice. *Nature Biotechnology* **25**, 1134-1143.

JAYE, D.L., GEIGERMAN, C.M., FULLER, R.E., AKYILDIZ, A. and PARKOS, C.A. (2004) Direct fluorochrome labeling of phage display library clones for studying binding specificities: applications in flow cytometry and fluorescence microscopy. *Journal of Immunological Methods* **295**, 119-127.

KARASSEVA, N.G., GLINSKY, V.V., CHEN, N.X., KOMATIREDDY, R. and QUINN, T.P. (2002) Identification and characterization of peptides that bind human ErbB-2 selected from a bacteriophage display library. *Journal of Protein Chemistry* **21**, 287-296.

KEHOE, J.W. and KAY, B.K. (2005) Filamentous phage display in the new millennium. *Chemical Reviews* **105**, 4056-4072.

KELLY, K., ALENCAR, H., FUNOVICS, M., MAHMOOD, U. and WEISSLEDER, R. (2004) Detection of invasive colon cancer using a novel, targeted, library-derived fluorescent peptide. *Cancer Research* **64**, 6247-6251.

KELLY, K.A., ALLPORT, J.R., TSOURKAS, A., SHINDE-PATIL, V.R., JOSEPHSON, L. and WEISSLEDER, R. (2005) Detection of Vascular Adhesion Molecule-1 expression using a novel multimodal nanoparticle. *Circulation Research* **96**, 327-336.

KELLY, K.A., BARDEESY, N., ANBAZHAGAN, R., GURUMURTHY, S., BERGER, J., ALENCAR, H., DEPINHO, R., MAHMOOD, U. and WEISSLEDER, R. (2008a) Targeted nanoparticles for imaging incipient pancreatic ductal adenocarcinoma. *PLoS Medicine* **5**, e85.

KELLY, K.A., CARSON, J., MCCARTHY, J.R. and WEISSLEDER, R. (2007) Novel peptide sequence ("IQ-tag") with high affinity for NIR fluorochromes allows protein and cell specific labeling for in vivo imaging. *PLoS ONE* **2**, e665.

KELLY, K.A., CLEMONS, P.A., YU, A.M. and WEISSLEDER, R. (2006a) High-throughput identification of phage-derived imaging agents. *Molecular Imaging* **5**, 24-30.

KELLY, K.A. and JONES, D.A. (2003) Isolation of a colon tumor specific binding peptide using phage display selection. *Neoplasia* **5**, 437-444.

KELLY, K.A., NAHRENDORF, M., YU, A.M., REYNOLDS, F. and WEISSLEDER, R. (2006b) In vivo phage display selection yields atherosclerotic plaque targeted peptides for imaging. *Molecular Imaging and Biology* **8**, 201-207.

KELLY, K.A., SETLUR, S.R., ROSS, R., ANBAZHAGAN, R., WATERMAN, P., RUBIN, M.A. and WEISSLEDER, R. (2008b) Detection of early prostate cancer using a hepsin-targeted imaging agent. *Cancer Research* **68**, 2286-2291.

KELLY, K.A., WATERMAN, P. and WEISSLEDER, R. (2006c) In vivo imaging of molecularly targeted phage. *Neoplasia* **8**, 1011-1018.

KIESEWETTER, D.O., KRAMER-MAREK, G., MA, Y. and CAPALA, J. (2008) Radiolabeling of HER2-specific Affibody molecule with F-18. *Journal of Fluorine Chemistry* **129**, 799-806.

KIMURA, R.H., CHENG, Z., GAMBHIR, S.S. and COCHRAN, J.R. (2009a) Engineered knottin peptides: a new class of agents for imaging integrin expression in living subjects. *Cancer Research* **69**, 2435-2442.

KIMURA, R.H., LEVIN, A.M., COCHRAN, F.V. and COCHRAN, J.R. (2009b) Engineered cystine knot peptides that bind $\alpha_v\beta_3$, $\alpha_v\beta_5$, and $\alpha_5\beta_1$ integrins with low-nanomolar affinity. *Proteins: Structure, Function, and Bioinformatics* **77**, 359-369.

KOIDE, S. and SIDHU, S.S. (2009) The importance of being tyrosine: lessons in molecular recognition from minimalist synthetic binding proteins. *ACS Chemical Biology* **4**, 325-334.

KOIVUNEN, E., ARAP, W., VALTANEN, H., RAINISALO, A., MEDINA, O.P., HEIKKILA, P., KANTOR, C., GAHMBERG, C.G., SALO, T., KONTTINEN, Y.T., SORSA, T., RUOSLAHTI, E. and PASQUALINI, R. (1999) Tumor targeting with a selective gelatinase inhibitor. *Nature Biotechnology* **17**, 768-774.

KOLONIN, M., PASQUALINI, R. and ARAP, W. (2001) Molecular addresses in blood vessels as targets for therapy. *Current Opinion in Chemical Biology* **5**, 308-313.

KRAG, D.N., SHUKLA, G.S., SHEN, G.P., PERO, S., ASHIKAGA, T., FULLER, S., WEAVER, D.L., BURDETTE-RADOUX, S. and THOMAS, C. (2006) Selection of tumor-binding ligands in cancer patients with phage display libraries. *Cancer Research* **66**, 7724-7733.

KRONQVIST, N., LOFBLOM, J., JONSSON, A., WERNERUS, H. and STAHL, S. (2008) A novel affinity protein selection system based on staphylococcal cell surface display and flow cytometry. *Protein Engineering, Design and Selection* **21**, 247-255.

KRUMPE, L. and MORI, T. (2006a) The use of phage-displayed peptide libraries to develop tumor-targeting drugs. *International Journal of Peptide Research and Therapeutics* **12**, 79-91.

KRUMPE, L.R.H., ATKINSON, A.J., SMYTHERS, G.W., KANDEL, A., SCHUMACHER, K.M., MCMAHON, J.B., MAKOWSKI, L. and MORI, T. (2006b) T7 lytic phage-displayed peptide libraries exhibit less sequence bias than M13 filamentous phage-displayed peptide libraries. *Proteomics* **6**, 4210-4222.

KUHNAST, B., BODENSTEIN, C., HAUBNER, R., WESTER, H.J., SENEKOWITSCH-SCHMIDTKE, R., SCHWAIGER, M. and WEBER, W.A. (2004) Targeting of gelatinase activity with a radiolabeled cyclic HWGF peptide. *Nuclear Medicine and Biology* **31**, 337-344.

KUMAR, S.R. and DEUTSCHER, S. (2008) ^{111}In-labeled galectin-3-targeting peptide as a SPECT agent for imaging breast tumors. *Journal of Nuclear Medicine* **49**, 796-803.

KUMAR, S.R., QUINN, T.P. and DEUTSCHER, S. (2007) Evaluation of an ^{111}In-radiolabeled peptide as a targeting and imaging agent for ErbB-2 receptor-expressing breast carcinomas. *Clinical Cancer Research* **13**, 6070-6079.

KUZMICHEVA, G.A., JAYANNA, P.K., SOROKULOVA, I.B. and PETRENKO, V.A. (2009) Diversity and censoring of landscape phage libraries. *Protein Engineering, Design and Selection* **22**, 9-18.

LAMBOY, J., TAM, P., LEE, L., JACKSON, P., AVRANTINIS, S., LEE, H., CORN, R. and WEISS, G. (2008) Chemical and genetic wrappers for improved phage and RNA display. *ChemBioChem* **9**, 2846-2852.

LANDON, L. and DEUTSCHER, S. (2003) Combinatorial discovery of tumor targeting peptides using phage display. *Journal of Cellular Biochemistry* **90**, 509-517.

LANDON, L., HARDEN, W., ILLY, C. and DEUTSCHER, S. (2004a) High-throughput fluorescence spectroscopic analysis of affinity of peptides displayed on bacteriophage. *Analytical Biochemistry* **331**, 60-67.

LANDON, L., ZOU, J. and DEUTSCHER, S. (2004b) Is phage display technology on target for developing peptide-based cancer drugs? *Current drug discovery technologies* **1**, 113-132.

LARBANOIX, L., BURTEA, C., LAURENT, S., VAN LEUVEN, F., TOUBEAU, G., ELST, L. and MULLER, R. (2008) Potential amyloid plaque-specific peptides for the diagnosis of Alzheimer's disease. *Neurobiology of Aging,* In press.

LAUMONIER, C., SEGERS, J., LAURENT, S., MICHEL, A., COPPEE, F., BELAYEW, A., ELST, L.V. and MULLER, R.N. (2006) A new peptidic vector for molecular imaging of apoptosis, identified by phage display technology. *Journal of Biomolecular Screening* **11**, 537-545.

Lee, S. and Chen, X. (2009) Dual-modality probes for in vivo molecular imaging. *Molecular Imaging* **8**, 87-100.

Lee, S., Hassan, M., Fisher, R., Chertov, O., Chernomordik, V., Kramer-Marek, G., Gandjbakhche, A. and Capala, J. (2008) Affibody molecules for in vivo characterization of HER2-positive tumors by near-infrared imaging. *Clinical Cancer Research* **14**, 3840-3849.

Lendel, C., Dogan, J. and Hard, T. (2006) Structural basis for molecular recognition in an affibody:affibody complex. *Journal of Molecular Biology* **359**, 1293-1304.

Li, P. and Roller, P.P. (2002) Cyclization strategies in peptide derived drug design. *Current Topics in Medicinal Chemistry* **2**, 325-341.

Li, Z.B., Niu, G., Wang, H., He, L., Yang, L., Ploug, M. and Chen, X. (2008) Imaging of urokinase-type plasminogen activator receptor expression using a ^{64}Cu-labeled linear peptide antagonist by microPET. *Clinical Cancer Research* **14**, 4758-4766.

Lipovsek, D. and Pluckthun, A. (2004) In-vitro protein evolution by ribosome display and mRNA display. *Journal of Immunological Methods* **290**, 51-67.

Liu, C.C., Mack, A.V., Tsao, M.L., Mills, J.H., Lee, H.S., Choe, H., Farzan, M., Schultz, P.G. and Smider, V.V. (2008) Protein evolution with an expanded genetic code. *Proceedings of the National Academy of Sciences of the United States of America* **105**, 17688-17693.

Liu, C.M., Jin, Q., Sutton, A. and Chen, L. (2005) A novel fluorescent probe: europium complex hybridized T7 phage. *Bioconjugate Chemistry* **16**, 1054-1057.

Liu, S. (2006) Radiolabeled multimeric cyclic RGD peptides as integrin $\alpha_v\beta_3$ targeted radiotracers for tumor imaging. *Molecular Pharmaceutics* **3**, 472-487.

Liu, Y., Adams, J., Turner, K., Cochran, F.V., Gambhir, S.S. and Soh, H.T. (2009) Controlling the selection stringency of phage display using a microfluidic device. *Lab on a Chip* **9**, 1033-1036.

Lonberg, N. (2008) Fully human antibodies from transgenic mouse and phage display platforms. *Current Opinion in Immunology* **20**, 450-459.

Magee, J.A., Araki, T., Patil, S., Ehrig, T., True, L., Humphrey, P.A., Catalona, W.J., Watson, M.A. and Milbrandt, J. (2001) Expression profiling reveals hepsin overexpression in prostate cancer. *Cancer Research* **61**, 5692-5696.

Mao, C., Liu, A. and Cao, B. (2009) Virus-based chemical and biological sensing. *Angewandte Chemie International Edition* **48**, 6790-6810.

Marks, K.M., Rosinov, M. and Nolan, G.P. (2004) In vivo targeting of organic calcium sensors via genetically selected peptides. *Chemistry and Biology* **11**, 347-356.

Mather, S. (2009) Molecular imaging with bioconjugates in mouse models of cancer. *Bioconjugate Chemistry* **20**, 631-643.

McCarthy, J.R. and Weissleder, R. (2008) Multifunctional magnetic nanoparticles for targeted imaging and therapy. *Advanced Drug Delivery Reviews* **60**, 1241-1251.

Medina, O.P., Kairemo, K., Valtanen, H., Kangasniemi, A., Kaukinen, S., Ahonen, I., Permi, P., Annila, A., Sneck, M., Holopainen, J.M., Karonen, S.L., Kinnunen, P.K. and Koivunen, E. (2005) Radionuclide imaging of tumor xenografts in mice using a gelatinase-targeting peptide. *Anticancer Research* **25**, 33-42.

Mier, W., Zitzmann, S., Kramer, S., Reed, J., Knapp, E.M., Altmann, A., Eisenhut, M. and Haberkorn, U. (2007) Influence of chelate conjugation on a newly identified tumor-targeting peptide. *Journal of Nuclear Medicine* **48**, 1545-1552.

Mishani, E., Abourbeh, G., Eiblmaier, M. and Anderson, C.J. (2008) Imaging of EGFR and EGFR tyrosine kinase overexpression in tumors by nuclear medicine modalities. *Current Pharmaceutical Design* **14**, 2983-2998.

Moore, S.J., Olsen, M.J., Cochran, J.R. and Cochran, F.V. (2009) Cell surface display systems for protein engineering. In *Protein Engineering and Design* (Park, S.J. and Cochran J.R. Eds), 23-50. CRC Press/Taylor and Francis, Boca Raton.

Mori, T. (2004) Cancer-specific ligands identified from screening of peptide-display libraries. *Current Pharmaceutical Design* **10**, 2335-2343.

Mukai, T., Suganuma, N., Soejima, K., Sasaki, J., Yamamoto, F. and Maeda, M. (2008) Synthesis of a β-tetrapeptide analog as a mother compound for the development of matrix metalloproteinase-2-imaging agents. *Chemical and Pharmaceutical Bulletin* **56**, 260-265.

Nahrendorf, M., Jaffer, F.A., Kelly, K.A., Sosnovik, D.E., Aikawa, E., Libby, P. and Weissleder, R. (2006) Noninvasive vascular cell adhesion molecule-1 imaging identifies inflammatory activation of cells in atherosclerosis. *Circulation* **114**, 1504-1511.

Newton, J. and Deutscher, S. (2008) Phage peptide display. *Handbook of Experimental Pharmacology* **185 Pt 2** 145-163.

Newton, J.R. and Deutscher, S. (2009) In vivo bacteriophage display for the discovery of novel peptide-based tumor-targeting agents. *Methods in Molecular Biology* **504**, 275-290.

Newton, J.R., Kelly, K.A., Mahmood, U., Weissleder, R. and Deutscher, S. (2006) In vivo selection of phage for the optical imaging of PC-3 human prostate carcinoma in mice. *Neoplasia* **8**, 772-780.

Newton, J.R., Miao, Y., Deutscher, S. and Quinn, T.P. (2007) Melanoma imaging with pretargeted bivalent bacteriophage. *Journal of Nuclear Medicine* **48**, 429-436.

Newton-Northup, J.R., Figueroa, S.D., Quinn, T.P. and Deutscher, S. (2009) Bifunctional phage-based pretargeted imaging of human prostate carcinoma. *Nuclear Medicine and Biology* **36**, 789-800.

Nishiyama, K., Takakusagi, Y., Kusayanagi, T., Matsumoto, Y., Habu, S., Kuramochi, K., Sugawara, F., Sakaguchi, K., Takahashi, H., Natsugari, H. and Kobayashi, S. (2009) Identification of trimannoside-recognizing peptide sequences from a T7 phage display screen using a QCM device. *Bioorganic & Medicinal Chemistry* **17**, 195-202.

Nord, K., Gunneriusson, E., Ringdahl, J., Stahl, S., Uhlen, M. and Nygren, P.A. (1997) Binding proteins selected from combinatorial libraries of an alpha-helical bacterial receptor domain. *Nature Biotechnology* **15**, 772-777.

Nordberg, A. (2009) The future: new methods of imaging exploration in Alzheimer's disease. *Frontiers of Neurology and Neuroscience* **24**, 47-53.

Nygren, P.A. (2008) Alternative binding proteins: affibody binding proteins developed from a small three-helix bundle scaffold. *The FEBS Journal* **275**, 2668-2676.

Orlova, A., Feldwisch, J., Abrahmsen, L. and Tolmachev, V. (2007a) Affibody molecules for molecular imaging and therapy for cancer. *Cancer Biotherapy and Radiopharmaceuticals* **22**, 573-584.

Orlova, A., Magnusson, M., Eriksson, T.L., Nilsson, M., Larsson, B., Hoiden-Guthenberg, I., Widstrom, C., Carlsson, J., Tolmachev, V., Stahl, S. and Nilsson,

F.Y. (2006) Tumor imaging using a picomolar affinity HER2 binding affibody molecule. *Cancer Research* **66**, 4339-4348.

Orlova, A., Tolmachev, V., Pehrson, R., Lindborg, M., Tran, T., Sandstrom, M., Nilsson, F.Y., Wennborg, A., Abrahmsen, L. and Feldwisch, J. (2007b) Synthetic affibody molecules: a novel class of affinity ligands for molecular imaging of HER2-expressing malignant tumors. *Cancer Research* **67**, 2178-2186.

Overoye-Chan, K., Koerner, S., Looby, R.J., Kolodziej, A.F., Zech, S.G., Deng, Q., Chasse, J.M., McMurry, T.J. and Caravan, P. (2008) EP-2104R: A fibrin-specific gadolinium-based MRI contrast agent for detection of thrombus. *Journal of the American Chemical Society* **130**, 6025-6039.

Ozawa, M.G., Zurita, A., Dias-Neto, E., Nunes, D.N., Sidman, R.L., Gelovani, J.G., Arap, W. and Pasqualini, R. (2008) Beyond receptor expression levels: the relevance of target accessibility in ligand-directed pharmacodelivery systems. *Trends in Cardiovascular Medicine* **18**, 126-132.

Pasqualini, R. and Ruoslahti, E. (1996) Organ targeting in vivo using phage display peptide libraries. *Nature* **380**, 364-366.

Pepper, L.R., Cho, Y.K., Boder, E.T. and Shusta, E.V. (2008) A decade of yeast surface display technology: Where are we now? *Combinatorial Chemistry and High Throughput Screening* **11**, 127-134.

Perillo, N.L., Marcus, M.E. and Baum, L.G. (1998) Galectins: versatile modulators of cell adhesion, cell proliferation, and cell death. *Journal of Molecular Medicine* **76**, 402-412.

Perkins, A.C. and Missailidis, S. (2007) Radiolabelled aptamers for tumour imaging and therapy. *The Quarterly Journal of Nuclear Medicine and Molecular Imaging* **51**, 292-296.

Petrenko, V.A. (2008) Evolution of phage display: from bioactive peptides to bioselective nanomaterials. *Expert Opinion on Drug Delivery* **5**, 825-836.

Pettersen, E.F., Goddard, T.D., Huang, C.C., Couch, G.S., Greenblatt, D.M., Meng, E.C. and Ferrin, T.E. (2004) UCSF Chimera - A visualization system for exploratory research and analysis. *Journal of Computational Chemistry* **25**, 1605-1612.

Pilch, J., Brown, D.M., Komatsu, M., Jarvinen, T.A.H., Yang, M., Peters, D., Hoffman, R.M. and Ruoslahti, E. (2006) Peptides selected for binding to clotted plasma accumulate in tumor stroma and wounds. *Proceedings of the National Academy of Sciences of the United States of America* **103**, 2800-2804.

Pini, A., Viti, F., Santucci, A., Carnemolla, B., Zardi, L., Neri, P. and Neri, D. (1998) Design and use of a phage display library: Human antibodies with subnanomolar affinity against a marker of angiogenesis eluted from a two-dimensional gel. *Journal of Biological Chemistry* **273**, 21769-21776.

Pini, A., Giuliani, A., Ricci, C., Runci, Y. and Bracci, L. (2004) Strategies for the construction and use of peptide and antibody libraries displayed on phages. *Current Protein and Peptide Science* **5**, 487-496.

Ploug, M., Ostergaard, S., Gardsvoll, H., Kovalski, K., Holst-Hansen, C., Holm, A., Ossowski, L. and Dano, K. (2001) Peptide-derived antagonists of the urokinase receptor. Affinity maturation by combinatorial chemistry, identification of functional epitopes, and inhibitory effect on cancer cell intravasation. *Biochemistry* **40**, 12157-12168.

Presta, L.G. (2008) Molecular engineering and design of therapeutic antibodies. *Current*

Opinion in Immunology **20**, 460-470.

RAE, J., CREIGHTON, C., MECK, J., HADDAD, B. and JOHNSON, M. (2007) MDA-MB-435 cells are derived from M14 Melanoma cells—a loss for breast cancer, but a boon for melanoma research. *Breast Cancer Research and Treatment* **104**, 13-19.

RAO, J., DRAGULESCU-ANDRASI, A. and YAO, H. (2007) Fluorescence imaging in vivo: recent advances. *Current Opinion in Biotechnology* **18**, 17-25.

REN, G., ZHANG, R., LIU, Z., WEBSTER, J.M., MIAO, Z., GAMBHIR, S.S., SYUD, F.A. and CHENG, Z. (2009) A 2-helix small protein labeled with ^{68}Ga for PET imaging of HER2 expression. *Journal of Nuclear Medicine* **50**, 1492-1499.

REUBI, J.C. and MAECKE, H.R. (2008) Peptide-based probes for cancer imaging. *Journal of Nuclear Medicine* **49**, 1735-1738.

ROMER, J., NIELSEN, B.S. and PLOUG, M. (2004) The urokinase receptor as a potential target in cancer therapy. *Current Pharmaceutical Design* **10**, 2359-2376.

ROSENBERG, A., GRIFFIN, K., STUDIER, F.W., MCCORMICK, M., BERG, J., NOVY, R. and MIERENDORF, R. (1996) T7Select phage display system: a powerful new protein display system based on bacteriophage T7. *inNovations* **6**, 1–6.

ROSSIN, R., BERNDORFF, D., FRIEBE, M., DINKELBORG, L.M. and WELCH, M.J. (2007) Small-animal PET of tumor angiogenesis using a ^{76}Br-labeled human recombinant antibody fragment to the ED-B domain of fibronectin. *Journal of Nuclear Medicine* **48**, 1172-1179.

ROZINOV, M.N. and NOLAN, G.P. (1998) Evolution of peptides that modulate the spectral qualities of bound, small-molecule fluorophores. *Chemistry and Biology* **5**, 713-728.

RUOSLAHTI, E. (2000) Targeting tumor vasculature with homing peptides from phage display. *Seminars in Cancer Biology* **10**, 435-442.

RUSCKOWSKI, M., GUPTA, S., LIU, G., DOU, S. and HNATOWICH, D.J. (2004) Investigations of a ^{99m}Tc-labeled bacteriophage as a potential infection-specific imaging agent. *Journal of Nuclear Medicine* **45**, 1201-1208.

SANTIMARIA, M., MOSCATELLI, G., VIALE, G.L., GIOVANNONI, L., NERI, G., VITI, F., LEPRINI, A., BORSI, L., CASTELLANI, P., ZARDI, L., NERI, D. and RIVA, P. (2003) Immunoscintigraphic detection of the ED-B domain of fibronectin, a marker of angiogenesis, in patients with cancer. *Clinical Cancer Research* **9**, 571-579.

SCHMIDT, M.M. and WITTRUP, K.D. (2009) A modeling analysis of the effects of molecular size and binding affinity on tumor targeting. *Molecular Cancer Therapeutics* **8**, 2861-2871.

SEGERS, J., LAUMONIER, C., BURTEA, C., LAURENT, S., ELST, L.V. and MULLER, R.N. (2007) From phage display to magnetophage, a new tool for magnetic resonance molecular imaging. *Bioconjugate Chemistry* **18**, 1251-1258.

SERGEEVA, A., KOLONIN, M.G., MOLLDREM, J.J., PASQUALINI, R. and ARAP, W. (2006) Display technologies: Application for the discovery of drug and gene delivery agents. *Advanced Drug Delivery Reviews* **58**, 1622-1654.

SIDHU, S.S., Ed (2005). *Phage Display in Biotechnology and Drug Discovery*. CRC Press/Taylor & Francis, Boca Raton.

SIMBERG, D., DUZA, T., PARK, J.H., ESSLER, M., PILCH, J., ZHANG, L., DERFUS, A.M., YANG, M., HOFFMAN, R.M., BHATIA, S., SAILOR, M.J. and RUOSLAHTI, E. (2007) Biomimetic amplification of nanoparticle homing to tumors. *Proceedings of the National Academy of Sciences of the United States of America* **104**, 932-936.

Slootweg, E.J., Keller, H.J., Hink, M.A., Borst, J.W., Bakker, J. and Schots, A. (2006) Fluorescent T7 display phages obtained by translational frameshift. *Nucleic Acids Research* **34**, e137.

Smith, G.P. and Petrenko, V.A. (1997) Phage display. *Chemical Reviews* **97**, 391-410.

Smith, G.P. and Scott, J.K. (1993) Libraries of peptides and proteins displayed on filamentous phage. *Methods in Enzymology* **217**, 228-257.

Sosnovik, D.E. and Weissleder, R. (2007) Emerging concepts in molecular MRI. *Current Opinion in Biotechnology* **18**, 4-10.

Souza, G.R., Christianson, D.R., Staquicini, F.I., Ozawa, M.G., Snyder, E.Y., Sidman, R.L., Miller, J.H., Arap, W. and Pasqualini, R. (2006) Networks of gold nanoparticles and bacteriophage as biological sensors and cell-targeting agents. *Proceedings of the National Academy of Sciences of the United States of America* **103**, 1215-1220.

Souza, G.R., Yonel-Gumruk, E., Fan, D., Easley, J., Rangel, R., Guzman-Rojas, L., Miller, J.H., Arap, W. and Pasqualini, R. (2008) Bottom-up assembly of hydrogels from bacteriophage and Au nanoparticles: the effect of cis- and trans-acting factors. *PLoS ONE* **3**, e2242.

Sprague, J.E., Li, W.P., Liang, K., Achilefu, S. and Anderson, C.J. (2006) In vitro and in vivo investigation of matrix metalloproteinase expression in metastatic tumor models. *Nuclear Medicine and Biology* **33**, 227-237.

Spuentrup, E., Botnar, R., Wiethoff, A., Ibrahim, T., Kelle, S., Katoh, M., Ozgun, M., Nagel, E., Vymazal, J., Graham, P., Gunther, R. and Maintz, D. (2008) MR imaging of thrombi using EP-2104R, a fibrin-specific contrast agent: initial results in patients. *European Radiology* **18**, 1995-2005.

Steffen, A., Orlova, A., Wikman, M., Nilsson, F.Y., Stahl, S., Adams, G.P., Tolmachev, V. and Carlsson, J. (2006) Affibody-mediated tumour targeting of HER-2 expressing xenografts in mice. *European Journal of Nuclear Medicine and Molecular Imaging* **33**, 631-638.

Steffen, A.-C., Wikman, M., Tolmachev, V., Adams, G.P., Nilsson, F.Y., Stahl, S. and Carlsson, J.r. (2005) In vitro characterization of a bivalent anti-HER-2 affibody with potential for radionuclide-based diagnostics. *Cancer Biotherapy and Radiopharmaceuticals* **20**, 239-248.

Steiner, D., Forrer, P. and Pluckthun, A. (2008) Efficient selection of DARPins with sub-nanomolar affinities using SRP phage display. *Journal of Molecular Biology* **382**, 1211-1227.

Steiner, D., Forrer, P., Stumpp, M.T. and Pluckthun, A. (2006) Signal sequences directing cotranslational translocation expand the range of proteins amenable to phage display. *Nature Biotechnology* **24**, 823-831.

Tavitian, B., Haberkorn, U. (2009) Darwinian molecular imaging. *European Journal of Nuclear Medicine and Molecular Imaging* **36**, 1475-1482.

Thapa, N., Kim, S., So, I., Lee, B., Kwon, I., Choi, K. and Kim, I. (2008) Discovery of a phosphatidylserine-recognizing peptide and its utility in molecular imaging of tumour apoptosis. *Journal of Cellular and Molecular Medicine* **12**, 1649-1660.

Tijink, B.M., Perk, L.R., Budde, M., Stigter-van Walsum, M., Visser, G.W.M., Kloet, R.W., Dinkelborg, L.M., Leemans, C.R., Neri, D. and van Dongen, G.A.M.S. (2009) Influence of valency and labelling chemistry on in vivo targeting using radioiodinated

HER2-binding Affibody molecules. *European Journal of Nuclear Medicine and Molecular Imaging* **36**, 1235-1244.

TOLMACHEV, V. (2008) Imaging of HER-2 overexpression in tumors for guiding therapy. *Current Pharmaceutical Design* **14**, 2999-3019.

TOLMACHEV, V., FRIEDMAN, M., SANDSTROM, M., ERIKSSON, T.L., ROSIK, D., HODIK, M., STAHL, S., FREJD, F. and ORLOVA, A. (2009a) Affibody molecules for epidermal growth factor receptor targeting in vivo: aspects of dimerization and labeling chemistry. *Journal of Nuclear Medicine* **50**, 274-283.

TOLMACHEV, V., MUME, E., SJOBERG, S., FREJD, F. and ORLOVA, A. (2009b) Influence of valency and labelling chemistry on in vivo targeting using radioiodinated HER2-binding Affibody molecules. *European Journal of Nuclear Medicine and Molecular Imaging* **36**, 692-701.

TOLMACHEV, V., ORLOVA, A., NILSSON, F.Y., FELDWISCH, J., WENNBORG, A. and ABRAHMSEN, L. (2007) Affibody molecules: potential for in vivo imaging of molecular targets for cancer therapy. *Expert Opinion on Biological Therapy* **7**, 555-568.

TRAN, T.A., EKBLAD, T., ORLOVA, A., SANDSTROM, M., FELDWISCH, J., WENNBORG, A., ABRAHMSEN, L., TOLMACHEV, V. and ERIKSSON KARLSTROM, A. (2008) Effects of lysine-containing mercaptoacetyl-based chelators on the biodistribution of ^{99m}Tc-labeled anti-HER2 Affibody molecules. *Bioconjugate Chemistry* **19**, 2568-2576.

TREPEL, M., PASQUALINI, R. and ARAP, W. (2008) Screening phage-display peptide libraries for vascular targeted peptides. *Methods in Enzymology* **445**, 83-106.

VAN DE WIELE, C. and OLTENFREITER, R. (2006) Imaging probes targeting matrix metalloproteinases. *Cancer Biotherapy and Radiopharmaceuticals* **21**, 409-417.

VITI, F., TARLI, L., GIOVANNONI, L., ZARDI, L. and NERI, D. (1999) Increased binding affinity and valence of recombinant antibody fragments lead to improved targeting of tumoral angiogenesis. *Cancer Research* **59**, 347-352.

WANG, H., CAI, W., CHEN, K., LI, Z., KASHEFI, A., HE, L. and CHEN, X. (2007a) A new PET tracer specific for vascular endothelial growth factor receptor 2. *European Journal of Nuclear Medicine and Molecular Imaging* **34**, 2001-2010.

WANG, W., KE, S., KWON, S., YALLAMPALLI, S., CAMERON, A.G., ADAMS, K.E., MAWAD, M.E. and SEVICK-MURACA, E.M. (2007b) A new optical and nuclear dual-labeled imaging agent targeting interleukin 11 receptor alpha-chain. *Bioconjugate Chemistry* **18**, 397-402.

WANG, W., SHAO, R., WU, Q., KE, S., MCMURRAY, J., LANG, F., CHARNSANGAVEJ, C., GELOVANI, J.G. and LI, C. (2009) Targeting gelatinases with a near-infrared fluorescent cyclic His-Try-Gly-Phe peptide. *Molecular Imaging and Biology* **11**, 424-433.

WEBSTER, J., ZHANG, R., GAMBHIR, S., CHENG, Z. and SYUD, F.A. (2009) Engineered two-helix small proteins for molecular recognition. *ChemBioChem* **10**, 1293-1296.

WEISSLEDER, R. and PITTET, M. (2008) Imaging in the era of molecular oncology. *Nature* **452**, 580-589.

WELLER, G.E.R., WONG, M.K.K., MODZELEWSKI, R.A., LU, E., KLIBANOV, A.L., WAGNER, W.R. and VILLANUEVA, F.S. (2005) Ultrasonic imaging of tumor angiogenesis using contrast microbubbles targeted via the tumor-binding peptide arginine-arginine-leucine. *Cancer Research* **65**, 533-539.

WIKMAN, M., STEFFEN, A.C., GUNNERIUSSON, E., TOLMACHEV, V., ADAMS, G.P., CARLSSON, J. and STAHL, S. (2004) Selection and characterization of HER2/neu-binding affibody ligands. *Protein Engineering, Design and Selection* **17**, 455-462.

WU, A.M. (2009) Antibodies and antimatter: the resurgence of immuno-PET. *Journal of Nuclear Medicine* **50**, 2-5.

YE, F., JEONG, E.K., JIA, Z., YANG, T., PARKER, D. and LU, Z.R. (2008) A peptide targeted contrast agent specific to fibrin-fibronectin complexes for cancer molecular imaging with MRI. *Bioconjugate Chemistry* **19**, 2300-2303.

ZANUY, D., CURCO, D., NUSSINOV, R. and ALEMAN, C. (2009) Influence of the dye presence on the conformational preferences of CREKA, a tumor homing linear pentapeptide. *Peptide Science* **92**, 83-93.

ZHAO, P., GRABINSKI, T., GAO, C., SKINNER, R.S., GIAMBERNARDI, T., SU, Y., HUDSON, E., RESAU, J., GROSS, M., VANDE WOUDE, G.F., HAY, R. and CAO, B. (2007) Identification of a met-binding peptide from a phage display library. *Clinical Cancer Research* **13**, 6049-6055.

ZHOU, Y., DRUMMOND, D.C., ZOU, H., HAYES, M.E., ADAMS, G.P., KIRPOTIN, D.B. and MARKS, J.D. (2007) Impact of single-chain Fv antibody fragment affinity on nanoparticle targeting of epidermal growth factor receptor-expressing tumor cells. *Journal of Molecular Biology* **371**, 934-947.

ZITZMANN, S., KRAMER, S., MIER, W., HEBLING, U., ALTMANN, A., ROTHER, A., BERNDORFF, D., EISENHUT, M. and HABERKORN, U. (2007) Identification and evaluation of a new tumor cell-binding peptide, FROP-1. *Journal of Nuclear Medicine* **48**, 965-972.

ZOU, J., DICKERSON, M.T., OWEN, N.K., LANDON, L. and DEUTSCHER, S. (2004) Biodistribution of filamentous phage peptide libraries in mice. *Molecular Biology Reports* **31**, 121-129.

ZOU, J., GLINSKY, V.V., LANDON, L., MATTHEWS, L. and DEUTSCHER, S. (2005) Peptides specific to the galectin-3 carbohydrate recognition domain inhibit metastasis-associated cancer cell adhesion. *Carcinogenesis* **26**, 309-318.

Biotechnology and Genetic Engineering Reviews - Vol. 27, 95-114 (2010)

Co-immobilized coupled enzyme systems in biotechnology

LORENA BETANCOR[1*] AND HEATHER R. LUCKARIFT[2,3*]

[1]Madrid Institute for Advanced Studies, Campus Universitario de Cantoblanco, c/ Einstein, 13 Pabellon C 1ºPlanta, E-28049, Madrid, Spain. [2]Air Force Research Laboratory, AFRL/RXQL, Microbiology and Applied Biochemistry, Tyndall Air Force Base, Florida 32403. [3]Universal Technology Corporation, 1270 N. Fairfield Road, Dayton, Ohio, 45432, USA

Abstract

The development of coimmobilized multi-enzymatic systems is increasingly driven by economic and environmental constraints that provide an impetus to develop alternatives to conventional multistep synthetic methods. As in nature, enzyme-based systems work cooperatively to direct the formation of desired products within the defined compartmentalization of a cell. In an attempt to mimic biology, coimmobilization is intended to immobilize a number of sequential or cooperating biocatalysts on the same support to impart stability and enhance reaction kinetics by optimizing catalytic turnover.

There are three primary reasons for the utilization of coimmobilized enzymes: to enhance the efficiency of one of the enzymes by the *in-situ* generation of its substrate, to simplify a process that is conventionally carried out in several steps and/or to eliminate undesired by-products of an enzymatic reaction. As such, coimmobilization provides benefits that span numerous biotechnological applications, from biosensing of molecules to cofactor recycling and to combination of multiple biocatalysts for the synthesis of valuable products.

---

*To whom correspondence may be addressed (lorena.betancor@uam.es or heather.luckarift.ctr@tyndall.af.mil)

Abbreviations: ATP- adenosine triphosphate; EDTA- ethylenediaminetetraacetic acid; CALB- Candida antarctica B lipase; GOX- glucose oxidase; HRP- horse radish peroxidase; ABTS- 2,2'-azino-bis(3-ethylbenzthiazoline-6-sulphonic acid); NADPH- Nicotinamide adenine dinucleotide phosphate reduced form; NADH- Nicotinamide adenine dinucleotide reduced form; CoA- Coenzyme A; FAD- flavin adenine dinucleotide; FADH2- flavin adenine dinucleotide reduced form; GTP- Guanosine-5'-triphosphate; GDP- Guanosine-5'-diphosphate; DNA- Deoxyribonucleic acid

Introduction

The best example of coimmobilized enzymes collaborating *in situ* is in living cells. Enzymes constitute the basis of metabolism in all living beings and, in a perfectly concerted succession of catalytic steps, form a network of reactions that make life possible. Enzymes that function to form a complete process such as metabolic cycling are physically associated so as to ensure substrate channeling without diffusion limitations. This metabolic organization is intrinsic to cells and although whole cells may be visualized as a random mix of enzymes, recent studies suggest a specific intracellular organization of enzymes specifically to allow for channeling of enzyme intermediates (Beeckmans *et al.*, 1993; Huang *et al.*, 2001). Many of the enzymes of the citric acid cycle, for example, are inhibited by their reactants, products, intermediates or even cofactors involved in the cycle; functioning as a continuous unit in which inhibitory intermediates are immediately removed, maintains enzyme functionality throughout the cycle.

In an attempt to mimic nature, science has benefited from the high efficiency, selectivity and specificity of coupled enzyme systems for detection, diagnosis or synthesis of industrially relevant molecules (Schoevaart and Kieboom, 2001; Bruggink *et al.*, 2003). One primary advantage of using multi-enzyme systems in biocatalysis is the ability to convert a starting material into a desired product without the need to separate or isolate intermediate products. Moreover, some intermediates may not be available or stable if added *ex situ*. In this instance, coimmobilization of enzymes can make a biocatalytic pathway more efficient by limiting the diffusion of unstable intermediates into the surrounding media. Biocatalysis can also offer a significant advantage over chemical catalysis by eliminating the defined reaction environments required in organic synthesis. The modification of penicillin to 6-aminopenicillanic acid, for example, is now predominantly manufactured by an enzymatic process using immobilized penicillin acylases in a single-step biocatalytic reaction that replaces the traditional three-step chemical reaction and eliminates the need for harsh solvents and the cost of operating the reaction at extreme temperatures (-40°C) (*Figure 1*) (Averill *et al.*, 1999; Zaks, 2001).

Herein, we will review what we consider the most current relevant examples of coupled-enzyme systems used for biosensing or biocatalytic purposes. The immobilization strategies will be described with special emphasis on those that have improved the process in which they were applied and that could potentially be used with other enzyme species. Particular attention will be given to new nanoscale architectures that are increasingly emerging as interesting supports for nanoscale biotechnological applications. This review is not intended as an exhaustive summary of all coimmobilized or cascade enzyme systems but rather aims to demonstrate the breadth of innovative immobilization strategies, the diverse range of applications that may benefit from immobilized biocatalysis and highlight pertinent examples that exemplify the technology.

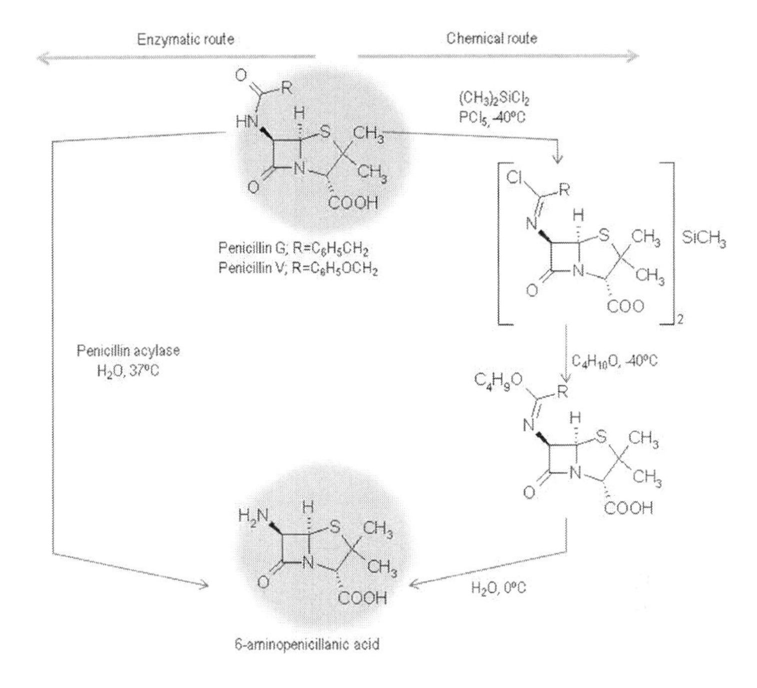

Figure 1. Enzymatic and chemical routes to 6-aminopenicillanic acid from penicillin G or V.

Fundamentals of immobilized enzymes

Inherently, the soluble nature of enzymes presents specific restrictions for biotechnological applications: specifically poor stability, difficulty of separation, product contamination and limited reuse. Enzyme immobilization is therefore commonly used to overcome the limitations of enzyme utilization, as stabilization is often provided against heat, organic solvents and/or changes in pH (Grazú *et al.*, 2005; Irazoqui *et al.*, 2007; Mateo *et al.*, 2007b; Montes *et al.*, 2007; Kim *et al.*, 2008; Hanefeld *et al.*, 2009). Furthermore, immobilized enzymes can be reused, often through many cycles, thereby minimizing costs and time of analysis and in certain applications facilitating the continuous use of the biocatalyst (Berne *et al.*, 2006; Filice *et al.*, 2009). Stabilization may also fortuitously improve enzyme properties by locking the protein structure in a configuration that enhances substrate specificity and reduces the effect of inhibitors (Mateo *et al.*, 2007b; Sheldon, 2007).

A wide variety of techniques are now available for enzyme attachment to a variety of supports (Girelli and Mattei, 2005; Mateo *et al.*, 2007b; Betancor and Luckarift, 2008). Immobilization techniques generally include chemical or physical mechanisms. Chemical immobilization methods mainly include enzyme attachment to the matrix by covalent bonds or other interactions and cross-linking between the enzyme and the matrix. Physical methods involve the entrapment of the enzymes within an insoluble matrix. A combination of chemical and physical methods has facilitated in certain circumstances for the immobilization of different enzyme species in the same composite (Kreft *et al.*, 2007). The requirements of different enzymes are inherently varied and specific conditions are often needed for a defined application. Unfortunately, there is at the present time, no generic method for enzyme stabilization that will be optimal for all enzyme systems, but a toolbox of versatile methodologies is now well documented in the literature and provides examples in which cooperating enzymes have been immobilized for various applications (Nahalka *et al.*, 2003; Berne *et al.*, 2006; Salinas–Castillo *et al.*, 2008).

Considerations for coimmobilization of enzymes

The limitations of biocatalysis are particularly evident when attempting to utilize a multitude of enzyme activities in concert. Coimmobilization of cooperating enzymes requires specific optimization to balance the catalytic components and therefore necessitates screening of suitable immobilization methods, design and preparation of the appropriate immobilization carriers, and analysis of the relevant reaction kinetics and mass transfer characteristics to determine the optimum reaction conditions (El–Zahab *et al.*, 2004; Lopez–Gallego *et al.*, 2005; Sun *et al.*, 2009). Immobilization of more than one enzyme on the same support, however, is especially challenging as it needs to preserve the catalytic activity of all the enzymes involved in the system and ideally improve stability. An ideal immobilization design should confer an overall operational stabilization to each of the enzymes involved; otherwise the half-life of the composite will be limited by the most unstable catalytic component.

The beauty of combined biological enzymes over purely chemical cascades is that enzymes inherently function in the same physiological environments, i.e., aqueous solvents, moderate temperature and defined pH. An elegant example of pH control of enzymes was utilized in the four-step enzymatic conversion of glycerol into a heptose in which pH switching was used to temporarily control the on/off catalysis of enzymes involved within the cascade (Schoevaart *et al.*, 2001).

Immobilization of sequentially acting enzymes within a confined space increases the catalytic efficiency of conversion due to a dramatic reduction in the diffusion time of the substrate. Moreover, the *in-situ* formation of substrates generates high local concentrations that lead to kinetic enhancements and can equate to substantial cost savings (Van De Velde *et al.*, 2000; El–Zahab *et al.*, 2004). The interest in reducing diffusion limitations and maximizing the functional surface area to increase enzyme loading has prompted the emergence of new nanoscaffolds that could potentially support enzyme immobilization (Kim *et al.*, 2008). Among them, the bioinspired formation of silica nanoparticles provides a versatile new technology for enzyme immobilization with several inherent advantages: inexpensive, rapid, mild, robust and stabilizing for the entrapped enzymes (Betancor and Luckarift, 2008; Luckarift, 2008; Vamvakaki *et al.*, 2008). This particular immobilization support has been used to couple sequentially acting enzymes with very good results (Luckarift *et al.*, 2007). The application utilized individual enzymes encapsulated in silica, packed into microfluidic chips and then connected in series to allow the flow of reaction products from one step to the next for the synthesis of a metabolite relevant to antibiotic synthesis. This type of sequential processing has numerous applications in catalysis including the ability to change the direction and ordering of the reaction sequence (Lee *et al.*, 2003; Ku *et al.*, 2006; Logan *et al.*, 2007). Logan *et al.*, demonstrated an elegant example of spatially separated multi-enzyme reactions by patterning enzymes onto monoliths using covalent attachment. Glucose oxidase and horseradish peroxidase were immobilized sequentially in a flow-through system whereby glucose oxidase converts glucose to gluconolactone, liberating hydrogen peroxide, which is subsequently utilized by horseradish peroxidase to oxidize amplex red to the red fluorescent product, resorufin. Interestingly, red fluorescence was observed only when glucose and amplex red were flushed in a forward direction. Reversing the flow essentially eliminated fluorescence and confirmed the correct sequential ordering of the catalytic steps. In an additional step, invertase was added prior to the reaction scheme to allow the *in-situ* hydrolysis of sucrose to glucose (and fructose) and extend the substrate range of the reaction system (Logan *et al.*, 2007).

Coimmobilization of enzymes in biosensing

Generally, coimmobilized enzymes serve one of two primary purposes: to channel an intermediate reaction product directly to a secondary enzyme and reduce the loss of intermediates as a result of instability or diffusion. Alternatively, a coimmobilized enzyme may be required to recycle a cofactor to maintain catalytic turnover and eliminate the need to continually add cofactor to the reaction (*Figure 2*).

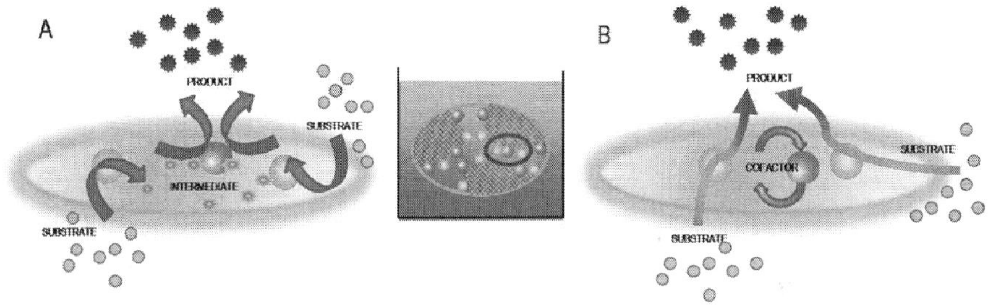

Figure 2. Coimmobilized enzyme systems for *in-situ* generation of an intermediate (A) or recycling of a cofactor (B).

In sensing applications, the use of more than one enzyme species often allows the sensitivity of the analytical method to be increased, expanding the range of applications for detection of numerous substrates at otherwise undetectable concentrations (Salinas–Castillo *et al.*, 2008). Numerous bi-enzyme systems have been reported for glucose detection, for example, by combining a secondary enzyme to enhance or improve the sensitivity and selectivity of the signal. Peroxidase and glucose oxidase, were immobilized onto carbon nanotubes using polypyrrole or Nafion® to provide a basis for bi-enzymatic glucose biosensors (Zhu *et al.*, 2007; Jeykumari and Narayanan, 2008). Glucose oxidase and horseradish peroxidase with fluorogenic detection by resorufin is a combination often used in bi-enzyme sensing systems due to the sensitivity of peroxide production coupled with fluorescence, which significantly increases the detection of the initial reaction product (hydrogen peroxide).

Recently, researchers investigated the ability to define the spatial orientation of enzymes by utilizing specific 'capture' oligomers to tag enzymes with a short nucleotide sequence and align the proteins in a specific location on a DNA backbone (Muller and Niemeyer, 2008). Using this method glucose oxidase and horseradish peroxidase were coimmobilized onto microplates. The catalytic activity was hampered by steric interactions as a result of specific organization of the enzymes within the system, but, when organized to limit steric affects, demonstrated the potential for enhanced detection specificity of glucose. Signal detection was based on direct coupling of the conversion of glucose (and the concomitant production of hydrogen peroxide) with the fluorescence based detection of resorufin. Similarly, the coimmobilized orientation of luciferase and oxidoreductase was used to catalyze flavin mononucleotide reduction and aldehyde oxidation (Niemeyer *et al.*, 2002).

Not all coimmobilization strategies, however, rely on catalytic cooperativity. Wang *et al.*, reported the coimmobilization of glucose oxidase and heparin by electropolymerization into a polymeric film (Wang *et al.*, 2000). The immobilized glucose oxidase provides an amperometric measure of glucose concentration in blood with application to needle-type implantable glucose biosensors. Implantable sensors, however, are susceptible to fouling upon continuous exposure to biological fluids. The inclusion of heparin (as an anticoagulant) extends the biocompatibility and hence reusability of the device.

Silica sol–gels have provided a broad and versatile basis for many examples of immobilized multi-enzyme systems but have limitations associated with drying and cracking. In an alternate design for amperometric detection of glucose, a hybrid silica sol–gel was used to encapsulate glucose oxidase and glucose-6-phosphate dehydrogenase. The silica sol–gel was formed from the hydrolysis of a mixture of silane precursors to create a three-dimensional structure that limited the cracking problems associated with conventional sol–gels (Liu and Sun, 2007). The addition of glucose-6-phosphate provides a competitive catalytic sink that utilizes a stoichiometric amount of ATP and results in a detection method for both glucose and ATP. The addition of ATP to the system is a typical example of one of the limitations of enzyme systems and particularly of multi-enzyme systems in that cofactors have to be continuously added or an additional enzyme included, for the cofactor to be recycled during catalysis. The conversion of glucose into riboflavin, for example, can be performed by six catalytic enzymes working in concert, but an additional two enzymes are required to recycle cofactors during synthesis. The eight-enzyme reaction functions entirely in an aqueous system as a random mixture of enzymes and, as such, the reusability of the enzymes within the system is limited (Romisch *et al.*, 2002).

Coimmobilization of enzymes in biocatalysis

The majority of multi-enzyme cascades in biocatalysis have been developed for carbohydrate synthesis or sugar conversions as enzymatic oligosaccharide synthesis using recombinant glycosyl transferases overcomes many of the hurdles associated with chemical synthesis. Many enzymes of the nucleotide biosynthetic pathway have now been well studied and recombinantly expressed. The attachment of a hexahistidine "tail" to the required biocatalytic enzymes allows for affinity binding to a metal-coated support (Nahalka *et al.*, 2003). By varying the number of enzymes in the coimmobilization step, four-enzyme (Superbeads I) or seven-enzyme cascades (Superbeads II) have been demonstrated that allow for efficient biocatalysis of a versatile range of oligosaccharides, depending upon the starting saccharide units.

Often, coimmobilization of enzymes helps to prevent inactivation that may arise due to high localized concentrations of intermediates or reaction products that may act as inhibitors upon catalytic activity. Limiting the local concentration of hydrogen peroxide, for example, is a common goal to preserve enzyme activity. Coimmobilization of peroxidase and glucose oxidase in a polyurethane foam, for example, resulted in the *in situ* generation of hydrogen peroxide from glucose and glucose oxidase but at a low internal concentration; sufficient to allow the catalytic turnover of the peroxidase enzyme without inactivation that arises from direct addition of peroxide (Van De Velde *et al.*, 2000). Coimmobilization may also provide an added benefit and create an apparent change in the enzymatic activity of a protein, due to reaction synergy. Soybean peroxidase, thus acts as an apparent oxygen transferase when immobilized with glucose oxidase (Van De Velde *et al.*, 2000).

In the conversion of dextran, the enzyme dextransucrase must be protected from dextranase for the two enzymes to work together. As dextran forms it remains associated with the dextransucrase enzyme and can be inactivated by dextranase. Successful coimmobilization of the two enzymes, however, was achieved by preliminary absorption of dextranase onto hydroxyapatite before coimmobilization with dextransucrase in alginate microbeads (Erhardt *et al.*, 2000). This method of compartmentalization is a common theme in coimmobilized systems to spatially separate conflicting catalytic activities. Compartmentalization, for example, was used to separate glucose oxidase and peroxidase in a shell-in-shell microcapsule (Kreft *et al.*, 2007). Polyelectrolyte layers deposited onto calcium carbonate microcapsules can later be dissolved with EDTA to leave an empty shell-in-shell structure with peroxidase in the inner shell and glucose oxidase in the outer shell. Hydrogen peroxide is generated in the outer shell (by the catalytic mechanism of glucose oxidase) and diffuses into the inner compartment, where peroxidase utilizes the peroxide in the conversion of amplex red to resorufin. The red fluorescence of the resulting resorufin can be imaged to demonstrate the architecture of the shell structure and confirm the multistep biocatalytic reaction. Similarly, hemoglobin and glucose oxidase were combined to create microcapsules through a similar layer-by-layer deposition to produce microcapsules that were responsive to the concentration of glucose (Qi *et al.*, 2009).

Initial studies for compartmentalization of enzymes relied on the use of phospholipid liposomes, which mimic the natural cell structure but are difficult to handle due to mechanical fragility; however, vesicles prepared by layer-by-layer techniques as described above have overcome some of the mechanical limitations. There are some examples in the literature that follow strategies for enzyme coimmobilization that resemble cell-like conditions, for instance, to control the order in which enzymes react or to protect one of them from the action of degrading byproducts. van Dongen *et al* have worked on the compartmentalization of sequentially acting enzymes immobilized in the same structure (van Dongen *et al.*, 2009). In an effort to mimic cell-like enzymatic cascades, these researchers developed spherical aggregates called polymersomes to spatially distribute enzymes acting in tandem (*Figure 3*).

There are still specific considerations in this approach, namely that the size of the enzymes may limit expression in certain compartments. The resulting "nanoreactors" contain glucose oxidase (in the lumen), lipase (in the membrane bilayer) and horseradish peroxidase (on the surface) (van Dongen *et al.*, 2009). Glucose oxidase, was included in the polymersome lumen as it was thought that its size might disrupt the structure if included in the membrane bilayer. Similarly, the hydrophobicity and structural affinity of the enzymes was also considered in respect to positioning. Specifically, azido-functionalization of the surface was required to enable active attachment of horseradish peroxide at the surface. Although the assembly did not provide a catalytic advantage over the use of the soluble enzyme mixture, this strategy can be adapted for future applications with enzyme systems that otherwise could not be coimmobilized.

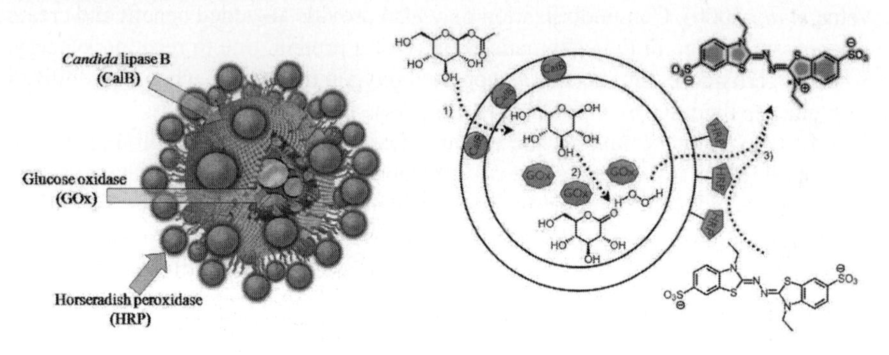

Figure 3. Compartmentalization of enzymes in a polymersome. Acetate-protected glucose is deprotected by *Candida* lipase B (CalB) at the polymersome membrane to give glucose, which is oxidized by glucose oxidase (GOx) in the inner compartment and generates hydrogen peroxide which is used by horseradish peroxidase (HRP) to oxidize ABTS [2,2'-Azino-bis(3-ethylbenzthiazoline-6-sulfonic acid) diammonium salt] at the polymersome surface.[van Dongen *et al.*, A three-enzyme cascade reaction through positional assembly of enzymes in a polymersome nanoreactor. Chemistry – A European Journal, 2009, 15, 1101. Copyright Wiley–VCH Verlag GmbH & Co. KGaA, reproduced with permission].

One similar consideration for coimmobilization is in preferential binding, particularly if the method of immobilization relies on covalent attachment. Lipase, trypsin and α-amylase, for example, were bound to fabrics using covalent fixation, but preferential binding was observed for trypsin over α-amylase due to the differing reactivity of the enzymes towards the activated support (Nouaimi-Bachmann *et al.*, 2007). Alternatively, non-sequential encapsulation favors a much more random organization, which may fortuitously favor protein–protein interaction. Nitrobenzene nitroreductase and glucose-6-phosphate dehydrogenase, for example, were coimmobilized by encapsulation in silica spheres that were formed by a polymer-templated silicification reaction (Betancor *et al.*, 2006). Nitrobenzene nitroreductase was used to catalyze the hydroxylation of nitrobenzene, a reaction that requires β-nicotinamide adenine dinucleotide phosphate (NADPH) as a cofactor. Glucose-6-phosphate dehydrogenase was co-encapsulated as a catalytic sink to allow the continuous conversion of $NADP^+$ to NADPH and provide a continuous supply of NADPH to the system (*Figure 4*). The resulting coimmobilized system was able to convert nitrobenzene at millimolar concentrations continuously (~ 8 hours) into the resulting hydroxylaminobenzene with excellent efficiency (>90%). In the absence of glucose-6-phosphate, conversion of nitrobenzene was minimal as NADPH became rapidly depleted from the system.

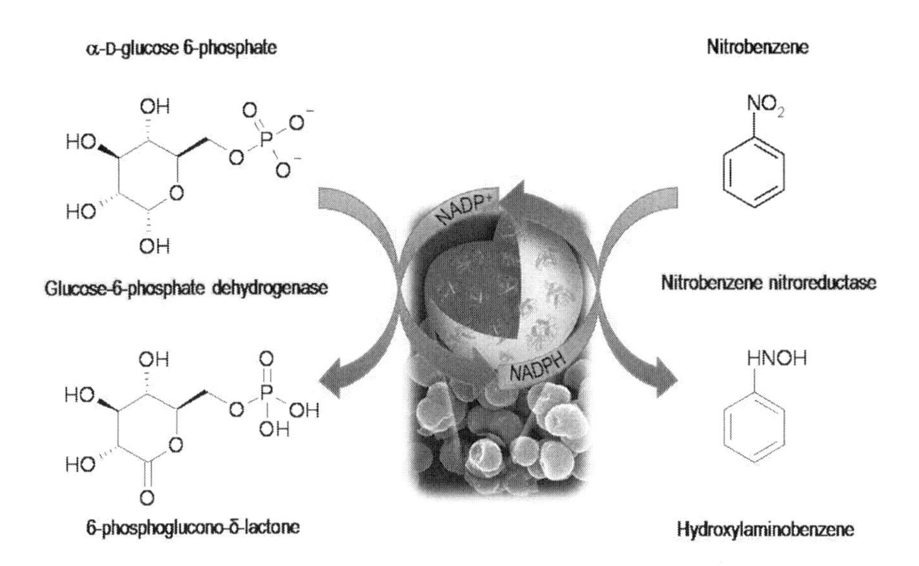

Figure 4. Schematic for the enzymatic hydroxylation of nitrobenzene with cofactor recycling via coimmobilization of enzymes in silica nanospheres.

Application of coimmobilized multi-enzyme systems

Multi-enzyme combinations now range from simple bi-enzymatic systems to complex multi-enzyme systems that mimic biochemical cycles (Table 1). There are numerous real world applications to these devices such as the determination of glycerol in wines (Gamella *et al.*, 2008). An amperometric biosensor, for example, based on glycerol dehydrogenase and diaphorase in which diaphorase recycles NAD^+ to NADH acts as a bi-enzyme system. An alternate tri-enzyme system uses glycerol kinase, glycerol-3-phosphate oxidase and peroxidase. Both systems cause the reduction and oxidation of tetrathiafulvalene, which causes a redox response that is directly related to the concentration of glycerol. Similarly, determination of acetaldehyde in alcoholic beverages can be monitored by coimmobilization of NADH oxidase with aldehyde dehydrogenase (Ghica *et al.*, 2007). Cross-linking the enzymes with glutaraldehyde, or entrapment in sol–gel, both provide stabilization to the enzyme but with a variation in the sensitivity of measurements dependent upon the immobilization strategy (up to 60 μM for sol–gel encapsulation and 100 μM for glutaraldehyde immobilization). The bi-enzymatic sensors showed improvement over aldehyde dehydrogenase alone with NAD^+ added exogenously to the reaction.

Future directions

LEARNING FROM CELLS

There is a wealth of knowledge that we can learn from cells regarding how enzymes function in a concerted manner; compartmentalization strategies (e.g., differences in

Table 1. Examples of coimmobilized enzymatic reactions

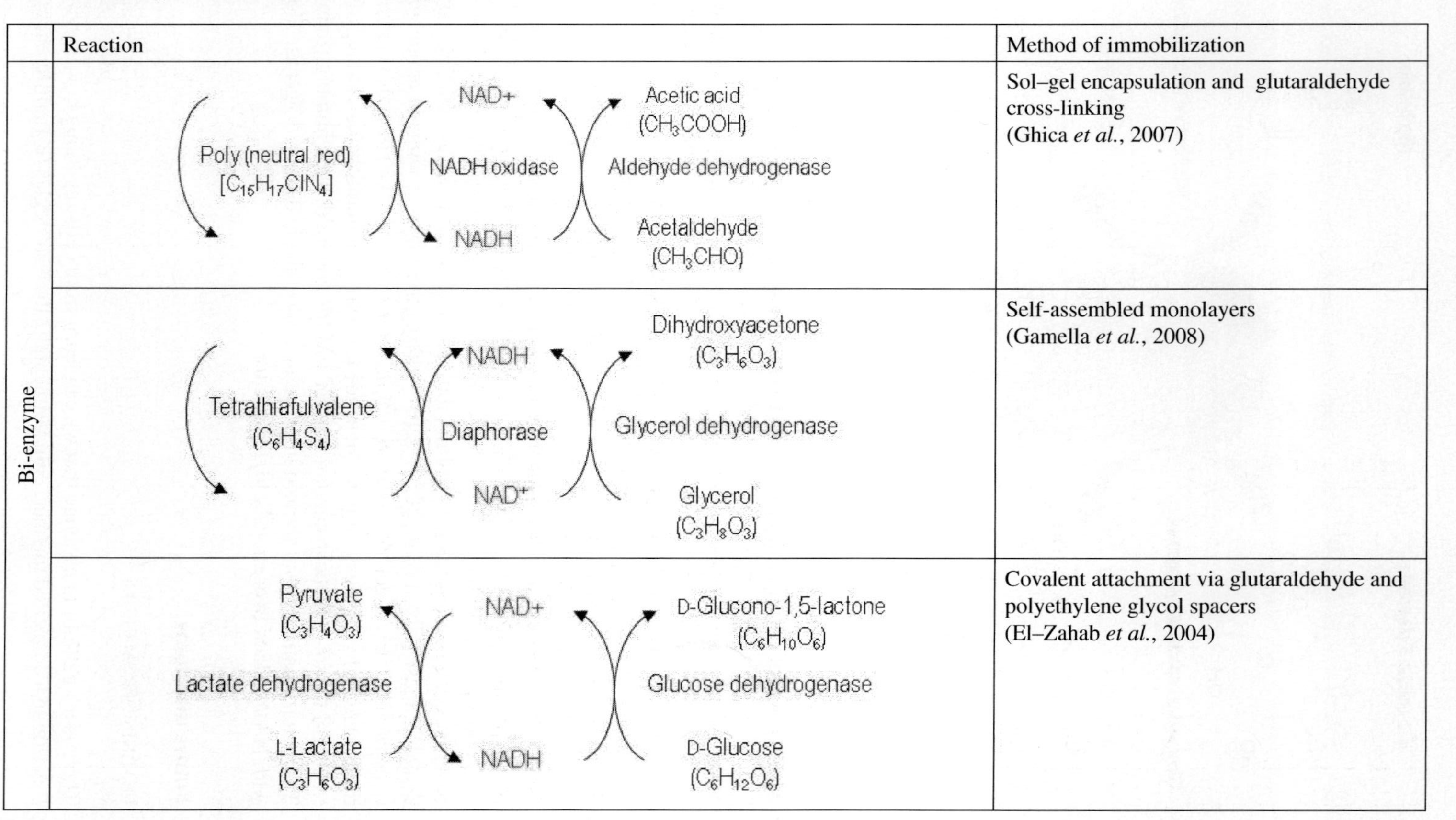

	Reaction	Method of immobilization
Bi-enzyme	Poly (neutral red) $[C_{15}H_{17}ClN_4]$; NAD+; NADH oxidase; NADH; Acetic acid (CH_3COOH); Aldehyde dehydrogenase; Acetaldehyde (CH_3CHO)	Sol–gel encapsulation and glutaraldehyde cross-linking (Ghica *et al.*, 2007)
	Tetrathiafulvalene $(C_6H_4S_4)$; NADH; Diaphorase; NAD^+; Dihydroxyacetone $(C_3H_6O_3)$; Glycerol dehydrogenase; Glycerol $(C_3H_8O_3)$	Self-assembled monolayers (Gamella *et al.*, 2008)
	Pyruvate $(C_3H_4O_3)$; Lactate dehydrogenase; L-Lactate $(C_3H_6O_3)$; NAD+; NADH; D-Glucono-1,5-lactone $(C_6H_{10}O_6)$; Glucose dehydrogenase; D-Glucose $(C_6H_{12}O_6)$	Covalent attachment via glutaraldehyde and polyethylene glycol spacers (El–Zahab *et al.*, 2004)

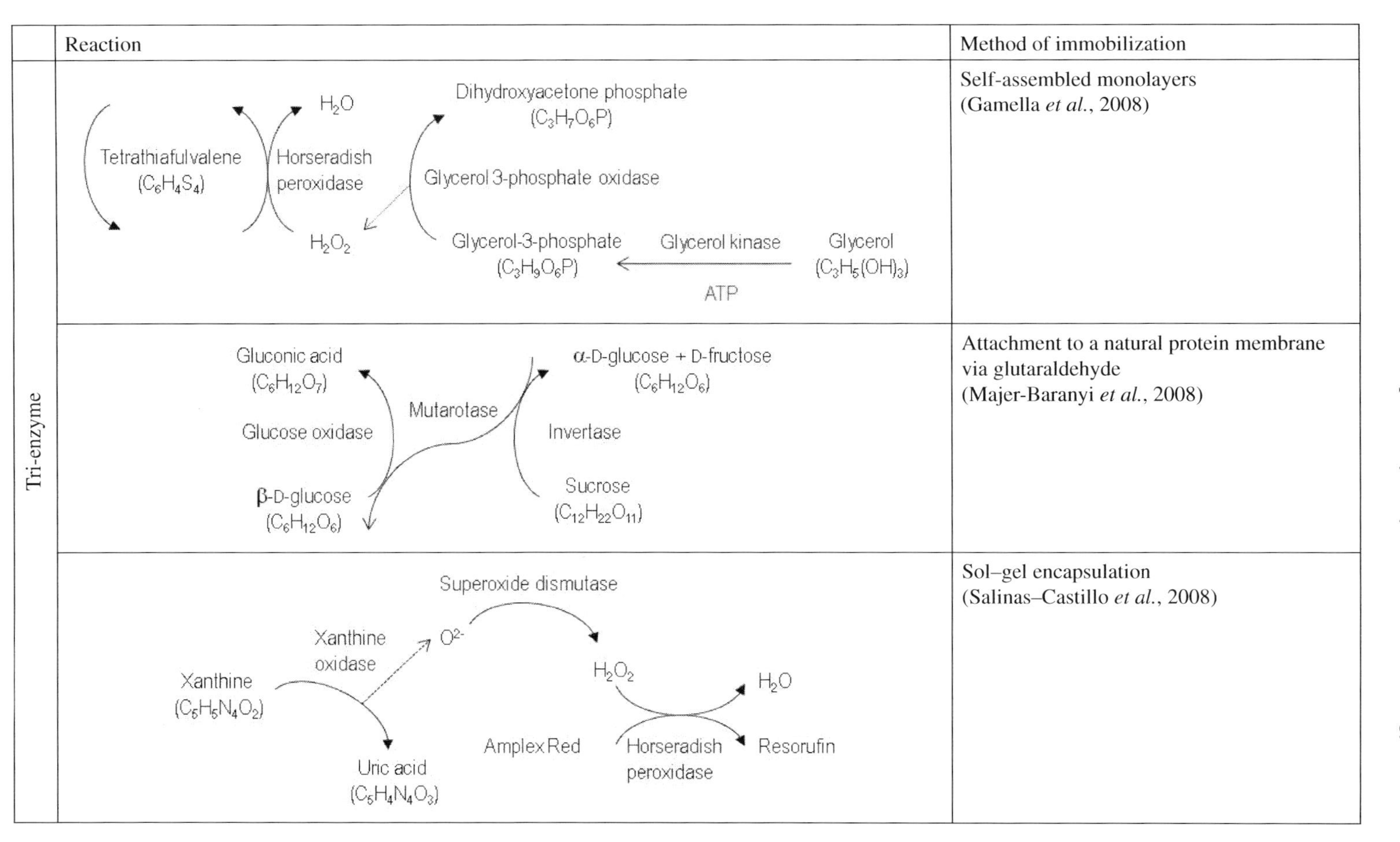

	Reaction	Method of immobilization
Tri-enzyme	Tetrathiafulvalene ($C_6H_4S_4$) Horseradish peroxidase H_2O H_2O_2 Dihydroxyacetone phosphate ($C_3H_7O_6P$) Glycerol 3-phosphate oxidase Glycerol-3-phosphate ($C_3H_9O_6P$) Glycerol kinase ATP Glycerol ($C_3H_5(OH)_3$)	Self-assembled monolayers (Gamella *et al.*, 2008)
	Gluconic acid ($C_6H_{12}O_7$) Glucose oxidase β-D-glucose ($C_6H_{12}O_6$) Mutarotase α-D-glucose + D-fructose ($C_6H_{12}O_6$) Invertase Sucrose ($C_{12}H_{22}O_{11}$)	Attachment to a natural protein membrane via glutaraldehyde (Majer-Baranyi *et al.*, 2008)
	Xanthine ($C_5H_5N_4O_2$) Xanthine oxidase Uric acid ($C_5H_4N_4O_3$) O^{2-} Superoxide dismutase H_2O_2 Amplex Red Horseradish peroxidase H_2O Resorufin	Sol–gel encapsulation (Salinas–Castillo *et al.*, 2008)

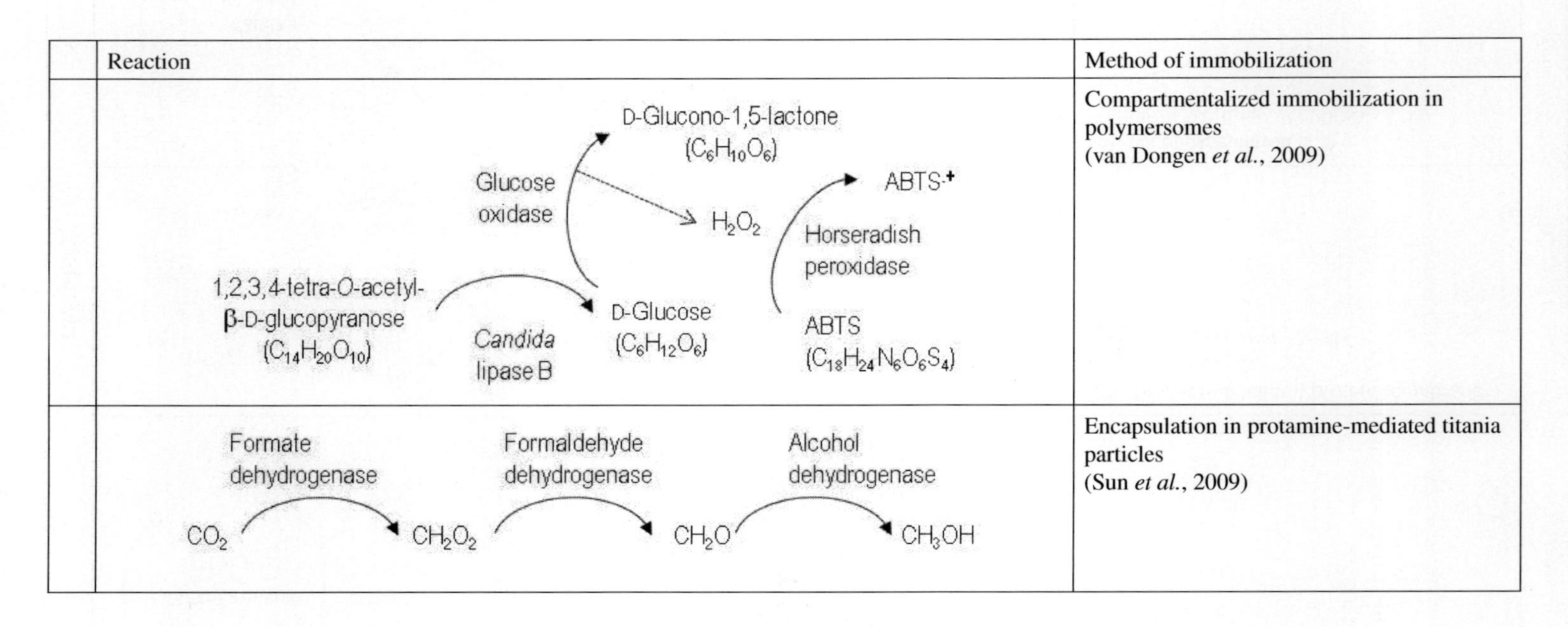

	Reaction	Method of immobilization
		Compartmentalized immobilization in polymersomes (van Dongen *et al.*, 2009)
		Encapsulation in protamine-mediated titania particles (Sun *et al.*, 2009)

	Reaction	Method of immobilization
Multi-enzyme	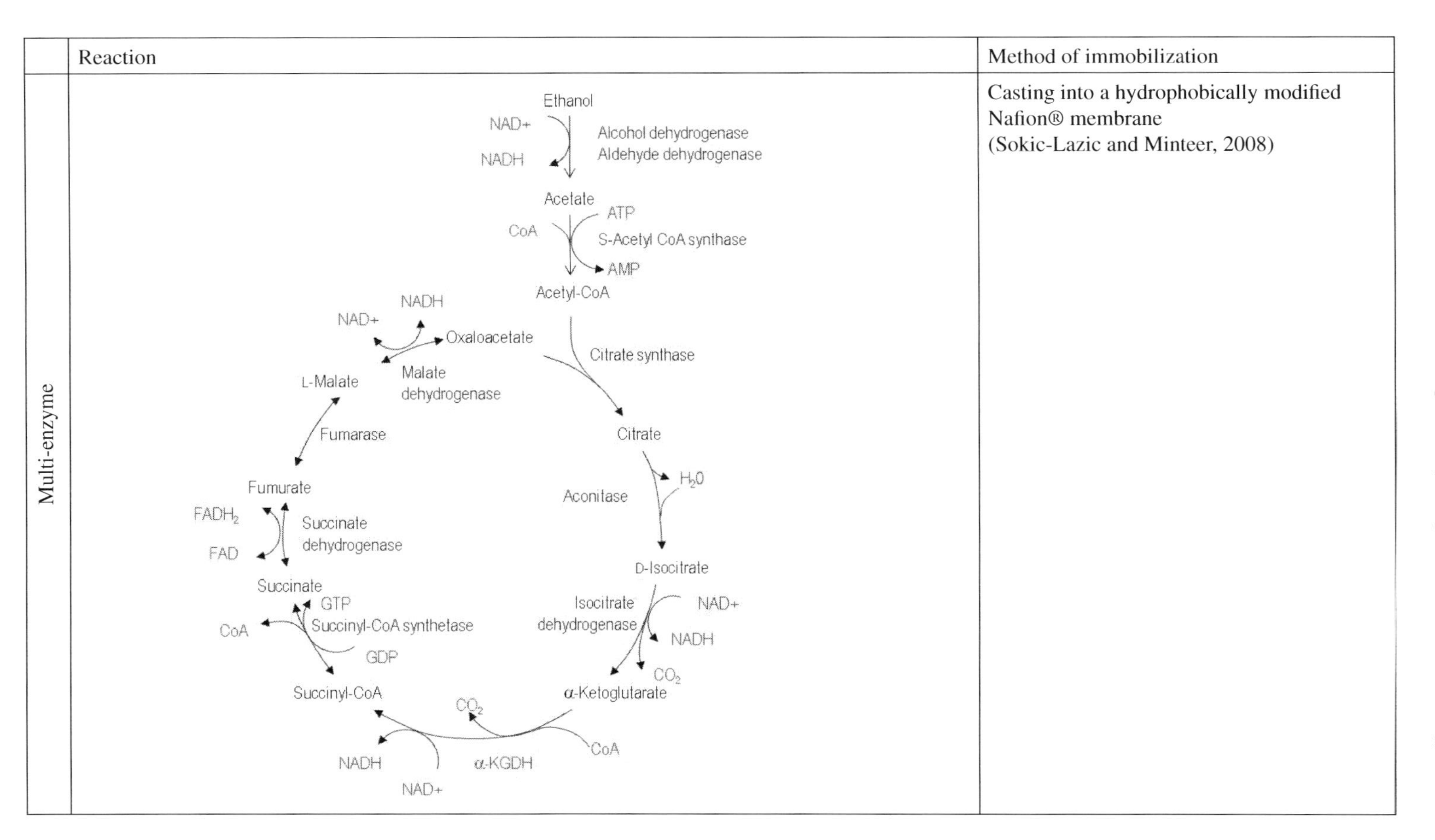	Casting into a hydrophobically modified Nafion® membrane (Sokic-Lazic and Minteer, 2008)

viscosity that contribute a one-sided partition of high-molecular-weight macromolecules and evenly distributed low-molecular-weight molecules) and the use of multi-enzyme complexes that avoid diffusion problems, balancing stability and function (e.g., intolerance to impurities) against substrate specificity (Chakrabarti *et al.*, 2003; Bhattacharya *et al.*, 2004). However, in many ways, cells still hold the key to how their complex enzymatic cascades function so perfectly. Efforts currently being pursued in the "omics" sciences (genomics, transcriptomics, proteomics, metabolomics, etc.) will certainly unveil at least some of these details from which the design of future bioprocesses will benefit significantly. A better understanding of cell function and communication between enzyme molecules within metabolic networks will undoubtedly increase the productivity of known processes and pave the way to new enzymatic syntheses with coupled biocatalysts. Exciting developments in the use of multi-enzyme systems that rely on signaling crosstalk in enzyme-based biomolecular computing (using Boolean style logic) relies on a fascinating utilization of multi-enzyme signaling to process information, but the state of the art immobilizes enzymes individually (Pita *et al.*, 2008).

THE ARCHETYPE OF COOPERATING ENZYMES

Polyketide synthases (PKS), non-ribosomal peptide synthases and fatty acid synthases constitute the paradigm of sequentially acting enzymes. As in an industrial assembly line, substrates are handed from one functional domain to the next one (sometimes housed in the same polypeptide) to produce a wealth of structurally diverse molecules. Such molecules comprise metabolites and pharmaceutically important natural products including antibiotics, immunosuppressants, antiparasitics and anticancer compounds (Weissman and Leadlay, 2005; Weissman, 2008). The literature on the immobilization of these multi-enzymatic and multi-domain megasynthase enzymes, however, is scarce. To our knowledge the few reports on the immobilized use of such enzymes have been contributed by Dordick's research group (Srinivasan *et al.*, 2004; Ku *et al.*, 2006; Kwon *et al.*, 2007; Kim *et al.*, 2009). The use of a microfluidic reactor with immobilized Type III PKS (single iterative domains) coupled to a second immobilized peroxidase reactor, for example, produces a variety of flavonoids and pyrone derivatives (Kim *et al.*, 2009). However, the use of large modular multi-domain synthases (eg., Type I PKS) still remains a challenge. One of the reasons for the infrequent use of these valuable biocatalysts may lie in the difficulty of expression of such giant multi-enzymes and the hurdle of using such enzyme complexes *in vitro* (poor stability, low activity, difficulty to identify the products, etc.) (Staunton and Weissman, 2001; Betancor *et al.*, 2008). Future studies on the immobilization of megasynthases may set the basis for the formation of attractive molecules for a range of potential applications. The results would be not only of academic importance but also vital for more applied purposes such as the production of novel compounds by rational reconfiguring of these synthases or subtle modification of the enzyme structure by immobilization.

NEW AGE CARRIERS

The availability of new carriers for enzyme immobilization may be crucial for designing new processes involving sequentially acting enzymes. Material science is constantly providing us with new or improved supports. Recently, De Geest *et al* published the use of self-exploding beads that release microcarriers (De Geest *et al.*, 2008). Upon variations in the pH of the medium and osmotic changes the outer layers of these polyelectrolyte-coated gel beads disintegrate, ejecting the inner microcapsules into the surrounding medium. Although the authors propose a possible and indeed useful application for the delivery of antigen-containing microcapsules within the body, the possibility to tailor the time of explosion could be of interest for the use of sequentially acting enzymes in certain applications (eg., when a certain amount of substrate is needed for the second enzyme to act or deleterious by-products need to be removed before the action of a sensitive subsequent enzyme).

Additionally, DNA scaffolds represent the ultimate example of new materials for enzyme immobilization with exciting future opportunities. Self-assembled single-stranded nucleic acids are able to form three-dimensional structures that can anchor DNA-tethered enzymes within the resulting scaffold. These structures have been proposed for the coimmobilization of enzyme cascades as they provide the possibility of controlling the reactivity of the system through the design of the individual DNA strips that form the scaffold (Wilner *et al.*, 2009).

ENZYME TECHNOLOGY IN *IN VITRO* SYSTEMS

As discussed, the advantages of using *in vitro* immobilized sequentially acting enzymes for bioprocesses are numerous. However, there are some problems inherent to the enzymes involved which still need to be overcome if we want the system to function repeatedly. Avoiding inactivation brought about by unfavourable reaction conditions could for instance greatly increase the economic feasibility of a process. In this regard, many scientific disciplines have contributed to provide an ever-increasing toolbox for the improvement of enzymatic properties. The design of tailor-made enzyme immobilization protocols, for example, has not only increased the stability of industrially relevant catalysts but also helps avoid inhibition problems and provide solutions for issues of enantioselectivity (Mateo *et al.*, 2007b). Protein aging can also be a significant reason for an enzyme-based system to cease functioning. The rate of some aging reactions can be reduced by carefully selecting proper reaction conditions such as oxygen content and pH. However, as the rate with which protein inactivation by covalent modifications proceeds is highly dependent on the specific sequence, the half-life of such proteins could potentially be improved by engineering the amino acid sequence (Hold and Panke, 2009). Similarly, the use of directed evolution has also been used to improve the activity recovery and stability of enzymes after immobilization to enhance biocatalysts (Ansorge-Schumacher *et al.*, 2006; Mateo *et al.*, 2007a). Moreover, site-directed mutagenesis of protein surfaces seems to be a powerful tool to greatly improve the immobilization and properties of the final immobilized biocatalyst and more effort may be expected in the next years in this regard. These are only a few examples of the many tools available to engineer putative immobilized enzymatic networks. Undoubtedly, the combined use of these tools will contribute to the rational design of integrated immobilized systems providing additional advantages over the use of engineered living systems.

Acknowledgements

Dr. Betancor gratefully recognizes funding from Ministerio di ciencia y technología (MCyT, Madrid; Ramón y Cajal research fellowship). Dr. Luckarift acknowledgements research funding from the Air Force Office of Scientific Research (program manager, Walt Kozumbo) and the Air Force Research Laboratory, Materials Science Directorate, Biotechnology Program.

References

ANSORGE-SCHUMACHER, M. B., SLUSARCZYK, H., SCHUMERS, J., HIRTZ, D. (2006). Directed evolution of formate dehydrogenase from Candida boidinii for improved stability during entrapment in polyacrylamide. *FEBS Journal,* **273**, 3938-3945.

AVERILL, B. A., LAANE, N. W. M., STRAATHOF, A. J. J., TRAMPER, J. (1999). Biocatalysis. In *Catalysis: An integrated approach*, p. 346. Elsevier Science, Amsterdam, The Netherlands.

BEECKMANS, S., VAN DRIESSCHE, E., KANAREK, L. (1993). Immobilized enzymes as tools for the demonstration of metabolon formation. A short overview. *Journal of Molecular Recognition,* **6**(4), 195-204.

BERNE, C., BETANCOR, L., LUCKARIFT, H. R., SPAIN, J. C. (2006). Application of a microfluidic reactor for screening cancer prodrug activation using silica-immobilized nitrobenzene nitroreductase. *Biomacromolecules,* **7**(9), 2631-2636.

BETANCOR, L., BERNE, C., LUCKARIFT, H. R., SPAIN, J. C. (2006). Coimmobilization of a redox enzyme and a cofactor regeneration system. *Chemical communications,* (34), 3640-3642.

BETANCOR, L., FERNANDEZ, M. J., WEISSMAN, K. J., LEADLAY, P. F. (2008). Improved catalytic activity of a purified multienzyme from a modular polyketide synthase after coexpression with Streptomyces chaperonins in *Escherichia coli*. *Chembiochem,* **9**(18), 2962-2966.

BETANCOR, L. AND LUCKARIFT, H. R. (2008). Bioinspired enzyme encapsulation for biocatalysis. *Trends in Biotechnology,* **26**(10), 566-572.

BHATTACHARYA, S., SCHIAVONE, M., GOMES, J., BHATTACHARYA, S. K. (2004). Cascade of bioreactors in series for conversion of 3-phospho-D-glycerate into D-ribulose-1,5-bisphosphate: kinetic parameters of enzymes and operation variables. *Journal of Biotechnology,* **111**(2), 203-217.

BRUGGINK, A., SCHOEVAART, R., KIEBOOM, T. (2003). Concepts of nature in organic synthesis: Cascade catalysis and multistep conversions in concert. *Organic Process Research & Development,* **7**(5), 622-640.

CHAKRABARTI, S., BHATTACHARYA, S., BHATTACHARYA, S. K. (2003). Biochemical engineering: cues from cells. *Trends in Biotechnology,* **21**(5), 204-209.

DE GEEST, B. G., MCSHANE, M. J., DEMEESTER, J., DE SMEDT, S. C., HENNINK, W. E. (2008). Microcapsules ejecting nanosized species into the environment. *Journal of the American Chemical Society,* **130**(44), 14480-14482.

EL–ZAHAB, B., JIA, H., WANG, P. (2004). Enabling multienzyme biocatalysis using nanoporous materials. *Biotechnology and Bioengineering,* **87**(2), 178-183.

ERHARDT, F. A., KUGLER, J., CHAKRAVARTHULA, R. R., JORDENING, H.-J. (2000). Co-immobilization of dextransucrase and dextranase for the facilitated synthesis of isomalto-oligosaccharides: Preparation, characterization and modeling. *Biotechnology and Bioengineering,* **100**, 673-683.

FILICE, M., VANNA, R., TERRENI, M., GUISAN, J. M., PALOMO, J. M. (2009). Lipase-catalyzed regioselective one-step synthesis of penta-*O*-acetyl-3-hydroxylactal. *European*

Journal of Organic Chemistry, **2009**(20): 3327-3329.

Gamella, M., Campuzano, S., Reviejo, A. J., Pingarron, J. M. (2008). Integrated multienzyme electrochemical biosensors for the determination of glycerol in wines. *Analytica Chimica Acta,* **609**(2), 201-209.

Ghica, M. E., Pauliukaite, R., Marchand, N., Devic, E., Brett, C. M. (2007). An improved biosensor for acetaldehyde determination using a bienzymatic strategy at poly(neutral red) modified carbon film electrodes. *Analytica Chimica Acta,* **591**(1), 80-86.

Girelli, A. M. AND Mattei, E. (2005). Application of immobilized enzyme reactor in on-line high performance liquid chromatography: A review. *Journal of Chromatography B: Analytical Technologies in the Biomedical and Life Sciences,* **819**(1), 3-16.

Grazú, V., Abian, O., Mateo, C., Batista-Viera, F., Fernández-Lafuente, R., Guisán, J. M. (2005). Stabilization of enzymes by multipoint immobilization of thiolated proteins on new epoxy-thiol supports. *Biotechnology and Bioengineering,* **90**(5), 597-605.

Hanefeld, U., Gardossi, L., Magner, E. (2009). Understanding enzyme immobilisation. *Chemical Society reviews,* **38**(2), 453-468.

Hold, C. AND Panke, S. (2009). Towards the engineering of in vitro systems. *Journal of the Royal Society Interface,* **6**(Supp 4), S507-S521.

Huang, X., Holden, H. M., Raushel, F. M. (2001). Channeling of substrates and intermediates in enzyme-catalyzed reactions. *Annual Review of Biochemistry,* **70**, 149-180.

Irazoqui, G., Giacomini, C., Batista-Viera, F., Brena, B. M. (2007). Hydrophilization of immobilized model enzymes suggests a widely applicable method for enhancing protein stability in polar organic co-solvents. *Journal of Molecular Catalysis B: Enzymatic,* **46**(1-4), 43-51.

Jeykumari, D. R. AND Narayanan, S. S. (2008). Fabrication of bienzyme nanobiocomposite electrode using functionalized carbon nanotubes for biosensing applications. *Biosensors and Bioelectronics,* **23**(11), 1686-1693.

Kim, J., Grate, J. W., Wang, P. (2008). Nanobiocatalysis and its potential applications. *Trends in Biotechnology,* **26**(11), 639-646.

Kim, M. I., Kwon, S. J., Dordick, J. S. (2009). In vitro precursor-directed synthesis of polyketide analogues with coenzyme A regeneration for the development of antiangiogenic agents. *Organic Letters,* **11**(17), 3806-3809.

Kreft, O., Prevot, M., Mohwald, H., Sukhorukov, G. B. (2007). Shell-in-shell microcapsules: A novel tool for integrated, spatially confined enzymatic reactions. *Angewandte Chemie; International Edition,* **46**(29), 5605-5608.

Ku, B., Cha, J., Srinivasan, A., Kwon, S. J., Jeong, J. C., Sherman, D. H., Dordick, J. S. (2006). Chip-based polyketide biosynthesis and functionalization. *Biotechnology Progress,* **22**(4), 1102-1107.

Kwon, S. J., Lee, M. Y., Ku, B., Sherman, D. H., Dordick, J. S. (2007). High-throughput, microarray-based synthesis of natural product analogues via in vitro metabolic pathway construction. *American Chemical Society, Chemical Biology,* **2**(6), 419-425.

Lee, M. Y., Srinivasan, A., Ku, B., Dordick, J. S. (2003). Multienzyme catalysis in microfluidic biochips. *Biotechnology and Bioengineering,* **83**(1), 20-28.

Liu, S. AND Sun, Y. (2007). Co-immobilization of glucose oxidase and hexokinase on silicate hybrid sol-gel membrane for glucose and ATP detections. *Biosensors and Bioelectronics,* **22**, 905-911.

LOGAN, T. C., CLARK, D. S., STACHOWIAK, T. B., SVEC, F., FRECHET, J. M. (2007). Photopatterning enzymes on polymer monoliths in microfluidic devices for steady-state kinetic analysis and spatially separated multi-enzyme reactions. *Analytical Chemistry,* **79**(17), 6592-6598.

LOPEZ–GALLEGO, F., BETANCOR, L., HIDALGO, A., MATEO, C., FERNANDEZ–LAFUENTE, R., GUISAN, J. M. (2005). One-pot conversion of cephalosporin C to 7-aminocephalosporanic acid in the absence of hydrogen peroxide. *Advanced Synthesis & Catalysis,* **347**(14), 1804-1810.

LUCKARIFT, H. R., KU, B. S., DORDICK, J. S., SPAIN, J. C. (2007). Silica-immobilized enzymes for multi-step synthesis in microfluidic devices. *Biotechnology and Bioengineering,* **98**(3), 701-705.

LUCKARIFT, H. R. (2008). Silica-immobilized enzyme reactors. *Journal of Liquid Chromatography and Related Technologies,* **31**(11-12), 1568-1592.

MAJER-BARANYI, K., ADANYI, N., VARADI, M. (2008). Investigation of a multienzyme based amperometric bisensor for determination of sucrose in fruit juices. *European Food Research Technology,* **228**, 139-144.

MATEO, C., GRAZÚ, V., PESSELA, B. C. C., MONTES, T., PALOMO, J. M., TORRES, R., LOPEZ-GALLEGO, F., FERNANDEZ-LAFUENTE, R., GUISAN, J. M. (2007a). Advances in the design of new epoxy supports for enzyme immobilization-stabilization. *Biochemical Society Transactions,* **35**, 1593-1601.

MATEO, C., PALOMO, J. M., FERNANDEZ-LORENTE, G., GUISAN, J. M., FERNANDEZ-LAFUENTE, R. (2007b). Improvement of enzyme activity, stability and selectivity via immobilization techniques. *Enzyme and Microbial Technology,* **40**(6), 1451-1463.

MONTES, T., GRAZÚ, V., MANSO, I., GALÁN, B., LÓPEZ-GALLEGO, F., GONZÁLEZ, R., HERMOSO, J. A., GARCÍA, J. L., GUISÁN, J. M., FERNÁNDEZ-LAFUENTE, R. (2007). Improved stabilization of genetically modified penicillin G acylase in the presence of organic cosolvents by co-immobilization of the enzyme with polyethyleneimine. *Advanced Synthesis and Catalysis,* **349**(3), 459-464.

MULLER, J. AND NIEMEYER, C. M. (2008). DNA-directed assembly of artificial multienzyme complexes. *Biochemical and Biophysical Research Communications,* **377**(1), 62-67.

NAHALKA, J., LIU, Z., CHEN, X., WANG, P. G. (2003). Superbeads: immobilization in "sweet" chemistry. *Chemistry,* **9**(2), 372-377.

NIEMEYER, C. M., KOEHLER, J., WUERDEMANN, C. (2002). DNA-directed assembly of bienzymic complexes from in vivo biotinylated NAD(P)H:FMN oxidoreductase and luciferase. *Chembiochem,* **3**(2-3), 242-245.

NOUAIMI-BACHMANN, M., SKILEWITSCH, O., SENHAJI-DACHTLER, S., BISSWANGER, H. (2007). Co-immobilization of different enzyme activities to non-woven polyester surfaces. *Biotechnology and Bioengineering,* **96**(4), 623-630.

PITA, M., KRAMER, M., ZHOU, J., POGHOSSIAN, A., SCHONING, M. J., FERNANDEZ, V. M., KATZ, E. (2008). Optoelectronic properties of nanostructured ensembles controlled by biomolecular logic systems. *American Chemical Society, Nano,* **2**(10), 2160-2166.

QI, W., YAN, X., DUAN, L., CUI, Y., YANG, Y., LI, J. (2009). Glucose-sensitive microcapsules from glutaraldehyde cross-linked hemoglobin and glucose oxidase.

Biomacromolecules, **10**, 1212-1216.

ROMISCH, W., EISENREICH, W., RICHTER, G., BACHER, A. (2002). Rapid one-pot synthesis of riboflavin isotopomers. *Journal of Organic Chemistry,* **67**(25), 8890-8894.

SALINAS–CASTILLO, A., PASTOR, I., MALLAVIA, R., MATEO, C. R. (2008). Immobilization of a trienzymatic system in a sol–gel matrix: A new fluorescent biosensor for xanthine. *Biosensors and Bioelectronics,* **24**(4), 1053-1056.

SCHOEVAART, R. AND KIEBOOM, T. (2001). Combined catalytic reactions—Nature's way. *Chemical Innovation,* **31**(12), 33-39.

SCHOEVAART, R., VAN RANTWIJK, F., SHELDON, R. A. (2001). A four-step enzymatic cascade for the one-pot synthesis of non-natural carbohydrates from glycerol. *Journal of Organic Chemistry,* **66**(1), 351.

SHELDON, R. A. (2007). Enzyme immobilization: The quest for optimum performance. *Advanced Synthesis and Catalysis,* **349**(8-9), 1289-1307.

SOKIC-LAZIC, D. AND MINTEER, S. D. (2008). Citric acid cycle biomimic on a carbon electrode. *Biosensors and Bioelectronics,* **24**(4), 945-950.

SRINIVASAN, A., BACH, H., SHERMAN, D. H., DORDICK, J. S. (2004). Bacterial P450-catalyzed polyketide hydroxylation on a microfluidic platform. *Biotechnology and Bioengeering,* **88**(4), 528-535.

STAUNTON, J. AND WEISSMAN, K. J. (2001). Polyketide biosynthesis: a millennium review. *Natural Product Reports,* **18**(4), 380-416.

SUN, Q., JIANG, Y., JIANG, Z., ZHANG, L., SUN, X., LI, J. (2009). Green and efficient conversion of CO_2 to methanol by biomimetic coimmobilization of three dehydrogenases in protamine-templated titania. *Industrial & Engineering Chemistry Research,* **48**(9), 4210-4215.

VAMVAKAKI, V., HATZIMARINAKI, M., CHANIOTAKIS, N. (2008). Biomimetically synthesized silica-carbon nanofiber architectures for the development of highly stable electrochemical biosensor systems. *Analytical Chemistry,* **80**(15), 5970-5975.

VAN DE VELDE, F., LOURENCO, N. D., BAKKER, M., VAN RANTWIJK, F., SHELDON, R. A. (2000). Improved operational stability of peroxidases by coimmobilization with glucose oxidase. *Biotechnology and Bioengineering,* **69**(3), 286-291.

VAN DONGEN, S. F. M., NALLANI, M., CORNELISSEN, J. J. L. M., NOLTE, R. J. M., VAN HEST, J. C. M. (2009). A three-enzyme cascade reaction through positional assembly of enzymes in a polymersome nanoreactor. *Chemistry-A European Journal,* **15**, 1107-1114.

WANG, J., CHEN, L., HOCEVAR, S. B., OGOREVC, B. (2000). One-step electropolymeric co-immobilization of glucose oxidase and heparin for amperometric biosensing of glucose. *The Analyst,* **125**(8), 1431-1434.

WEISSMAN, K. J. AND LEADLAY, P. F. (2005). Combinatorial biosynthesis of reduced polyketides. *Nature Reviews: Microbiology,* **3**(12), 925-936.

WEISSMAN, K. J. (2008). Taking a closer look at fatty acid biosynthesis. *ChemBioChem,* **9**(18), 2929-2931.

WILNER, O. I., WEIZMANN, Y., GILL, R., LIOUBASHEVSKI, O., FREEMAN, R., WILNER, I. (2009). Enzyme cascades activated on topologically programmed DNA scaffolds. *Nature Nanotechnology,* **4**(4), 249-254.

ZAKS, A. (2001). Industrial biocatalysis. *Current Opinion in Chemical Biology,* **5**(2), 130-136.

ZHU, L., YANG, R., ZHAI, J., TIAN, C. (2007). Bienzymatic glucose biosensor based on coimmobilization of peroxidase and glucose oxidase on a carbon nanotube electrode. *Biosensors and Bioelectronics,* **23**, 528-535.

Biotechnology and Genetic Engineering Reviews - Vol. 27, 115-134 (2010)

Carbon, food and fuel security – will biotechnology solve this irreconcilable trinity?

WAYNE MARTINDALE

Centre for Food Innovation, Sheffield Hallam University, Sheffield, S1 1WB, UK

Abstract

The emergence of food security as a key policy issue in developed nations has been concomitant with the need to reduce greenhouse gas emissions and the implementation of Environmental Management Systems in primary industries. Biotechnological interventions such as biorefinery platforms that produce chemicals and fuels provide opportunities to reconcile the security and environmental sustainability criteria increasingly sought after by governments. Indeed, sustainable and more carbon neutral options have been positively benchmarked against scenarios based solely on petrochemical feedstocks. Notably, biotechnology companies are beginning to use Environmental Management Systems employed by other industries to advocate the benefits of green technologies that employ GM, industrial enzymes and bio-materials. Management systems such as Life Cycle Analysis are providing a powerful means to measure benefits and augment change in the biotechnology sector. These methods are discussed here in the context of the emergent 21st Century debates on security. The evidence presented leads to a conclusion where biotechnologies are likely to offer increasingly high impact options for sustainability and security criteria required for food and fuel supply.

---

*To whom correspondence may be addressed (w.martindale@shu.ac.uk)

Abbreviations: BRIC, Brazil, Russia, India, China; GHG, Greenhouse Gas; FAO, Food and Agricultural Organisation; DM, Dry Matter; GJ, Giga Joule; GM, Genetically Modified; Ha, Hectare; IAASTD, International Assessment of Agricultural Science and Technology for Development; LCA, Life Cycle Analysis; NGO, Non governmental Organisation; SCAR, Standing Committee on Agricultural Research

Introduction

The United Nations Climate Change Conference hosted in Copenhagen during December 2009 reaffirmed one of the summed impact of our modern lifestyle will be the continued and irreversible influence of climate change on environmental systems. Many reports now suggest general disappointment with the Copenhagen meeting stating its failure to change the status-quo in the development of a new global climate change policy arena (European Commission, 2010). However, Copenhagen exposed the clear and urgent problem of solving the difficult and unruly relationship between increased greenhouse gas (GHG) emissions and the need to improve lifestyles globally (Cohen, Rau and Brüning, 2009). Indeed, the problems of matching the aspiration of many to improve lifestyles against the need to reduce GHG emissions were the very issues that hindered the Kyoto Protocol framework of the last 12 years and stunted possible agreement December 2009 (Kerr, 2009; McCarthy, 2010).

Crutzen and others suggest our current situation represents a need to define this human impact as a transition in geological time from the Holocene geological period to the Anthropocene (Crutzen and Stoermer, 2000). Such creative terminology is suggestive of a move from evidence gathering culture to one that has a need to be more responsive to environment change. The impact of 9 to 12 billion carbon and energy intensive lifestyles has been quantified and the stabilisation scenarios for GHG emissions and population growth to 2050 are established (McCarthy, 2010). There remains the question of what technical and management systems will be used to affect a positive change and stabilise this growth. Indeed, this is the very place that biotechnologies hold a central position for these visions because of the pivotal role of food and biofuel systems have in recycling GHG emissions and providing security (Martindale and Trewavas, 2008). The deployment of biotechnologies in agricultural systems holds an important position because of their potential to provide sustainable food and fuel production while minimising energy inputs and GHG emissions associated with the production of biomass for food and industrial feedstock. Indeed energy intensity and GHG emissions are areas of modern farming that are commonly criticised by NGOs who suggest a less intensive agriculture is the only option for reducing energy inputs in agriculture (Niggli et al., 2009). The evidence reviewed here clearly shows that not to be the case.

The potential for future biofuel development presents a huge current opportunity for biotechnology because evidence suggests these can ameliorate GHG emissions and provide sustainable food and fuel for the expected 2050 global population of 9 to 12 billion people (Sommerville, 2006). Current international GHG trading has evolved within the Kyoto Protocol where historically there has been an erroneous dismissal of the importance of agricultural systems in the achieving a capping GHG emissions at nation state scales. This is despite agricultural systems providing some of the Kyoto Protocols greatest GHG trading successes. The use of the Clean Development Mechanism (CDM) to successfully help the development of the Brazilian bioethanol industry whose feedstock for bioethanol is sugar cane (See Martindale and Trewavas, 2008) Significant evidence supports the use of liquid biofuel on the basis of it recycling carbon dioxide emissions to provide a more carbon neutral fuel compared to a gasoline baseline (Wassenaar and Kay, 2008). The CDM has been successfully used within the international GHG trading infrastructure between the European Union and Brazil on this basis. This process has raised important issues regarding the traceability of biofuels and the ethical costs associated with their production (European Commission, 2009a). The current focus for ethical issues is the currently unknown effect of biofuel production on Indirect Land Use Changes (Wassenaar and Kay, 2008). Indeed, the GHG trading debate around ethically sourced biofuel has become a frictional one with emergent BRIC (Brazil, Russia, India and China) nations increasingly dominating the direction of debate (European Commission, 2009b).

More generally, biotechnologies applied to agronomic decision making processes can provide the potential to develop the more efficient use of energy in agricultural production systems across two specific areas of agronomy. These are the more efficient use of plant nutrients and diesel fuel on farms whose energy input can make up to 60% of the energy balance for typical farming operations (Hülsbergen and Kalk, 2001). The reduction of energy inputs into food systems will also reduce GHG emissions associated with food production tackling an important criticism applied to modern intensive agriculture that it uses more energy than organic and other production systems to produce food products (Küstermann and Hülsbergen, 2008). The evidence often cited is the intensive use of mineral nitrogen fertiliser (ammonia derived), crop protection inputs and soil cultivation far outweigh the energy required to produce animal manures and the establishment of legume nitrogen fixing crop rotations. Indeed the term intensive has been often used critically despite long term studies showing the optimization of nitrogen intensity in farming systems can reduce energy inputs and consequently GHG emissions (Pimental et al., 2005). The UK Royal Society has also recently published a report that provides a new meaning for intensity in agriculture by advocating the use of sustainable intensive agriculture (Royal Society, 2009). While these words mark a change in viewpoints from the policy arena because of the immediate pressures of food and fuel security, the evidence to support continued progress to increased intensification has been established for decades that has been evident in long term field trials (Rasmussen et al., 1998). The continued intensification of agriculture results in more efficient use of inputs and a conservation of land area used for agriculture. However, the intensification process must be balanced with market trends, product development and due attention to the external costs associated with environmental and health impacts as recently reported by the Standing Committee for Agricultural Research in Europe (SCAR, 2008) and the UK government Office for Science (2010).

A traditional dismissal of the external costs of food and biofuel products by the agricultural sector as resulted in a slowing of the application of agricultural biotechnologies on European farms (House of Commons, 2009). This is evidenced as a lack of will in deploying Genetically Modified (GM) crops where current external costs are viewed negatively (Trewavas, 2008). They are directly associated with intellectual property of crop varieties and crop protection chemicals and these are very much external to the agronomic production process on farm, they are very much associated with the agricultural input supply industry, not the consumption of food or fuel by consumers (Trewavas, 2004). However, many biotechnologies currently employed by agronomists regarding nitrogen management and energy use in producing optimal biomass are very much intrinsic to the agronomic process on farm and are not determined by external supply, they are concerned with how efficient farmers can use crop protection and plant nutrient inputs. The development of more-efficient practice on farm with regard to the use of nitrogen and fuel presents some of our greatest challenges in developing farms that can meet the lifestyle requirements of 9-12 billion.

This situation is further compounded by the growth of a resistance to industrial agriculture in popular culture. The resulting resistance to GM, is focused on traits that have been dominated by soybean and maize crops and herbicide resistance, this has surprised the research community and resulted in a lower profile role for agricultural biotechnologies in the sustainable development folio. The evidence base for this overlooking of GM has been scant and a result of what Trewavas has recently called the 'cult of the amateur' (Trewavas, 2008). The negative consideration of biotechnological applications in agriculture is evident in a recent International Agricultural Assessment for Science Technology and Development (IAASTD) published by the World Bank and UN Food and Agriculture Organisation (FAO) (see International Assessment of Agricultural Science and Technology for Development IAASTD; http://www.agassessment.org/). The IAASTD has received criticism for focusing on the negative issues of GM and biotechnology from many parts of the agricultural sector (Trewavas, 2008). The IAASTD has questioned the used of biotechnology with respect to ownership of genetic material and traceability of GM crops in food and biofuel feedstocks. It is likely that the deployment of GM and biotechnology in agriculture will now be applied across the whole agricultural supply chain from farm to shopper because the maintenance of food and fuel security will demand it. What is now critical to the development of agriculture is we do not develop low intensity farming at the expense of applying biotechnologies that can help achieve GHG stabilisation and security scenarios to 2050 (Royal Society, 2009).

The issue of fuel security and liquid biofuel deserves a specific consideration because the USA and European Union have committed additions of biofuel to transport fuels with clear deadlines that must be met by 2020 (Kay and Loudjani, 2009). Enhancing the deployment and performance of the technologies that will deliver these targets from the food system while improving the consumer experience have often been communicated by high profile projects led by visionary institutions like the Venter Institute (Venter, 2008). The issues have been previously reviewed by Martindale and Trewavas (2008) but the technological visions are often focused on development of breakthrough innovations that are speculative and of very high risk nature. As such, they are in their infancy and they are concerned with post second generation, that's is beyond cellulosic digestion of biomass into simple sugars for fermentation to biofuel, to the so-called synthetic genomes that directly synthesise fine chemicals and fuel from whole cell culture systems. These post second generation fuels do not yet consider the reality that we will need to deploy these biotechnologies as product choices for 9-12 billion shoppers and it is this very detail of supplying products that must become a vibrant area of research in the biotechnology arena. The recent emergent demand for food and fuel security in developed nations adds impetus to initiating such understanding and quantification of the value biotechnology across food and biofuel supply chains from the farm to the shopper. Indeed this entails calculating the external costs and benefits of biotechnology using standard industrial procedures such as Life Cycle Analysis (LCA) and Environmental Management System (EMS) techniques that already exist for agricultural commodities and food products (Küsters 1999; Hülsbergen and Kalk, 2001; Nielsen et al., 2003; Wallén, Brandt and Wennersten, 2004; Martindale et al., 2008).

The requirement for robust food security has been highlighted by recent variability in food purchase prices, the demand of improved lifestyle in BRIC nations, a diversion of food into fuel supply chains and intensified financial speculation on future crop prices (FAO, 2009). These market impacts have all provided insight to how biotechnologies might stabilise variability and provide security within supply chains (Deloitte, 2007). A shopper and consumer focus for future near market biotechnology is essential because current political will often responds to pressure groups, media and NGOs that have developed and gained status with consumers that Trewavas aptly calls the 'cult of the amateur' (Trewavas, 2008). The cult is often a result of poor communication, mis-understanding issues of security and the ability to make high profile statements that are backed by scant evidence (for example, Friends of the Earth, 2009). The requirement to communicate scientific evidence to shoppers is of increased importance and scientists are now aware of management systems that can optimise communication from the stage of new product development through to product.

This review provides three areas of current work that will result in the deployment of biotechnologies through agricultural supply chains, they are biofuel production, food production and sensory science. They are explored here because they reflect much of what is changing with regard to how consumers interact with biotechnology. Previously, pharmaceutical and therapeutic technologies have delivered low volume high value commodities and now we are beginning to see a transition for high volume lower value commodities such as food and fuel. This market packaged by policy makers through an agenda of security and climate change not individuals and free markets as was the case for healthcare (see UK Government Office for Science, 2010). The approach has changed but the requirement is acute with a clear timeframe of ten years in Europe (Barosso, 2009) and 40 years at a global scale (House of Commons, 2009), the impact of these policy moves will be felt in coming years when we will know whether it will succeed or not.

A growing need for security and protection of citizens

There have been notable recent changes in the perception of the application of biotechnology in food and fuel supply chains by policy makers. At a pan-European level a recent statement made by the European Commission marking the start of the Spanish presidency for the next five years. The statement made by the new President of the European Commission, Senor Barosso of Spain clearly focuses the funding associated with the Common Agricultural Policy to environmental and food security benefits for European citizens (Barosso, 2009). The UK Parliament has now published responses to a large-scale review and stakeholder study of UK food security to 2050 states that although there is a requirement to double food production there is a need to consider food supply trends and technologies that will enhance the efficiency of the food supply chain (House of Commons 2009). These technologies will be applied in operations such as preservation techniques, storage design and inventory or management planning for high impact. This type of response has been supported through a UK Royal Society review of the agri-food research arena that calls for 'sustainable intensification' of agriculture (Royal Society, 2009). While intensification of agriculture can conserve land area and maximise production completely new models of understanding what is sustainable agriculture and sustainable diet must be deployed, the current Royal Society report does not goes as far as stating a new approach but suggests this will require significant investment over the next decade (2 Bn pounds sterling) and this has already impacted within the USA where over 1 Bn US Dollars per annum is likely to be committed to agri-food research (Stockstad, 2009). The case studies used in this review suggest that it will not be enough to use previous models of sustainable agriculture because it is clear that biotechnology and genetic modification hold a central role to achieving a halving of resource use, that includes nitrogen and diesel use on farms, and doubling outputs, as food and biofuel products by 2050 (House of Commons, 2009).

The policy arena associated with agriculture clearly shows a requirement to develop a food and fuel supply chain that will both meet the demands of improving lifestyles and lowering GHG emissions for decreased environmental impact. This scenario has been developed by the reports already cited here and those of the European Commission's Standing Committee on Agicultural Research offer additional evidence for a need to be pro-active across food and biofuel supply chains from farm to shopper (SCAR, 2008). The task of meeting feedstock demands for food and fuel supply chains while reducing the environmental impact of production is not possible using previous models of agricultural improvement that were successfully deployed in the 20th Century. It is clear that we are in a time of transition that is poignantly marked by the death of Norman Borlaug the visionary and architect of the 20th Century Green Revolution that defied the previous scenarios of population time bombs cited by Ehrlich and others in the late 1960s. The Green Revolution resulted in the removal of the scourge of hunger for billions of people but those same population time bomb scenarios of the late 1960's have re-emerged in the last decade with a further proviso of reducing environmental impact. The issues are compounded by BRIC economies in transition where there are requirements from citizens that go beyond providing basic resources for life and extend to enhancing quality of lifestyles. The three analyses of energy balance in agriculture, production of food and production of biofuel presented here use scenarios for developing these markets sustainably. Their potential success is critical at a micro-regional scale for the future development of biotechnology and GM in agriculture. They are also central to the development of economies in transition on which so much depends if we are to meet international targets for reducing the impact of climate change. Indeed, they form the basis of a discussion for the formulation of a new model of sustainable agriculture that will maintain 9 to 12 billion people (McCarthy, 2010). At the very core of this new model must exist a whole supply chain approach of sustainablilty from farm to shopper and need to measure the efficiency of taking biomass and feedstock to food or fuel product. The cited case studies here will begin to provide insight into that process.

Case study 1: Balancing energy in the agricultural system and the limits to efficiency

Critics of modern agriculture often cite intensive energy agricultural inputs as evidence of unsustainable production techniques and increased dependence on fossil fuel (Pimental et al., 2005). However, long-term agricultural experiments have shown extremely favourable energy balances for cereal and root crop rotations where energy outputs of 180-200 GJ/ha require energy inputs of 20-30 GJ/ha/yr (Hulsbergen and Kalk, 2001). Increased nitrogen inputs are a major energy input but are required at applications of 90-140 kg N/ha (as opposed to minimised and organic systems at 60-100 kg N/ha) to obtain optimal yields (Hülsbergen and Kalk, 2001). Organic and integrated enterprises show little difference in energy intensity because organic systems use less fossil fuel energy resulting in lower energy recovery due to decreased crop yields. Soil cultivation and harvesting are some of the most intensive on-farm operations accounting for some 2000 MJ/ha (Küsters, 1999; Hülsbergen and Kalk, 2001). The total energy input for cereals is around 20 GJ/ha/yr rising to 25-30 GJ/ha/yr for root crops. A major proportion of this energy input, up to 25%, is associated with nitrogen use. A further 25% of energy input for efficient cropping systems is accounted for by diesel consumed in ploughing and this can be reduced by utilising minimal soil cultivation methods (Küsters, 1999). The use nitrogenous fertilisers present a specific problem and the source of many current concerns with the GHG N_2O. Long term agronomic experiments show how modern agricultural production can operate close to the possible maximum nitrogen use efficiency defined by plant metabolic and industrial synthesis limitations. For example, the nitrogen nutrient budget for the typical wheat crop with a yield for grain at 8t/ha will result in a nitrogen off-take at 160 kg (N at 2% DM), the crop might typically have a Leaf Area Index (LAI) of 3 or a leaf area of 30 000 m^2 per hectare that will contain 84-109.2 kg N/ha nitrogen for C_3 leaves (typical analysis 200-260 mmol N/m^2 or 6000-7800 moles N/ha) in the canopy that is largely transferred into grain as leaves senescence (Ehleringer and Monson, 1993). The typical fertiliser applications for wheat currently range from 100-180 kg/ha and this clearly means agronomic systems can operate close to maximum efficiency. Thus, there is limited potential to reduce and extremely small emission of 5 kg N_2O/ha by manipulating optimum nitrogen applications at this Nitrogen Use Efficiency in wheat agronomic systems. Such scenarios are likely to be similar for other agronomic systems that are managed responsibly using integrated agricultural techniques (Leake, 2000).

The requirement to reduce the energy requirement for the industrial synthesis of ammonia is not an necessary outcome for the application of biotechnology because it is close to the theoretical minimum. The energy required to fix nitrogen as ammonia has decreased from around 200 MJ/kg ammonia manufactured in the 1900's to 30 MJ/kg using 1990's natural gas steam reforming techniques (the theoretical minimum for synthesis is 23 MJ/kg) (Smil, 1999). Furthermore, Smil (2002) has demonstrated that 23% of all nitrogen consumed by humans is derived from industrial fixation because of vastly improved agricultural energy balances. This industry has been the gearing of previous Green Revolutions to enable the development of highly productive crop varieties, agri-mechanical innovation and specific crop protection inputs that respond to energy efficient mineral and limited organic (as manures) nitrogen inputs. Thus coupling increased production efficiency and reduced GHG emissions is a significant challenge because many agricultural inputs are lower than ever before and our most challenging emissions, such as N_2O and phosphate occur at very low amounts that must be reduced. Reducing inputs at this scale will not occur without innovative biotechnologies, management practices and transfer of optimal integrated management to mainstream agricultural practices.

Reductions in farm diesel consumption can be achieved if minimal soil cultivations can be used. Mminimal-till or no-till agriculture reduces the use of the plough resulting in a decrease in the energy intensity and diesel consumption taken up with soil cultivation (Leake, 2000). The plough has dominated soil cultivation because it is suited most soils and can reduce weed pressure. The adoption of herbicide resistant crops has resulted in a growth of min-till approaches because weed pressure can be effectively dealt with using total herbicides. This means non-inversion of soils to reduce weed pressure is possible and the comparatively superficial cultivation of discing top soil and stubbles will provide suitable seed beds. It has been suggested that min-till agriculture could sequester an additional 10.4 Tg of carbon per year for the UK, equivalent to 6.6% of 1990 UK carbon emissions (Smith et al., 2000).

While significant management methodologies such as LCA are established where the environmental impacts of production are quantified for agricultural and food products (for examples see, Nielsen et al., 2003; Wallén, Brandt and Wennersten, 2004), LCAs do not commonly measure the impact of utilising biotechnologies that may decrease environmental impacts. It is now of some importance that the biotechnology industry responds to this deficit and utilises LCA and energy balance methodologies to demonstrate the benefits of utilising GM and other biotechnologies in agricultural and food industry production systems. These LCA methodologies have been used to assess polymers derived from plant sources using biotechnological processes (Murphy, Bonin and Hillier, 2004). These approaches should be utilised more rigorously across the agri-food supply chain from farm to shopper. Furthermore studies and meta-analyses of GM crops and transgenes have shown no statistical difference between GM and wild type cultivars (Shewry et al., 2007) and no statistical difference in bioavailability during consumption between GM and wild type crops (Kristensen et al., 2008). The approach of assessing costs and benefits by using LCA methodologies could be utilised more extensively for biotechnologies and GM products in order to measure environmental benefit. This approach has been carried out for milk production from cows between 1944 and 2007 with the conclusion that modern diary production can achieve greater sustainability outcomes if modern production intensive methods are utilised to increase milk yield per cow (Capper, Cady and Bauman, 2009). These interventions require significant veterinary inputs derived from the biotechnological sectors. Indeed, Novozymes A/S company presented the Copenhagen Climate Conference (COP 15) in 2009 with policy recommendation to ameliorate climate change using biotechnologies (Novozymes A/S, 2009).

Case study 2: Developing a secure food system the requirement for whole supply chain approaches from farm to shopper

The European food system in Europe serves some 480 million people each day with food and drink (Raspor, Mckenna and de Vries, 2007). It is of intense current research interest to understand changes in food purchase and consumption trends because they will ultimately impact on resource use, climate change and public health (Deloitte, 2007). It is clear that the current food needs of customers in developed nations are becoming more complex with environmental impact, social responsibility, functional foods, nutraceuticals, obesity and food miles, amongst many other issues, driving new products and successful business development (UK Cabinet Office Strategy

Unit, 2008). Research that defines the complexity of purchase choice supports the continued need to supply low-prices, product variety and diet choice (Deloitte, 2007). These factors stimulate consumer demand for improvement and innovation in the food and system (Costa and Jongen, 2006). Consumer purchase decisions focussed on choice and variety are often associated with health and wellbeing attributes of food and drink products. Ethical, labelling and environmental concerns, while important, take a lesser role in the determination of product purchase. Convenience, ability to snack and out-of-home consumption are also a significant for purchase. Analyses of consumer purchase choice do offer opportunities to implement innovations that are able to track consumer trends. This review will highlight a number of food system developments that are enabling innovations in response to supply chain and consumer demands. Previous reviews have highlighted the importance of defining consumer led innovations in the food system (Linnemann et al., 2006). Further development of these previous analyses will provide demonstrations whole supply chain led innovation.

Mechanisms that clearly link production to dietary culture do however have a clear role in reducing consumption of more energy intensive livestock products. However, austere dietary policy approaches will not work because even at a global level the cultural interface of food and tastes are changing dramatically with production increases of garlic (230%), chillies (223%) and spinach (337%) between 1990 and 2005 (FAOStat national statistical data). Our current understanding of how taste and fragrance is perceived through the human gut and digestive system is being understood more by the molecular understanding of taste receptors (Breslin and Spector, 2008). Placing the molecular understanding of food tastes and flavours within database structures offers huge potential impact in understanding product development and the interactions between dietary choice and health, there are European databases that are attempting to develop these types of functions (Slimania et al., 2007). Developing databases of nutrients and flavours with those genomics databases offers much potential in the dietary arena. These typical of trends show how cultural issues can be as important as production when considering food supply and they are likely to form a component of much future planning for the global food system. Food choice is often in conflict with nutritional benefit and many national food regulatory agencies representative of industrial agricultural industries have used regulatory constraints that have forced manufacturers to improve healthiness of food. Ultimately, an outcome of efficient food production and increased product development capacity has been an improvement in quality of life for billions of people. However, there is a finely tuned balance between quantity, health and enjoyment will not be simply decided by producers. The Green Revolution that faces us will be dramatically different to those of the 20th Century and biotechnologies that enhance health and choice will be increasingly important in the development of new agricultural systems.

It is clear human health will continue to drive much innovation because it is clearly associated with enhanced pleasure and quality of life. Emergent technologies are providing nutritional solutions to obesity, malnutrition and calorific deficiency. For example, there is intense interest in providing products with enhanced delivery and proposed bioavailable mineral, vitamin and fatty acid nutrition's. Such developments will change our current innovation and NPD environments beyond recognition, if implemented across the food supply chain from using biofortification, novel manufacturing and improved preservation methodologies.

Ethical considerations of consumers are an emergent trend in market innovations for food and beverage products. Ethical values associated with food products are closely related to issues of health for the consumer, and, sustainability and Corporate Social Responsibility for food companies. There is no doubt poor nutrition is a major cause of ill health and premature death in many developing and developed countries. Although the figures are avidly debated, it is reported obesity is responsible for an estimated 9,000 premature deaths per year in the UK and an estimated treatment cost for ill health due to poor diet of at least £4 billion each year (UK Cabinet Office Strategy Unit 2008a). This 'obesity epidemic' seems in conflict with the increased use of functional foods in diets and the emergence of health and well-being led innovations in the food and beverage industry (UK Cabinet Office Strategy Unit 2008b). Such situations have tiers of the regulatory, policy and research expertise in the public health arena thoroughly confused about how balanced diet and nutrition should or could be communicated to consumers. Clearly, the interactions between lifestyle, diet and public health issues are not as simple as many commentators have thought and understanding the obesity epidemic in emerging economies and the developed nations is not straightforward. Communicating the importance of balanced diets is clearly a cornerstone of robust public health communication. The analysis presented here provides a means of linking dietary, nutritional and environmental impact and could be of significant value to the development of balanced eating and public health policy.

Developing supply chain approaches to stimulate food innovation requires the analysis of the supply chain from the farm to the fork. Innovation can be established throughout the food production, processing, manufacturing and retails supply chain. Ultimately supply is determined by activities that are pre-farm gate in supply chains. This is evident in the application of agronomic technologies that have dramatically increased food yield per unit agricultural area in North America and Europe (Smil, 1999; Evans, 2005). Rapid changes in the distribution of food and beverage processing infrastructure have accompanied improvements in primary production on farms. Indeed, significant urbanisation of populations in Europe during the Nineteenth Century initiated agronomic developments at Rothamsted and other agricultural research stations globally (Rasmussen et al., 1998). The limits in regional agricultural product supply have been ameliorated by efficient logistical infrastructure, consumer demand, preservation and packaging of food (Kumar, 2008,). These developments have hidden the full cost of not producing food regionally. We are now beginning to account for these limits to agronomic capacity with the emergence of assurance and environmental labelling schemes for food (Clements, Lazo and Martin, 2008). Current food security concerns, suggest increased synergy between agriculture, food manufacture and novel plant biotechnologies is required in the future food system. This is evident in the emergent biofortification technologies and agrifortification policies for recipe-ready (eliminating processing and preservation) and more nutritious crop feedstocks for food production (White and Broadley, 2005). Such innovative activities can eliminate or establish the potential for minimal processing and preservation of foods (Allende , Toma´s-Barberan and Gil, 2006; Martýn-Dian et al., 2007). These activities impact clearly on the ability to produce food that has lower allergenicity demanded by regulators in response to consumers purchase choices that are based on 'free-from' ingredient labels (Singh and Bhalla, 2008). There are a range of applications that can enhance nutritional quality of food including novel materials and nutrient delivery mechanisms utilising nano-materials that encapsulate nutrients for release in specific gut environments (Graveland-Bikker and de Kruifa, 2006). The nutritional content of agri-products is an area of intense interest in the food and drink industry where improvement in the consumption of calcium, zinc, iron and selenium are recognised public health criteria (Morris et al., 2008). Exciting and continued development of agrifortification and biofortication technologies will improve dietary consumption of these nutrients and subsequently the health of billions of people where crop breeding and agronomic management have a central role in closing yield gaps and increasing nutritive quality (Al-Babili and Beyer, 2005). Such technological innovations will provide a means to develop sustainable supply chains while meeting consumer and regulator requirements.

Case study 3: Managing the GHG and carbon cycle with biofuel

It is becoming clear that the development of biofuel and biomass feedstocks can provide solutions for obtaining agricultural systems that have lower GHG emissions at an individual farm scale (Farrell et al., 2006; Hill et al., 2006). Biofuel and biomass can increase the energy balance of the farm with positive carbon balances (i.e. sequestration of GHG) with willow or Miscanthus crops grown for biomass fuel (heat and power) and carbon neutrality achieved for starch to bioethanol crops (Bertilsson, 1992; Hulsbergen and Kalk, 2001). Global agricultural land use pressures and conflict between fuel and food chains will be ameliorated by biotechnological, agronomic and supply chain interventions. Fledgling European biorefineries will provide an important test for liquid biofuels (European Environment Agency, 2007) and the associated changes in land use and agri-production within and outside of the EU (Harberl et al., 2007; Searchinger et al., 2008). Indeed, these developments could at their most innovative, mark a turning point in an acceptance of these manufacturing solutions by critics who have previously shown lack of vision for the application of biotechnology in supply chains.

The regional challenge in the UK is currently focussed on the development of bioethanol refineries within the Yorkshire and Humber and neighbouring regions and their impact on the Yorkshire and Humber agri-food system. The construction of refineries within and outside the Region has resulted in the need for an impact assessment to be used by food manufacturers utilising local wheat and those farmers who produce premium wheat for the food system (principally bread wheat). There are also priorities concerning regional food security that must be considered so that robust and sustainable biofuel-food policy is developed with food manufacturers, farmers and bioethanol producers. Research provides the basis for delivering these producer, manufacturer and processor driven (supply chain) frameworks (Martindale, 2009). Sensitivities regarding global land use change associated with biofuel production may stimulate the development of a completely UK-based biofuel industry. Most importantly, the results from current research lead us to consider the impact of biofuel crops on other crops in the Region. This relationship is often misrepresented at global scales (as reported by Wassenaar and Kay, 2008) and it is important that similar misrepresentation is not developed regionally in the UK.

The bioethanol manufacturing and processing plants that are projected to become operational in 2009-2010 in the UK will require farmers and food manufacturers to have efficient forecasting methodologies for both land use planning and the determination of business risks associated with investing in biofuel and/or food crops. The Vivergo fuels refinery near Kingston upon Hull will require 1.1 million tonnes of wheat each year. If we were to consider a conservative wheat yield for the Region (average for the UK) at 8.5 tonnes of grain per hectare, this will equate to a land-use requirement for almost 130 000 hectares of wheat within Yorkshire and Humber region. This is nearly 60% of the current Region's cultivated area of wheat (Defra Agricultural and Horticultural Survey (AHS) statistics, 2007). The impact of using wheat and other starch crops on food supply is not currently known. What is known is there is a demand for local wheat grain as a feedstock for bioethanol refineries because it will minimise transport costs and greenhouse gas (GHG) emissions associated with grain haulage. Current research presented and shown in Figure 1 has developed a spatial analysis of crop production and bioethanol refineries. This study has used the Agcensus Database (http://www.edina.ac.uk) to plot 2004 agricultural land use datasets spatially using GIS analyses and cross referenced them with the Defra AHS 2007. Land use capacities were defined within 50 km radii of the Vivergo, Ensus and Cargill plants using MapInfo Professional 9.5 software tools and the Agcensus database published by EDINA. Figure 1 shows a GIS analysis for the three biorefineries including Cargill PLC (at Trafford Park near Manchester UK), Vivergo Biofuels Ltd (Saltend, near Kingston upon Hull UK) and Ensus (Wilton, Teeside UK) that will be of most importance in forecasting impacts on the Yorkshire and Humber grain chain. The concentric circles shown are 5 km wide circles with a total radius of 50 km from the biorefinery. The background grid (light grey) shows the Agcensus plots for each 2 km grid, the grey-scale grids show the intensity of land used for wheat production for 2km radii (lightest grey shade at 510 hectares ; to darkest grey shade at 0). Black circles show the location of major bakeries from the Yorkshire Forward Food and Drink Cluster businesses database.

Using experimental data from field trials that have shown a hectare of wheat cultivation is associated with 1.5 tonnes of CO_2 emissions from agri-inputs and cultivation (Cowell and Clift, 1995); and, Life Cycle Assessment data that show 1 tonne of wheat is equivalent to 0.71 t CO_2eq (Nielsen, Nielsen et al., 2009) we can provide GHG emissions associated with this intensity of regional wheat production. A further environmental consideration is water use by feedstock crops. Using the equivalence measure of 750g water transpired to produce 1g C3 biomass provided by the review of Ehleringer and Monson, (1993), we have calculated the water utilised by this intensity of regional wheat production.

The research shows that the 1.1 million tonnes of wheat capacity of the Vivergo refinery can be met within 50 km of the refinery if optimal agronomic yields are achieved. Over 0.163 million hectares of wheat are grown within this radius. If optimal yields approach those of the national average the refinery capacity will be reached with a near 40% reserve on the 1.1 Mt of grain required. The Ensus Refinery has 40% of the wheat area capacity of the Vivergo refinery and the Cargill starch refinery has 14% of the wheat area capacity of the Vivergo refinery. The Vivergo Fuel refinery will also produce 0.5 million tonnes of animal feed each year. The value of these co-products in new and existing feed and fine/speciality chemical chains is currently untested and the impact of wheat grown for bioethanol on current feed chains might be minimal. The relationship between feedstock, fuel, feed and impact on crop management is unknown and future research will define these relationships.

This study has made an estimate of the regional bakery demand for local grain based of the local grain requirements of a specific bakery and extrapolated these findings to 36 other bakeries in the Region using companies reported annual financial turnovers. This estimate of local grain demand by regional bakeries has suggested a land requirement of 0.023 million hectares to produce 0.195 million tonnes of wheat for the 37 regional bakeries that have a total turnover of £214 million and 8 071 employees.

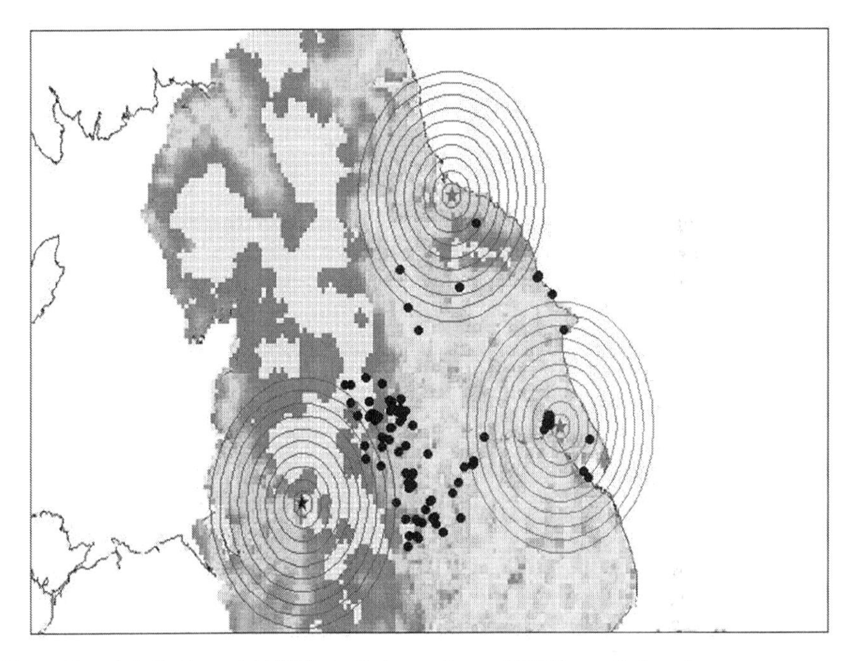

Figure 1. The GIS analysis for the three biorefineries including Cargill PLC (Trafford Park), Vivergo Biofuels Ltd (Saltend, Kingston upon Hull) and Ensus (Wilton).

An important consideration in utilising wheat for bioethanol is to consider benchmarking gross margins of wheat production against other biomass (cellulosic) crops that are utilised for fibre, heat and power. The National Non-Food Crops Centre (NNFCC) has developed a calculator for this purpose (NNFCC, 2009). The calculator shows that wheat produced at a yield of 8.25 t/ha needs to reach a price of £110-120/t to compete with a tonne of hemp fibre, £100-110/t for a tonne of Miscanthus and £90-100/t for a tonne Short Rotation Coppice (SRC, willow). The price of feed wheat per tonne in the UK has varied between £90 and £110 per tonne in the 2008/2009 period (reported HGCA delivered prices). The Region's food manufacturers and farmers who have invested in local procurement of wheat require a forecasting tool that will allow them to assess the impact of diverting wheat grain into fuel and associated feed chains. Wheat grown within the 50 km radius of the Vivergo plant represents 50% of the cropping land in this area.

The impact of lower nitrogen wheat will also be assessed. There is the potential to develop food technology solutions (enzyme and protein addition) to enable the use of lower protein wheat in bread making. These are innovative and if proven, the techniques will be of international importance. Indeed competition between bread and fuel wheat needs to be determined because the conflict between food and fuel may be manageable using current land use scenarios presented here. Evidence derived from existing biofuel crop systems show GHG emissions associated with less intensive agricultural production (lower nitrogen fertiliser inputs) are likely to be lower than those of high quality wheat production (bread making wheat). There is an immediate requirement for a study of the Vivergo Fuel refinery impact on the growing of higher nitrogen bread wheat in the region.

The demand for local wheat by millers and bakers in the Yorkshire and Humber Region is not accurately recorded and the food chain data in this study has been extrapolated using limited data. We are currently developing more robust GIS models and scenarios for the production and supply of feedstock into food, biofuel and co-product supply chains. The water, GHG and land use impacts of these production scenarios will be benchmarked and ranked with respect to agri-sustainability criteria. A key element of agri-sustainability will be economic outcomes and the NNFCC biofuel crops calculator shows how the current relationship between food and fuel feedstock is only just providing economic support for the use wheat as a biofuel crop. This economic calculator demonstrates wheat will have to reach prices of over £100 per tonne to be competitive with the current prices obtained for Miscanthus and SRC. Currently, the price offered for delivered biomass feedstocks (for heat and power) might be considered stronger with the demand from power generators having been established on the basis of renewable energy obligation and carbon reduction targets. Liquid fuel (gasoline substitute and gasohols) security has been a significant driver in the development of biofuels and the volatility of wheat price will increasingly be a factor in determining how the liquid biofuels industry in the UK develops. Furthermore, farm management systems that can deliver increased gross margins for wheat destined for the biofuel, feed or food will be of increased importance in the emerging European biofuels industry.

Conclusions and forward look

In forward looking for the application of biotechnology in the agri-food arena it is perhaps pertinent to ask what the role and value of understanding plant metabolism is. Further more, what is the value of continuing to understand the molecular basis for improving crop varieties and how this relates to agronomic decisions made on farm. We must also wonder why the issue of matching population growth and the sustainable use of natural resources has never been suitably dealt with. We now stand in a position where industrialised agriculture has had a substantial trial of at least 150 years since the start of the field trials initiated at Rothamsted in the UK (Rasmussen et al., 1998). These trials have shown that the issues of conserving resources and land can be achieved using intensive and energy efficient agriculture. However, it is apparent that all of the problems associated global food supply has not been solved because technological advances on farms have not been linked to food supply and consumer requirements effectively. It is clear that energy balance, LCA and EMS methodologies can provide evidence and a measure of the efficiency and benefits of technological interventions in food and agricultural supply chains. If GM or biotechnologies are utilised their impact can be effectively measured using these management systems. It is becoming critically important that we communicate these measurements to shoppers in appropriate ways. Increasingly, knowledge of sensory and consumer studies associated with food purchase and consumption will become of value to products that utilise biotechnology in their development.

Furthermore, the genetic base we have used for providing energy and protein globally over the last 10 000 years is surprisingly small. There are perhaps around 120 crop types that provide most of our dietary and fibre needs covering just over 10 taxonomic plant families. We are now able to improve crop yields and quality attributes by targeting gene expression and protein synthesis within plant cells using biotechnologies that do not necessarily require field observation. These technologies have developed the genomics arena where there is currently debate on how these datasets can be best utilised for greatest impact (Field et al., 2009). They will not mean the field trial is dying out because ultimately crop production must be interfaced with management in the field, these 'omics technologies will potentially take GM crops from limited herbicide and insect resistance traits and a handful of crops to an applied arena for the whole food supply chain. the result will be a greater understanding of determinants of quality and taste so that they can be manipulated more accurately on farm by management and in crop breeding trials by genetics.

The use of agronomy and our land resources to produce fine chemicals for a range of markets that cross the energy, manufacturing, medical and food sectors offers huge potential in developing new and novel markets for agricultural produce. In Europe we have seen the emergence of oil seed rape (*Brassica napus*) otherwise known as canola in the last 25 years to become a mainstream crop that is familiar in the countryside of North Western Europe with diverse markets including the manufacturing, foods and medical sectors. We must view our traditional agri-landscapes as the chemical processing industry of the future. Our view of agricultural and food supply chains has changed radically in the last decade and the influence particularly with regard to the diversification of agricultural products for feedstocks in fuel and fine chemical supply chains. This represents a present challenge that has been defined by GHG emission targets at nation state level and represents opportunity for many biotechnologies to impact in society as their intervention in reducing GHG emissions will begin to unfold.

References

Al-Babili, S. and Beyer, P. (2005) Golden Rice – five years on the road – five years to go? *Trends in Plant Science* **10**, 565-573

Allende A., Toma´s-Barberan F.A. and Gil M.I. (2006) Minimal processing for healthy traditional foods. *Trends in Food Science and Technology* **17,** 513–519

Barroso, J.M. (2009) Political guidelines for the next Commission. *European Commission* http://ec.europa.eu/commission_barroso/president/pdf/press_20090903_EN.pdf

Bertilsson, G. (1992) Environmental consequences of different farming systems using good agricultural practices. *Proceedings of the Fertiliser Society*. (December) pp.1-27.

Breslin, P.A.S. and Spector, A.C. (2008) Mammalian taste perception. *Current Biology* **18** , 148-155

Capper, J.L., Cady, R.A. and Bauman, D.E. (2009) The environmental impact of dairy production: 1944 compared with 2007. *Journal of Animal Science* **87,** 2160-2167

Clements, M.D., Lazo, R.M. and Martin, S.K. (2008) Relationship connectors in NZ fresh produce supply chains. *British Food Journal* **110.** 346-360

Cohen, J., Rau, A. and Brüning, K. (2009) Bridging the Montreal-Kyoto Gap, *Science* **326,** 140-141

Costa, A.I.A. and Jongen, W.M.F. (2006) New insights into consumer-led food product development. *Trends in Food Science and Technology* **17**, 457–465

Crutzen, P.J., Mosier, A.R., Smith, K.A. and Winiwarter W. (2007) N_2O release from agrobiofuel production negates global warming by replacing fossil fuels. *Atmospheric Chemistry and Physics Disscussions* **7,** 11191-11205

Crutzen P.J. and Stoermer E.F. (2000), The 'Anthropocene. *Global Change Newsletter* **41,** 17-18

Deloitte Touche Tohmatsu. (2007) An appetite for change, food and beverage 2012, a commercial report presented by DTT available at: http://www.deloitte.com/assets/Dcom-UnitedKingdom/Local%20Assets/Documents/An%20appetite%20for%20change(2).pdf

Ehleringer J.R. and Monson R.K. (1993) Evolutionary and ecological aspects of photosynthetic pathway variation. *Annual Review Ecology Systematics* **24,** 411–439

EuroFIR (European Food Information Resource Network) (2008) http://www.eurofir.net

European Commission (2010) EU sticks to climate pledges amid 'soft' UN deadline concerns http://www.euractiv.com/en/climate-environment/eu-sticks-climate-pledges-amid-soft-un-deadline-concerns/article-189133

European Commission (2009a) EU biofuel sustainability criteria 'inconsistent' http://www.euractiv.com/en/climate-environment/eu-biofuel-sustainability-criteria-inconsistent/article-188224

European Commission (2009b) Brazil warns EU on biofuel sustainability http://www.euractiv.com/en/energy/brazil-warns-eu-biofuel-sustainability/article-188445

European Environment Agency (2007) Estimating the environmentally compatible bioenergy potential from agriculture. EEA technical report No 12/2007

Evans, L.T. (2005) The changing context for agricultural science. *Journal of Agricultural Science* **143,** 7-10

F.A.O. (2009) The State of Food and Agriculture 2008 http://www.fao.org/publications/sofa/en/

Farrell, A.E., Plevin, R.J., Turner, B.T., Jones, A.D., O'Hare, M. and Kammen, D.M. (2006) Ethanol can contribute to energy and environmental goals. *Science* **311**, 506-508

Field D., Sansone S-A., Collis A., Booth T., Dukes P., Gregurick S.K., Kennedy K., Kolar P., Kolker E., Maxon M., Millard S., Mugabushaka A-M., Perrin N., Remacle J.E., Remington K., Rocca-Serra P., Taylor C.F., Thorley M., Tiwari B, and Wilbanks J. (2009) 'Omics Data Sharing. *Science* **326,** 234-236

Friends of the Earth (2009) Biofuels policy doubles CO_2 emissions - new research http://www.foe.co.uk/resource/press_releases/biofuels_double_carbon_emissions_15042009.html

GRAVELAND-BIKKER, J.F. AND DE KRUIFA, C.G. (2006) Unique milk protein based nanotubes: Food and nanotechnology meet. *Trends in Food Science and Technology* **17,** 196–203

HABERL H., ERB K.H., KRAUSMANN F., GAUBE V., BONDEAU A., PLUTZAR C., GINGRICH S., LUCHT W. AND FISCHER-KOWALSKI M. (2007) Quantifying and mapping the human appropriation of net primary production in earth's terrestrial ecosystems. *Proceedings of the National Academy of Science U S A.* **104,** 12942-12947.

HILL, J., NELSON, E., TILMAN, D., POLASKY, S. AND TIFFANY D. (2006) Environmental, economic and energetic costs and benefits of biodiesel and ethanol biofuels. *Proceedings of the National Academy of Science U S A*, **103,** 11206-11210.

HÜLSBERGEN, K-J. AND KALK, W-D. (2001) Energy Balances in Different Agricultural Systems - Can they be Improved? *The International Fertiliser Society* - Proceeding **476**

HOUSE OF COMMONS (2009) Securing food supplies up to 2050: Government response to the Committees fourth report of session 2008-2009 HC1022

KAY, S. AND LOUDJANI, P. (2009) Future measures for European and global agriculture. *Aspects of Applied Biology* **95,** 65-71

KERR, R.A. (2009) Amid worrisome signs of warming, 'Climate Fatigue' sets in. *Science* **326,** 926-928

KRISTENSEN, M., ØSTERGAARD, L.F., HALEKOH, U., JØRGENSEN, H., LAURIDSEN, C., BRANDT, K, AND BÜGEL S. (2008) Effect of plant cultivation methods on content of major and trace elements in foodstuffs and retention in rats. *Journal of the Science of Food and Agriculture* **88,** 2161-2172

KUMAR, S. (2008) A study of the supermarket industry and its growing logistics capabilities. *International Journal of Retail and Distribution Management* **36,** 192-211

KÜSTERS, J. (1999) Life cycle Approach to nutrient and energy efficiency in European Agriculture. *Proceedings of the International Fertiliser Society* No. **438**

KÜSTERMANN, B. AND HÜLSBERGEN, K.J. (2008) Emission of Climate-Relevant Gases in Organic and Conventional Cropping Systems. *16th IFOAM Organic World Congress, Modena, Italy, June 16-20, 2008.* Archived at http://orgprints.org/12813/1/12813.pdf

LEAKE, A.R. (2000) Climate change, Farming systems and Soils. *Aspects in Applied Biology* **62,** 253-259

LINNEMANN, A.R., BENNER, M., VERKERK, R. AND VAN BOEKEL, A.J.S. (2006) Consumer-driven food product development. *Trends in Food Science and Technology* **17,** 184-190

MARTINDALE, W. (2009) Co-development of bioethanol, feed and food supply chains that meet European agricultural sustainability criteria. *Aspects of Applied Biology* **95**, 79-84

MARTINDALE, W., MCGLOIN, R., JONES, M. AND BARLOW, P. (2008) The carbon dioxide emission footprint of food products and their application in the food system. *Aspects of Applied Biology*, Proceedings of the Association of Applied Biologists ISSN 0265-1491, Greening the Food Chain **86,** 55-60.

MARTINDALE, W. AND TREWAVAS, A. (2008) Fuelling the 9 Billion, *Nature Biotechnology*

26, 1068 - 1070 (2008) doi:10.1038/nbt1008-1068

Martýn-Dian, A.B., Rico, D., Frýas, J.M., Barat, J.M., Henehan, G.T.M. and Barry-Ryan, C. (2007) Calcium for extending the shelf life of fresh whole and minimally processed fruits and vegetables: a review. *Trends in Food Science and Technology* **18,** 210-218

McCarthy, J.J. (2010) Reflections on: our planet and its life, origins, and futures. Science **326,** 1646-1655

Morris, J., Hawthorne, K.M., Hotze, T., Abrams, S.A. and Hirsch, K.D. (2008) Nutritional impact of elevated calcium transport activity in carrots. *Proceedings of the National Academy of Science U S A.* **105,** 1431-1435

Murphy, R.J., Bonin, M. and Hillier, W.R. (2004) Life cycle assessment of potato starch based packaging trays. Report (draft) for STI Project Sustainable GB Potato Packaging, Imperial College London, contact r.murphy@imperial.ac.uk

Nielsen, P.H., Nielsen, A.M., Weidema, B.P., Dalgaard, R. and Halberg, N. (2003) LCA food data base. www.lcafood.dk

Niggli, U., Fliessbach, A., Hepperly, P. and Scialabba, N. (2009) Low Greenhouse Gas Agriculture: Mitigation and Adaptation Potential of Sustainable Farming Systems. *FAO*, April 2009, Rev. 2 – 2009 ftp://ftp.fao.org/docrep/fao/010/ai781e/ai781e00.pdf

Novozymes A/S. (2009) Novozymes' policy recommendations for COP15 http://report2009.novozymes.com/Menu/Novozymes+Report+2009/Report/Letter+from+the+Board+of+Directors and Novozymes fights climate change today http://www.novozymes.com/en/MainStructure/Climate+in+focus/Novozymes+fights+climate+change+today/

NNFCC (2009) UK Energy Crops Calculator http://www.nnfcc.co.uk/

Pimental, D., Hepperly, P., Hanson, J., Douds, D. and Seidel, R. (2005) Environmental energetic and economic comparisons of organic and conventional farming systems. *Bioscience* 55, 573-582.

Raspor, P., McKenna, B. and de Vries, H.S.M. (2007) Food processing: food quality, food safety, technology in ESF-COST (2007) Forward Look: European Food Systems in a Changing World [Au:Please clarify publication]

Rasmussen, P.E., Goulding, K.W.T., Brown, J.R., Grace, P.R., Janzen, H.H. and Kõrschens, M. (1998) Long-term agroecosystem experiments: assessing agricultural sustainability and global change. *Science* **282,** 893-896 (1998).

Royal Society (2009) Reaping the benefits Science and the sustainable intensification of global agriculture. RS Policy document 11/09 ISBN: 978-0-85403-784-1

SCAR (Standing Committee for Agricultural Research) (2008) New challenges for agricultural research: climate change, food security, rural development, agricultural knowledge systems. *Report from the 2nd Foresight Exercise.*

Searchinger, T., Heimlich, R., Houghton, R.A., Dong, F., Elobeid, A., Fabiosa, J., Tokgoz, S., Hayes, D. and Yu, T.H. (2008) Use of U.S. croplands for biofuels increases greenhouse gases through emissions from land-use change *Science,* **319,** 1238-1240

Searchinger, T.D., Hamburg, S.P., Melillom J., Chameides, W., Havlik, P., Kammen, D.M., Likens, G.E., Lubowski, R.N., Obersteiner, M., Oppenheimer, M., Robertson, P., Schlesinger, W.H. and Tilman, D. (2009) Fixing a critical climate accounting error. *Science* **326,** 527-528

SHEWRY, P.R., BAUDO, M., LOVEGROVE, A., POWERS, S., NAPIER, J.A., WARD, J.L., BAKER, J.M. AND BEALE, M.H. (2007) Are GM and conventionally bred cereals really different? *Trends in Food Science and Technology* **18,** 201-209

SINGH, M.B. AND BHALLA, P.L. (2008) Genetic engineering for removing food allergens from plants. *Trends in Plant Science* **13,** 257-260

SLIMANIA, N., DEHARVENG, G., UNWIN, I., VIGNAT, J., SKEIE, G., SALVINI, S., MØLLER, A., IRELAND, J., BECKER, W. AND SOUTHGATE, D.A.T. (2007) Standardisation of an European end-user nutrient database for nutritional epidemiology: what can we learn from the EPIC Nutrient Database (ENDB) Project? *Trends in Food Science and Technology* **18**, 407-419

SMIL, V. (2002) Nitrogen and food production: Proteins for human diets. *Ambio* **31**,126-131

SMIL, V. (1999) Detonator of the population explosion. *Nature* **401,** 415

SMITH, B.G. (2008) Sourcing from a more sustainable agriculture. *Aspects of Applied Biology*, Proceedings of the Association of Applied Biologists ISSN 0265-1491, Greening the Food Chain series, **87,** 45-54

SMITH, P., POWLSON, D.S., SMITH, J.U., FALLOON, P. AND COLEMAN, K. (2000) Meeting the UK's climate change commitments: options for carbon mitigation on agricultural land. *Soil Use and Management* **16,** 1-11

SOMERVILLE, C. (2006) The billion ton biofuels vision. *Science* **312,** 1277

STOCKSTAD, E. (2009) Agricultural Science Gets More Money, New Faces. *Science* **326,** 216

TREWAVAS. A. J. (2008) The cult of the amateur in agriculture threatens food security. Trends in *Biotechnology* **26**, 475-479

TREWAVAS, A.J. (2004) A critical assessment of organic farming and food assertions with particular respect to the UK and the potential environmental benefits of no-till agriculture. *Crop Protection* **23,** 757-781

UK CABINET OFFICE STRATEGY UNIT (2008a) Food: an analysis of the issues, discussion paper http://www.cabinetoffice.gov.uk/strategy/work_areas/food_policy.aspx

UK CABINET OFFICE STRATEGY UNIT (2008b) Food matters: towards a strategy for the 21st Century http://www.cabinetoffice.gov.uk/~/media/assets/www.cabinetoffice.gov.uk/strategy/food/food_matters1%20pdf.ashx

UK GOVERNMENT OFFICE FOR SCIENCE (2010) UK Cross government food research and innovation strategy. A copy of the full report is available at: www.bis.gov.uk/GO-Science

VENTER, C. (2008) On the verge of creating synthetic life. *Technology Entertainment Design - an on-line global community and collection of lectures and talks available at* http://www.ted.com/index.php/talks/craig_venter_is_on_the_verge_of_creating_synthetic_life.html

WALLÉN, A., BRANDT, N. AND WENNERSTEN, R. (2004) Does the Swedish consumer's choice of food influence greenhouse gas emissions? *Environmental Science and Policy* **7**, 525–535

WASSENAAR, T. AND KAY, S. (2008) Biofuels: One of Many Claims to Resources *Science* **321,** 201

WHITE, P.J. AND BROADLEY, M.R. (2005) Adding nutritional value to food ingredients by biofortication *Trends in Plant Science* **10,** 586-593

Biotechnology and Genetic Engineering Reviews - Vol. 27, 135-158 (2010)

Targeted deep resequencing of the human cancer genome using next-generation technologies

SAMUEL MYLLYKANGAS[1], HANLEE P. JI[1]*

[1]*Stanford Genome Technology Center and Division of Oncology, Department of Medicine, Stanford University School of Medicine, CCSR 3215, 269 Campus Drive, Stanford, California 94305, USA*

Abstract

Next-generation sequencing technologies have revolutionized our ability to identify genetic variants, either germline or somatic point mutations, that occur in cancer. Parallelization and miniaturization of DNA sequencing enables massive data throughput and for the first time, large-scale, nucleotide resolution views of cancer genomes can be achieved. Systematic, large-scale sequencing surveys have revealed that the genetic spectrum of mutations in cancers appears to be highly complex with numerous low frequency bystander somatic variations, and a limited number of common, frequently mutated genes. Large sample sizes and deeper resequencing are much needed in resolving clinical and biological relevance of the mutations as well as in detecting somatic variants in heterogeneous samples and cancer cell sub-populations. However, even with the next-generation sequencing technologies, the overwhelming size of the human genome and need for very high fold coverage represents a major challenge for up-scaling cancer genome sequencing projects. Assays to target, capture, enrich or partition disease-specific regions of the genome offer immediate solutions for reducing the complexity of the sequencing libraries. Integration of targeted DNA capture assays and next-generation deep resequencing improves the ability to identify clinically and biologically relevant mutations.

*To whom correspondence may be addressed (samuel.myllykangas@gmail.com)

Abbreviations: MIP, molecular inversion probe; BAC, bacterial artificial chromosome; SMART, Spacer Multiplex Amplification Reaction

Introduction

Polymorphisms, rare variants and mutations in genes contribute to increased malignancy predisposition, cancer development (Kinzler *et al.*, 1991) and tumor behavior such as metastatic spread (Vogelstein and Kinzler, 1992). This has been substantiated with the completion of recent large-scale cancer genome resequencing surveys where it is clear that for any given tumor there are critical somatic mutations in specific sets of cancer genes (Ding *et al.*, 2008; 2008; Sjöblom *et al.*, 2006; Wood *et al.*, 2007). These same mutations may be useful as genetic prognostic and predictive biomarkers with clinical application (Di Fiore *et al.*, 2007; Liévre *et al.*, 2008). Fortunately, innovative high-throughput DNA sequencing technologies have greatly increased DNA sequencing throughput and thus expanded one's ability to survey cancer genomes for mutations and other genomic aberrations at a scale once considered impossible (Bentley *et al.*, 2008; Bentley, 2006; Harris *et al.*, 2008; Margulies *et al.*, 2005; Shendure *et al.*, 2005). Typically, these "next-generation" platforms are based on parallelization of the cyclic sequencing reactions on a solid surface, where the ascertainment of the nucleotide order of a random selection of DNA fragments occurs (Margulies *et al.*, 2005; Shendure *et al.*, 2005). Although, these novel technologies produce gigabases of sequence data with a single run, the higher error rates and short read lengths compared to Sanger sequencing require substantially increased sequencing fold coverage to reliably identify mutations from an entire cancer genome (Smith *et al.*, 2008). For these reasons, whole cancer genome resequencing is not cost-effective for translational and clinical studies requiring large sample sizes or deep sequencing of individual samples. To overcome the throughput limitations of sequencing large number of cancer genomes, new approaches are needed that target, capture, enrich or partition disease-specific regions of the genome (Albert *et al.*, 2007; Bau *et al.*, 2009; Dahl *et al.*, 2007; Gnirke *et al.*, 2009; Hodges *et al.*, 2007; Krishnakumar *et al.*, 2008; Okou *et al.*, 2007; Porreca *et al.*, 2007). These genomic "snap shot" resequencing approaches open new avenues to disentangle the complex genetic variation and somatic mutations in cancer.

Next-generation DNA sequencing technologies

Next-generation of DNA sequencing dramatically increases throughput by orders of magnitude compared to Sanger sequencing (Margulies *et al.*, 2005; Shendure *et al.*, 2005) (Reviewed by (Shendure and Ji, 2008) and (Holt and Jones, 2008)). Immobilizing template DNA fragments on a solid support enables mass parallelization of cyclic sequencing reactions with subsequent high-resolution imaging (Bentley *et al.*, 2008; Bentley, 2006; Harris *et al.*, 2008; Margulies *et al.*, 2005; Shendure *et al.*, 2005). Commercial product lines for next-generation DNA sequencers are currently available (Table 1). The Genome Sequencer GS20 (454 Life Sciences) was the first commercially available next-generation sequencing platform and was recently acquired by Roche Applied Science (Basel) (Margulies *et al.*, 2005). The first commercial short read sequence (< 50 bp) technology was developed by Solexa, which was subsequently acquired by Illumina (Hayward, CA) (Bentley *et al.*, 2008; Bentley, 2006). The SOLiD Sequencer is a sequencing platform by Applied Biosystems (Foster City, CA) and applies uniquely labeled oligonucleotide panel and ligation chemistry

Table 1. Overview of the next-generation DNA sequencing platforms.

Platform	Vendor	Clonal amplification	Sequencing application	Sequencing feature	Extension chemistry	Read length	Accuracy	Sequence per run	Separate reactions	Multiplexing capacity	Run time
Genome Sequencer FLX Titanium	Roche	Emulsion PCR	Synthesis	Pyrophosphate labeled nucleotides	Asynchronous	500	99 %	1 Gb	16	12	10 h
Genome Analyzer II	Illumina	Bridge PCR	Synthesis	Fluorescent terminators	Synchronous	~100	>99%	10 Gb	8	12	5 days
SOLiD 3	Applied Biosystems	Emulsion PCR	Ligation	Semi-degenerate, fluorescent octamers with two coding bases	Synchronous	50	99.94 %	10 Gb	16	16	7 days
HeliScope	Helicos	Single molecule	Synthesis	Fluorescent terminators	Asynchronous	35	N/A	28 Gb	50	>2,000	14 days
Polonator G.007	Dover Systems and Harvard Medical School	Emulsion PCR	Ligation	Fluorescent nonamers	Synchronous	13	N/A	5 Gb	32	N/A	4 days

in sequencing. Sequencing technology conceived by the Quake lab (2003) is based on detection of single molecule sequencing-by-synthesis events and does not require clonal DNA template amplification for sequencing feature generation (Braslavsky *et al.*, 2003). This technology is available commercially through HeliScope Single Molecule Sequencer (Harris *et al.*, 2008) as manufactured by Helicos (Cambridge, MA, USA). In addition, Dover Systems (Salem, NH) and Harvard Medical School (Cambridge, MA) are developing an open-source platform called Polonator G.007 for high-throughput DNA sequencing (Shendure *et al.*, 2005).

SEQUENCING LIBRARY PREPARATION

The next-generation sequencers typically use random fragmentation of genomic DNA from a given sample with subsequent ligation of the adaptor oligonucleotide sequences to the ends of the DNA fragments. This preparation is frequently referred to as a "sequencing library". Sonication, physical interruption, chemical or enzyme-based methods have been utilized in fragmentation of the genomic DNA sample. Size selected DNA fragments are tagged with common adaptor oligonucleotides or, if multiple samples are analyzed in the same reaction volume, barcoded multiplex adaptors that allow mapping of the sequencing reads to specific samples (Binladen *et al.*, 2007; Meyer *et al.*, 2007; Nielsen *et al.*, 2006). Furthermore, methods have been developed to sequence paired ends of a DNA fragment of known size with next-generation platforms (Campbell *et al.*, 2008; Korbel *et al.*, 2007). Paired-end sequencing improves mapping of short reads and facilitates detection of structural variation, insertions, deletions, and rearrangements, in the genome. Other than 454, the expense of sequencing paired-ends is still double compared to single end sequencing.

IN VITRO CLONAL AMPLIFICATION

In the majority of the next-generation platforms, the detection of individual enzymatic reaction is not sensitive enough to record single DNA molecule events. Thus, DNA templates in the sequencing library require *in vitro* clonal amplification to form "magnified" sequencing features that become detectable by imaging. The main approaches in generating the clonal amplicons from the sequencing libraries include emulsion PCR (Dressman *et al.*, 2003; Ghadessy *et al.*, 2001; Margulies *et al.*, 2005; Tawfik and Griffiths, 1998), applied in 454, SOLiD and Polonator platforms, and bridge PCR (Adessi *et al.*, 2000; Fedurco *et al.*, 2006), which is utilized in Illumina sequencing (Table 1). Since single DNA molecules are interrogated in the HeliScope sequencing process, clonal amplification is not required.

In emulsion PCR, entrapment of a single template DNA molecule and a clone collection bead inside an aqueous bubble dispersed in immiscible liquid is achieved by limiting dilutions. Then, PCR in emulsion clonally amplifies the entrapped DNA molecule and the beads are employed to collect the newly synthesized DNA molecules (Adessi *et al.*, 2000; Dressman *et al.*, 2003; Tawfik and Griffiths, 1998). In the Illumina system, bridge PCR is performed *in vitro* on solid support. Briefly, immobilized DNA fragment bends towards the flow cell surface until the head of the arched DNA molecule hybridizes with an adjacent anchor oligonucleotide to stabilize the molecular bridge structure (Adessi *et al.*, 2000; Fedurco *et al.*, 2006). Then, synthesis reaction is primed by the adaptor oligonucleotide tethering the arched head of the template DNA molecule. The bridge-PCR process can generate clusters of ~1,000 clonally amplified DNA molecules deriving from single immobilized DNA molecules. All clonal amplification methods result in localized and retractable DNA objects, either collection of beads covered with DNA or bridge PCR clusters.

PARALLELIZED SEQUENCING

Next-generation DNA sequencing is dependent on multiple processes. One of the critical components of the system are the fluid dispensers that portion out accurate volumes of sequencing reagents through solid phase support devices supporting the template DNA. By immobilizing an array of DNA templates on a solid surface, enzymatic, cyclic sequencing-by-synthesis or sequencing-by-ligation reactions can be parallelized in single reaction volume. High-performance cameras and sophisticated image processing are integrated with the chemistry core of the sequencing systems for data acquisition.

In Illumina sequencing, DNA templates are immobilized on the surface of a "flow cell". Illumina flow cells are planar, fluidic devices that contain eight separate lanes that can be flushed with sequencing reagents. A field of oligonucleotide anchors that are complementary to the adaptor sequences covers the flow cell surface. Adaptor-tagged DNA fragments are bound to the flow cell surface by hybridization between the anchor and the adaptor sequences. Scattered and immobilized single DNA molecules on the surface of a flow cell are used as substrates for bridge PCR. Over a hundred million randomly scattered clusters, clonal bridge PCR amplicons, can be generated on the surface of an Illumina flow cell.

In 454, SOLiD and Polonator systems, the clonally amplified DNA templates are bound to the surface of nano beads. These DNA covered nano beads can be fixed on porous plates or glass slides. The 454 sequencing platform uses a "PicoTiterPlate", a solid phase support containing over a million picoliter wells (Margulies *et al.*, 2005). The dimensions of the wells are such that only one bead is able to enter each position on the plate. Sequencing chemistry flows through the plate and insular sequencing reactions take place inside the wells. The PicoTiterPlate can be compartmentalized up to 16 separate reaction entities using different gaskets. In SOLiD and Polonator sequencing, clonal bead populations with 3' end modified DNA fragments can be covalently bound to a glass slide or coverslip. High-density arrays with hundreds of millions of beads can be prepared in the SOLiD System and glass slides can be further segmented into eight chambers to facilitate up scaling of the number of analyzed samples. The Polonator strategy involves arraying DNA beads on an amine-functionalized coverslip glass slide (Shendure *et al.*, 2005). The coverslip is then overlaid on a fluidics grid, also called a flow cell, that permits sequencing biochemistry cycles to contact with the DNA covered beads that are attached to the coverslip sealing. In the HeliScope single molecule sequencing technology, billions of individual DNA fragments amended with poly A-tails are randomly attached on a glass surface of the flow cell by hybridization with covalently anchored poly d(T) oligonucleotides (Braslavsky *et al.*, 2003). The HeliScope Sequencer can analyze two flow cells, each containing 25 discrete channels, simultaneously.

ENZYMATIC SEQUENCING REACTIONS

Current next-generation sequencing is based on enzymatic sequencing-by-synthesis or sequencing-by-ligation reactions (Table 1). The process of sequencing-by-synthesis involves cycles of DNA polymerase catalyzing the incorporation of labeled nucleotides to the synthesized DNA chains and chemistry that allows the cleavage of the label before proceeding to the next cycle (Braslavsky *et al.*, 2003; Brenner *et al.*, 2000; Margulies *et al.*, 2005). In systems provided by Illumina, newly synthesized DNA polymer is synchronously extended by one nucleotide at a time using reversible terminators (Bentley *et al.*, 2008; Braslavsky *et al.*, 2003), whereas 454 and Helicos technologies rely on asynchronous extension chemistry. In asynchronous approach, there is no termination moiety that would prevent addition of multiple bases in a homopolymer during one cycle. As result, accurate sequencing through homopolymer stretches (i.e., AAA) represents a challenging technical issue for 454 and Helicos systems (Harris *et al.*, 2008).

Sequencing-by-ligation utilizes the ligase enzyme, which attaches labeled oligonucleotides to the 5' end of the sequencing DNA strand (Brenner *et al.*, 2000; Shendure *et al.*, 2005). A pool of uniquely labeled sequencing oligonucleotides contains all possible variations of the complementary bases for the template sequence. Oligonucleotides are annealed to the template and perfectly matching sequences are ligated. SOLiD technology applies partially degenerate, fluorescently labeled, DNA octamers with dinucleotide complement sequence recognition core. Since ligation based method in the SOLiD system requires complex panel of reporter oligonucleotides and sequencing proceeds by off-set steps, the interpretation of the raw data is not as straightforward as in sequencing-by-synthesis platforms and requires complicated algorithms. However, the SOLiD system achieves higher accuracy due to redundant sequencing of each base by octamer sequencing oligonucleotides with dinucleotide detection cores. Polonator technology uses fully degenerate nonamers in which each nucleotide is labeled using separate fluorophores. These nonamers selectively ligate to anchor primers providing a fluorescence signal that corresponds to a specific base in the interrogated genomic DNA.

HIGH-PERFORMANCE IMAGING AND DATA ACQUISITION

Sequencing reactions are detected by fluorescent labels or by a process called pyrosequencing as utilized by the 454 platform (Table 1) (Margulies *et al.*, 2005). In pyrosequencing, DNA polymerase catalyzed incorporation of an individual nucleotide leads to the release of a pyrophosphate. ATP sulfurylase and luciferase enzymes convert the pyrophosphate into a visible burst of light, which is detected by a CCD imaging system. Other sequencing platforms use fluorescent labels conjugated to nucleotides as specific sequence reporters (Bentley *et al.*, 2008; Braslavsky *et al.*, 2003). There are a variety of fluorescent dyes with different absorption and emission properties, which allows sequencing with multiple labeled nucleotides or oligonucleotides in the same reaction volume during one cycle, while in pyrosequencing, each nucleotide needs to be incorporated in separate cycles.

As result of this cyclic imaging acquisition, next-generation sequencing technologies generate huge amount of raw image data as the full array of synthesis reactions and each labeled nucleotide or oligonucleotide sequence composition are imaged at each cycle. High-performance computational infrastructure is required for organizing and storing the raw image data. These periodical "star field-like" images need to be interpreted into data files that contain the information about the spatial positions of the sequencing features and their signal intensities in tabulated format. Base calling software is then used to combine and analyze these data matrices and transcribe intensity values of spatially localized sequencing features into DNA sequence reads.

Except for the HeliScope, next-generation sequencing processes require clonal amplification of the DNA templates. Since sequence signal is generalized from multiple clonally amplified DNA molecules, off-phasing effects (strand extension by synthesis or ligation does not proceed in concert in all molecules but some molecules are lagging behind or run ahead) become an issue. Off-phasing can be corrected to some extent but eventually it causes sequence quality to drop as cycles proceed and, therefore, majority of these technologies can generate short reads between 25 - 100 nucleotides in length (Table 1). The exception is 454, which represents long read technology that is capable of producing reads of five hundred nucleotides.

Advances continue to be made in the next-generation sequencing field with new sequencing technologies emerging commercially. Companies like Complete Genomics, Oxford Nanopore Technologies and Pacific Biosciences are developing ultra high-density DNA nano-arrays (Complete Genomics, 2009), single molecule, nanopore-based, label-free DNA sequencing applications (Clarke *et al.*, 2009; Stoddart *et al.*, 2009), and single molecule real-time sequencing technologies (Eid *et al.*, 2009; Korlach *et al.*, 2008; Korlach *et al.*, 2008; Levene *et al.*, 2003; Lundquist *et al.*, 2008) that are anticipated to lower the cost of whole genome sequencing by increasing sequencing throughput and read length.

Sequencing the cancer genome

Human cancer represents a heterogeneous group of diseases originating from different tissues. Each type of cancer has its own complement of genetic and genomic aberrations, which contribute to the neoplastic process and affect the cancer phenotype. What is common to all cancers is that explicit changes in the genomic DNA sequence of the cancer cells underlie their pathogenesis. The initiating events can be inherited genetic variants, somatic mutations, copy number aberrations and epigenetic changes. Other critical mutations and genomic aberrations follow afterwards, contributing to the unhampered proliferation of the cancer cells and defining complex cancer phenotypes such as the metastatic spread. For the past decades, central objective of cancer research has been to identify the "cancer genes", mutated genes that are causally implicated in oncogenesis (Futreal *et al.*, 2004) and translate that knowledge into clinically relevant applications. The emergence of next-generation DNA sequencing technologies have made high-throughput identification of somatically acquired genetic alterations in cancer possible. However, the requirement to deep resequence cancer genomes to obtain complete catalogues of somatic mutations has limited the number of analyzed samples to few, although, at the same time, clinical validation and identification of driver mutations with low prevalence necessitate sequencing of hundreds of cancer genomes.

DEEP RESEQUENCING OF THE CANCER GENOME

The next-generation DNA sequence data represents a compilation of individual DNA molecules – multitude of independent, tiny fractions of the entirety of the genome. Yet, the objective of cancer genome sequencing is to reconstruct the original nucleotide order and identify alterations within. *De novo* genome assembly is challenging because combining overlapping short reads is often ambiguous. However, "resequencing" approaches can be efficiently applied to reconstruct genomes, even from fractionated collection of short reads, by aligning them against reference sequences from the human genome. Furthermore, the digital nature of the next-generation DNA sequence data makes it possible to compile independent reads with partially common DNA content. The obtained redundancy in sequence reads, the number of times that a given nucleotide has been sequenced, corresponds to sequencing "depth" or "fold coverage". Although next-generation sequencing platforms have higher error rates and generally, shorter read lengths (Table 1) than Sanger sequencing, increased fold coverage leads to improved accuracy for variant detection.

Practically, "deep resequencing" approach has two contexts. First, it is often applied to identify the candidate genes or regions from large populations. For example, the 1000 Genomes Project is an effort to produce a catalog of variants that are present at one percent or greater frequency in the human population across most of the genome, and down to 0.5 percent or lower within genes (Kuehn, 2008; Wise, 2008). One thousand individuals will have their full genomes sequenced to identify these rare variants. Nonetheless, whole genome sequencing requires high fold coverage. For example, sequencing a normal diploid human genome using Illumina technology required an average sequencing fold coverage of thirty (Bentley *et al.*, 2008) or thirty-six (Wang *et al.*, 2008). The high fold coverage was required for improving sequencing accuracy and compensating for short read length. In comparison, complete human genome sequencing using Sanger (Levy *et al.*, 2007) and 454 (Wheeler *et al.*, 2008) sequencing required 7.5 and 7.4-fold coverage respectively.

Another context for deep resequencing is the application towards cancer genomes where increased sequencing fold coverage improves the sensitivity of detection of somatic cancer mutations. Previously, discovering cancer mutations has relied on direct Sanger sequencing of PCR products with typically a two-pass approach using the forward and reverse strands. However, direct sequencing has disadvantages. First, multiple handling steps involved in PCR increase the rate of amplicon failures, generate artifacts or induce contamination. Second, sequence analysis for detecting mutations is at a disadvantage with heterogeneous tumor samples, which contain a mixture of normal stroma and tumor. For example, the background of normal genomic DNA originating from normal stromal components oftentimes obscures mutations found in primary tumors. At its lowest threshold of sensitivity, direct sequencing requires approximately 30% of mutant allele from the primary sample (Ohnishi *et al.*, 2006). Therefore, any representation of tumor where the cancer cells make up less than 50% of the gross tumor tissue results in obscured cancer mutation detection. Next-generation sequencing technology with its high redundancy of sequence reads generated from the same genomic region can discriminate somatic mutations that would otherwise be impossible to detect with traditional Sanger sequencing (Thomas *et al.*, 2006). For example, Thomas et al. (2005) demonstrated by sequencing simplex PCR products amplifying the *EGFR* gene in lung adenocarcinoma samples, that oversampling of PCR amplicons with a 454 system enabled detection of mutations, which were missed by Sanger sequencing of those same amplicons (Thomas *et al.*, 2005).

LARGE SCALE CANCER GENOME SEQUENCING EFFORTS

Recently, several groups sequenced large number of coding regions from small sample subsets of individual tumors. All of the cited studies used Sanger sequencing. For example, Wood et al. (2008) sequenced exons of 20,857 transcripts, representing 18,191 genes from 11 breast and 11 colorectal cancers and reported 1,718 genes with somatic non-silent mutations (Wood *et al.*, 2007). By evaluating the mutation data they identified 140 candidate cancer genes in both tumor types that were significantly represented above background mutations (Wood *et al.*, 2007). Their study revealed only few frequently mutated genes among abundance of rare variants and the identified gene mutations clustered key biological pathways such as PI3K-mediated signal transduction (Wood *et al.*, 2007). Jones et al. (2008) sequenced 20,661 protein-coding genes in 24 pancreatic cancers and identified 1,327 genes with one mutation and 148 genes with two or more mutations (Jones *et al.*, 2008). An average of 63 genetic alterations were identified in the 24 pancreatic cancers surveyed that clustered in 12 generally defined cellular signaling pathways and biological processes. Similarly, Parsons et al. (2008) sequenced 20,661 protein coding genes in 22 glioblastoma multiforme samples (Parsons *et al.*, 2008). Importantly, they identified frequent (12%) mutations in the *IDH1* (isocitrate dehydrogenase enzyme) gene that were associated with young patients with secondary disease and increased overall survival. Overall, these large-scale investigations show that most of the somatic mutations in cancer genomes are primarily background mutations, otherwise known as passengers and only a handful of genes are commonly mutated.

The results from preexisting studies can be used to curate biologically and clinically relevant compilations of genes that have been previously connected with the disease phenotype. For example, Futreal et al. (2004) conducted a census of published cancer genes based on presence of germline and somatic mutations and found that several hundred genes fulfilled these criteria (Futreal *et al.*, 2004). Although, there is definitely uncharted biology to be discovered in the non-coding human genome sequences as well as in the genes that have not yet been implicated in cancer, this *a priori* knowledge about biological relevance and connection with cancer is valuable when cost-benefit functions of the genomic research designs are being optimized. By reducing the sequenced portion of the genome to the fraction of sequences that represent the majority of critical cancer mutation drivers, it is possible to expand resequencing efforts, covering a larger number of cancer cases and stay within a reasonable budget. For example, using a candidate gene approach, Ding et al. (2008) sequenced 623 cancer genes in 188 primary lung adenocarcinomas and discovered 1,013 non-synonymous somatic mutations in 163 tumors but 26 genes were mutated at statistically significantly high frequencies (Ding *et al.*, 2008). Similarly, the Cancer Genome Atlas consortium sequenced 601 genes in 91 glioblastoma multiforme samples. Out of the total, 224 of these genes contained mutations in at least one sample. However, only eight genes and their mutations were identified to be statistically overrepresented compared to background mutations (The Cancer Genome Atlas Research Network, 2008). Noteworthy, the *IDH1* gene, which was indentified in the sequencing of all coding regions of 22 glioblastoma multiforme samples (Parsons *et al.*, 2008), was omitted in the study conducted by the Cancer Genome Atlas Research Network - since it was not among the list of candidates - underlining the importance of the comprehensive, all coding regions covering mutation screens in identifying the candidate genes. In proportion, using the candidate gene approach and multi-dimensional genomic characterization such as integration with transcriptome and copy number variation surveys, new biological and clinical paradigms of gliomas could be revealed (The Cancer Genome Atlas Research Network, 2008). These complementary efforts to characterize glioblastoma multiforme genomes demonstrate the usefulness of integrating large scale mutation screens and candidate gene approaches in identification of cancer genes.

THE COST OF WHOLE GENOME SEQUENCING IN CANCER – IS IT PRACTICAL?

The first complete human genome sequence was estimated to cost ~$300 million. In 2008, cost of sequencing diploid genome of an individual human using contemporary Sanger capillary electrophoresis technology was ~$10 million (Bentley *et al.*, 2008). Next-generation technologies have dramatically reduced the cost of sequencing a human genome even more. However, whole genome resequencing remains costly with next-generation systems. As mentioned, this continuing high cost with next-generation technology, is attributable to the overall sequence size of the human genome, higher error rates associated with next-generation technology, its short reads and necessity of multi-fold sequencing coverage. For example, sequencing of a diploid genome using Illumina technology was quoted to $250,000 (Bentley *et al.*, 2008). Furthermore, to resequence comprehensively a cancer genome, one needs to also sequence the matched normal DNA to distinguish acquired somatic mutations from the germline variation. Therefore, one needs to resequence over 100 billion bases for a single cancer genome (Stratton *et al.*, 2009).

An example of this whole genome approach and the entailed cost is the recent publication of a cancer genome from an acute myelogeneous leukemia (AML) sample with normal cytogenetics (e.g. diploid chromosome complement without aneuploidy) (Ley *et al.*, 2008). For this cancer genome, the leukemia sample required 32.7-fold sequencing coverage, equivalent to 98 billion bases, and 13.9-fold coverage (41.8 billion bases) for the normal skin sample. Of the 2,647,695 well-supported single nucleotide variants (SNVs) found in the tumor genome, 2,584,418 (97.6%) were also detected in the patient's skin genome, limiting the number of variants that required further studies. The majority of mutations (5) were bystanders and not confirmed in other AML samples. Two mutations were known AML mutations that had previously been discovered. The cost was described as being over $150,000 for the two samples, which obviously limits the application of this type of sequencing approach to large clinical populations.

Ultimately, the discovery and validation of specific cancer mutations as prognostic or predictive diagnostics requires multiple studies be conducted. As a case in point, the detection of *KRAS* mutations from primary colorectal cancers has become standard-of-care diagnostic in clinical oncology (Amado *et al.*, 2008; Di Fiore *et al.*, 2007; Khambata-Ford *et al.*, 2007; Lievre *et al.*, 2008; Personeni *et al.*, 2008). The adoption of *KRAS* mutation testing represents a dramatic example of the enormous utility of a genetic diagnostic as applied to a large clinical population, namely individuals with metastatic colorectal cancer. To reach this point required the combined analysis of nearly 1,500 patients in total among more than six studies (Amado *et al.*, 2008; Di Fiore *et al.*, 2007; Khambata-Ford *et al.*, 2007; Lievre *et al.*, 2008; Personeni *et al.*, 2008). Only then, have clinicians been insured that the test had been sufficiently validated to warrant its use clinically. This example is particularly instructive in showing the case numbers required for clinical validation. Basic researchers may find this level of validation to be surprising, but the entire nature of clinical trials with the inherent complexities of heterogeneous populations requires this level of redundancy.

Targeted resequencing of the cancer genome: a future approach for cancer mutation discovery

As a potential solution to the cost and technical limitations of a whole genome sequencing approach for cancer mutation discovery, strategies for targeting, capturing or partition specific regions of genomic DNA will prove to be increasingly useful and cost-effective, particularly when studying large sample collections of tumors of variable quality. These strategies can be readily integrated with current next-generation sequencing technologies for variant discovery. One way of categorizing these strategies is based on the phase in which the targeted genomic DNA hybridization event takes place. Solid phase technology, as used in microarray-based methods, relies on hybridization between the oligonucleotide probes and the target DNA on a solid-phase support such as a microarray surface. For example, a probe covalently bound to microarray surface will hybridize to its specific target. This enables washing off the non-hybridized genomic material and eluting out the target sequences. Respectively, solution-based technologies apply nested multiplex PCR or oligonucleotide constructs that conjugate designated primer pairs or select genomic DNA by circularization. The enrichment of the target sequences occurs in aqueous solution where DNA polymerase amplifies the genomic target DNA. Hybrid-phase technologies utilize bead technology and DNA hybridization to capture target sequences in a solution. When evaluating and comparing the target DNA capture methods, important parameters to assess are multiplex capacity, specificity, and uniformity. Multiplexing potential of the assay can be evaluated by the size of the targeted genomic region, specificity of the technique can be measured by calculating the proportion of sequences that originate from the targeted DNA and uniformity refers to the evenness and coverage of the targeted regions.

SOLID-PHASE TARGET DNA CAPTURE

Microarray hybridization has been used to selectively capture genomic targets (Albert *et al.*, 2007; Bau *et al.*, 2009; Hodges *et al.*, 2007; Okou *et al.*, 2007) (Figure 1). In each method, the sequencing library preparation is performed by random fragmentation of genomic DNA followed by adaptor ligation. Adapted DNA fragments are then hybridized on a DNA microarray containing probes that correspond to targeted sequences of the human genome. Hodges et al. (2007) applied a set of seven Nimblegen arrays, each containing 385,000 probes positioned in 20 bp intervals and spanning all hu-

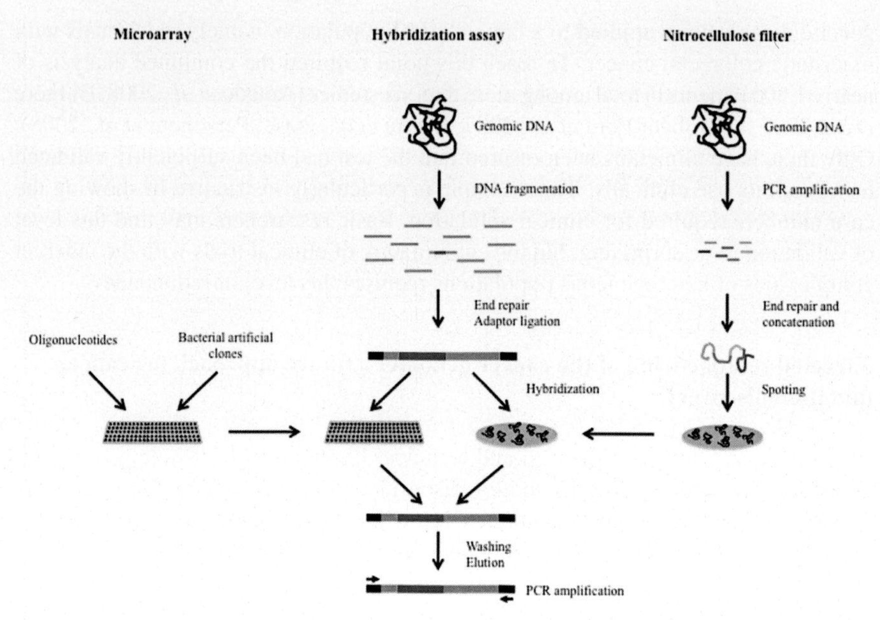

Figure 1. Solid-phase target DNA capture. Oligonucleotides, bacterial artificial clones or PCR products associated with the targeted genomic regions are immobilized on solid surface, either on microarray glass slide or nitrocellulose filter. Target capture is executed by DNA hybridization.

man exons (Hodges *et al.*, 2007) (Table 2). Albert et al. (2007) utilized a single 385k NimbleGen microarray to capture 6,726 genomic regions from 660 genes scattered throughout the human genome (Albert *et al.*, 2007), while Okou et al. (2007) captured X chromosome-linked genomic regions using a bacterial artificial chromosome (BAC) array (Okou *et al.*, 2007). Bau et al. (2009) applied microfluidic microarrays with 12,498 and 13,296 probes (50-mer) and 12-20 nucleotide tiling design to enrich 6,776 features from *BRCA1*, *BRCA2*, and *TP53* genes' sense strands (Bau *et al.*, 2009).

In a variation of the array hybridization assay, Herman et al. (2009) utilized nitrocellulose filter-based capture using PCR products to target genomic subsets (Herman *et al.*, 2009). First, targets representing 47 genes and 115 kb of human genome implicated in cardiovascular disease were amplified using PCR and 154 primer pairs. PCR amplicons were pooled and ligated into large DNA concatemers and isothermally amplified using ϕ29 DNA polymerase. Concatenated and amplified target DNA was bound to nitrocellulose filter, which served as a trap to capture genomic subsets corresponding to the immobilized PCR products. Genomic DNA samples were prepared by shearing and adaptor ligation and hybridized on the filter trap. Non-targeted DNA was washed and target DNA libraries were eluted from the filters. After PCR amplification, target libraries were sequenced using Illumina sequencers.

Microarray-based capture assays offer high level of multiplexity, reaching to cover all coding exons. Hybridization-based capture assays require large quantities of starting material, while the amount of DNA captured using hybridization-based methods are minute. Therefore, amplification steps are often required before hybridization experiment and sequencing. This may not become an impediment for these technologies if amplification methods continue to improve and requirements

Table 2. Overview and performance of the target DNA capture technologies.

Capture phase	Capture application	Number of capture elements	Targeted genomic features	Targeted genomic region	Reads mapping to targeted regions	Targeted bases covered	Reference
Solid	Microarray	2,685,000	200,000 exons	44 Mb	36 – 55 %	40 – 78 %	Hodges *et al.*, 2007
Solid	Microarray	385,000	6,726 exons 660 genes	5 Mb	65 – 77 %	93 – 96 %	Albert *et al.*, 2007
Solid	Microarray	385,000	*FMR1*, *FMR1NB*, *AFF2*	304 kb 1,7 Mb	N/A	N/A	Okou *et al.*, 2007
Solid	Microfluidic microarray	25,794	*BRCA1*, *BRCA2*, *TP53*	185 kb	7 %	46 - 84 %	Bau *et al.*, 2009
Solid	Nitrocellulose filter	154	47 genes implicated in cardiovascular disease	115 kb	58-67%	89 - 94%	Herman *et al.*, 2009
Solution	Nested patch PCR	N/A	Six cancer genes	21.6 kb	90 %	95.7 %	Varley and Mitra, 2008
Solution	Gene-Collector	170	Coding sequences of ten cancer genes	N/A	58 %	90 %	Fredriksson *et al.*, 2007
Solution	Targeted genomic circularization technology	508	Coding sequences of ten cancer genes	49 kb	90 %	74 – 93 %	Dahl *et al.*, 2007
Solution	Molecular inversion probe	55,000	55,000 human exons	6.7 Mb	98 %	18 %	Porreca *et al.*, 2007
Solution	Molecular inversion probe	55,000	55,000 human exons	6.7 Mb	99 %	91 %	Turner *et al.*, 2009
Solution	SMART probe	557	485 Exons 30 kinase genes	N/A	N/A	90 %	Krishnakumar *et al.*, 2008
Hybrid	Biotin-RNA / Streptavidin beads	22,000	15,565 exons 1,900 genes	3.7 Mb	58 %	88 %	Gnirke *at al.*, 2009

of template DNA concentration of the sequencing libraries ease. However, the sample processing, which involves library preparation, clonal amplification, selection of small fraction of the material and another amplification step is likely to induce a phenomenon called "molecular bottlenecking". Molecular bottlenecking refers to a situation where identical DNA sequence reads emerge that correspond to identical DNA molecules, presumably originating from clonal amplification products rather than independent genomic counterparts. Therefore, identical sequences are usually filtered from the data to ensure that each sequence represents unique genomic component. The amount of retained DNA is proportional to the hybridization efficiency and molarity of the arrayed probes, suggesting that high-resolution microarrays capture less material than arrays with larger spotted probe features. The specificity of microarray-based assays as determined by percentage of reads mapping to targeted sequences, varied between 7% and 77% (Table 2). Evidently, off target sequences are carried over since hybridization-based capture harvests DNA fragments that may contain adjacent genomic regions on top of the targeted sequences. Like Hodges et al. (2007) discovered, majority of the off target sequences in exon-wide capture mapped to close proximity of the protein coding sequences (Hodges *et al.*, 2007). In addition, significant portion of off-target sequence reads is likely to be related to non-specific hybridization. Uniformity of the microarray-based assays varied between 40 and 96 % (Table 1). Optimizing the probe composition and oligonucleotide design of the capture microarrays can reduce biases in capture, extend the coverage and improve balance between different genomic regions.

Microarray-hybridization based capture can be multiplexed to hybridize multiple samples as a way of increasing capacity. However, even with this strategy, which remains largely untested for larger number of samples, the cost for array-based targeted capture is still high for individual laboratories to conduct studies on thousands of samples. The Sequence Capture Human Exome 2.1M Array (Roche Nimblegen), which targets 180,000 human protein-coding exons and 700 micro RNA exons, 34 Mb of genomic DNA in total, and has been optimized to integrate with the 454-sequencing platform, costs $2,000 in the US. Microarrays can be stripped for reuse but there is a possibility of residual contaminating genomic material remaining and the performance may vary after stripping. For these reasons, these technologies will probably attract large cancer genome consortia rather than small research units.

IMPROVEMENTS OF MULTIPLEXED PCR FOR TARGETED SELECTION

For large scale projects such as cancer genome resequencing, conventional simplex or multiplex PCR to select and amplify genomic sequences has many limitations. The production of template libraries for sequencing requires large scale synthesis of oligonucleotides and handling of enormous number of simplex PCR reactions. For example, the study by Sjöblom et al. (2006) involved synthesis of 135,483 oligonucleotides and 3 million PCR reactions to produce 465 Mb of sequence (Sjöblom *et al.*, 2006). This approach is dependent on automation infrastructure (e.g. robotics, laboratory information management systems, etc.) and is neither applicable nor cost-effective in individual laboratories. Nonetheless, direct multiplex PCR amplification produces non-specific PCR products that result from non-designated pairing of the primers (Fan *et al.*, 2006) and is often complicated by primer interactions which lead to extensive artifactual products (Han *et al.*, 2006).

As a potential solution to the issues of PCR, there are number of approaches that have been developed. Varley and Mitra introduced a nested patch PCR method to reduce the mispriming effects of multiplex PCR (Varley and Mitra, 2008) (Figure 2A). The nested patch PCR assay has two primer pairs for each amplified locus. First, low cycled PCR reaction using outer marginal primers, where thymidine has been substituted by uracil, is carried out to define the amplicon ends. Then, primer overhangs are cleaved from amplicons by uracil DNA glycosylase. Next, nested patch oligonucleotides are annealed and universal primers are ligated to the amplicon ends. Finally, selected DNA fragments are simultaneously amplified using universal primers. Nested patch PCR was applied to amplify 94 exons from six cancer related genes with high specificity and coverage (Table 2); however, it remains to be seen what level of multiplexing this method can achieve.

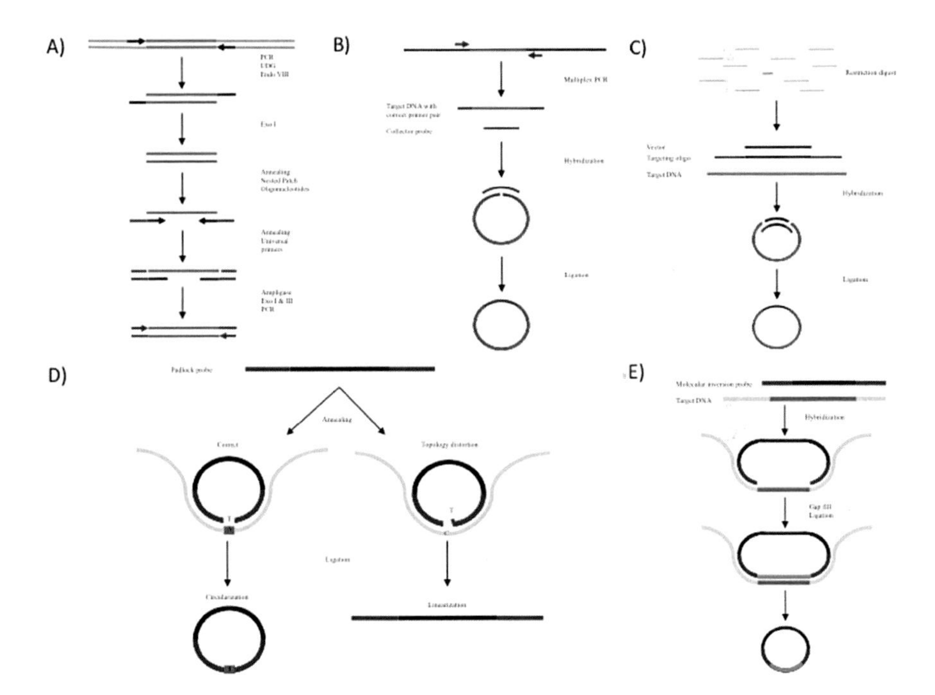

Figure 2. Multiplex PCR and solution-phase target DNA capture. A) Nested Patch PCR utilizes two sets of primers and subsequent PCR reactions to amplify specific target regions. B) The Gene-Collector assay is based on collector oligonucelotide probe, which selectively circularizes correctly amplified PCR products. C) Targeted genomic circularization technology uses "Vector" and "Targeting" oligonucleotides to circularize genomic DNA fragments that have been digested with restriction enzymes. D) Padlock probe and E) Molecular inversion probe assays induce circular DNA molecules by cap filling between specific recognition sites.

Fredriksson et al. (2007) developed the gene-collector assay, which utilizes targeted circularization of the correct multiplex PCR products based on cognate primer pairing (Fredriksson *et al.*, 2007) (Figure 2B). The assay relies on traditional multiplex PCR, after which, the collector probe oligonucleotides are hybridized with ends of the accurately amplified PCR amplicon. A ligase enzyme seals the circular DNA constructs. Non-cognate products remain linear and can be degraded with exonucleases while amplification products of matched primer pairs are circularized and unaffected by degradation. The intact circular target DNA molecules are amplified using random priming and rolling circle amplification that generate linear, concatenated DNA molecules. This PCR circularization assay successfully enriched 90 % of target sequences with close to 60 % specificity. The main limitation for multiplexing the Gene-Collector assay is the number of primer pairs initiating the PCR amplification.

IN SOLUTION BASED TARGETING APPROACHES

Targeted genomic circularization technology involves restriction enzyme mediated linearization and genomic target DNA circularization using specific oligonucleotide constructs (Dahl *et al.*, 2005; Dahl *et al.*, 2007; Fredriksson *et al.*, 2007). The process is initiated by restriction enzyme fragmentation of the genomic DNA (Figure 2C). Targeted genomic circularization oligonucleotides (~80 nucleotides) contain two nucleotide sequences (~ 20 nucleotides each), designed to hybridize with the restriction enzyme-cut ends of the targeted DNA fragments and a connector sequence (~ 40 nucleotides) that is complementary with a linker sequence. Hybridization between targeted genomic circularization oligonucleotide and the target DNA fragment result in circular molecule that can be completed using a general linker sequence (~ 40 nucleotides) and a ligase enzyme activity. Circularized targets can be amplified using one universal primer pair specific for the linker sequence. Dahl et al. (2007) amplified and sequenced 10 genes (177 exons) from colorectal cancer samples using 508 targeted genomic circularization oligonucleotides (Dahl *et al.*, 2007) (Table 2). Targeted genomic circularization assay provided high coverage of the targeted genomic regions as the drop-out rate was only 10 %. This proof-of-concept study shows that the targeted genomic circularization technology can be reliably used to genotype cancer samples and suggests that comprehensive multiplexing is potential.

As another in-solution approach, padlock or molecular inversion probes (MIP) - long oligonucleotides that target specific genomic regions - are used to reduce non-specific PCR priming by conjugating primer pairs and generating circular DNA elements that are protected from exonuclease activity (Figure 2). A padlock probe is composed of target complementary sequences that flank a linker sequence and can be circularized in the presence of a target DNA sequence and a ligase enzyme (Nilsson *et al.*, 1994) (Figure 2D). Padlock probes have been utilized to multiplex single nucleotide polymorphism genotyping (Hardenbol *et al.*, 2003; Hardenbol *et al.*, 2005) but, instead of single nucleotide, the target site can be designed to capture larger genomic region. As a variation of the padlock probe strategy, MIPs are oligonucleotides that contain two 20-nucleotide elements that hybridize to sequences flanking the genomic region of interest, a restriction site for probe linearization and universal primers for inverted target amplification (Akhras *et al.*, 2007) (Figure 2E). Porreca et al. (2007) utilized array-based synthesis approach and generated 55,000 oligonucleotides (100-mers) that were restriction digested to release 70-mer "capture probes", each targeting a unique human exon (Porreca *et al.*, 2007). Then, MIP mediated circularization and target DNA gap extension by a polymerase were used to amplify 6.7 Mb of genomic DNA in single reaction volume.

This first generation of highly multiplex MIP assay suffered from poor uniformity as detection of 80 % of the target nucleotides failed (Porreca *et al.*, 2007) (Table 2). Turner et al. (2009) optimized the MIP assay by increasing hybridization and gap-fill incubation times as well as oligonucleotide and ligase concentrations and succeeded in detecting 91 % of the targeted exons (Turner *et al.*, 2009). Moreover, improvements in the protocol resulted in equal representation of heterozygous alleles, which subsequently led to improve sensitivity for variant detection. Krishnakumar et al. (2008) modified the MIP assay by lengthening the probes to over 300-mer nucleotide length and optimizing the reaction conditions during the gap fill step (Krishnakumar *et al.*, 2008). Referred to as the SMART (Spacer Multiplex Amplification Reaction) amplification strategy, they demonstrated improvements in the uniformity and increased target coverage compared to shorter oligonucleotides. For their study, they analyzed 485 exons using 557 SMART probes and succeeded to capture over 90 % of the targeted DNA sequences based on sequencing microarray hybridization.

Overall, solution-phase technologies provide good coverage of the targeted sequences, while having high specificity and low fraction of off-target sequences. Solution-phase assays are based on polymerase or ligase driven amplification of the selected genomic components. It remains to be seen if different solution-phase strategies differ in characteristics of the molecular bottlenecks. Targeted genomic circularization design is dependent on genomic distribution of restriction sites, whereas MIP-based methods offer more design flexibility. Furthermore, the number of restriction enzymes used in the design determines the number of reactions in the targeted genomic circularization protocol, while MIP targeting can be executed in a single volume. Uniformity of the captured DNA - high variation between molarities of captured sequences - as well as low multiplexing capacity remains major challenges for solution-phase assays. Novel circularization strategies and improvements in the oligonucleotide production capabilities may offer more scalability in the next-generation of the solution-phase assays. Major advantages of solution-phase assays are that they can be easily customized and after initial investment of generating oligonucleotide libraries, experiment costs are low, potentially reaching $10 per sample.

HYBRID-PHASE TARGET DNA CAPTURE

Hybrid-phase targeting methods take advantage of the multiplexing capabilities of the hybridization-based approach and the flexibility provided by the solution-based applications (Figure 3). Direct genomic selection using biotin-labeled DNA fragments and streptavidin coated microbeads has been introduced to capture genomic target DNA (Bashiardes *et al.*, 2005). Gnirke et al. (2009) modified this approach by utilizing 170-mer biotin-labeled RNA baits (Gnirke *et al.*, 2009) instead of BAC clones, applied by Bashiardes et al. (2005) (Bashiardes *et al.*, 2005). Gnirke and co-workers targeted 15,000 coding exons and four genomic regions using 22,000 baits. The baits were generated by synthesizing 200-mer pre-bait oligonucleotides in parallel on an Agilent microarray production platform. Pre-bait oligonucleotide constructs contained primer sequences of 15 bases separated by a 170-mer bait sequence. Target-specific core sequences were amplified, labeled and transcribed to single strand, biotinylated RNA using PCR, *in vitro* transcription and biotin-UTPs. Using array-based oligonucleotide production enabled large-scale bait oligonucleotide production. Hybrid-phase strategy can be multiplexed to cover sub-set of human exon sequences and manages to capture target DNA with high specificity and coverage (Table 2). Agilent Technologies (Santa Clara, CA) has commercialized the hybrid-phase target DNA capture assay and recently launched SureSelect Target Enrichment System, which is based on biotinylated RNA library of 55,000 baits (120 bp) and hybridization and bead collection in solution. The SureSelect assay has been streamlined to connect with the Illumina sequencing system and captures a subset of human exons (3 Mb) with a cost of $1,100 per sample.

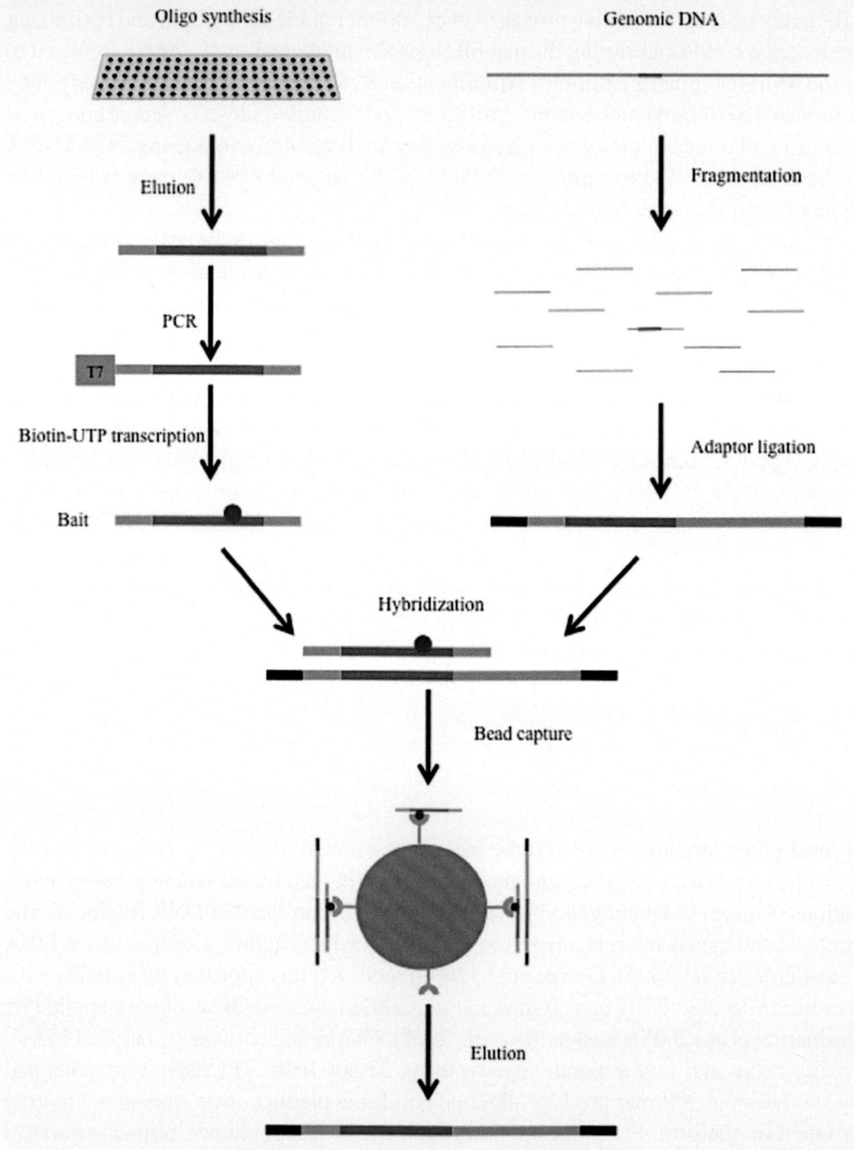

Figure 3. Hybrid-phase target DNA capture. Microarray production platform is used to generate oligonucleotides that are further processed to biotinylated RNA baits. Baits are hybridized with genomic DNA fragments that have been ligated with sequencing adaptors. Streptavidin coated beads are then used to collect selected DNA fragments.

ANALYSIS OF TARGETED RESEQUENCING DATA

Even at this early stage of commercial availability of next-generation sequencing technology, a variety of software tools are available for analyzing resequencing data from targeted sequences. Short read sequences pose a unique challenge in detecting cancer mutations, particularly when they come from single-end alignment (Levy *et al.*, 2007). Aligners like BLAST (Altschul *et al.*, 1997) or BLAT (Kent, 2002) are adequate for longer reads such as those generated by conventional Sanger sequencing, but are not optimal for handling short read sequence data and detecting insertions and deletions. An increasing number of alignment tools have been developed specifically for rapid alignment of large sets of short reads, while allowing for mismatches and/or gaps. Some of these tools take advantage of well-established alignment algorithms such as Smith-Waterman. To improve variant discovery, alignment software is increasingly taking into account the estimated quality of the underlying data in generating read-placements. This is the case with a number of short read sequence alignment tools for single read alignment have been specially developed for next-generation sequencers, including ELAND (Cox, unpublished software, Illumina), SeqMap (Jiang and Wong, 2008) and MAQ (Li *et al.*, 2008). Many of these alignment and variation discovery tools work with data from the Illumina and SOLiD platforms. Finally, identification of somatic mutations requires normal tissue sequencing to eliminate germline variants, which are omnipresent in either tumor or normal tissue.

Conclusions

Three integral parameters of cancer genome research designs are number of analyzed samples, sequencing depth and genomic coverage (Figure 4). Main factor in this cost-benefit equation is the enormous size of the human genome and cancer genome resequencing efforts are often hampered by inability to analyze large-enough sample cohorts or reach adequate sequencing depth to identify variants. Targeted resequencing approaches integrate the potential to capture, partition, index or target genomic DNA subsets with power of the next-generation sequencing technologies. This snap-shot resequencing offers cost-effective and feasible alternative to shot gun sequencing for many biomedical applications that require, from statistical perspective, high-throughput deep resequencing of large gene sets from thousands of samples. Although resequencing entire cancer genomes provides valuable information by systematically cataloguing mutations, reducing the complexity of the sequencing libraries and focusing the sequencing efforts towards genomic regions with highest clinical and biological interest, provides myriad of alternative approaches to sleuth the details of the cancer genomes. In the near future, targeted resequencing studies of large clinical sample cohorts will be accessible and, undoubtedly, advance our understanding of the molecular genetics and genomics of a variety of human cancers. Hopefully, that new insight and deeper knowledge will revolutionize the treatment of these grave diseases.

Acknowledgements

This work was supported by grants from the NIH (2P01HG000205 to HPJ, R21CA12848, HPJ), the Reddere Foundation (to HPJ), the Liu Bie Ju Cha and Family

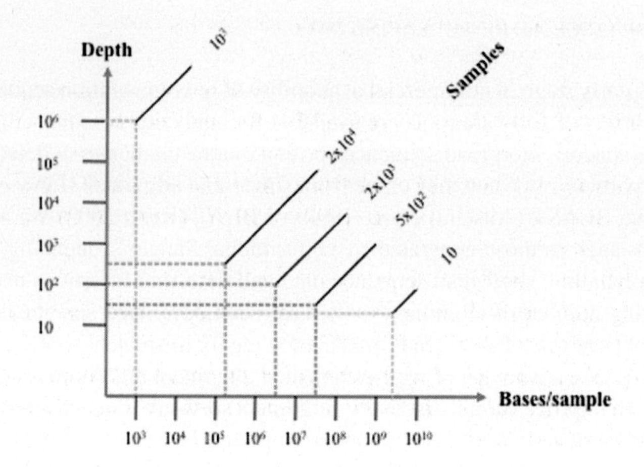

Figure 4. Experimental designs of 1 Tb sequencing projects. Experimental dimensions of next-generation sequencing projects of the whole cancer genome (green), protein coding exons (blue), cancer genes (red), 25 target genes (purple) and mutation biomarkers from circulating tumor cells (yellow) are illustrated in the figure. When sample number increase less stringent criteria for sample inclusion can be maintained, suggesting that more sequencing depth is required for large sample size studies. Moreover, sample multiplexing necessitates sequencing barcode sequences, which also is reflected in the required sequencing depth in up-scaled projects.

Fellowship in Cancer (to HPJ), the Wang Family Foundation (HPJ), Sigrid Jusélius Foundation (SM), Academy of Finland (SM), the Lymphoma and Leukemia Society (HPJ) and Howard Hughes Medical Foundation (HPJ).

References

ADESSI, C., MATTON, G., AYALA, G. *ET AL.* (2000). Solid phase DNA amplification: characterisation of primer attachment and amplification mechanisms. *Nucleic Acids Research* **28**, e87.

AKHRAS, M., UNEMO, M., THIYAGARAJAN, S. *ET AL.* (2007). Connector inversion probe technology: a powerful one-primer multiplex DNA amplification system for numerous scientific applications. *PLoS ONE* **2**, e915.

ALBERT, T., MOLLA, M., MUZNY, D. *ET AL.* (2007). Direct selection of human genomic loci by microarray hybridization. *Nature Methods* **4**, 903-5.

ALTSCHUL, S.F., MADDEN, T.L., SCHAFFER, A.A. *ET AL.* (1997). Gapped BLAST and PSI-BLAST: a new generation of protein database search programs. *Nucleic Acids Research* **25**, 3389-402.

AMADO, R.G., WOLF, M., PEETERS, M. *ET AL.* (2008). Wild-type KRAS is required for panitumumab efficacy in patients with metastatic colorectal cancer. *Journal of Clinical Oncology* **26**, 1626-34.

BASHIARDES, S., VEILE, R., HELMS, C. *ET AL.* (2005). Direct genomic selection. *Nature Methods* **2**, 63-9.

BAU, S., SCHRACKE, N., KRÄNZLE, M. *ET AL.* (2009). Targeted next-generation sequencing by specific capture of multiple genomic loci using low-volume microfluidic DNA arrays. *Analytical and Bioanalytical Chemistry* **393**, 171-5.

BENTLEY, D., BALASUBRAMANIAN, S., SWERDLOW, H. *ET AL.* (2008). Accurate whole human genome sequencing using reversible terminator chemistry. *Nature* **456**, 53-9.

BENTLEY, D.R. (2006). Whole-genome re-sequencing. *Current Opinion in Genetics & Development* **16**, 545-52.

BINLADEN, J., GILBERT, M.T., BOLLBACK, J.P. *ET AL.* (2007). The use of coded PCR primers enables high-throughput sequencing of multiple homolog amplification products by 454 parallel sequencing. *PLoS ONE* **2**, e197.

BRASLAVSKY, I., HEBERT, B., KARTALOV, E. AND QUAKE, S.R. (2003). Sequence information can be obtained from single DNA molecules. *Proceeding of the National Academy of Sciences of USA* **100**, 3960-4.

BRENNER, S., JOHNSON, M., BRIDGHAM, J. *ET AL.* (2000). Gene expression analysis by massively parallel signature sequencing (MPSS) on microbead arrays. *Nature Biotechnol*ogy **18**, 630-4.

CAMPBELL, P., STEPHENS, P., PLEASANCE, E. *ET AL.* (2008). Identification of somatically acquired rearrangements in cancer using genome-wide massively parallel paired-end sequencing. *Nature Genetics* **40**, 722-9.

CLARKE, J., WU, H.C., JAYASINGHE, L. *ET AL.* (2009). Continuous base identification for single-molecule nanopore DNA sequencing. *Nature Nanotechnology* **4**, 265-70.

COMPLETE GENOMICS (2009). Complete Human Genome Sequencing Technology Overview. http://www.completegenomics.com/pages/materials/CompleteGenomicsTechnologyPaper.pdf

DAHL, F., GULLBERG, M., STENBERG, J. *ET AL.* (2005). Multiplex amplification enabled by selective circularization of large sets of genomic DNA fragments. *Nucleic Acids Research* **33**, e71.

DAHL, F., STENBERG, J., FREDRIKSSON, S. *ET AL.* (2007). Multigene amplification and massively parallel sequencing for cancer mutation discovery. *Proceeding of the National Academy of Sciences of USA* **104**, 9387-92.

DI FIORE, F., BLANCHARD, F., CHARBONNIER, F. *ET AL.* (2007). Clinical relevance of KRAS mutation detection in metastatic colorectal cancer treated by Cetuximab plus chemotherapy. *British Journal of Cancer* **96**, 1166-9.

DING, L., GETZ, G., WHEELER, D. *ET AL.* (2008). Somatic mutations affect key pathways in lung adenocarcinoma. *Nature* **455**, 1069-75.

DRESSMAN, D., YAN, H., TRAVERSO, G. *ET AL.* (2003). Transforming single DNA molecules into fluorescent magnetic particles for detection and enumeration of genetic variations. *Proceeding of the National Academy of Sciences of USA* **100**, 8817-22.

EID, J., FEHR, A., GRAY, J. *ET AL.* (2009). Real-time DNA sequencing from single polymerase molecules. *Science* **323**, 133-8.

FAN, J.B., CHEE, M.S. AND GUNDERSON, K.L. (2006). Highly parallel genomic assays. *Nature Review Genetics* **7**, 632-44.

FEDURCO, M., ROMIEU, A., WILLIAMS, S. *ET AL.* (2006). BTA, a novel reagent for DNA attachment on glass and efficient generation of solid-phase amplified DNA colonies. *Nucleic Acids Research* **34**, e22.

FREDRIKSSON, S., BANÉR, J., DAHL, F. *ET AL.* (2007). Multiplex amplification of all coding sequences within 10 cancer genes by Gene-Collector. *Nucleic Acids Research* **35**, e47.

FUTREAL, P.A., COIN, L., MARSHALL, M. *ET AL.* (2004). A census of human cancer genes.

Nature Reviews Cancer **4**, 177-83.

Ghadessy, F.J., Ong, J.L. and Holliger, P. (2001). Directed evolution of polymerase function by compartmentalized self-replication. *Proceeding of the National Academy of Sciences of USA* **98**, 4552-7.

Gnirke, A., Melnikov, A., Maguire, J. *et al.* (2009). Solution hybrid selection with ultra-long oligonucleotides for massively parallel targeted sequencing. *Nature Biotechnology* **27**, 182-9.

Han, J., Swan, D.C., Smith, S.J. *et al.* (2006). Simultaneous amplification and identification of 25 human papillomavirus types with Templex technology. *Journal of Clinical Microbiology* **44**, 4157-62.

Hardenbol, P., Banér J., Jain, M. *et al.* (2003). Multiplexed genotyping with sequence-tagged molecular inversion probes. *Nature Biotechnology* **21**, 673-8.

Hardenbol, P., Yu, F., Belmont, J. *et al.* (2005). Highly multiplexed molecular inversion probe genotyping: over 10,000 targeted SNPs genotyped in a single tube assay. *Genome Research* **15**, 269-75.

Harris, T.D., Buzby, P.R., Babcock, H. *et al.* (2008). Single-molecule DNA sequencing of a viral genome. *Science* **320**, 106-9.

Herman, D.S., Hovingh, G.K., Iartchouk, O. *et al.* (2009). Filter-based hybridization capture of subgenomes enables resequencing and copy-number detection. *Nature Methods* **6**, 507-10.

Hodges, E., Xuan, Z., Balija, V. *et al.* (2007). Genome-wide in situ exon capture for selective resequencing. *Nature Genetics* **39**, 1522-7.

Holt, R. and Jones, S. (2008). The new paradigm of flow cell sequencing. *Genome Research* **18**, 839-46.

Jiang, H. and Wong, W.H. (2008). SeqMap : mapping massive amount of oligonucleotides to the genome. *Bioinformatics* **24**, 2395-6.

Jones, S., Zhang, X., Parsons, D.W. *et al.* (2008). Core signaling pathways in human pancreatic cancers revealed by global genomic analyses. *Science* **321**, 1801-6.

Kent, W.J. (2002). BLAT--the BLAST-like alignment tool. *Genome Research* **12**, 656-64.

Khambata-Ford, S., Garrett, C., Meropol, N. *et al.* (2007). Expression of epiregulin and amphiregulin and K-ras mutation status predict disease control in metastatic colorectal cancer patients treated with cetuximab. *Journal of Clinical Oncology* **25**, 3230-7.

Kinzler, K.W., Nilbert, M.C., Su, L.K. *et al.* (1991). Identification of FAP locus genes from chromosome 5q21. *Science* **253**, 661-5.

Korbel, J., Urban, A., Affourtit, J. *et al.* (2007). Paired-end mapping reveals extensive structural variation in the human genome. *Science* **318**, 420-6.

Korlach, J., Bibillo, A., Wegener, J. *et al.* (2008). Long, processive enzymatic DNA synthesis using 100% dye-labeled terminal phosphate-linked nucleotides. *Nucleosides, Nucleotides and Nucleic Acids* **27**, 1072-83.

Korlach, J., Marks, P., Cicero, R. *et al.* (2008). Selective aluminum passivation for targeted immobilization of single DNA polymerase molecules in zero-mode waveguide nanostructures. *Proceeding of the National Academy of Sciences of USA* **105**, 1176-81.

Krishnakumar, S., Zheng, J., Wilhelmy, J. *et al.* (2008). A comprehensive assay for targeted multiplex amplification of human DNA sequences. *Proceeding of the*

National Academy of Sciences of USA **105**, 9296-301.

KUEHN, B.M. (2008). 1000 Genomes Project promises closer look at variation in human genome. *Journal of the American Medical Association* **300**, 2715.

LEVENE, M., KORLACH, J., TURNER, S. *ET AL.* (2003). Zero-mode waveguides for single-molecule analysis at high concentrations. *Science* **299**, 682-6.

LEVY, S., SUTTON, G., NG, P. *ET AL.* (2007). The diploid genome sequence of an individual human. *PLoS Biology* **5**, e254.

LEY, T.J., MARDIS, E.R., DING, L. *ET AL.* (2008). DNA sequencing of a cytogenetically normal acute myeloid leukaemia genome. *Nature* **456**, 66-72.

LI, H., RUAN, J. AND DURBIN, R. (2008). Mapping short DNA sequencing reads and calling variants using mapping quality scores. *Genome Research* **18**, 1851-8.

LIÉVRE, A., BACHET, J., BOIGE, V. *ET AL.* (2008). KRAS mutations as an independent prognostic factor in patients with advanced colorectal cancer treated with cetuximab. *Journal of Clinical Oncology* **26**, 374-9.

LUNDQUIST, P., ZHONG, C., ZHAO, P. *ET AL.* (2008). Parallel confocal detection of single molecules in real time. *Optics Letters* **33**, 1026-8.

MARGULIES, M., EGHOLM, M., ALTMAN, W.E. *ET AL.* (2005). Genome sequencing in microfabricated high-density picolitre reactors. *Nature* **437**, 376-80.

MEYER, M., STENZEL, U., MYLES, S. *ET AL.* (2007). Targeted high-throughput sequencing of tagged nucleic acid samples. *Nucleic Acids Research* **35**, e97.

THE CANCER GENOME ATLAS RESEARCH NETWORK (2008). Comprehensive genomic characterization defines human glioblastoma genes and core pathways. *Nature* **455**, 1061-8.

NIELSEN, K.L., HOGH, A.L. AND EMMERSEN, J. (2006). DeepSAGE--digital transcriptomics with high sensitivity, simple experimental protocol and multiplexing of samples. *Nucleic Acids Research* **34**, e133.

NILSSON, M., MALMGREN, H., SAMIOTAKI, M. *ET AL.* (1994). Padlock probes: circularizing oligonucleotides for localized DNA detection. *Science* **265**, 2085-8.

OHNISHI, H., OHTSUKA, K., OOIDE, A. *ET AL.* (2006). A simple and sensitive method for detecting major mutations within the tyrosine kinase domain of the epidermal growth factor receptor gene in non-small-cell lung carcinoma. *Diagnostic Molecular Pathology* **15**, 101-8.

OKOU, D., STEINBERG, K., MIDDLE, C. *ET AL.* (2007). Microarray-based genomic selection for high-throughput resequencing. *Nature Methods* **4**, 907-9.

PARSONS, D., JONES, S., ZHANG, X. *ET AL.* (2008). An integrated genomic analysis of human glioblastoma multiforme. *Science* **321**, 1807-12.

PERSONENI, N., FIEUWS, S., PIESSEVAUX, H. *ET AL.* (2008). Clinical usefulness of EGFR gene copy number as a predictive marker in colorectal cancer patients treated with cetuximab: a fluorescent in situ hybridization study. *Clinical Cancer Research* **14**, 5869-76.

PORRECA, G., ZHANG, K., LI, J. *ET AL.* (2007). Multiplex amplification of large sets of human exons. *Nature Methods* **4**, 931-6.

SHENDURE, J. AND JI, H. (2008). Next-generation DNA sequencing. *Nature Biotechnology* **26**, 1135-45.

SHENDURE, J., PORRECA, G.J., REPPAS, N.B. *ET AL.* (2005). Accurate multiplex polony sequencing of an evolved bacterial genome. *Science* **309**, 1728-32.

SJÖBLOM, T., JONES, S., WOOD, L. *ET AL.* (2006). The consensus coding sequences of human breast and colorectal cancers. *Science* **314**, 268-74.

Smith, D.R., Quinlan, A.R., Peckham, H.E. *et al.* (2008). Rapid whole-genome mutational profiling using next-generation sequencing technologies. *Genome Research* **18**, 1638-42.

Stoddart, D., Heron, A.J., Mikhailova, E. *et al.* (2009). Single-nucleotide discrimination in immobilized DNA oligonucleotides with a biological nanopore. *Proceeding of the National Academy of Sciences of USA* **106**, 7702-7.

Stratton, M.R., Campbell, P.J. and Futreal, P.A. (2009). The cancer genome. *Nature* **458**, 719-24.

Tawfik, D.S. and Griffiths, A.D. (1998). Man-made cell-like compartments for molecular evolution. *Nature Biotechnol*ogy **16**, 652-6.

Thomas, R.K., Greulich, H., Yuza, Y. *et al.* (2005). Detection of oncogenic mutations in the EGFR gene in lung adenocarcinoma with differential sensitivity to EGFR tyrosine kinase inhibitors. *Cold Spring Harbor Symposia on Quantitative Biology* **70**, 73-81.

Thomas, R.K., Nickerson, E., Simons, J.F. *et al.* (2006). Sensitive mutation detection in heterogeneous cancer specimens by massively parallel picoliter reactor sequencing. *Nature Medicine* **12**, 852-855.

Turner, E.H., Lee, C., Ng, S.B. *et al.* (2009). Massively parallel exon capture and library-free resequencing across 16 genomes. *Nature Methods* **6**, 315-6.

Varley, K.E. and Mitra, R.D. (2008). Nested Patch PCR enables highly multiplexed mutation discovery in candidate genes. *Genome Research* **18**, 1844-50.

Vogelstein, B. and Kinzler, K.W. (1992). p53 function and dysfunction. *Cell* **70**, 523-6.

Wang, J., Wang, W., Li, R. *et al.* (2008). The diploid genome sequence of an Asian individual. *Nature* **456**, 60-5.

Wheeler, D., Srinivasan, M., Egholm, M. *et al.* (2008). The complete genome of an individual by massively parallel DNA sequencing. *Nature* **452**, 872-6.

Wise, J. (2008). Consortium hopes to sequence genome of 1000 volunteers. *British Medical Journal* **336**, 237.

Wood, L., Parsons, D., Jones, S. *et al.* (2007). The genomic landscapes of human breast and colorectal cancers. *Science* **318**, 1108-13.

Biotechnology and Genetic Engineering Reviews - Vol. 27, 159-184 (2010)

Quorum sensing: implications on Rhamnolipid biosurfactant production

DEVENDRA H. DUSANE[1], SMITA S. ZINJARDE[1], VAYALAM P. VENUGOPALAN[2], ROBERT J.C. MCLEAN[3], MARY M. WEBER[3] AND PATTANATHU K.S.M. RAHMAN[4*]

[1]Institute of Bioinformatics and Biotechnology, University of Pune, Pune-411 007 India, [2]Biofouling and Biofilm Processes Section, Water and Steam Chemistry Division, BARC Facilities, Kalpakkam-603 102 India, [3]Department of Biology, Texas State University-San Marcos, 601 University Drive, San Marcos, TX 78666, USA and [4]Chemical and Bioprocess Engineering Group, School of Science and Engineering, Teesside University, Middlesbrough-TS13BA, UK

Abstract

Quorum sensing (QS) has received significant attention in the past few decades. QS describes population density dependent cell to cell communication in bacteria using diffusible signal molecules. These signal molecules produced by bacterial cells, regulate various physiological processes important for social behavior and pathogenesis. One such process regulated by quorum sensing molecules is the production of a biosurfactant, rhamnolipid. Rhamnolipids are important microbially derived surface active agents produced by *Pseudomonas* spp. under the control of two interrelated quorum sensing systems; namely *las* and *rhl*. Rhamnolipids possess antibacterial, antifungal and antiviral properties. They are important in motility, cell to cell interactions, cellular differentiation and formation of water channels that

*To whom correspondence may be addressed (p.rahman@tees.ac.uk)

Abbreviations: QS: quorum sensing; AI: autoinducer; HSL: homoserine lactone; AHL: acyl homoserine lactone; PQS: pseudomonas quinolone signal; sRNA: small ribonucleic acid; RL: rhamnolipid; HAA: 3-(3-hydroxyalkanoyloxy) alkanoic acid; CTAB: cetyl trimethyl ammonium bromide; TLC: thin layer chromatography; GC: gas chromatography; MS: mass spectroscopy; FTIR: fourier transform infrared spectroscopy; ATR: attenuated total reflectance; NMR: nuclear magnetic resonance; SDS: sodium dodecyl sulfate; CMC: critical micelle concentration; LPS: lipopolysaccharide; TMV: tobacco mosaic virus; PAI: pseudomonas autoinducer; FAD: flavin adenine dinucleotide; NAD: nicotinamide adenine dinucleotide; TDP: thiamine di-phosphate; mN/m: milli-Newton per meter are characteristics of *Pseudomonas* biofilms. Rhamnolipids have biotechnological applications in the uptake of hydrophobic substrates, bioremediation of contaminated soils and polluted waters. Rhamnolipid biosurfactants are biodegradable as compared to chemical surfactants and hence are more preferred in environmental applications. In this review, we examine the biochemical and genetic mechanism of rhamnolipid production by *P. aeruginosa* and propose the application of QS signal molecules in enhancing the rhamnolipid production.

Introduction

Quorum sensing (QS) is the mechanism by which bacteria engage in cell-to-cell communication using diffusible molecules based on a critical cell density (Williams *et al.*, 2007). When the cell density increases these molecules referred to variously as autoinducers (Fuqua *et al.*, 1997, Kleerebezem *et al.*, 1997, Williams and Camara, 2009), pheromones or quoromones are produced that dictate the behavior of bacterial populations. QS signaling molecules, control diverse physiological processes; some of which are inter-related and under the control of multifaceted QS systems. For instance, in *P. aeruginosa,* exo-polysaccharide production (Davies *et al.*, 1998), antibiotic resistance (Bjarnsholt *et al.*, 2005) and biofilm formation (Davies *et al.*, 1998, Hentzer *et al.*, 2001) are all under the control of QS molecules. In addition to the aforementioned examples, certain *Pseudomonas* spp. also produces a surface active agent, viz. rhamnolipid, the production of which is regulated by QS molecules (Pearson *et al.*, 1997).

Rhamnolipids have been extensively studied due to their antibacterial, antifungal and antiviral properties (Haferburg *et al.*, 1987, Stanghellini and Miller, 1997, Syldatk *et al.*, 1985). They are important in bacterial cell motility, cell to cell interactions, cellular differentiation and formation of water channels that are characteristics of *Pseudomonas* biofilms. Rhamnolipids also enable *Pseudomonas* spp. to access poorly soluble hydrophobic carbon sources and thereby facilitate their uptake (Maier and Soberon-Chavez, 2000, Nealson *et al.*, 1970). These properties have encouraged the use of rhamnolipid compounds in environmental bioremediation of contaminated soils and polluted waters. In the medical scenario, they are important as antimicrobials, healing of wounds and in organ transplants (Tatjana and Goran, 2007). Apart from the above applications, rhamnolipids are also used in cosmetics, pesticide removal, pharmaceutical, oil sludge recovery, enhanced oil recovery, household cleaning, agriculture and food industry. Morever, rhamnolipids are biodegradable and less toxic than many synthetic surfactants, and hence their use is highly favored (Hommel, 1990, Volkering *et al.*, 1995).

In this review, we will focus on quorum sensing in detail and describe its role in rhamnolipid production, with particular reference to *Pseudomonas aeruginosa.* The role of quorum signaling in rhamnolipid biosynthesis, bacterial physiology and ecology is described. We have also discussed the application of quorum signaling molecules in enhancing the production of rhamnolipids.

Quorum Sensing

QUORUM SENSING MOLECULES

Quorum sensing has received a great deal of attention, primarily due to the diverse roles it plays in regulating bacterial physiology (Miller and Bassler, 2001, Waters and Bassler, 2005). QS implies that bacteria sense each other by detecting a threshold accumulation of the secreted signals. The signal molecules are well documented in both Gram positive and Gram negative bacterial species. However, there seems to be a significant difference in the signal molecules amongst these bacterial groups. In the Gram-positive bacteria, QS is associated with a number of linear and post-translationally modified peptide based signal molecules, such as the peptide lactones and peptide thiolactones which are found in *Bacillus subtilis*, *Enterococcus* spp., and *Staphylococcus aureus*. The chemical structures of the Gram-positive QS peptides vary greatly in the number of residues and the type of modifications. The biosynthesis pathways are however more complex in Gram-positive bacteria than the AHL molecules in Gram-negative bacteria, because of the post translational modifications of the peptides and their inability to diffuse across the membranes. Interestingly, to date the largest studied most complex peptide signal molecules produced by a few Gram-positive bacterial species are the lantibiotics. These molecules possess antimicrobial activity, as shown by nisin produced by *Lactobactococcus lactis* (Lubelski *et al.*, 2008). Another emerging class of compounds in *Staphylococcus aureus*; *Enterococcus faecalis*; *Listeria monocytogenes* and other *Staphylococci* are the type I autoinducing peptide. These QS molecules play an important role in the Gram-positive bacterial physiology (Miller and Bassler, 2001, Waters and Bassler, 2005).

In Gram-negative bacteria, the regulation of quorum sensing is under the control of the autoinducer (AI) molecules. These AI molecules belong to the biochemical class of acyl homoserine lactones, which are lipophilic in nature. Homoserine lactones are derived from S-adenosyl-methionine, which is one of the substrates for AHL synthesis and consists of a hydrophilic homoserine lactone head group and a hydrophobic acyl side chain that varies based on species. The side chain ranges from 4 to 18 carbons, with the most significant divergence in length and chemical composition occurring at the third carbon. These alterations in structure act to provide specificity to QS signals and facilitate communication between bacteria. The quorum sensing molecule, designated as autoinducer 1 (AI-1), includes *Lux* based quorum sensing systems present in Gram-negative bacteria such as *Agrobacterium tumefaciens*, *Pseudomonas aeruginosa* and *Vibrio fischeri* (Engebrecht *et al.*, 1983, Fuqua *et al.*, 1994). Quorum sensing was first characterized in the marine bacteria *Vibrio harveyi* and *Vibrio fischeri* (Nealson *et al.*, 1970, Nealson and Hastings, 1979). In *V. harveyi*, there are two types of density-dependent signaling systems that regulate bioluminescence activity consisting of autoinducer 1 and 2. The AI-1 (N-3-oxohexanoyl-L-homoserine lactone) molecule found in *V. fischeri* governs the induction of luminescence operon (Gilson *et al.*, 1995). *V. harveyi* and *V. cholerae* have been reported to use the AI-1 quorum sensing circuit for intra-species communication. The essential characteristics of AI-1 systems are the biosynthesis of acylated homoserine lactones (AHLs) by an AHL synthase, encoded by *luxI* in *V. fischeri* or *luxI* homologs in other bacteria; and an AHL response regulator, encoded by *luxR* (or luxR homologs). The N-octanoyl-L-homoserine lactone (AI-1) molecule in *V. fischeri* interacts with and activates the luminescence in *E. coli* via LuxR (Gilson *et al.*, 1995). The other QS signal molecules designated as autoinducer 2 (AI-2) is observed in both Gram-negative and Gram-positive bacterial species and is suggested to mediate communication among and between species (Bassler *et al.*, 1997, Schauder *et al.*, 2001). AI-2 signal production occurs in bacteria that possess a *luxS* homologue. The AI-2 molecule in *Vibrio harveyi* is currently believed to be furanosyl borate diesters (Chen *et al.*, 2002). Several other bacterial species can interact with the *Vibrio harveyi* AI-2 signaling pathway and the AI-2 modifying LuxS protein sequence is extremely conserved throughout the bacterial kingdom. The third type of autoinducer (AI-3) molecules are involved in cross talk and inter-kingdom signaling with the eukaryotic hormones (epinephrine/ norepinephrine). The AI-3 molecules are observed in *E. coli* O157:H7 and the host epinephrine cell signaling. This signaling activates transcription of virulence genes in enterohemorrhagic *E. coli* O157:H7 as well as intestinal cell actin rearrangement. The structure of AI-3 molecules is however yet not elucidated.

Apart from these autoinducer molecules, other non-AHL compounds such as indole, PQS, small RNA and secondary messengers are also involved in quorum sensing induction. Indole is produced and is reported to act as an extracellular signal in the induction of quorum in *E. coli* (Wang *et al.*, 2005). Another molecule heptyl-hydroxy-quinolone, designated the *Pseudomonas* quinolone signal (PQS) found exclusively in *Pseudomonas* spp. is a part of the quorum sensing hierarchy. PQS acts as a link between *las* and *rhl* systems (McKnight *et al.*, 2000). PQS is similar to AHLs with respect to size and its lipophilic nature. Most of the genes involved in the synthesis and regulation of PQS have been described in detail earlier, however the mechanism of activity is unknown (Cao *et al.*, 2001, Deziel *et al.*, 2004, Diggle *et al.*, 2003, Gallagher *et al.*, 2002). These molecules diffuse freely through the bacterial membrane and are internally sensed. Other molecules, such as the small RNAs also play a role in quorum sensing. It is becoming increasingly apparent that like other bacterial processes, integration of information by QS systems is regulated by noncoding small RNAs (sRNAs). These sRNAs are global regulators that act directly or indirectly to control gene expression by post-transcriptional mechanisms. sRNAs are important regulators involved in bacterial and eukaryotic developmental processes (Masse *et al.*, 2003, Wienholds and Plasterk, 2005). Bejerano-Sagie and Xavier (2007) have recently reviewed the crucial role of small noncoding RNAs in the regulation of bacterial QS. Regulation by sRNAs rather than by proteins is presumed to be beneficial when a rapid response is required, because of the short time required to synthesize or degrade sRNAs compared with synthesizing and degrading proteins.

QS systems govern a diverse set of microbial processes, including antibiotic biosynthesis, swarming, swimming and twitching motility, plasmid conjugal transfer, biofilm formation (Davies *et al.*, 1998), pathogenesis, production of biosurfactant, enzymes and other secondary metabolites (for reviews see Camara *et al.*, 2002, Fuqua and Greenberg, 2002, Lazdunski *et al.*, 2004, Miller and Bassler, 2004, Pappas *et al.*, 2004, Whitehead *et al.*, 2001). Recently, the study of QS systems has been extended to include implications in synthetic biology for population control (You *et al.*, 2004), band detection (Basu *et al.*, 2005) and predator-prey systems (Balagadde *et al.*, 2008). In this review, we focus especially on the aspects of QS involved in the production of rhamnolipid.

QUORUM SENSING IN MICROBIAL COMMUNITIES

QS activities have been documented in biofilms for some time (Davies *et al.*, 1998, McLean *et al.*, 1997), although the magnitude of their role in biofilms depends on the nutritional environment (Shrout *et al.*, 2006). As stated earlier, *P. aeruginosa* has several QS systems, including the AHL-mediated *las* and *rhl* systems, the PQS system (Mashburn and Whiteley, 2005); as well as the AI-2 QS system (Duan *et al.*, 2003). Very recently, small regulatory RNA molecules have also been shown to influence QS regulation (Tu *et al.*, 2008). These signals and many of the functions that they encode are quite important for biofilm development and bacterial interactions within microbial communities (Givskov *et al.*, 1996, Parsek and Greenberg, 2000). Several years ago, Singh *et al.* (2000) investigated the AHL expression levels in planktonic and biofilm grown *P. aeruginosa*. They found that the ratio of 3-oxo-dodecanoyl homoserine lactone (3-oxo-C12 HSL, produced by the *lasI* gene product) to N-butanoyl homoserine lactone (C4-HSL, produced by the *rhlI* gene product) in planktonic populations was approximately 3:1. In laboratory-grown and clinically-obtained biofilms (sputum samples of cystic fibrosis patients), the ratio was reversed with C4-HSL being the predominant AHL (Singh *et al.*, 2000). Among other things, C4-HSL regulates rhamnolipid biosynthesis (Ochsner and Reiser, 1995). Here, we focus on biofilm and microbial community features that are influenced by rhamnolipids.

Rhamnolipids function as biosurfactant molecules (Davey *et al.*, 2003). In this fashion they have been predicted to facilitate uptake of poorly soluble, hydrophobic compounds. Work from the laboratories of PA Holden (2002) and RM Miller (1994, 1995) have shown the emulsifying nature of rhamnolipids that enables *Pseudomonas* spp. to facilitate hydrocarbon utilization. Interestingly, one study showed that the *P. aeruginosa* outer membrane was removed in the presence of rhamnolipids, such that the hydrophobic membrane interior could bind directly to lipids (Al-Tahhan *et al.*, 2000).

During the process of biofilm development and maturation, surface-attached cells will aggregate into microcolonies that are surrounded by regions of few cells referred to as water channels (Davey and O'Toole, 2000, Sauer *et al.*, 2002). During the aggregation process, surface-colonized *P. aeruginosa* move across the substratum by a combination of twitching motility, which involves type IV pili (O'Toole and Kolter, 1998); and swarming, which involves cell elongation, hyper-flagellation and differentiation (Kohler *et al.*, 2000). Rhamnolipids play a role in the swarming process, acting both as surface wetting agents and as chemotaxis stimuli. In swarming but not swimming, rhamnolipids function as chemoattractants whereas the chemically related, hydroxy alkanoic acids function as chemorepellants (Tremblay *et al.*, 2007). Swarming can be blocked by branched chain fatty acids, which presumably compete with rhamnolipids (Inoue *et al.*, 2008).

Rhamnolipids are also important for the formation of water channels in mature biofilms as shown by Davey *et al.* (2003). During this study, *rhl* mutants, unable to synthesize rhamnolipids, formed biofilms lacking the characteristic architecture (micro-colonies and water channels). Co-culture of the *rhl* mutants with wild type *Pseudomonas* could partially rescue the biofilm structural phenotype. Overproduction of rhamnolipids caused an inhibition of biofilm formation, blocked cellular aggregation, and also blocked secondary colonization onto preformed biofilms by other planktonic bacteria (Davey *et al.*, 2003). Rhamnolipids have also been associated with cell dispersal from biofilms (Boles *et al.*, 2005, Pamp and Tolker, 2007).

One notable feature of biofilms is the protection that is offered to their component cells from antimicrobial agents and external forces including predation and the immune system (Costerton *et al.*, 1987). Rhamnolipids do play a role in the chemical ecology of biofilms. Rhamnolipid production within *P. aeruginosa* biofilms has been shown to cause the rapid killing of polymorphonuclear leukocytes during experimental lung infections of mice (Jensen *et al.*, 2007). From a microbial competition perspective, rhamnolipids, produced by *P. aeruginosa* have been shown to be able to disrupt preformed biofilms of *Bordetella bronchiseptica* (Irie *et al.*, 2005). Production of these biosurfactants is not always beneficial to *P. aeruginosa*. Kohler *et al.* (2007) showed that the action of the antibiotic, azithromycin, was enhanced in the presence of rhamnolipids, presumably as these compounds facilitated the transport of the antibiotic across the bacterial membrane. Although they do have varied roles within biofilms (Pamp and Tolker, 2007), rhamnolipids are an important component of *Pseudomonas* biofilm development, structure, and functions.

Biosurfactant

Biosurfactants are surface active agents that have been receiving increasing attention on account of their unique properties such as their mild production conditions, lower toxicity, and higher biodegradability, compared to their synthetic chemical counterparts (Rosenberg and Ron, 1999). Biosurfactants are produced by bacteria or yeasts from variety of sources such as sugars, glycerol, oils etc. Biosurfactants are classified as glycolipids, lipopeptides, phospholipids, fatty acids, neutral lipids, and polymeric or particulate compounds (Desai and Banat, 1997). The hydrophobic portion of the molecule may be long-chain fatty acids, hydroxyl fatty acids or α-alkyl-β-hydroxyl fatty acids. The hydrophilic moiety can be a carbohydrate, amino acid, cyclic peptide, phosphate, carboxylic acid or alcohol. One such biosurfactants that have been extensively studied is rhamnolipid (Lang and Wullbrandt, 1999).

RHAMNOLIPID BIOSURFACTANT

Rhamnolipid production is a characteristic of *P. aeruginosa* and was first described by Jarvis and Johnson (1949), however recently other *pseudomonads*, *P. putida* and *P. chlororaphis*, as well as *Burkholderia pseudomallei* have been reported to produce a variety of rhamnolipids (Gunther *et al.*, 2005, Haussler *et al.*, 1998, 2003, Tuleva *et al.*, 2002). The production of rhamnolipids is species specific. Some species produce a mono-rhamnolipid, others produce a di-rhamnolipid and yet others produce a mixture of rhamnolipids, all of which vary in lipid chain lengths. The rhamnolipids are composed of a polar head group and one or more non-polar tail. *P. aeruginosa* produces four types of rhamnolipids, including a mixture of homologous species of RL1 ($RhC_{10}C_{10}$), RL2 (RhC_{10}), RL3 ($Rh_2C_{10}C_{10}$) and RL4 (Rh_2C_{10}) (Rahman *et al.*, 2002). The length of the carbon chains found on the β-hydroxy portion of the rhamnolipid can vary significantly; however, in case of *P. aeruginosa* ten carbon molecule chains are the predominant form (Deziel *et al.*, 2000). Rhamnolipid as well as their precursor, 3-(3-hydroxyalkanoyloxy) alkanoic acids (HAAs), display tensioactive properties (Deziel *et al.*, 2003), which further facilitate their medical, environmental and industrial applications. Rhamnolipid induces a remarkably larger reduction in the surface tension of water from 72mN/m to values below 30mN/m and it also reduces the interfacial tension of water/oil systems from 43mN/m to values below 1mN/m. Rhamnolipids also show an excellent emulsifying activity with a variety of hydrocarbons and vegetable oils (Abalos *et al.*, 2001). Rhamnolipids solubilize hydrophobic molecules, such as long-chain hydrocarbons, and allow their use as a carbon source and in addition facilitate the interactions between cells by promoting aggregation (Herman *et al.*, 1997).

Rhamnolipids are extensively used in the production of fine chemicals, characterization of surfaces and surface coatings, as additives for environmental remediation and as biological control agents (Stanghellini and Miller, 1997). Rhamnolipids have been regarded as virulence factors (Kownatzki *et al.*, 1987) and antimicrobials (Abalos *et al.*, 2001), and are implicated in the development of biofilms (Davey *et al.*, 2003) and along with HAAs, have been documented to be crucial for *P. aeruginosa* swarming motility (Deziel *et al.*, 2003, Kohler *et al.*, 2000).

METHODS FOR RHAMNOLIPID DETECTION AND QUANTIFICATION

In the last few decades, extensive research has been conducted in the area of biosurfactants. This has lead to the development of various techniques to detect, quantify and enhance the production of biosurfactants. Various techniques, especially with respect to rhamnolipid biosurfactant, are mentioned here. Each of these methods has been described in a recent review by Heyd *et al.* (2008) and other references stated below.

i. Surface tension reduction (Guerra-Santos *et al.*, 1984, Haussler *et al.*, 1998).
ii. Hemolytic activity (Siegmund and Wagner, 1991).
iii. Colorimetric method using Cetyl Trimethyl Ammonium Bromide (CTAB) plate assay (Siegmund and Wagner, 1991).
iv. Methylene blue complexation method (Pinzon and Ju, 2009).
v. Rhamnolipid estimation using Anthrone reagent (Helbert and Brown, 1957).
vi. Rhamnolipid estimation using Orcinol method (Chandrasekaran and Bemiller, 1980, Koch *et al.*, 1991).
vii. Thin layer chromatography (TLC) (de Koster *et al.*, 1994, Rendell *et al.*, 1990).
viii. Gas Chromatography (GC) (Arino *et al.*, 1996, Van Dyke *et al.*, 1993).
ix. High performance liquid chromatography (HPLC) (Deziel *et al.*, 1999, Lepine *et al.*, 2002, Mata-Sandoval *et al.*, 1999).
x. Mass spectrometry (MS) (Deziel *et al.*, 1999).
xi. Fourier transform infrared spectroscopy (Borgund *et al.*, 2007, Gartshore *et al.*, 2000) and Attenuated Total Reflectance (ATR) FTIR (Leitermann *et al.*, 2008).
xii. NMR spectroscopy (Choe *et al.*, 1992, Monteiro *et al.*, 2007).

Combinations of these methods are generally used for the detection and estimation of rhamnolipid production.

Applications of rhamnolipid biosurfactant

Compared to chemical surfactants, biological biosurfactant possess numerous attributes that make them invaluable in both environmental and industrial settings (Hommel, 1990, Rahman and Gakpe, 2008, Volkering *et al.*, 1995). Noordman and Janssen (2002) claimed the degradation of hexadecane by rhamnolipid with rates higher as compared to other biosurfactants. Urum *et al.* (2006), on the other hand compared the effectiveness of biosurfactant rhamnolipid, saponin and sodium dodecyl sulfate (SDS). They found that rhamnolipid and saponin aided crude oil degradation almost equally, whilst SDS was found to be ineffective. Rhamnolipid is therefore the best biosurfactant since it is produced naturally via microbial activity, while SDS is a synthetic surfactant.

Rhamnolipids in particular are biotechnologically important due to their antibacterial, antifungal and antiviral activities (Haferburg *et al.*, 1987, Stanghellini and Miller, 1997, Syldatk *et al.*, 1985). Rhamnolipid biosurfactants increases membrane permeability of the bacterial cells thereby causing cell death. The biosurfactant probably forms molecular aggregates in surface bacterial membranes, leading to the formation of trans-membrane pores (King *et al.*, 1991). Studies conducted by Sotirova *et al.* (2008) showed the rhamnolipid biosurfactant complex termed PS mediates permeabilizing effects on Gram-positive and Gram-negative bacterial strains, namely *B. subtilis* and *P. aeruginosa*. They reported that at lower concentrations of rhamnolipid biosurfactant close to CMC, the growth of bacterial cells is not influenced, however concentrations greater than CMC exhibit toxic conditions for *B. subtilis* cells but not for *P. aeruginosa* as evident from the levels of extracellular proteins. The biosurfactant enhanced levels of extracellular protein in *B. subtilis* cells compared with those of *P. aeruginosa*, which confirmed the higher susceptibility of Gram-positive cells to the effect of the studied biosurfactant. It is evident that the outer membrane of Gram-negative bacteria have lipopolysaccharide (LPS), porin channels, and murein lipoprotein, all of which are absent in Gram-positive bacteria. Also, the outer membrane functions as an efficient permeability barrier that is able to exclude biosurfactant molecules. The permeability barrier property is largely caused by the presence of the LPS layer. Increased cell permeability induced by rhamnolipid biosurfactant was most likely caused by the release of LPS from the outer membrane (Al-Tahhan *et al.*, 2000, Sotirova *et al.*, 2007). Rhamnolipids have been used as emulsifying agents for the transport of drugs to the site of action. Rhamnolipids in combination with the antibiotic, azithromycin facilitated destruction of the bacterial cells by increasing the bacterial membrane permeability (Kohler *et al.*, 2007). *P. aeruginosa* rhamnolipid mixture was found to inhibit a majority of pathogenic bacteria such as *A. faecalis*, *E. coli*, *Micrococcus luteus*, *Mycobacterium phlei*, *Serratia marcescens* and *S. epidermidis*. The marine bacterium, *B. pumilus* cell adhesion and biofilm disruption was also achieved using rhamnolipids (unpublished data).

Rhamnolipids also show antifungal activity against *Aspergillus niger*, *Aureobasidium pullulans*, *Chaetonium globosum* and *Penicillum crysogenum* (Abalos *et al.*, 2001, Rahman and Gakpe, 2008). The zoosporicidal activity of mono and dirhamnolipids against phytopathogens is reported by Stanghellini and Miller (1997). Stanghellini and coworkers (1998) patented their work on rhamnolipid biosurfactants produced by *Pseudomonas* spp. able to rapidly kill zoospores by rupturing the plasma membrane of three representative plant pathogenic microorganisms; namely *Pythium apha-*

nidernatum, *Plasmopara lactucae-radicis*, and *Phytophthora capsici*. Rhamnolipid-producing strains also provide control of *Pseudoperonospora cubensis*, the causal agent of downy mildew of cucurbits, by intercalating into the plasma membrane and thereby destroying the cell structure. Rhamnolipids inhibit the growth of algal species including *Heterosigma akashivo* and *Protocentrum dentatum* (Wang *et al.*, 2005). Rhamnolipids also display antiviral properties against pathogens like TMV and potato X virus (Haferburg *et al.*, 1987). They can be used as adjuvants for vaccines. McClure and Schiller (1992) reported the enhancement of phospholipase C activity after addition of rhamnolipids. Tatjana and Goran (2007) patented their work on rhamnolipids as effective in wound healing, treating burn shock, atherosclerosis, organ transplants, depression, schizophrenia and cosmetics (Tatjana and Goran, 2007). Rhamnolipids are however also responsible for increasing the virulence factors secreted by *P. aeruginosa* that affect the structure of human airway epithelium in the early stages of infection (Zulianello *et al.*, 2006).

In environmental bioremediation, rhamnolipids play a significant role in the treatment of soils contaminated with industrial waste, crude oils, polyaromatic hydrocarbons, refinery products, pesticides and heavy metals (Rahman *et al.*, 2003). The ability of a rhamnolipid mixture produced by *P. aeruginosa* UG2 to solubilize the pesticides atrazine, trifluralin, and coumaphos was compared with a chemical surfactant Triton X-100. It was observed that the values of maximum micellar solubilization capacities for trifluralin and coumaphos in Triton X-100 were double those for the rhamnolipid mixture, whereas atrazine maximum micellar solubilization capacity value for the rhamnolipid biosurfactant was in the same range as that for the synthetic surfactant (Mata-Sandoval *et al.*, 2000).

Muller-Hurtig *et al.* (1993) and Finnerty (1994) reviewed the possible use of biosurfactants in soil remediation. Rhamnolipids can alter the physicochemical properties of oil, thereby facilitating the removal of oil from contaminated soils. Rhamnolipids have also shown to enhance the cell surface hydrophobicity of *P. aeruginosa* cells by inducing removal of lipopolysaccharides, thereby increasing the uptake of hydrophobic compounds by the cells (Al-Tahhan *et al.*, 2000). Heavy metals are included on the EPA's list of priority pollutants (Mulligan *et al.*, 2001). Heavy metals pose a persistent problem at many contaminated sites and are being added to soil, water, and air in increasing amounts from a variety of sources including industrial, agricultural and domestic effluents. Heavy metals can be removed from contaminated areas through direct binding of the rhamnolipids to the metals forming a stable complex, which is subsequently removed from the soil (Ochoa-Loza *et al.*, 2001).

In food industries, rhamnolipids serve as a good source of rhamnose sugar that acts as a precursor for high quality flavor components (Linhardt *et al.*, 1989). Van Haesendonck and Vanzeveren (2004) reported the application of rhamnolipid for volume enhancement and for texture modification in bakery and pastry products.

Rhamnolipids and quorum sensing

QUORUM SENSING MOLECULES IN RHAMNOLIPID PRODUCTION

P. aeruginosa possesses two interrelated QS systems, namely the *las* and *rhl,* (Gambello and Iglewski, 1991, Passador *et al.*, 1993, Toder *et al.*, 1991) that regulate different processes including rhamnolipid expression, enzyme production, pyocyanin pigment production and maintenance of biofilm architecture (Davies *et al.*, 1998, de Kievit and Iglewsky, 2000, Rumbaugh *et al.*, 2000, Smith and Iglewski, 2003). Production of rhamnolipid is governed by three QS molecules: the twelve carbon *Pseudomonas* autoinducer 1 (PAI-1) [N-(3-oxododecanoyl) homoserine lactone also known as 3-oxo-C_{12}-HSL] (Pearson *et al.*, 1994), *Pseudomonas* autoinducer 2 (PAI-2) [N-butyryl homoserine lactone known also as C_4-HSL] (Pearson *et al.*, 1995), and PQS, [2-heptyl-3-hydroxy-4-quinolone] (Pesci *et al.*, 1999) (Fig. 1).

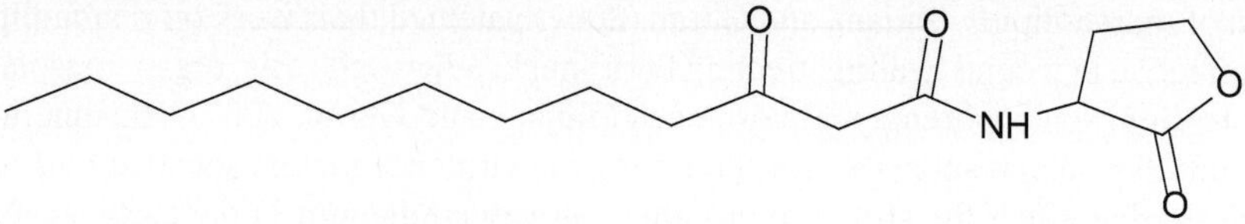

PAI-1 (3-oxo-C12- HSL)

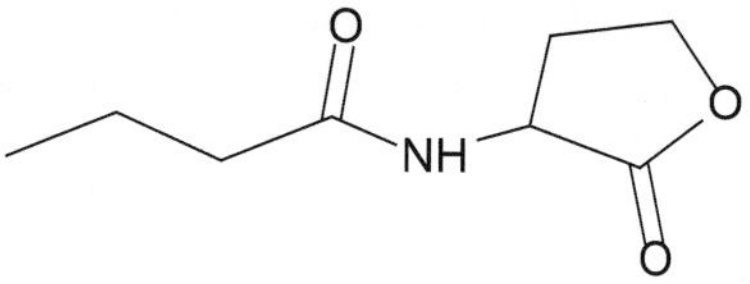

PAI-2 (C4-HSL)

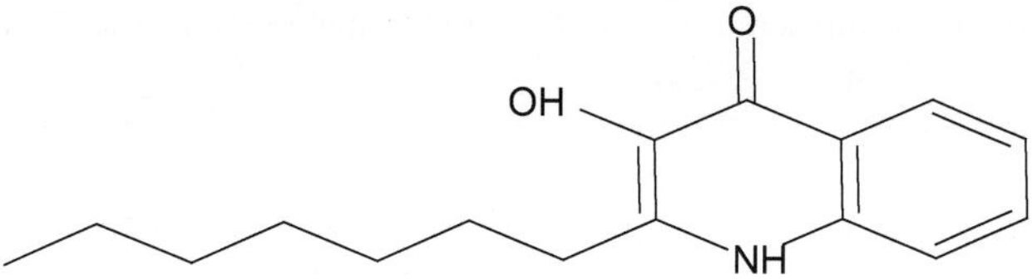

PQS

Figure 1. Quorum sensing molecules in *Pseudomonas aeruginosa* PAO1. The three quorum sensing molecules denoted as (a) *Pseudomonas* autoinducer PAI-1 [N-(3-oxododecanoyl) homoserine lactone] also known as 3-oxo-C_{12}-HSL; (b) *Pseudomonas* autoinducer PAI-2 [N-butyryl homoserine lactone] known as C_4-HSL and (c) *Pseudomonas* Quinolone Signal (PQS), [2-heptyl-3-hydroxy-4-quinolone] coordinates the cellular activities.

Genetic basis of rhamnolipid production

The production of rhamnolipids is under the control of two quorum sensing systems, namely the *las* and *rhl* QS systems. In *P. aeruginosa*, the *las* operon consists of two transcriptional activator proteins, the LasR and LasI, which directs the synthesis of N-3-oxododecanoyl homoserine lactone (PAI-1 or 3-oxo-C_{12}-HSL) autoinducer. Induction of the *lasB* gene that encodes the elastase enzyme and other virulence genes requires the expression of LasR and PAI-1 autoinducer. The production of rhamnolipid is regulated by the *rhl* system (Johnson and Boese-Marazzo, 1980). The synthesis of rhamnolipids takes place under the coordinated guidance of *rhlAB* genes that encodes a group of enzymes termed the rhamnosyltransferases (Ochsner *et al.*, 1995). Rhamnolipid is a complex synthesized by two enzymes namely, rhamnosyltransferase 1 and rhamnosyltransferase 2. The *rhl* system consists of transcriptional activator proteins RhlR and RhlI, which regulates the synthesis of a QS molecule, N-butyryl homoserine lactone (PAI-2 also called the C_4-HSL) (Ochsner *et al.*, 1994b, Pearson *et al.*, 1995). The transcriptional activator RhlR activates the transcription of *rhlAB* operon genes, coding for rhamnosyltransferase 1 (Ochsner *et al.*, 1994b), and another gene, *rhlC* encoding for the rhamnosyltransferase 2 (Rahim *et al.*, 2001). The genes involved in the production of rhamnolipid are mentioned in table 1. With increase in bacterial cell density, the induction of *las* quorum sensing system takes place resulting in an increase in the concentration of PAI-1 (3-oxo-C_{12}-HSL) autoinducer molecule. This quorum sensing molecule (PAI-1) then binds to the transcriptional activator site LasR and forms the LasR–PAI-1 complex. The LasR–PAI-1 complexes induces genes controlled by the *las* quorum-sensing system, including a negative regulator gene *rsaL*, *rhlR* (Ochsner *et al.*, 1994b, Pearson *et al.*, 1995) and *pqsH*, required for PQS production (Mashburn and Whiteley, 2005). The activity of these signals is dependent upon their ability to dissolve in and freely diffuse through the aqueous solution.

Table 1. Quorum sensing systems prevalent in *P. aeruginosa* suggesting the genes and functions assisting rhamnolipid production.

Quorum sensing system	Genes involved	Enzyme product
las system	*las*I	Autoinducer synthesis, LasI synthase
	*las*R	Transcriptional regulator, LasR synthase
	*las*A	LasA protease precursor
	*las*B	Elastase LasB
rhl system	*rhl*I	Autoinducer synthesis protein, RhlI synthase
	*rhl*R	Transcriptional regulator, RhlR synthase
	rhlAB	Rhamnosyltransferase 1
	*rhl*C	Rhamnosyltransferase 2
	*rhl*G	β-ketoacyl reductase
pqs system	*pqs*A	coenzymeA ligase
	*pqs*B	Homologous to β-keto acyl carrier protein synthase
	*pqs*C	Homologous to β-keto acyl carrier synthase (3-oxoacyl-[acyl-carrier protein])
	*pqs*D	3-oxoacyl-[acyl-carrier protein] synthase III
	*pqs*E	Quinolone signal response protein
	*pqs*H	Probable FAD-dependent monooxygenase
	*pqs*L	Probable FAD-dependent monooxygenase

P. aeruginosa produced rhamnolipid biosurfactant that enhances the solubility of PQS in aqueous solutions (Calfee *et al.*, 2005). However unlike other QS, the hydrophobic

PQS is transported primarily through outer membrane vesicles (Mashburn and Whiteley, 2005), the formation of which are PQS-induced (Mashburn-Warren *et al.*, 2008, Mashburn-Warren *et al.*, 2009). PQS (3, 4-hydroxy-2-heptylquinoline), as mentioned earlier acts as a link between *las* and *rhl* quorum sensing systems (Muller-Hurtig *et al.*, 1993). Using mutants deficient in the synthesis of PQS, the cells of *P. aeruginosa* make less rhamnolipid than the wild type strains (Diggle *et al.*, 2003). PQS here either directly or indirectly induces the *rhlI* gene which directs the production of PAI-2 (C_4-HSL) quorum sensing molecule that binds to and activates RhlR (McKnight *et al.*, 2000). The operon, *rhlAB* that encodes these enzymes responsible for rhamnolipid production is controlled at the transcriptional and translational levels by RhlR and C4-HSL (Ochsner and Reiser, 1995). The RhlR–PAI-2 complex induces genes controlled by the *rhl* quorum sensing system for the production of rhamnolipid. The *las* system controls the expression of transcriptional activator RhlR (Fig. 2). Along with this an important gene, *rhlG* is involved in the synthesis of β-hydroxyacid moiety of rhamnolipids (Campos-Garcia *et al.*, 1998). A QS hierarchy therefore exists in *P. aeruginosa las* and *rhl* systems for the synthesis of rhamnolipid.

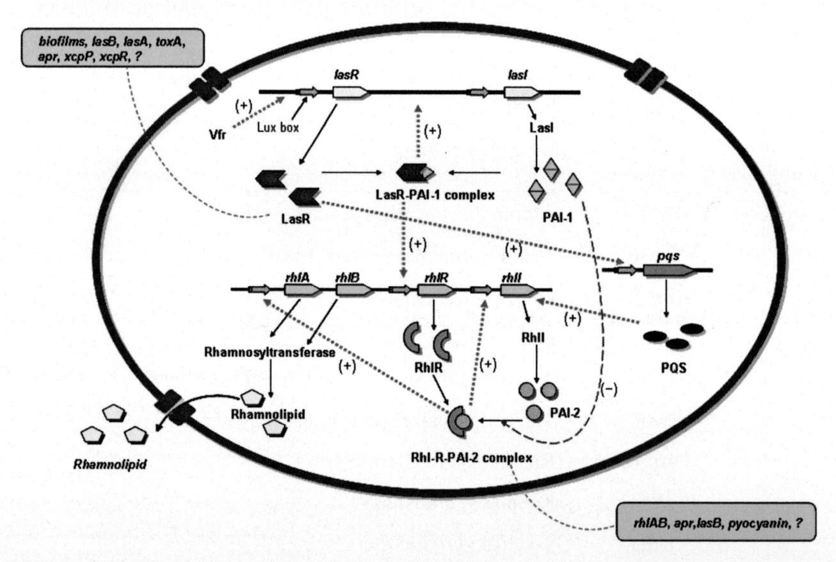

Figure 2. Schematic representation of the *las* and *rhl* genes and quorum sensing molecules in *Pseudomonas aeruginosa* for the production of rhamnolipid. The *las* system produces transcriptional activators, LasR and LasI (producing PAI-1). The *rhlA* and *rhlB* genes are arranged as an operon and are clustered with *rhlR* and *rhlI*. These genes, *rhlABRI* directs the synthesis of rhamnosyltransferase and transcriptional activators, RhlR and RhlI (producing PAI-2), which are responsible for the production of rhamnolipids. Vfr induces *lasR* and the concentration of PAI-1 increases where it binds to and activates LasR (Albus *et al.*, 1997). The autoinducer PAI-1 (encoded by *lasI*) binds with LasR and forms a LasR-PAI-1 complex. This complex regulates the transcription of *rhlR*. *rhlR* produces RhlR protein, which binds to the PAI-2 autoinducer resulting in RhlR-PAI-2 complex that interacts with the *rhlA* promoter (*lux* box) to begin transcription of the rhamnolipid producing *rhlAB* gene. Here (+) indicates transcriptional activation and (–) indicates transcriptional repression of the concerned genes.

Biochemical pathway of rhamnolipid production

Initiation of quorum sensing is contingent upon the accumulation of sufficient signal molecules. Due to this stringency, quorum sensing is only initiated once the population reaches a critical threshold (Fuqua *et al.*, 1994, Pierson *et al.*, 1994). In *P. aeruginosa,* rhamnolipid production initially depends on PAI-1 and PAI-2 diffusible molecules, which interact with the activators, LasR and RhlR at high bacterial cell densities. For more information refer to Fig. 3.

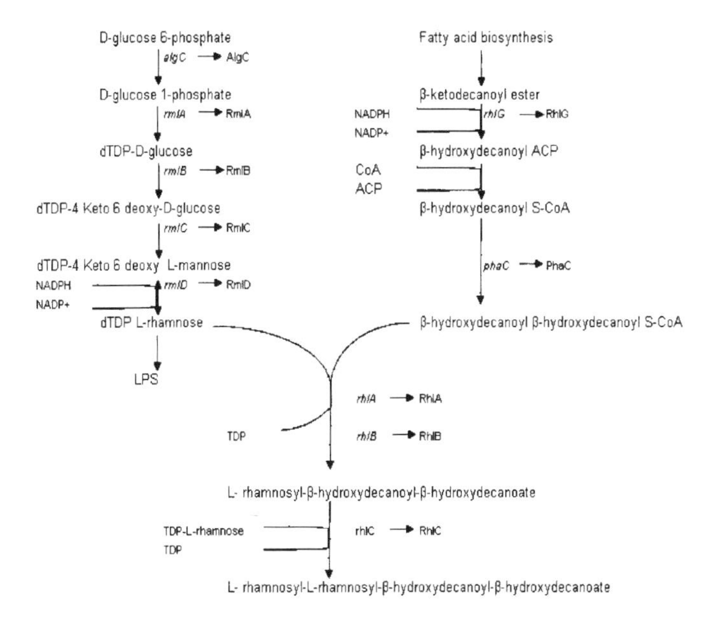

Figure 3. Biochemical pathway for rhamnolipid biosynthesis showing the genes and proteins involved in *Pseudomonas aeruginosa* PAO1 (Maier and Soberson-Chavez, 2000). The synthesis of rhamnolipid proceeds via the transfer of TDP-L-rhamnose. The 3-(3-hydroxyalkanoyloxy) alcanoic acid (HAA) is synthesized by RhlA enzyme and is then converted to mono-rhamnolipid by RhlB enzyme (Deziel *et al.*, 2003, Ochsner *et al.*, 1994b). The mono-rhamnolipid is converted to di-rhamnolipid by the RhlC enzyme (Rahim *et al.*, 2001). CoA-β-hydroxyacids are the precursors of rhamnolipids. The *rhlAB* operon and *rhlC* genes are induced by homoserine lactone activated RhlR and are thus under the control of QS system (Ochsner *et al.*, 1994b, Rahim *et al.*, 2001). RhlR protein is known to activate *rhlG* transcription for rhamnolipid production. The biochemical synthesis of rhamnolipid is shown in fig. 3. Many QS molecules are known to regulate the synthesis of rhamnolipids. An autoinducer, N-butyryl homoserine lactone (PAI-2) present in *P. aeruginosa* restores rhamnolipid production in *P. aeruginosa rhlI* mutant (Pearson et al., 1995, Winson *et al.*, 1995). PAI-2 and RhlR enhances the expression of *rhlI* in *E. coli* (Latifi *et al.*, 1996). These finding suggests the importance of *rhlR* and *rhlI* quorum sensing components required for the auto-induction of rhamnolipid biosynthesis genes *rhlA* and *rhlB* (Pearson *et al.*, 1997). The activator LasR-PAI1 complex induces the production of several virulence factors, such as the alkaline protease, exotoxin A and also regulates the expression of secretion proteins (Gambello *et al.*, 1993, Morihara and Homma, 1985, Toder *et al.*, 1991, 1994). The Rhl-PAI2 complex present in the biosynthesis of rhamnolipid induces expression of the LasA and LasB proteases as well as the secretion proteins (Xcp). In addition, the complex also controls expression of *rhlAB* and *rhlG* genes responsible for rhamnosyltransferase (Burger *et al.*, 1963) and synthesis of hydroxyalkanoate for rhamnolipid production respectively (Campos-Garcia *et al.*, 1998).

Enhancement of rhamnolipid production

P. aeruginosa is an opportunistic pathogen associated with cystic fibrosis and infections associated with severe burns. This bacterium is known for its resistance towards a variety of antibiotics and is one of the leading causes of nosocomial infections (Tummler *et al.*, 1991). Numerous studies on *P. aeruginosa* and rhamnolipid biosynthesis have improved the understanding of methods for the enhancement of rhamnolipid production.

NUTRITIONAL AND ENVIRONMENTAL CONDITIONS

In *P. aeruginosa,* rhamnolipid production occurs typically in late exponential or stationary phase (Guerra-Santos *et al.*, 1986). The presence of nutrients, such as carbon and nitrogen, also play an important role in the production of rhamnolipids (Wu *et al.*, 2008). *P. aeruginosa* growth and rhamnolipid production can occur using a range of different primary carbon sources. The highest level of rhamnolipid production in *P. aeruginosa* occurs when using vegetable based oils as carbon sources including soybean oil (Lang and Wullbrandt, 1999), corn oil (Linhardt *et al.*, 1989), canola oil (Sim *et al.*, 1997), and olive oil (Robert *et al.*, 1989). Rhamnolipid production is dependent upon environmental and nutritional conditions. Environmental factors play a crucial role in influencing the productivity and efficacy of rhamnolipids. In general, as a biosurfactant, rhamnolipid activity is controlled by environmental conditions such as pH, salinity and temperature (Ilori *et al.*, 2005, Inakollu *et al.*, 2004, Jirasripongpun, 2002). Ilori *et al* (2005) pointed out that the chemical structure of biosurfactant gives benefit for hydrocarbon degradation and very unlikely to be disrupted due to extreme temperature and pH. Benka-Coker and Ekundayo (1996) highlighted the amount of oil might affect the biodegradation rate as well, due to poor aeration and lack of oxygen. The action of *Pseudomonas* in degrading oil is accelerated by the formation of rhamnolipid. The size and structure of hydrocarbon substrates may however slow down this oil degradation process.

SUPPLEMENTING EXOGENOUS QS MOLECULES

In mutants, unable to produce rhamnolipids, external addition of autoinducer molecules, N-acyl homoserine lactones regains the phenotype of rhamnolipid production. Kuniho *et al* (1998) found that autoinducer activity increased approximately ten-fold in fed batch system which strongly correlated with increased rhamnolipid production. Enhancement of rhamnolipid production occurred in the presence of either N-(3-oxohexanoyl)-L-homoserine lactone (OhDHL) or N-(3-oxododecanoyl)-L-homoserine lactone (OdDHL). Overall, the presence of exogenous autoinducer increased rhamnolipid production five-fold, with maximal yields occurring during stationary phase of growth. Construction of rhamnolipid mutants has allowed for the identification of several genes that are essential for rhamnolipid biosynthesis. The first, identified in *P. aeruginosa,* comprises the *rhlAB* operon; here both the genes play an essential role in rhamnolipid production (Ochsner *et al.*, 1994a). The *rhlAB* genes encode rhamnosyltransferase 1 complex involved in the formation of the RhlAB heterodimer (Ochsner *et al.*, 1994b) and defects in either gene result in deficiencies in rhamnolipid production (Deziel *et al.*, 2003). In addition to the two aforementioned genes, *rhlC* (encoding rhamnosyltransferase 2, which adds the second rhamnosyl group to form RL2) is essential for the production of RL2, but is not essential for RL1 production (Rahim *et al.*, 2001). A stable mutant strain would be a great advantage for rhamnolipid production by fermentation (Wang *et al.*, 2007).

GENETIC ENGINEERING

Screening high rhamnolipid-producing microorganism from the natural environment is a good strategy; however engineering strains for rhamnolipid production is another alternative. Rhamnolipid production could also be effectively enhanced by cloning the wild-type *rhlI* gene into a suitable strain such as *E. coli*, or by the addition of *P. aeruginosa* cell-free spent supernatant containing the autoinducer molecules (Ochsner and Reiser, 1995). Rhamnolipid production has been shown to be transcriptionally regulated by quorum-sensing circuitry (Ochsner and Reiser, 1995, Heurlier *et al.*, 2004). In a recent study, Cha *et al* (2008) were able to successfully increase rhamnolipid production by cloning both the *rhlAB* rhamnosyltransferase genes and the *rhlRI* quorum sensing system into *P. putida* to enhance rhamnolipid production.

In another study carried out by Wang and co-workers (2007) the novel transposome biotechnique was used. They integrated successfully the key genes of rhamnolipid biosynthesis into the chromosome of *P. aeruginosa* and *E. coli* cells, which were originally devoid of rhamnolipid production and the engineered strains, thus produced rhamnolipid. This technique would allow one to create a stable insertion mutation in a wide range of bacteria (Hoffman *et al.*, 2000). Unlike plasmid-based engineered strain, transposon-based strains could exist stably under no drug-selection pressure, and the integration site of the targeted gene(s) would easily be confirmed by inverse-PCR, DNA sequencing, and alignment with a vast repository of genome information available from public database. The mechanism of gene regulation enables controlled production of rhamnolipid. Ochsner *et al* (1994b) reported that *rhl*R gene is essential in synthesizing rhamnolipids since the interruption at this locus contributed to the formation of rhamnolipid deficient mutants.

By using a suitable medium with the addition of QS molecules (AHL) at an early stage of bacterial growth along with genetically modified bacterial strains could be used for enhanced synthesis of rhamnolipid. A recent development in synthetic biology where synthetic molecules of quorum sensing are used for induction has shown numerous applications. Better understanding of the QS based synthetic networks is useful and has been applied in studies related to programming cell death in *E. coli* (Balagadde *et al.*, 2005), constructing microbial consortia (Brenner *et al.*, 2007), building of artificial intercellular communication and quorum-sensing behavior in prokaryotes (Bulter *et al.*, 2004) and eukaryotes (Chen and Weiss, 2005). The topic is of significant interest and there is a need to explore it in great detail.

Conclusion

Rhamnolipids are effective biosurfactants with numerous applications. The production of rhamnolipids is under the control of quorum sensing. Over the past decade, significant strides have been made towards understanding the cell to cell communication, especially in the production of rhamnolipids biosurfactant. Evidence suggests that knowledge of cell to cell communication molecules and their role in biosurfactant production could be exploited to industrial scale production. There are numerous methods of enhancing rhamnolipids, however knowledge of the genes required for biosurfactant production can be critical for application in industry.

Currently, biosurfactants are unable to compete economically with chemically synthesized compounds in the market due to high production costs. Once the genes required for biosurfactant production have been identified, they can be placed under the regulation of strong promoters in nonpathogenic, heterologous hosts to enhance production. The production of rhamnolipids could be increased by cloning both the *rhlAB* rhamnosyltransferase genes and the *rhlRI* quorum sensing system into a suitable bacterium such as *E. coli* or *P. putida* and facilitate rhamnolipid production. Biosurfactants can also be genetically engineered for different industrial applications assuming there is a strong understanding of both the genetics and the structure-function relationships of each component of the molecule. Genetic engineering of surfactin has already been reported, with recent papers describing the creation of novel peptide structures from the genetic recombination of several peptide synthetases. Recent application of dynamic metabolic engineering strategies for controlled gene expression could lower the cost of fermentation processes by increasing the product formation. Therefore, by integrating a genetic circuit into applications of metabolic engineering the biochemical production can be optimized. Furthermore, novel strategies could be designed on the basis of information obtained from the studies of quorum sensing and biosurfactants produced suggesting enormous practical applications.

Acknowledgements

DHD would like to acknowledge Bhabha Atomic Research Centre (BARC) and University of Pune (UoP) collaborative research programme. PKSMR wishes to thank UK- Bioscience for Business KTN for the award of FROPTOP fund to explore the biocatalytic study of biosurfactant production from renewable resources. RJCM is funded by the Norman Hackerman Advanced Research Program of the Texas Higher Education Coordinating Board (003615-0037-2007).

References

ABALOS, A., PINAZO, A., INFANTE, M.R., CASALS, M., GARCIA, F. *et al.* (2001) Physicochemical and antimicrobial properties of new rhamnolipids produced by *Pseudomonas aeruginosa* AT10 from soybean oil refinery wastes. *Langmuir* **17**, 1367–1371

ALBUS, A.M., PESCI, E.C., RUNYEN-JANECKY, L.J., WEST, S.E.H. AND IGLEWSKI, B.H. (1997) Vfr controls quorum sensing in *Pseudomonas aeruginosa. Journal of Bacteriology* **179**, 3928–3935

AL-TAHHAN, R.A., SANDRIN, T.R., BODOUR, A.A. AND MAIER, R.M. (2000) Rhamnolipid-induced removal of lipopolysaccharide from *Pseudomonas aeruginosa*: effect on cell surface properties and interaction with hydrophobic substrates. *Applied and Environmental Microbiology* **66**, 3262–3268

ARINO, S., MARCHAL, R. AND VANDECASTEELE, J.P. (1996) Identification and production of a rhamnolipidic biosurfactant by a *Pseudomonas* species. *Applied Microbiology and Biotechnology* **45**, 162–168

BALAGADDE, F.K., YOU, L., HANSEN, C.L., ARNOLD, F.H. AND QUAKE, S.R. (2005) Longterm monitoring of bacteria undergoing programmed population control in a microchemostat. *Science* **309**, 137–140

BALAGADDE, F.K., SONG, H., OZAKI, J., COLLINS, C.H., BARNET, M. *et al.* (2008) A synthetic *Escherichia coli* predator–prey ecosystem. *Molecular and Systematic Biology* **4**, 1-8

BASSLER, B.L., GREENBERG, E.P. AND STEVENS, A.M. (1997) Cross-species induction of luminescence in the quorum-sensing bacterium *Vibrio harveyi*. *Journal of Bacteriology* **179**, 4043–4045

BASU, S., GERCHMAN, Y., COLLINS, C.H., ARNOLD, F.H. AND WEISS, R. (2005) A synthetic multicellular system for programmed pattern formation. *Nature* **434**, 1130–1134

BEJERANO-SAGIE, M. AND **XAVIER**, K.B. (2007) The role of small RNAs in quorum-sensing. *Current Opinion in Microbiology* **10**, 189-98

BENKA-COKER, M.O. AND EKUNDAYA, J.A. (1996) Applicability of evaluating the ability of microbes isolated from an oil spill site to degrade oil. *Environment Monitor Assessment* **45**, 259–272

BJARNSHOLT, T., JENSEN, P.O., BURMOLLE, M., HENTZER, M., HAAGENSEN, J.A.J *et al.* (2005) *Pseudomonas aeruginosa* tolerance to tobramycin, hydrogen peroxide and polymorphonuclear leukocytes is quorum-sensing dependent. *Microbiology* **151**, 373–383

BOLES, B.R., THOENDEL, M. AND SINGH, P.K. (2005) Rhamnolipids mediate detachment of *Pseudomonas aeruginosa* from biofilms. *Molecular Microbiology* **57**, 1210-1223

BORGUND, A.E., ERSTAD, K. AND BARTH, T. (2007) Normal phase high performance liquid chromatography for fractionation of organic acid mixtures extracted from crude oils. *Journal of Chromatography: A* **1149**, 189–196.

BRENNER, K., KARIG, D.K., WEISS, R. AND ARNOLD, F.H. (2007) Engineered bidirectional communication mediates a consensus in a microbial biofilm consortium. *Proceedings of the National Academy of Sciences (USA)* **104**, 17300–17304

BULTER, T., LEE, S.G., WONG, W.W., FUNG, E., CONNER, M.R. *et al.* (2004) Design of artificial cell–cell communication using gene and metabolic networks. *Proceedings of the National Academy of Sciences (USA)* **101**, 2299–2304

BURGER, M.M., GLASER, L. AND BURTON, R.M. (1963) The enzymatic synthesis of a rhamnose-containing glycolipid by extracts of *Pseudomonas aeruginosa*. *Journal of Biological Chemistry* **238**, 2595–2602

CALFEE, M.W., SHELTON, J.G., MCCUBREY, J.A. AND PESCI, E.C. (2005) Solubility and bioactivity of the *Pseudomonas* quinolone signal are increased by a *Pseudomonas aeruginosa*-produced surfactant. *Infection Immunity* **73**, 878–882

CAMARA, M., WILLIAMS, P. AND HARDMAN, A. (2002) Controlling infection by tuning in and turning down the volume of bacterial small-talk. *Lancet* **2**, 667-676

CAMPOS-GARCIA, A., JESU, S., CARO, A.D., JERA, R., MILLER-MAIER, R.M. *et al.* (1998) The *Pseudomonas aeruginosa rhlG* gene encodes an NADPH dependent β-ketoacyl reductase which is specifically involved in rhamnolipid synthesis. *Journal of Bacteriology* **180**, 4442–4451

CAO, H., KRISHNAN, G., GOUMNEROV, B., TSONGALIS, J., TOMPKINS, R. *et al.* (2001) A quorum sensing-associated virulence gene of *Pseudomonas aeruginosa* encodes a LysR-like transcription regulator with a unique self regulatory mechanism. *Proceedings of the National Academy of Sciences (USA)* **98**, 14613–14618

CHA, M., LEE, N., KIM, M., KIM, M. **AND** LEE, S. (2008) Heterologous production of *Pseudomonas aeruginosa* EMS1 biosurfactant in *Pseudomonas putida*. *Bioresource*

Technology **99**, 2192–2199

CHANDRASEKARAN, E.V. AND BEMILLER, J.N. (1980) Constituent analysis of glycosaminoglycans. In *Methods in Carbohydrate Chemistry*. Eds. R. L. Whistler and M. L. Wolform, pp 89-96. Academic Press, New York

CHEN, X., SCHAUDER, S., POTIER, N., VAN DORSSELAER, A., PELCZER, I. *et al.* (2002) Structural identification of a bacterial quorum-sensing signal containing boron. *Nature* **415**, 545–549

CHEN, M.T. AND WEISS, R. (2005) Artificial cell–cell communication in yeast *Saccharomyces cerevisiae* using signaling elements from *Arabidopsis thaliana. Nature Biotechnology* **23**, 1551–1555

CHOE, B.Y., KRISHNA, N.R. AND PRITCHARD, D.G. (1992) Proton NMR study on rhamnolipids produced by *Pseudomonas aeruginosa*. *Magnetic Resonance Chemistry* **30**, 1025–1026

COSTERTON, J.W., CHENG, K.J., GEESEY, G.G., LADD, T.I., NICKEL, J.C. *et al.* (1987) Bacterial biofilms in nature and disease. *Annual Reviews of Microbiology* **41**, 435-464

DAVEY, M.E. AND O'TOOLE, G.A. (2000) Microbial biofilms: from ecology to molecular genetics. *Microbiology and Molecular Biology Reviews* **64**, 847–867

DAVEY, M.E., CAIAZZA, N.C. AND O'TOOLE, G.A. (2003) Rhamnolipid surfactant production affects biofilm architecture in *Pseudomonas aeruginosa* PAO1. *Journal of Bacteriology* **185**, 1027–1036

DAVIES, D.G., PARSEK, M.R., PEARSON, J.P., IGLEWSKI, B.H., COSTERTON, J.W. *et al.* (1998) The involvement of cell-cell signals in the development of a bacterial biofilm. *Science* **280**, 295–298

DE KIEVIT, T.R. AND IGLEWSKY, B.H. (2000) Bacterial quorum sensing in pathogenic relationships. *Journal of Bacteriology* **68**, 4839–4849

DE KOSTER, C.G., VOS, B., VERSLUIS, C., HEERMA, W. AND HAVERKAMP, J. (1994) High-performance TLC/fast atom bombardment (tandem) mass spectrometry of *Pseudomonas* rhamnolipids. *Biological Mass Spectroscopy* **23**, 179–185

DESAI, J.D. AND BANAT, I.M. (1997) Microbial production of surfactants and their commercial potential, *Microbiology and Molecular Biology Reviews* **61**, 47–64

DEZIEL, E., LEPINE, F., DENNIE, D., BOISMENU, D., MAMER, O.A. *et al.* (1999) Liquid chromatography/mass spectrometry analysis of mixtures of rhamnolipids produced by *Pseudomonas aeruginosa* strain 57RP grown on mannitol or naphthalene. *Biochimica Biophysica Acta* **1440**, 244–252

DEZIEL, E., LEPINE, F., MILOT, S. AND VILLEMUR, R. (2000) Mass spectrometry monitoring of rhamnolipids from a growing culture of *Pseudomonas aeruginosa* strain 57RP. *Biochimica Biophysica Acta* **1485**, 145–152

DEZIEL, E., LEPINE, F., MILOT, S. AND VILLEMUR, R. (2003) *rhlA* is required for the production of a novel biosurfactant promoting swarming motility in *Pseudomonas aeruginosa*: 3-(3- hydroxyalkanoyloxy) alkanoic acids (HAAs), the precursors of rhamnolipids. *Microbiology* **149**, 2005–2013

DEZIEL, E., LEPINE, F., MILOT, S., HE, J., MINDRINOS, M.N. *et al.* (2004) Analysis of *Pseudomonas aeruginosa* 4-hydroxy-2-alkylquinolines (HAQs) reveals a role for 4-hydroxy-2-heptylquinoline in cell-to-cell communication. *Proceedings of the National Academy of Sciences (USA)* **101**, 1339–1344

DIGGLE, S.P., WINZER, K., CHHABRA, S.R., WORRALL, K.E., CAMARA, M. *et al.* (2003) The

Pseudomonas aeruginosa quinolone signal molecule overcomes the cell density-dependency of the quorum sensing hierarchy, regulates *rhl*-dependent genes at the onset of stationary phase and can be produced in the absence of LasR. *Molecular Microbiology* **50**, 29–43

DUAN, K., DAMMEL, C., STEIN, J., RABIN, H. AND SURETTE, M.G. (2003) Modulation of *Pseudomonas aeruginosa* gene expression by host microflora through interspecies communication. *Molecular Microbiology* **50,** 1477–1491

ENGEBRECHT, J., NEALSON, K. AND SILVERMAN, M. (1983) Bacterial bioluminescence: isolation and genetic analysis of functions from *Vibrio fischeri. Cell* **32**, 773–781

FINNERTY, W.R. (1994) Biosurfactants in environmental biotechnology. *Current Opinion in Biotechnology* **5**, 291–295

FUQUA, W.C., WINANS, S.C. AND GREENBERG, E.P. (1994) Quorum sensing in bacteria: the LuxR-LuxI family of cell density-responsive transcriptional regulators. *Journal of Bacteriology* **176**, 269–275

FUQUA, C. AND GREENBERG, E.P. (2002) Listening in on bacteria: acyl-homoserine lactone signalling. *Nature Reviews in Molecular and Cell Biology* **3,** 685–695

GALLAGHER, L.A., MCKNIGHT, S.L., KUZNETSOVA, M.S., PESCI, E.C. AND MANOIL, C. (2002) Functions required for extracellular quinolone signaling by *Pseudomonas aeruginosa. Journal of Bacteriology* **184**, 6472–6480

GAMBELLO, M.J. AND IGLEWSKI, B.H. (1991) Cloning and characterization of the *Pseudomonas aeruginosa lasR* gene, a transcriptional activator of elastase expression. *Journal of Bacteriology* **173**, 3000–3009

GAMBELLO, M.J., KAYE, S. AND IGLEWSKI, B.H. (1993) LasR of *Pseudomonas aeruginosa* is a transcriptional activator of the alkaline protease gene (*apr*) and an enhancer of exotoxin A expression. *Infection and Immunity* **61**, 1180–1184

GARTSHORE, J., LIM, Y.C. AND COOPER, D.G. (2000) Quantitative analysis of biosurfactants using Fourier Transform Infrared (FT-IR) spectroscopy. *Biotechnology Letters* **22**, 169–172

GILSON, L., KUO, A. AND DUNLAP, P.V. (1995) AinS and a new family of autoinducer synthesis proteins. *Journal of Bacteriology* **177**, 6946–6951

GIVSKOV, M., DE NYS, R., MANEFIELD, M., GRAM, L., MAXIMILIEN, R. *et al.* (1996) Eukaryotic intereference with homoserine lactone mediated prokaryotic signalling. *Journal of Bacteriology* **178**, 6618–6622

GUERRA-SANTOS, L., KAPPELI, O. AND FIECHTER, A. (1984) *Pseudomonas aeruginosa* biosurfactant production in continuous culture with glucose as carbon source. *Applied and Environmental Microbiology* **48**, 301–305

GUERRA-SANTOS, L., KAPPELI, O. AND FIECHTER, A. (1986) Dependence of *Pseudomonas aeruginosa* continuous culture biosurfactants production on nutritional and environmental factors. *Applied Microbiology and Biotechnology* **24**, 443–448

GUNTHER, N.W. IV., NUNEZ, A., FETT, W. AND SOLAIMAN, D.K. (2005) Production of rhamnolipids by *Pseudomonas chlororaphis*, a nonpathogenic bacterium. *Applied Environmental Microbiology* **71**, 2288–2293

HAFERBURG, D., HOMMEL, R., KLEBER, H.P., KLUGE, S., SCHUSTER, G. *et al.* (1987) Antiphytoviral activity of a rhamnolipid from *Pseudomonas aeruginosa. Acta Biotechnology* **7**, 353–356

HAUSSLER, S., NIMTZ, M., DOMKE, T., WRAY, V. AND STEINMETZ, I. (1998) Purification and characterization of a cytotoxic exolipid of *Burkholderia pseudomallei. Infection and*

Immunity **66**, 1588–1593

HAUSSLER, S., ZIEGLER, I., LOTTEL, A., VON GOTZ, F., ROHDE, M. *et al.* (2003) Highly adherent small-colony variants of *Pseudomonas aeruginosa* in cystic fibrosis lung infection. *Journal of Medical Microbiology* **52**, 295–301

HELBERT, J.R. AND BROWN, K.D. (1957) Color reaction of anthrone with monosaccharide mixtures and oligo- and polysaccharides containing hexuronic acids. *Analytical Chemistry* **29**, 1464–1466

HENTZER, M., TEITZEL, G.M., BALZER, G.J., HEYDORN, A., MOLIN, S. *et al.* (2001) Alginate over-production affects *Pseudomonas aeruginosa* biofilm structure and function. *Journal of Bacteriology* **183**, 5395–5401

HERMAN, D.C., ZHANG, Y. AND MILLER, R.M. (1997) Rhamnolipid (biosurfactant) effects on cell aggregation and biodegradation of residual hexadecane under saturated flow conditions. *Applied and Environmental Microbiology* **63**, 3622–3627

HEURLIER, K., WILLIAMS, F., HEEB, S., DORMOND, C., PESSI, G. *et al.* (2004) Positive control of swarming, rhamnolipid synthesis, and lipase production by the posttranscriptional RsmA/RsmZ system in *Pseudomonas aeruginosa* PAO1. *Journal of Bacteriology* **186**, 2936–2945

HEYD, M., KOHNERT, A., TAN, T.H., NUSSER, M., KIRSCHHOFER, F. *et al.* (2008) Development and trends of biosurfactant analysis and purification using rhamnolipids as an example. *Analytical and Bioanalytical Chemistry* **391**, 1579–1590

HOFFMAN, L.M., JENDRISAK, J.J., MEIS, R.J., GORYSHIN, I.Y. AND REZNIKOF, S.W. (2000) Transposome insertional mutagenesis and direct sequencing of microbial genomes. *Genetica* **108**, 19–24

HOLDEN, P.A., LAMONTAGNE, M.G., BRUCE, A.K., MILLER, W.G. AND LINDOW, S.E. (2002) Assessing the role of *Pseudomonas aeruginosa* surface-active gene expression in hexadecane biodegradation in sand. *Applied and Environmental Microbiology* **68**, 2509–2518

HOMMEL, R. (1990) Formation and physiological role of biosurfactants produced by hydrocarbon utilizing microorganisms. *Biodegradation* **1**, 107–119

ILORI, M.O., AMOBI, C.J. AND ODOCHA, A.C. (2005) Factors affecting the production of oil degrading *Aeromonas* sp. isolated from a typical environment. *Chemosphere* **61**, 985-992

INAKOLLU, S., HUNG, H.C. AND SHREVE, G.S. (2004). Biosurfactant enhancement of microbial degradation of various structural classes of hydrocarbon in mixed water systems. *Environmental Engineering Science* **21**, 463–469

INOUE, T., SHINGAKI, R. AND FUKUI, K. (2008) Inhibition of swarming motility of *Pseudomonas aeruginosa* by branched-chain fatty acids. *FEMS Microbiology Letters* **281,** 81–86

IRIE, Y., O'TOOLE, G. AND YUK, M.H. (2005) *Pseudomonas aeruginosa* rhamnolipids disperse *Bordetella bronchiseptica* biofilms. *FEMS Microbiology Letters* **250**, 237–243

JARVIS, F.G. AND JOHNSON, M.J.A. (1949) Glycolipid produced by *Pseudomonas aeruginosa*. *Journal of Americal Chemical Society* **71**, 4124–4126

JENSEN, P.O., BJARNSHOLT, T., PHIPPS, R., RASMUSSEN, T.B., CALUM, H. *et al.* (2007) Rapid necrotic killing of polymorphonuclear leukocytes is caused by quorum-sensing-controlled production of rhamnolipid by *Pseudomonas aeruginosa*. *Microbiology* **153**, 1329–1338

JIRASRIPONGPUN, K. (2002) The characterization of oil-degrading microorganisms from lubricating oil contaminated (scale) soil. *Letters in Applied Microbiology* **35**, 296–300

JOHNSON, M.K. AND BOESE-MARAZZO, O. (1980) Production and properties of heat-stable extracellular hemolysin from *Pseudomonas aeruginosa. Infection and Immunity* **29**, 1028–1033

KING, A.T., DAVEY, M.R., MELLOR, I.R., MULLIGAN, B.J. AND LOWE, K.C. (1991) Surfactant effects on yeast cells. *Enzyme and Microbial Technology* **13**, 148–153

KLEEREBEZEM, M., QUADRI, L.E.N., KUIPERS, O.P. AND DEVOS, W.M. (1997) Quorum sensing by peptide pheromones and two component signal transduction systems in Gram positive bacteria. *Molecular Microbiology* **24**, 895–904

KOCH, A.K., KAPPELI, O., FIECHTER, A. AND REISER, K. (1991) Hydrocarbon assimilation and biosurfactant production in *Pseudomonas aeruginosa* mutants. *Journal of Bacteriology* **173**, 4212–4219

KOHLER, T., CURTY, L.K., BARJA, F., VAN DELDEN, C. AND PECHERE, J.C. (2000) Swarming of *Pseudomonas aeruginosa* is dependent on cell-to-cell signaling and requires flagella and pili. *Journal of Bacteriology* **182**, 5990–5996

KOHLER, T., DUMAS, J.L. AND VAN DELDEN, C. (2007) Ribosome protection prevents azithromycin-mediated quorum sensing modulation and stationary phase killing of *Pseudomonas aeruginosa. Antimicrobial Agents Chemotherapy* **51**, 4243–4248

KOWNATZKI, R., TUMMLER, B. AND DORING, G. (1987) Rhamnolipid of *Pseudomonas aeruginosa* in sputum of cystic fibrosis patients. *Lancet* **1**, 1026–1027

KUNIHO, N., YOSHIMOT, A. AND YAMADA, Y. (1998) Correlation between autoinducers and rhamnolipids production by *Pseudomonas aeruginosa* IFO 3924. *Journal of Fermentation Bioengineering* **86**, 608–610

LANG, S. AND WULLBRANDT, D. (1999) Rhamnose lipids-biosynthesis, *microbial* production and application potential. *Applied Microbiology and Biotechnology* **51**, 22–32

LATIFI, A., FOGLINO, M., TANAKA, K., WILLIAMS, P. AND LAZDUNSKI, A. (1996) A hierarchical quorum-sensing cascade in *Pseudomonas aeruginosa* links the transcriptional activators LasR and RhIR (VsmR) to expression of the stationary-phase sigma factor RpoS. *Molecular Microbiology* **21**, 1137–1146

LAZDUNSKI, A.M., VENTRE, I. AND STURGIS, J.N. (2004) Regulatory circuits and communication in gram-negative bacteria. *Nature Reviews in Microbiology* **2**, 581-592

LEITERMANN, F., SYLDATK, C. AND HAUSMANN, R. (2008) Fast quantitative determination of microbial rhamnolipids from cultivation broths by ATR-FTIR Spectroscopy. *Journal of Biological Engineering* **2**, 13 doi:10.1186/1754-1611-2-13

LEPINE, F., DEZIEL, E., MILOT, S. AND VILLEMUR, R. (2002) Liquid chromatographic/mass spectrometric detection of the 3-(3-hydroxyalkanoyloxy)alkanoic acid precursors of rhamnolipids in *Pseudomonas aeruginosa* cultures. *Journal of Mass Spectrometry* **37**, 41–46

LINHARDT, R.J., BAKHIT, R., DANIELS, L., MAYERL, F. AND PICKENHAGEN, W. (1989) Microbially produced rhamnolipid as a source of rhamnose. *Biotechnology and Bioengineering* **33, 365–368**

LUBELSKI, J., RINK, R., KHUSAINOV, R., MOLL, G.N. AND KUIPERS, O.P. (2008) Biosynthesis, immunity, regulation, mode of action and engineering of the model lantibiotic nisin.

Cell and Molecular Life Sciences **65**, 455–476

Maier, R.M. and Soberon-chavez, G. (2000) *Pseudomonas aeruginosa* rhamnolipids: biosynthesis and potential applications. *Applied Microbiology and Biotechnology* **54**, 625–633

Mashburn, L.M. and Whiteley, M. (2005) Membrane vesicles traffic signals and facilitate group activities in a prokaryote. *Nature* **437,** 422–425

Mashburn-warren, L., Howe, J., Garidel, P., Richter, W., Steiniger, F. *et al.* (2008) Interaction of quorum signals with outer membrane lipids: insights into prokaryotic membrane vesicle formation. *Molecular Microbiology* **69**, 491–502

Mashburn-warren, L., Howe, J., Brandenburg, K. and Whiteley, M. (2009) Structural requirement of the *Pseudomonas* quinolone signal for membrane vesicle stimulation. *Journal of Bacteriology* **191**, 3411–3414

Masse, E., Majdalani, N. and Gottesman, S. (2003) Regulatory roles for small RNAs in bacteria. *Current Opinion in Microbiology* **6**, 120–124

Mata-sandoval, J.C., Karns, J. and Torrents, A. (1999) HPLC method for the characterization of rhamnolipid mixtures produced by *P. aeruginosa* UG2 on corn oil. *Journal of Chromatography* A **864**, 211–220

Mata-sandoval, J.C., Karns, J. and Torrents, A. (2000) Effect of rhamnolipids produced by *Pseudomonas aeruginosa* UG2 on the solubilization of pesticides. *Environmental Science and Technology* **34**, 4923–4930

Mcclure, C.D. and Schiller, N.L. (1992) Effects of *Pseudomonas aeruginosa* rhamnolipids on human monocyte derived macrophages. *Journal of Leukocyte Biology* **51**, 97–102

Mcknight, S.L., Iglewski, B.H. and Pesci, E.C. (2000) The *Pseudomonas* quinolone signal regulates *rhl* quorum sensing in *Pseudomonas aeruginosa. Journal of Bacteriology* **182**, 2702–2708

Mclean, R.J.C., Whiteley, M., Stickler, D.J. and Fuqua, W.C. (1997) Evidence of autoinducer activity in naturally occurring biofilms. *FEMS Microbiology Letters* **154**, 259–263

Miller, M.B. and Bassler, B.L. (2001) Quorum sensing in bacteria. *Annual Reviews of Microbiology* **55**, 165–199

Monteiro, S.A., Sassaki, G.L., De souza, L.M., Meira, J.A., De araujo, J.M. *et al.* (2007) Molecular and structural characterization of the biosurfactant produced by *Pseudomonas aeruginosa* DAUPE 614. *Chemical and Physical Lipids* **147**, 1–13

Morihara, K. and Homma, J.Y. (1985) Pseudomonas protease. In *Bacterial Enzymes and Virulence* eds. I.A. Holder, pp 41–79. Boca Raton, FL, CRC Press.

Muller-hurtig, R., Blaszczyk, R., Wagner, F. and Kosaric, N. (1993) Biosurfactants for environmental control. In *Biosurfactant* eds. N. Kosaric, pp 447–469. New York, Marcel Dekker Inc.

Mulligan, C.N., Yong, R.N. and Gibbs, B.F. (2001) Heavy metal removal from sediments by biosurfactants. *Journal of Hazardous Materials* **85**, 111–125

Mulligan, C.N. (2005) Environmental applications for biosurfactants. *Environmental Pollution* **133**, 183–198

Nealson, K.H., Platt, T. and Hastings, J.W. (1970) Cellular control of the synthesis and activity of the bacterial luminescent system. *Journal of Bacteriology* **104**, 313–322

Nealson, K.H. and Hastings, J.W. (1979) Bacterial bioluminescence: its control and ecological significance. *Microbiology Reviews* **43**, 496–518

NOORDMAN, W.H. AND JANSSEN, D.B. (2002) Rhamnolipid stimulates uptake of hydrophobic compounds by *Pseudomonas aeruginosa*. *Applied and Environmental Microbiology* **68**, 4502–4508

OCHOA-LOZA, F.J., ARTIOLA, J.F. AND MAIER, R.M. (2001) Stability constants for the complexation of various metals with a rhamnolipid biosurfactant. *Journal of Environmental Quality* **30**, 479–485

OCHSNER, U.A., FIECHTER, A. AND REISER, J. (1994) Isolation, characterization, and expression in *Escherichia coli* of the *Pseudomonas aeruginosa rhlAB* genes encoding a rhamnosyltransferase involved in rhamnolipid biosurfactant synthesis. *Journal of Biological Chemistry* **269**, 19787–19795

OCHSNER, U.A., KOCH, A.K., FIECHTER, A. AND REISER, J. (1994) Isolation and characterization of a regulatory gene affecting rhamnolipid biosurfactant synthesis in *Pseudomonas aeruginosa*. *Journal of Bacteriology* **176**, 2044–2054

OCHSNER, U.A. AND REISER, J. (1995) Autoinducer-mediated regulation of rhamnolipid biosurfactant synthesis in *Pseudomonas aeruginosa*. *Proceedings of the National Academy of Sciences (USA)* **92**, 6424–6428

OCHSNER, U., REISER, J., FIECHTER, A. AND WITHOLT, B. (1995) Production of *Pseudomonas aeruginosa* rhamnolipid biosurfactants in heterologous hosts. *Applied and Environmental Microbiology* **61**, 3503–3506

O'TOOLE, G.A. AND KOLTER, R. (1998) Flagellar and twitching motility are necessary for *Pseudomonas aeruginosa* biofilm development. *Molecular Microbiology* **30**, 295–304

PAMP, S.J. AND TOLKER-NIELSEN, T. (2007) Multiple roles of biosurfactants in structural biofilm development by *Pseudomonas aeruginosa*. *Journal of Bacteriology* **189**, 2531–2539

PAPPAS, K., WEINGART, C.L. AND WINANS, S.C. (2004) Chemical communication in proteobacteria: biochemical and structural studies of signal synthases and receptors required for intercellular signaling. *Molecular Microbiology* **53**, 755-769

PARSEK, M.R. AND GREENBERG, E.P. (2000) Acyl-homoserine lactone quorum sensing in Gram-negative bacteria: a signaling mechanism involved in associations with higher organisms. *Proceedings of the National Academy of Sciences (USA)* **97**, 8789–8793

PASSADOR, L., COOK, J.M., GAMBELLO, M.J., RUST, L. AND IGLEWSKI, B.H. (1993) Expression of *Pseudomonas aeruginosa* virulence genes requires cell-to-cell communication. *Science* **260**, 1127–1130

PEARSON, J.P., GRAY, K.M., PASSADOR, L., TUCKER, K.D., EBERHARD, A. *et al.* (1994) Structure of the autoinducer required for expression of *P. aeruginosa* virulence genes. *Proceedings of the National Academy of Sciences (USA)* **91**, 197–201

PEARSON, J.P., PASSADOR, L., IGLEWSKI, B.H. AND GREENBERG, E.P. (1995) A second *N*-acyl homoserine lactone signal produced by *Pseudomonas aeruginosa*. *Proceedings of the National Academy of Sciences (USA)* **92**, 1490–1494

PEARSON, J.P., PESCI, E.C. AND IGLEWSKI, B.H. (1997) Roles of *Pseudomonas aeruginosa* las and rhl quorum-sensing systems in control of elastase and rhamnolipid biosynthesis genes. *Journal of Bacteriology* **179**, 5756–5767

PESCI, E.C., MILBANK, J.B., PEARSON, J.P., MCKNIGHT, S., KENDE, A.S. *et al.* (1999) Quinolone signaling in the cell-to-cell communication system of *Pseudomonas aeruginosa. Proceedings of the National Academy of Sciences (USA)* **96**, 11229–11234

PIERSON, L.S., KEPPENNE, V.D. AND WOOD, D.W. (1994) Phenazine antibiotic biosynthesis in *Pseudomonas aureofaciens* is regulated by PhzR in response to cell density. *Journal of Bacteriology* **176**, 3966–3974

PINZON, N.M. AND JU, L.K. (2009) Analysis of rhamnolipid biosurfactants by methylene blue complexation. *Applied Microbiology and Biotechnology* **82**, 975–981

RAHIM, R., OCHSNER, U.A., OLVERA, C., GRANINGER, M., MESSNER, P. *et al.* (2001) Cloning and functional characterization of the *Pseudomonas aeruginosa* rhlC gene that encodes rhamnosyltransferase 2, an enzyme responsible for dirhamnolipid biosynthesis. *Molecular Microbiology* **40**, 708–718

RAHMAN, K.S.M., RAHMAN, T.J., MCCLEAN, S., MARCHANT, R. AND BANAT, I.M. (2002) Rhamnolipid biosurfactant production by strains of *Pseudomonas aeruginosa* using low-cost raw materials. *Biotechnology Progress* **18**, 1277–1281

RAHMAN, K.S.M., RAHMAN, T.J., KOURKOUTAS, Y., PETSAS, I., MARCHANT, R. *et al.* (2003) Enhanced bioremediation of n-alkanes in petroleum sludge using bacterial consortium amended with rhamnolipid and micronutrients. *Bioresource Technology* **90**, 159–168

RAHMAN, K.S.M. AND GAKPE, E. (2008) Production, characterization and applications of biosurfactants - Review. *Biotechnology* **7**, 360–370

RENDELL, N.B., TAYLOR, G.W., SOMERVILLE, M., TODD, H., WILSON, R. *et al.* (1990) Characterization of *Pseudomonas* rhamnolipids. *Biochimica Biophysica Acta* **1045**, 189–193

ROBERT, M., MERCADE, M.E., BOSCH, M.P., PARRA, J.L., ESPUNY, M.J. *et al.* (1989) Effect of carbon source on biosurfactant production by *Pseudomonas aeruginosa* 44T1. *Biotechnology Letters* **11**, 871–874

ROSENBERG, E. AND RON, E.Z. (1999) High and low molecular mass microbial surfactants. *Applied Microbiology and Biotechnology* **52**, 154–162

RUMBAUGH, K.P., GRISWOLD, J.A. AND HAMOOD, A.N. (2000) The role of quorum sensing in the in vivo virulence of *Pseudomonas aeruginosa. Microbes and Infection* **2**, 1721–1731

SAUER, K., CAMPER, A.K., EHRLICH, D., COSTERTON, J.W. AND DAVIES, D.G. (2002) *Pseudomonas aeruginosa* displays multiple phenotypes during development as a biofilm. *Journal of Bacteriology* **184**,1140–1154

SCHAUDER, S., SHOKAT, K., SURETTE, M.G. AND BASSLER, B.L. (2001) The LuxS family of bacterial autoinducers: biosynthesis of a novel quorum-sensing signal molecule. *Molecular Microbiology* **41**, 463–476

SHROUT, J.D., CHOPP, D.L., JUST, C.L., HENTZER, M., GIVSKOV, M. *et al.* (2006) The impact of quorum sensing and swarming motility on *Pseudomonas aeruginosa* is nutritionally conditional. *Molecular Microbiology* **62**, 1264–1277

SIEGMUND, I. AND WAGNER, F. (1991) New method for detecting rhamnolipids excreted by *Pseudomonas* species during growth on mineral agar. *Biotechnology Techniques* **5**, 265–268

SIM, L., WARD, O.P. AND LI, Z.Y. (1997) Production and characterization of a biosurfactant isolated form *Pseudomonas aeruginosa* UW-1. *J Ind Microbiol Biotechnol* **19**, 232–238.

SINGH, P.K., SCHAEFER, A.L., PARSEK, M.R., MONINGER, T.O., WELSH, M.J. *et al.* (2000)

Quorum-sensing signals indicate that cystic fibrosis lungs are infected with bacterial biofilms. *Nature* **407**, 762–764

SMITH, R.S. AND IGLEWSKI, B.H. (2003) *P. aeruginosa* quorum-sensing systems and virulence. *Current Opinion in Microbiology* **6**, 56–60

SOTIROVA, A.V., SPASOVA, D.I., VASILEVA-TONKOVA, E. AND GALABOVA, D.N. (2007) Effects of rhamnolipid-biosurfactant on cell surface of *Pseudomonas aeruginosa. Microbiology Research* **164**, 297–303

SOTIROVA, A.V., SPASOVA, D.I., GALABOVA, D.N., KARPENKO, E. AND SHULGA, A. (2008) Rhamnolipid biosurfactant permeabilizing effects on Gram positive and Gram negative bacterial strains. *Current Microbiology* **56**, 639–644

STANGHELLINI, M.E. AND MILLER, R.M. (1997) Biosurfactants: their identity and potential efficacy in the biological control of zoosporic plant pathogens. *Plant Disease* **81**, 4–12

STANGHELLINI, M.E., MILLER, R.M., RASMUSSEN, S.L., KIM, D.H. AND ZHANG, Y. (1998) Microbially produced rhamnolipids (biosurfactants) for the control of plant pathogenic zoosporic fungi. United States Patent 5767090

SYLDATK, C., LANG, S., MATULOVIC, U. AND WAGNER, F. (1985) Production of four interfacial active rhamnolipids from *n*-alkanes or glycerol by resting cells of *Pseudomonas aeruginosa* species DSM 2874. *Z Naturforsch C* **40**, 61–67

TATJANA, P. AND GORAN, P. (2007) Use of rhamnolipids in wound healing, treating burn shock, atherosclerosis, organ transplants, depression, schizophrenia and cosmetics. US Patent 7262171

TODER, D.S., GAMBELLO, M.J. AND IGLEWSKI, B.H. (1991) *Pseudomonas aeruginosa* LasA: a second elastase gene under transcriptional control of LasR. *Molecular Microbiology* **5**, 2003–2010

TODER, D.S., FERRELL, S.J., NEZEZON, J.L., RUST, L. AND IGLEWSKI, B.H. (1994) *lasA* and *lasB* genes of *Pseudomonas aeruginosa*: analysis of transcription and gene product activity. *Infection and Immunity* **62**, 1320–1327

TREMBLAY, J., RICHARDSON, A.P., LEPINE, F. AND DEZIEL, E. (2007) Self-produced extracellular stimuli modulate the *Pseudomonas aeruginosa* swarming motility behaviour. *Environmental Microbiology* **9**, 2622–2630

TU, K.C., WATERS, C.M., SVENNINGSEN, S.L. AND BASSLER, B. (2008) A small RNA-mediated feedback loop controls quourm-sensing dynamics in *Vibrio harveyi. Molecular Microbiology* **70**, 896–907

TULEVA, B.K., IVANOV, G.R. AND CHRISTOVA, N.E. (2002) Biosurfactant production by a new *Pseudomonas putida* strain. *Z Naturforsch C* **57**, 356–360

TUMMLER, B., KOOPMANN, U., GROTHUES, D., WEISSBRODT, H., STEINKAMP, G. *et al.* (1991) Nosocomial acquisition of *Pseudomonas aeruginosa* by cystic fibrosis patients. *Journal of Clinical Microbiology* **29**, 1265-1267

URUM, K., GRIGSON, S., PEKDEMIR, T. AND MCMENAMY, S. (2006) A comparison of the efficiency of different surfactants for removal of crude oil from contaminated soils. *Chemosphere* **62**, 1403–1410

VAN DYKE, M.I., COUTURE, P., BRAUER, M., LEE, H. AND TREVORS, J.T. (1993) *Pseudomonas aeruginosa* UG2 rhamnolipid biosurfactants: structural characterization and their use in removing hydrophobic compounds from soil. *Canadian Journal of Microbiology* **39**, 1071–1078

VAN HAESENDONCK, INGRID PAULA, H., VAN ZEVEREN, EMMANUEL CLAUDE, A.L.B. (2004)

Rhamnolipids in bakery products PURATOS NV (BE) EP1415538.

VOLKERING, F., BREURE, A.M., VAN ANDEL, J.G. AND RULKENS, W.H. (1995) Influence of nonionic surfactants on bioavailability and biodegradation of polycyclic aromatic hydrocarbons. *Applied and Environmental Microbiology* **61**, 1699–1705

WANG, D., DING, X. AND RATHER, P.N. (2001) Indole can act as an extracellular signal in *Escherichia coli. Journal of Bacteriology* **183**, 4210–4216

WANG, X., GONG, L., LIANG, S., HAN, X., ZHU, C. *et al.* (2005) Algicidal activity of rhamnolipid biosurfactants produced by *Pseudomonas aeruginosa. Harmful Algae* **4**, 433–443

WANG, Q.H., FANG, X., BAI, B.J., LIANG, X.L., SHULER, P.L. *et al.* (2007) Engineering bacteria for production of rhamnolipid as an agent for enhanced oil recovery. *Biotechnology and Bioengineering* **98**, 842–853

WATERS, C.M. AND BASSLER, B.L. (2005) Quorum sensing: cell-to-cell communication in bacteria. *Annual Reviews in Cell Developmental Biology* **21**, 319–346

WHITEHEAD, N.A., BARNARD, A.M., SLATER, H., SIMPSON, N.J. AND SALMOND, G.P. (2001) Quorum-sensing in gram-negative bacteria. *FEMS Microbiology Review* **25**, 365–404

WIENHOLDS, E. AND PLASTERK, R.H. (2005) MicroRNA function in animal development. *FEBS Letters* **579**, 5911–5922

WILLIAMS, P., WINZER, K., CHAN, W.C. AND CAMARA, M. (2007) Look who's talking: communication and quorum sensing in the bacterial world. *Philosophical Transactions of the Royal Society London B: Biological Sciences* **362**, 1119-1134

WILLIAMS, P. AND CAMARA, M. (2009) Quorum sensing and environmental adaptation in *Pseudomonas aeruginosa*: a tale of regulatory networks and multifunctional signal molecules. *Current Opinion in Microbiology* **12**,182-91

WINSON, M.K., CAMARA, M., LATIFI, A., FOGLINO, M., CHHABRA, S.R. *et al.* (1995) Multiple N-acyl-Lhomoserine lactone signal molecules regulate production of virulence determinants and secondary metabolites in *Pseudomonas aeruginosa. Proceedings of the National Academy of Sciences (USA)* **92**, 9427–9431

WU, J.Y., YEH, K.L., LU, W.B., LIN, C.L. AND CHANG, J.S. (2008) Rhamnolipid production with indigenous *Pseudomonas aeruginosa* EM1 isolated from oil-contaminated site. *Bioresource Technology* **99**, 1157–1164

YOU, L., COX, R.S., WEISS, R. AND ARNOLD, F.H. (2004) Programmed population control by cell–cell communication and regulated killing. *Nature* **428**, 868–871

ZHANG, Y. AND MILLER, R.M. (1994) Effect of a *Pseudomonas* rhamnolipid biosurfactant on cell hydrophobicity and biodegradation of octadecane. *Applied and Environmental Microbiology* **60**, 2101–2106

ZHANG, Y. AND MILLER, R.M. (1995) Effect of rhamnolipid (biosurfactant) structure on solubilization and biodegradation of n-alkanes. *Applied and Environmental Microbiology* **61**, 2247–2251

ZULIANELLO, L., CANARD, C., KOHLER, T., CAILLE, D., LACROIX, J.S. *et al.* (2006) Rhamnolipids are virulence factors that promote early infiltration of primary human airway epithelia by *Pseudomonas aeruginosa. Infection and Immunity* **74**, 3134–3147

Biotechnology and Genetic Engineering Reviews - Vol. 27, 185-216 (2010)

Advances in nanopatterned and nanostructured supported lipid membranes and their applications

ERIK REIMHULT*, MARTINA K. BAUMANN, STEFAN KAUFMANN, KARTHIK KUMAR AND PHILIPP R. SPYCHER

Department for Materials Science, ETH Zurich, 8093 Zurich, Swizerland

Abstract

Lipid membranes are versatile and convenient alternatives to study the properties of natural cell membranes. Self-assembled, artificial, substrate-supported lipid membranes have taken a central role in membrane research due to a combination of factors

*To whom correspondence may be addressed (erik.reimhult@mat.ethz.ch)

Abbreviations: αHL: alpha-Hemolysin; AFM: Atomic force microscopy; AHP cells: Adult hippocampal stem/progenitor *cells;* AMP: Antimicrobial peptides; ATR-FTIR: Attenuated total reflection Fourier transform infrared spectroscopy; BSA: Bovine serum albumin; CcO: Cytochrome c oxidase; CPP: Cell penetrating peptides; DiynePC: 1,2-bis(10,12-tricosadiynoyl)-sn-glycero-3-phosphocholine; DNA: Deoxyribonucleic acid; DMPC: 1,2-dimyristoyl-*sn*-glycero-3-phosphocholine; DMPS: 1,2-dimyristoyl-*sn*-glycero-3-phospho-L-serine; DODAP: 1,2-dioleoyl-3-dimethylammonium-propane; DOPE: 1,2-dioleoyl-sn-glycero-3-phosphoethanolamine; DPhyPC: 1,2-Diphytanoyl-*sn*-Glycero-3- Phosphocholine; DPhyTL: 2,3-di-O-phytanyl-*sn*-glycerin-1-tetraethylenglycol-lipoic acid ester; DPPE: 1,2-dihexadecanoyl-sn-glycero-3-phosphoethanolamine; DPS: 1,2-dimyristoyl phosphatidlethanolamine-N-(polyethylene glycol-triethoxysilane); DSPE: 1,2-distearoyl-sn-glycero-3-phosphoethanolamine; DSPE–PEG3400–NHS: 1,2-Distearoyl-*sn*-glycero-3-phosphoethanolamine–poly(ethylene glycol)-N-hydroxysuccinimide; Egg PC: L-α-phosphatidylcholine (Egg, Chicken); EIS: Electrochemical impedance spectroscopy; FLIC: Fluorescence interference contrast microscopy; FRAP: Fluorescence recovery after photobleaching; GFP: Green fluorescent protein; HA: Hyaluronic acid; hDOR: human delta-opioid receptor; HF: Hydrofluoric acid; ITO: Indium tin oxide; LB/VF: Langmuir-Blodgett/vesicle fusion technique; l_d: liquid disordered; l_o: iquid ordered; LUV: large unilamellar vesicles; LSPR: Localised surface plasmon resonance; NBD-PE: 1,2-dioleoyl-*sn*-glycero-3-phosphoethanolamine-N-(7-nitro-2-1,3-benzoxadiazol-4-yl); NIPAAM: N-isopropylacrylamide; Ni-NTA-DOGS: 1,2-Dioleyl-*sn*-glycero-3-[*N*-(5-amino-1-carboxypentyliminodiacetic acid) succinyl] nickel salt; NTA: *N*-(5-Amino-1-carboxypentyl)iminodiacetic Acid; OTS: Octadecyltrichlorosilane; OWLS: Optical waveguide light mode spectroscopy; PDMS: Polydimethylsiloxane; PEG: Poly(ethylene glycol); PEG-DMA: poly(ethylene glycol) dimethacrylate: PEMA: poly(ethylene-*alt*-maleic anhydride); PLL-*g*-PEG: poly(L-Lysine)-*graft*-poly(ethylene glycol); POMA: poly(octadecene-*alt*-maleic anhydride); POPC: Palmitoyl-oleoyl-phosphatidyl-choline; POPG: 1-palmitoyl-2-oleoyl-sn-glycero-3-phospho-(1'-rac-glycerol); PPMA: poly(propene-*alt*-maleic anhydride); PPy: Polypyrrole; QCM-D: Quartz crystal microbalance with dissipation monitoring; R_{max}: Maximum roughness; R_{min}: Minimum roughness; R_{rms}: Root-mean-square roughness; RICM: Reflection interference contrast microscopy; SAM: self-assembled monolayer; SFG: Sum frequency generation spectroscopy; SLB: Supported lipid bilayer; SM: Sphingomyelin; SOPC: 1-stearoyl-2-oleoyl-*sn*-glycero-3-phosphocholine; SPR: Surface plasmon resonance; TCPS: hydrophilic tissue culture polystyrene; TEM: Transmission electron microscopy; TEOS: Tetraethyl orthosilicate; TIRF: Total internal reflection fluorescence microscopy; UV: Ultraviolet such as ease of creation, control over complexity, stability and the applicability of a large range of different analytical techniques. While supported lipid bilayers have been investigated for several decades, recent advances in the understanding of the assembly of such membranes from liposomes have spawned a renaissance in the field.

Supported lipid bilayers are a highly promising tool to study transmembrane proteins in their native state, an application that could have tremendous impact on, e.g. drug discovery, development of biointerfaces and as platforms for glycomics and probing of multivalent binding which requires ligand mobility. Parallel advances in microfluidics, biosensor design, micro- and nanofabrication have converged to bring self-assembled supported lipid bilayers closer to a versatile and easy to use research tool as well as closer to industrial applications. The field of supported lipid bilayer research and application is thus rapidly expanding and diversifying with new platforms continuously being proposed and developed.

In order to use supported lipid bilayers for such applications several advances have to be made: decoupling of the membrane from the support while maintaining it close to the surface, making use of biologically relevant lipid compositions, patterning of lipid membranes into arrays, and application to nanostructured substrates and sensors. This review summarizes recent advances in the field which addresses these challenges.

Introduction

Lipid membranes are versatile and convenient alternatives to study the properties of natural cell membranes. Self-assembled, artificial, substrate-supported lipid membranes have taken a central role in membrane research due to a combination of factors such as ease of formation, control over complexity, stability and the applicability of a large range of different analytical techniques, including highly sensitive surface probes. In particular research on membrane platforms has been stimulated by the possibility of studying membrane proteins in a near-native environment using label free surface sensitive techniques such as quartz crystal microbalance with dissipation monitoring (QCM-D), surface plasmon resonance (SPR), waveguide spectroscopy, electrochemical impedance spectroscopy (EIS) and scanning probe microscopy. This is especially attractive for elucidating protein binding and other interactions as fluorescent labels might alter interaction sites and protein function.

While so-called supported lipid bilayers (SLBs), which denote planar lipid membranes resting on or near a supporting inorganic or organic substrate, have been investigated for several decades recent advances in the understanding of the assembly of such membranes from liposomes have spawned a renaissance in the field. Formation of SLBs using, for example, liposome fusion combined with sensing on planar substrates, has been extensively discussed in the literature (Richter *et al.*, 2006; Rossetti *et al.*, 2005; Merz *et al.*, 2008; Janshoff and Steinem, 2006; Tanaka and Sackmann 2005). However, the number of inorganic supports which can be used for SLB formation by self-assembly from liposomes is still mostly restricted to a select few oxide surfaces (Richter and Brisson, 2005; Rossetti *et al.*, 2005; Reimhult *et al.*, 2003; Merz *et al.*, 2008; Kumar *et al.*, 2009). To extend that range, other metal or metal oxide substrates can be modified with hydrophobic tethers or polymers which have been developed to promote SLB formation and impart additional advantages such as stability and additional hydrated spacer volume from the substrate (Tanaka and Sackmann, 2005; Sinner and Knoll, 2001). However, this often complicates the assembly of the system and the sensor readout. Parallel advances in microfluidics, biosensor design, micro- and nanofabrication have combined to bring self-assembled SLBs closer to a versatile and easy to use research tool as well as closer to implementation in important applications in biosensing, drug discovery, membrane electrophoresis and biointerfaces. The field of SLB research and application is thus rapidly expanding and diversifying with new platforms continuously being proposed and developed.

This review summarizes recent advances in some key areas, which will allow SLBs to be applied on a larger scale to screen membrane processes in increasingly biomimetic membranes but does not attempt the impossible feat to cover all advances and trace all areas of artificial membrane research. The focus is on recent advances in polymer spacer supports, which will allow incorporation of large membrane proteins, substrate topography as a means to control membrane curvature and its influence on SLB self-assembly, patterning of lipid membranes for creation of arrays and some recent steps that have been taken to move beyond simple model lipid mixtures to specifically recreate natural lipid membrane compositions and asymmetries in SLB systems.

Biomimicking supported lipid bilayers

Biological membranes are differentiated through unique lipid compositions as well as a plethora of associated molecules and proteins for different organelles, leading to diverse physical properties and thus biological functions. Studying events at the cell membrane *in vivo* is quite challenging due to the high turnover and number of its components as well as complexity of the molecules involved. Therefore, simplified (reduced) systems with control over all components present a very appealing platform to study processes such as protein-lipid interactions, and unique properties of the membrane components itself such as raft (ordered domain) formation. The precision with which lipid mixtures can be adjusted to mimic the desired aspect of the biological question at hand is an advantage as complexity faced in living cells can be reduced to address only one interaction at a time. However, biological membranes display complex asymmetries and heterogeneities, which are still challenging to recreate in supported lipid bilayers. In the following section recent advances in using self-assembled SLBs to mimic native cell membranes will be described.

SLBS TO MIMIC DIFFERENT TYPES OF CELL MEMBRANES

SLBS TO STUDY CELL SURFACE INTERACTIONS

Getting closer to the native membrane composition in SLBs in order to study specific interactions of key species in the membrane has been an important goal for several years. In order to apply surface sensitive techniques also to native biological membranes several groups have aimed to transfer native cellular membranes to a solid support. Tanaka and coworkers have transferred human erythrocyte membranes to polymer coated substrates in an orientation selective manner (Tanaka *et al.*, 2001). To probe native cell membranes in an SLB platform Vogel and coworkers have detached the plasma membrane of HEK-293 cells and stabilized it on a poly-L-lysine (PLL)-coated glass slide (Perez *et al.*, 2006). Properties of the original cell membranes such as fluidity of the leaflets as well as protein composition were found to be conserved by FRAP microscopy. It was shown that recombinant membrane receptors involved in signal transduction pathways (α1b-adrenergic receptor, a receptor of the GPCR and a ligand gated ion channel (LGIC) the 5HT3) were present in the detached membrane sheets and could be visualized as fusion complexes with fluorescently labeled proteins (Perez *et al.*, 2006). Functionality of the GFP labeled α1b-adrenergic receptor was tested with a fluorescently labeled (Bodipy) antagonist and co-localization of the fluorescent signals from receptors and binding partners were observed. Due to dramatic background fluorescence reduction in the supported membrane system as compared to live cells single molecule tracking of labeled lipid probes could be achieved as well.

In order to reduce complexity and achieve detailed control over the composition as well as lateral homogeneity of SLB functionalized biosensors, several groups have instead pursued assembly of synthetic or reconstituted systems at the solid interface. Merz and coworkers have demonstrated formation of SLBs mimicking aspects of *E.coli* lipid membranes using vesicle fusion on standard biosensor surfaces (glass, QCM-D and waveguide sensors with SiO_2, TiO_2 or indium tin oxide (ITO) coating) (Merz *et al.*, 2008). They developed protocols using Ca^{2+} containing buffers to tune vesicle adsorption, SLB formation and membrane aggregation with bacteria mimicking lipid compositions ranging from POPC/POPG to *E.coli* total lipid extract (Merz *et al.*, 2008). By a combined analysis with QCM-D, OWLS and FRAP of the SLBs they found that the compositionally most complex SLBs formed from *E.coli* total lipid extract were laterally connected but non-planar. Therefore they propose to always use a combination of techniques for adequate characterization of SLBs from complex lipid mixtures (Merz *et al.*, 2008).

Signaling and activation of proteins through special membrane lipids such as phosphatidylinositol-4,5-bisphosphate (PIP2) has received increasing interest in recent years. These lipids have rapid turnover according to demand *in vivo*. Thus, developing platforms with a controlled and stable number of e.g PIP2 is vital for studying interactions and signaling cascades occurring at the cell membranes in detail. Using POPC lipid bilayers with 10 mol% PIP2 formed by fusion of large unilamellar vesicles (LUVs) on silica substrates, Steinem, Janshoff and co-workers have studied the activation of the actin binding protein ezrin via PIP2 (Janke *et al.*, 2008; Herrig *et al.*, 2006). They showed that ezrin is activated upon changing conformation after binding to PIP2 and thus able to interact with F-actin. Force-distance measurements using F-actin functionalized colloidal probes were performed to quantitatively assess the maximal adhesion forces and the work of adhesion of the ezrin - F-actin interface (Janke *et al.*, 2008).

Barfoot and colleagues developed an SLB platform with an attached F-actin network to mimic the eukaryotic cell membrane. The developed minimal model system of a eukaryotic cell surface can be used to study events at cell membranes which require a cytoskeleton in a controlled SLB environment allowing the use of surface sensitive techniques (Barfoot *et al.*, 2008). They reconstituted the transmembrane protein ponticulin in egg-PC supported lipid bilayers and EO3-cholesterol tethered bilayers and observed similar bilayer properties as for pure egg-PC bilayers with SPR, QCM-D, AFM and FRAP. F-actin cytoskeletal fibers were then attached through the high affinity link to ponticulin (Barfoot *et al.*, 2008). Specific interactions of F-actin with ponticulin were measured with several complementary techniques. Physical characteristics of the F-actin filaments adsorbed on SLB were mapped with AFM in agreement with small-angle X-ray scattering for the height of the fibers as well as the cross sections

Kiessling and colleagues developed a supported double membrane platform to mimic for example the periplasm of bacteria and mitochondria or the pre- and postsynaptic neuronal membrane cleft (Murray *et al.*, 2009). First, a supported SLB was formed by the Langmuir-Blodgett/vesicle fusion (LB/VF) technique with POPC lipids and 1 mol% biotin-PEG-DPPE in the distal layer. After exposure of the thus formed SLB to streptavidin the SLB was incubated with vesicles containing 0.1 mol% biotin-PEG-DPPE (at 0.1 mM total lipid concentration). Vesicle binding was found to be saturated after 3 h, and unbound vesicles were washed out after 4 h of incubation. To verify that the vesicles had fused to form a second bilayer, FRAP experiments were performed. The binding of the streptavidin to the proximal layer as well as the binding of vesicles to the proximal SLB was monitored with TIRF microscopy. The distance between the two supported double membranes was analyzed using FLIC microscopy (Lambacher, 1996, 2002) and found to be between 16 and 24 nm. They explain the larger experimental distances (expected distance between the membranes for the system (PEG_{2000}-biotin-streptavidin) is 14 nm) with bilayer undulations, partial extension of the polymer, or a small fraction of unfused vesicles on top of the second bilayer. Single particle tracking experiments in the second bilayer were performed with labeled syntaxin-1A. The planar nature of the developed platform allows studying biological events in close proximity of two membranes using advanced detection and imaging techniques (Murray *et al.*, 2009).

SLBS TO STUDY ANTIMICROBIAL PEPTIDE INTERACTIONS WITH MEMBRANES

SLBs with cell mimicking compositions have also been established to study the interaction of peptides (e.g. antimicrobial peptides (AMPs) and cell penetrating peptides (CPPs)) with cell membranes. Nguyen and coworkers investigated the interaction and orientation of the AMP magainin 2 with POPC (as mammalian cell membrane mimic) and POPG (as bacterial cell membrane mimic) SLBs *in situ* using sum frequency generation (SFG) vibrational spectroscopy and attenuated total reflectance-Fourier transform infrared spectroscopy (ATR-FTIR) (Nguyen *et al.*, 2009a,b). The SLBs were formed with the Langmuir-Blodgett and Langmuir-Schäfer (LB/LS) method directly on a prism. The interaction experiments were then performed with the bilayer continuously immersed in a small reservoir equipped with a magnetic stir bar to ensure homogenous concentration distribution of the peptide in the subphase below the bilayer. They found that at low magainin 2 concentration (200 nM) the peptide orients parallel to the POPG SLB surface while at higher concentration (800 nM) the peptide inserts into the lipid bilayer with an angle of 20° to the lipid orientation direction. For the interaction with POPC even at much higher concentration (2 μM) no ATR-FTIR signal was detected but SFG experiments suggested parallel orientation to the POPC SLB surface with 75° to the surface normal (Nguyen *et al.*, 2009a). By these results they also demonstrated that SFG has a better detection limit for studying interfacial molecules with low surface coverage than ATR-FTIR.

Shaw and colleagues investigated the antimicrobial activity, haemolytic toxicity and membrane domain remodeling of cationic peptides on membranes composed of

DSPC/DOPC and SM/cholesterol/DOPC by *in situ* AFM microscopy (Shaw *et al.*, 2007). These lipid compositions are of interest since SM and cholesterol are two major lipid species found in the outer leaflet of the erythrocyte plasma membrane. Unilammellar vesicles were prepared by ultrasonication and SLBs were formed from preheated vesicle solutions (65°C, Hepes buffer with 4 mM Ca^{2+}) on freshly cleaved muscovite mica directly in the AFM fluid cell. For all peptides investigated (indocilin, ILA, ILF, mellitin and Tat) at low peptide concentration (~ 0.5 μM) coalescence of circular cholesterol-SM l_o domains to larger irregularly shaped domains was observed as a consequence of reduced line tension. At higher peptide concentration the interaction with the membrane was found to be peptide specific and dependent on the hydrophobicity of the peptides (Shaw *et al.*, 2007). Cholesterol was shown to prevent membrane disruption of the l_o domains for most of the investigated peptides due to the high lipid mobility granted by cholesterol in the l_o domains (Shaw *et al.*, 2007).

SLBS TO STUDY CELL ADHESION AND REPULSION

SLBs can also be adapted to modify interfaces to immobilize and study cells and in particular cell adhesion (Sengupta *et al.*, 2003). By incorporation of receptors and membrane bound signalling molecules, ligand-cell interactions can be closely probed on an SLB modified surface (Thid *et al.*, 2007). Thid and colleagues demonstrated how an SLB platform can be functionalized with a short laminin-derived peptide (IKVAV) to yield a cell culture substrate for rat-derived hippocampal progenitor (AHP) cells. They prepared SLBs by adsorption and fusion of POPC vesicles doped with up to 10% of maleimido-terminated lipids for further peptide functionalization on SiO_2 surfaces. They were able to demonstrate that AHP cells could attach and grow on substrates prepared in this way without induction or differentiation (Thid *et al.*, 2007).

To study the properties of hyaluronic acid (HA) attachment to cell membranes, SLBs with immobilized HA layers have been established (Sengupta *et al.*, 2003; Benz *et al.*, 2004; Richter *et al.*, 2007). Sengupta and colleagues prepared a supported lipid bilayer via the Langmuir-Schäfer technique. In this way the first monolayer consisted of 1,2-dimyristoyl-*sn*-glycero-3-phosphocholine (DMPC) while the top monolayer had additionally chelating lipids with a Ni-NTA-DOGS (nitriloacetic acid (dioleoyl-sn-glycero-3-[N-(5-amino-1carboxypentyl-iminodiacetic acid)succinyl])) functionality incorporated (Sengupta *et al.*, 2003). By linking p32, a recombinant intracellular HA binding protein via His-tag to the nickel moiety, an HA film could be bound to the bilayer surface. The resultant HA cushion could then be used as a platform to study cell adhesion (Sengupta *et al.*, 2003). Richter *et al.* (2007) have also demonstrated a hyaluronan grafted SLB platform prepared via adsorption of liposomes incorporating biotin binding sites. After exposure of the formed SLBs to streptavidin, biotinylated HA was attached and the density of the grafted layer was varied from the mushroom to the brush regime and then characterized by colloidal probe reflection interference microscopy (RICM) and reflectometry (Richter *et al.*, 2007). These researchers further tested the permeability of the hyaluronan films by biotinylated probes and monitored the process by QCM-D. The permeability of the HA films was found to be dependent on the grafting ratio density.

SLBS TO STUDY RAFT FORMATION/MEMBRANE PROPERTIES

One aspect of biological membranes that has been studied in detail by use of SLBs is ordered domain formation. For some time it has been known that proteins may assemble into functional clusters on the cell membrane and that the lipids themselves may laterally organize into separate phase domains. An advantage of artificial membranes such as SLBs is that the domains can be much larger than domains occurring in native systems and thus can be analyzed by biophysical techniques. Tamm and colleagues have investigated different lipid compositions to isolate and understand the dynamics of liquid ordered (l_o) vs. liquid disordered (l_d) phases in membranes with asymmetrical lipid distributions in the two leaflets. The distal layer lacking lipids normally associated with domain formation and therefore mimicking the inner plasma membrane, was produced on top of a monolayer withlipids found in l_o phases (sphingomyelin (SM) and cholesterol) mimicking the outer plasma membrane. They discovered that to maintain asymmetry in the SLB one should use the LB/VF technique, tether the LB layer to the substrate with e.g DPS (1,2-dimyristoyl phosphatidlethanolamine-N-[polyethylene glycol-triethoxysilane]) and use equal amounts of cholesterol in both layers (Kalb *et al.*, 1992; Crane *et al.*, 2005; Crane and Tamm, 2007; Kiessling *et al.*, 2006). They further report that stable asymmetric SLBs could be more easily formed with natural lipid mixtures mimicking the inner-leaflet composition than with purely synthetic lipid mixtures with fewer components. They assign this behavior to more degrees of freedom available to the natural compositions due to a larger variation in chain length compared to the synthetic mixtures, and thus in melting temperature heterogeneity (Wan *et al.*, 2008). This finding helps explain the large variety of different lipid species found in natural membranes (Wan *et al.*, 2008; Van Meer, 2005). This research was also recently reviewed by Kiessling *et al.* (2009).

Polymer-supported lipid bilayers

A classical supported lipid bilayer has the lipids in one of the leaflets in direct contact with the supporting substrate. This close proximity impairs the ability to incorporate large membrane proteins as direct surface interactions can lead to denaturation of the protein structure. The problem can be overcome by placing the lipid bilayer on a highly hydrated polymer support where it is either anchored to the surface separated by a polymer spacer or deposited on a polymer cushion (Tanaka and Sackmann, 2005 Sinner and Knoll, 2001). Tethered lipid bilayers make use of a covalently anchored amphiphile which normally contains a lipidic part, a spacer molecule and an anchor responsible for covalent attachment to the substrate (Sinner and Knoll, 2001). On polymer cushions, in contrast to polymer tethered systems, the lipid bilayer as a whole is connected to the polymer support through many weak interactions instead of being covalently attached through single lipids. The advantage is that all lipids are mobile. However the balance of attractive and repulsive interactions between the polymer and the lipid bilayer has to be adjusted precisely to ensure the formation of a continuous bilayers with mobile lipids (Tanaka and Sackmann, 2005). In addition the hydration, the roughness and the thickness of the polymer cushion all play an important role. Depending on the particular application, the system characteristics that

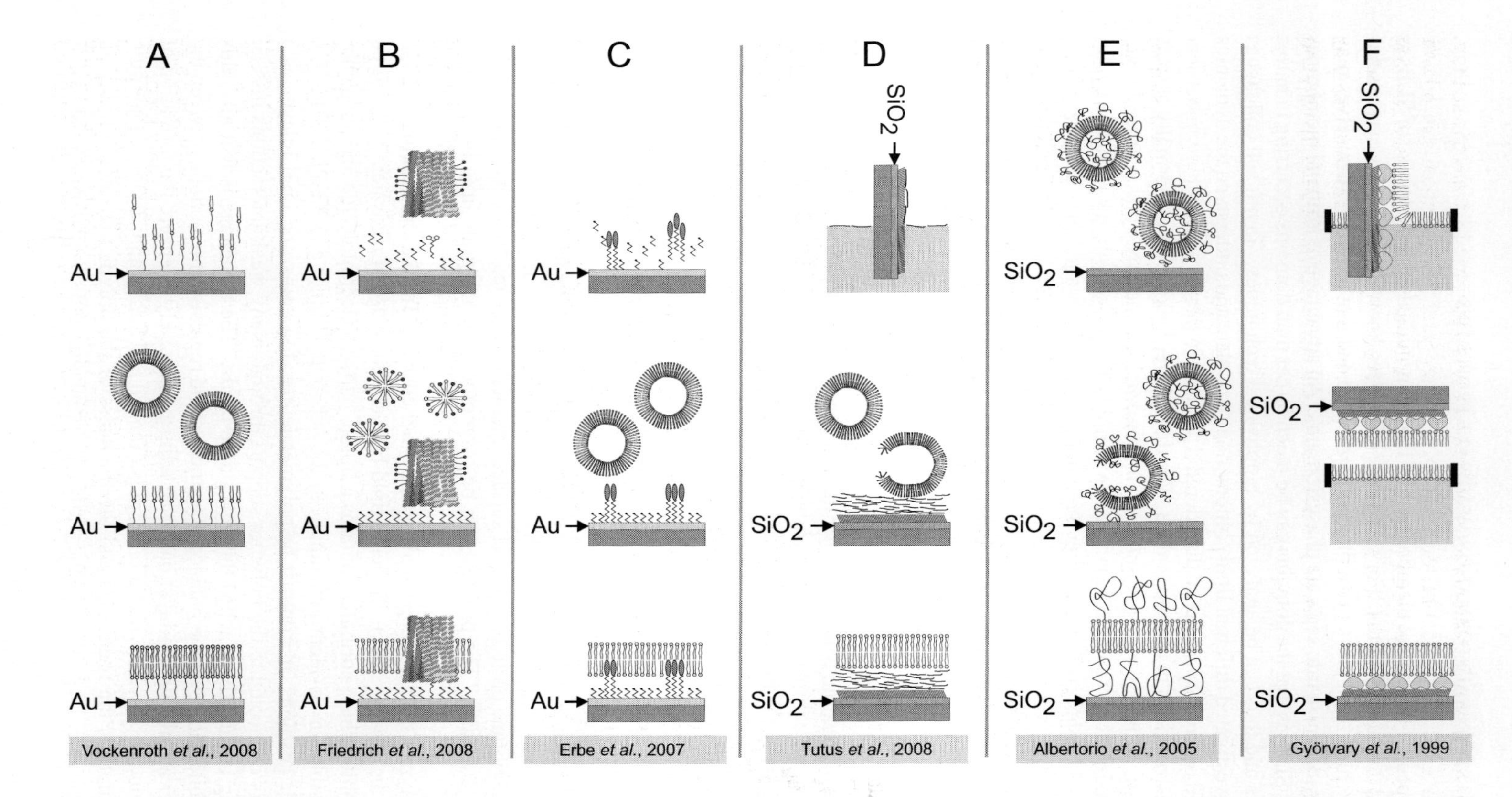

Figure 1. Schematics of the different advances in polymer supported lipid bilayers. (A) Tehered lipid bilayer. (B) Protein-tethered lipid bilayer. (C) Cholesterol-tethered lipid bilayer. (D) Cellulose cushioned lipid bilayer. (E) PEG-lipid bilayer). (F) S-layer supported lipid bilayer

are sought for polymer-supported lipid bilayers are as follows: high electric sealing, to allow electro-chemical measurements of single ion channels; good lipid bilayer stability, to perform long-term measurements and to reduce the membrane fragility for storage and shipping; low tether density, sufficient bilayer substrate separation and high membrane fluidity, so that membrane proteins and lipids diffuse freely as they would in a native environment; flexible choice of substrate, so that the system is not restricted to a specific surface material.

ADVANCES IN FORMATION OF POLYMER-SUPPORTED LIPID BILAYERS

Gold is a desired substrate for SLBs due to its extraordinary optical and electrical properties for sensing, although it is a material on which it has been difficult to self-assemble traditional SLBs of high quality. However polymer-supported lipid bilayers can be formed using thiols to anchor tethered membranes. *Figure 1A* shows the approach presented by Koeper and coworkers where a lipid monolayer of the anchor lipid DPhyTL (two phytanyl chains coupled via a glycerol linker to a tetraethylene oxide spacer (Atanasov *et al.*, 2006)) is grafted by sulfur-gold bonds to an ultraflat gold substrate (Vockenroth *et al.*, 2008). Fusion of small unilamelar (1,2-Diphytanoyl-sn-Glycero-3- Phosphocholine) DPhyPC vesicles completes the second leafleat of the lipid bilayer. The use of template-stripped ultraflat gold was found necessary for high quality SLBs to be formed. The group of Naumann modified a template-stripped gold surface with NTA and could reversibly bind cytochrome c oxidase (CcO) proteins with a histidine tag (Friedrich *et al.*, 2008) after NTA chelation with Ni^{2+} as shown in *Figure 1B*. A DPhyPC lipid bilayer was formed around the tethered proteins by *in situ* dialysis.

Figure1C shows the approach presented by Jeuken and coworkers (Erbe *et al.*, 2007). Mono-layers of EO3-cholesteryl and 6-mercapto-hexanol are self-assembled on a gold substrate and in a subsequent step EggPC lipid vesicles are fused on the modified gold surface forming a tethered lipid membrane.

Deme and Marchal have presented a platform based on a porous alumina support where they formed a PEG cushion or cushions bearing a lipid anchor (DSPE-PEG) in the mushroom regime. On top of this PEG cushion a lipid bilayer was formed by liposome fusion (Deme and Marchal, 2005).

SLBs can also be formed on top of preformed polymer cushions without a tether into the membrane. Tanaka and coworkers showed the formation of a regenerated cellulose film on glass substrates from cellulose powder (Tanaka *et al.*, 2001). Layers of cellulose are deposited on hydrophobized substrates by the Langmuir-Blodgett technique and to create a hydrophilic interface the hydrophobic trimethylsilyl side chains are cleaved by exposing the film to vapor of concentrated HCl. Onto this hydrophilic regenerated cellulose (1-Stearoyl-2-oleyl-*sn*-glycero-3-phosphatidylcholin) SOPC vesicles were fused to form a supported lipid membrane on the polymer cushion (Tutus *et al.*, 2008) as shown in *Figure 1D*.

Renner *et al.* (2008) examined the formation of lipid bilayers on three different polymer cushions, POMA (poly(octadecene-*alt*-maleic anhydride)), PPMA (poly(propene-*alt*-maleic anhydride)) and PEMA(poly(ethylene-*alt*-maleic anhydride)) with vesicles prepared from egg PC, PS, PE, cholesterol and NBD-PE (5:2:1:2:0.1). At pH 7.2 the supported lipid bilayer formation by vesicle fusion failed due to electrostatic repulsion of the negatively charged polymers and vesicles containing negatively charged lipids. The charge density of the polymers decreases at pH 4 and vesicle fusion proceeds. The diffusion coefficient was shown to increase with increasing hydrophilicity of the polymer cushion.

Recently, Reimhult *et al.* self-assembled poly(L-Lysine)-*graft*-poly(ethylene glycol) (PLL-*g*-PEG) end-functionalized with quaternary ammonium compounds to tether liposomes assembled on the hydrophilic brush. The tethered liposomes were at sufficiently high density ruptured osmotically by injection of a solution of PEG to form a supported planar lipid bilayer with an order of magnitude higher lipid mobility than for lipid bilayers supported on glass (Ye *et al.*, 2009).

Additionally, protein layers have been used to supporttethered SLBs. Bourdillon and coworkers used the biotin - streptavidin linkage and biotinylated lipids to form a tethered lipid bilayer from inner membrane of HepG2 mitochondria (Elie-Caille *et al.*, 2005) and prevented non-specific binding of protruding proteins to the substrate by forming a PEG layer between the biotin linkers. The tethered, immobilized proteoliposomes were ruptured osmotically by injection of a solution of PEG.

A different approach to both using covalent tethers and forming SLBs on top of preformed polymer layers has been followed by the group of Cremer. They demonstrated the formation of SLBs from EggPC vesicles containing PEG-PE lipopolymers, which are lipids with PEG-chains attached to the lipid head group (Albertorio *et al.*, 2005). These vesicles fused spontaneously on a glass surface forming a PEG-supported lipid bilayer as shown in *Figure 1E*, where all components are fluid including the polymer cushion which is not attached to the substrate.

Sleytr and coworkers have demonstrated the versatility of S-layer protein lattices for supporting lipid bilayers. S-layers are monomolecular crystalline bacterial glycoprotein lattices which self-assemble in two-dimensional films on various surfaces. S-layers, containing water, act as tethers and an ionic reservoir: they maintain the fluidity of the membrane even in an S-layer sandwich-like structure and they enhance the long-term stability of membranes (Gyorvary *et al.*, 1999; Gufler *et al.*, 2004; Schuster *et al.*, 2004). For example on silicon, as shown in *Figure 1F*, an S-layer is self-assembled on a silanized silicon substrate with subsequent deposition of the first monolayer by the Langmuir-Blodgett and the second monolayer by the Langmuir-Schaefer techniques (Gyorvary *et al.*, 1999). The functionality of S-layer supported lipid bilayers was proven by single pore recordings with αHL and incorporation of membrane active peptides alamethicin, gramicidin and valinomycin (Schuster *et al.*, 2001; Gufler *et al.*, 2004). In addition, the interaction of S-layer proteins with specific bacterial secondary cell wall biopolymers allows adjusting the separation between the membrane and the solid support (Sara *et al.*, 1998; Egelseer *et al.*, 1998).

FLEXIBLE CHOICE OF SUBSTRATE

The general approach for flexibility in the choice of the substrate is the formation of a SAM on the respective surface with a suitable anchor. This approach was successfully used on gold (thiols) and silicon oxide (silanes) surfaces where for instance Koeper and coworkers presented a molecular toolbox to produce lipid anchors directly on gold and silicon oxide surfaces (Atanasov *et al.*, 2005, 2006). Similarly, in the group of Chopineau substrates were modified in a two step approach by first forming a self-assembled monolayer of amines and then binding vesicles containing DSPE-PEG3400-NHS lipids (Deniaud *et al.*, 2007; Rossi *et al.*, 2007). Membranes supported on regenerated cellulose cushions, polyacrylamide brush layers and polyelectrolyte layers are described on glass surfaces but these systems should in principlebe possible to transfer to other substrates (Tanaka *et al.*, 2001; Smith *et al.*, 2005; Fischlechner *et al.*, 2008). S-layer lattices form on a variety of substrates as for example on silicon, glass, metals polymers and lipid structures and therefore provide a flexible platform for supported lipid bilayer systems (Pum and Sleytr, 1995; Weygand *et al.*, 2002; Gufler *et al.*, 2004; Schuster *et al.*, 2004). The PLL-*g*-PEG based approach by Reimhult is readily applicable to any negatively charged substrate from typical oxides to TCPS and oxidized PDMS (Ye *et al.*, 2009). Systems based on self-assembly of vesicles with PEG-lipids with subsequent fusion to PEG-tethered lipid bilayers are principally not restricted to specific surfaces. The attractive interaction with the substrate, however, has to be sufficient to induce the rupture of vesicles (Albertorio *et al.*, 2005), and tuning of this interaction is often delicate as is known from self-assembly of supported lipid bilayers (Reimhult *et al.*, 2003; Richter *et al.*, 2005).

BILAYER FLUIDITY

Membrane fluidity is important to conserve membrane physical properties and for membrane proteins to retain their natural mobility and functionality. This has been shown to be achieved by minimizing the interactions between (protruding) proteins and lipids with the substrate. Wagner and Tamm (2000) and Chopineau and coworkers have presented the formation of a lipid bilayer which is tethered by PEG-lipids that are covalently bound to a glass or amine modified surface. Highly mobile membranes were obtained at low polymer concentration in the mushroom regime ($0.8 – 1.2 \times 10^{-8}$ cm^2/s and $2.8 – 3.0$ x 10^{-8} cm^2/s and 2.5 - 3.6 x 10^{-8} cm^2/s respectively) (Deniaud *et al.*, 2007; Rossi *et al.*, 2007). Cremer and coworkers (Diaz *et al.*, 2008) have used non-covalently bound PEG-lipids in low concentrations to incorporate annexin V which showed high mobility (3 x 10^{-8} cm^2/s with a protein mobile fraction of about 75 %) and this represents a significant improvement compared to the systems with covalently attached PEG chains.

The increase in mobility of annexin V (Diaz *et al.*, 2008) was shown to be dependent on the molecular weight of PEG, which strongly indicates that a larger separation from the substrate in this range results in higher mobility as also shown quantitatively for lipids (Kaufmann *et al.*, 2009). The approach of non-covalent bound PEG-lipids was however shown to be restricted to low PEG-lipid concentrations (Kaufmann *et al.*, 2009). Smith *et al.* (2005) used polyacrylamide brush layers of 2.5 5 and 10 nm thickness, grown on fused silica using atom-transfer radical polymerization. Lipid bilayers of POPC were formed on these polymer brushes by vesicle fusion and they showed that the optimal polymer thickness is a trade-off between decoupling from the substrate and increasing roughness of the polymer layer (fastest diffusion coefficient (2 x 10^{-7} cm^2/s) and highest recovery fraction (99 %) for 5 nm layer). The hDOR receptor mobility was shown for a layer thickness between 2.5 – 5.0 nm to be 3 x 10^{-8} cm^2/s with decreasing values for thicker polymer layers (2.8 x 10^{-8} cm^2/s (lipid), 2.0 x 10^{-10} cm^2/s (protein), 97% recovered fraction) due to increasing roughness. The thick but smooth PEG spacer layer (10 nm) with sparse and small tethers used by Reimhult *et al.* also resulted in a diffusion coefficient of ~1 x 10^{-7} cm^2/s (Ye *et al.*, 2009).

Kunding and Stamou (2006) used a sparsely PEG-tetheredlipid bilayer system (0.1 mol % PEG tethers) with negatively charged lipids POPG (10 mol%) and POPC by coupling biotinylated PEG lipids (PEG_{2000}-biotin) to immobilized biotinylated bovine serum albumin via avidin. The immobilized giant unilamelar vesicles after binding to the protein functionalized surface were ruptured upon change of solvent from physiological buffer to MilliQ water. Addition of NaCl shielded the attractive electrostatic forces between the negatively charged bilayer and the positively charged avidin causing a stretching of the PEG tethers The lipid mobility was found to increase upon stretching of the tethers.

Hwang *et al.* (2008) used a benzophenone-modified substrate and attached a DODA-poly(D-glucose-2-propenoate) telechelic lipopolymer (GH24000) using UV illumination for which the fluidity of the membrane was decreasing for increasing tether concentration. Sackmann, Tanaka and coworkers have demonstrated that integrin receptors showed long range lateral mobility on regenerated cellulose cushions (Goennenwein *et al.*, 2003). Sleytr, Knoll and coworkers showed that S-layer supported lipid bilayers have diffusion coefficients of the order of ~ 2 - 3 × 10^{-8} cm^2/s possessing higher lipid mobilities than lipid monolayers supported on alkylsilanes and lipid bilayers on dextran (Gyorvary *et al.*, 1999).

ELECTRIC SEALING

Electric resistances needed for single-channel recordings are on the order of 10 GΩ (Sakmann, 1995), however for the comparison of different platforms the actual area should be taken into account using the *specific membrane resistance* (Ω cm^2). Supported lipid membranes with a high electric sealing appropriate for electrochemical measurements of ion channels have been reported for tethered lipid bilayers. Resistances of a few kΩ cm^2 to MΩ cm^2 have been measured on gold surfaces, by for example the groups of Vogel, Knoll and Evans (Lang *et al.*, 1994; Atanasov *et al.*, 2006; Jeuken *et al.*, 2007). The approach of S-layer supported DPhyPC membranes on gold substrates has resulted in high membrane resistances of ~ 80 MΩcm^2 (Gufler *et al.*, 2004). On silicon surfaces tethered lipid bilayers could be formed using silane anchor moieties with resistances in the MΩ cm^2 regime as well (Atanasov *et al.*, 2005).

Electrochemical measurements of polymer cushion supported lipid bilayers have, by contrast, been far more infrequent, although there are some good examples. For instance, the formation of DMPC and DMPS lipid bilayers by vesicle fusion on polypyrrole (PPy) a conducting polymer has been studied by the research team of Dong. An increase in the measured impedance indicated the formation of the membrane, although no exact specific membrane resistance is given (Shao *et al.*, 2005). Searson and coworkers showed the formation of a supported lipid bilayer using PEG-lipids (1,2-dipalmitoyl-sn-glycero-3-phosphoethanolamine-N-[methoxy-(polyethylene glycol)-2000]) and Langmuir-Blodgett deposition of the lower leaflet with completion by liposome fusion of the second leaflet (Nikolov *et al.*, 2007). The specific membrane resistance was found highly reproducible but very low at about 10 kΩ cm^2. Summarising, platforms having achieved high specific membrane resistances of a few MΩ cm^2 on areas of approximately 0.2 cm^2 lead to resistances of some MΩ but not yet reaching the required GΩ for single-channel recordings and it is likely that for most polymer cushion based platforms further improvement will be necessary.

BILAYER STABILITY

The stability of lipid membranes can be increased by the formation of a protecting hygroscopic polymer layer surrounding the bilayer Schmidt and coworkers presented a stable platform for single-channel measurements by stabilizing a lipid bilayer spanning a 200 μm Teflon aperture filled by a PEG-DMA hydrogel (Jeon *et al.*, 2006). The spanning bilayer showed a high resistance (> 10 GΩ) and mechanical stability for 2-5 days. Later this PEG-encapsulated lipid bilayer platform was further developed by covalent anchoring of the lipids to the hydrogel which enhances the stability of the membrane capacitance under applied pressures and prevents solvent drainage (Malmstadt *et al.*, 2008). Recently, the groups of Bayley and of Gu presented similar platforms where a lipid bilayer crossing an aperture was encapsulated between two layers of agarose gel (Kang *et al.*, 2007; Shim and Gu, 2007). These mechanically robust membrane devices could be stored with fully functional αHL for at least for 3 weeks. DPhyPC membranes supported on an S-layer lattice on gold and on porous microfiltration membranes showed high long-term stability of maximum 46 h and 17 h respectively (Schuster *et al.*, 2001; Gufler *et al.*, 2004). Koeper and coworkers demonstrated the long-term stability of DPhyTL tethered lipid bilayer systems over a time period of 3 months (resistance > 10 MΩ cm^2) and the stability of the system to application of high electric fields. These membranes were also protected against delamination upon air exposure by encapsulation in a poly(NIPAAM) hydrogel. After 24 hours in dry state the membrane was still electrically insulating (≈ 1.5 MΩ cm^2) (Vockenroth *et al.*, 2008). Similarly, Parikh and coworkers have shown that the preparation of lipid bilayers on trehalose films by liposome fusion from a 1 M trehalose phosphate buffered saline solution results in protection of the membranes against dehydration, possibly by encapsulation (Oliver *et al.*, 2008). Finally, Cremer and coworkers (Albertorio *et al.*, 2005; Diaz *et al.*, 2008) have demonstrated that supported lipid bilayers from fused vesicles with PEG-lipids (PEG-DPPE, PEG-DSPE, PEG-DOPE), mixed in a low concentration with egg PC show a stability enhancement upon drying for increased PEG-lipid concentration.

Physical surface modification for the sensing of supported lipid bilayers

Structured surfaces, i.e. substrates with micro- or nanostructures on the surface, allow novel sensor concepts based on nanoscale elements such as nanoplasmonics and zero-mode waveguides to be applied to increase the sensitivity and lateral resolution of membrane measurements. Such structures reduce the amount of analyte necessary and may also allow the implementation of combinatorial sensing techniques.

Micro- and nanostructures on the surface also allow the in-depth study of the effect of curvature in membranes. Lipids and membrane proteins are known to control membrane curvature, but induced curvature may also control the distribution of lipids and membrane proteins (Parthasarathy and Groves, 2007). To date, vesicles have been used in most research to study the influence of curvature on lipid distribution, protein binding and function. However, membrane curvature could also be controlled by the shape of surface structures to which it adheres with the advantage that the size and shape of such structures can be much better controlled and be more uniform than with the curvature of liposomes. Still, the interplay of lipids with nanostructures and the formation of SLBs on nanostructures in particular are not well explored.

We will broadly classify structured surfaces into two categories for the purposes of this review, namely coarse surfaces and nanostructured surfaces. *Coarse surfaces* refer to those where the texturing on the surface is random, and the surface can be defined in terms of roughness measures such as R_{max}, R_{min} and R_{rms}. These surfaces are typically not described as being porous and consist of the same material everywhere. Nanostructured surfaces are those where structures are fabricated into the substrate to provide a clear height and shape distinction and, in most cases a variation of chemistry on different surfaces within the structure.

FORMATION OF SLBS ON COARSE SURFACES

Coarse surfaces have only seldom been used as sensing substrates for SLBs. Melzak and Gizeli (2002) introduced a method of condensing TEOS on various sensor surfaces to create both coarse and smooth SiO_2 surfaces using a sol-gel. The coarse surfaces had a roughness of R_{rms} ~ 10 nm, while the smooth surfaces had an R_{rms} ~ 0 nm. Vesicle adsorption was observed on the coarse surfaces, while formation of SLBs was observed on the smooth surfaces, as investigated by the surface acoustic wavetechnique and fluorescence recovery after photobleaching (FRAP). This result has been contradicted by Lee and coworkers (Lee *et al.*, 2009) using the same kind of substrates and with the same roughness. They found that SLBs form both on the coarse and smooth substrates, being formed faster on the coarse surfaces rather than on the smooth surfaces.

Hovis and coworkers (Seu *et al.*, 2007) presented an in-depth study on the effect of surface treatment on the quality of the SLBs formed. Different surface treatments changed the effective surface roughness. SLBs were formed from the rupture of vesicles on solid glass supports which were subsequently either etched in pirahna solution for various times or annealed at 450 °C. Those researchers showed that where the surface R_{rms} roughness varies by as little as 2.5% from smooth substrate (with R_{rms} ~ 0.4 nm), a change in diffusion coefficient of the SLB by as much as 20% was observed.

FORMATION OF SLBS ON NANOSTRUCTURED SURFACES

In order to quantify the influence of nanoscale topography on the formation of SLBs, several groups have also resorted to nanostructuring of substrates or studying the interaction of membranes and nanoparticles. Shreve and coworkers (Werner *et al.*, 2009) have studied the formation dynamics and the kinetics of SLBs on a substrate patterned with grooves produced by interferometric lithography. The grooves had a depth of 320 nm and width of 160 nm. SLBs were observed by AFM and fluorescence microscopy to follow the substrate closely. An interesting finding was that the diffusion coefficient along the patterns within the grooves was similar to that of the flat substrate, while the diffusion coefficient across the grooves was lower even after taking the geometry of the surface into account. It was proposed that the SLB conformed tightly to the surface. The SLBs across the grooves had a lower diffusion coefficient since they had to traverse a larger distance in the field of view than the SLBs along the grooves. This decoupling step could make FRAP experiments far more descriptive in reflecting the state of the underlying substrate.

Zach and coworkers (Pfeiffer *et al.*, 2008) fabricated 110 nm and 190 nm diameter pits, 25 nm deep in SiO_2 sputtered on gold-coated QCM-D crystals. Vesicles of 30 nm and 110 nm diameter were allowed to adsorb to the surface and monitored using QCM-D and AFM. SLBs were formed in all cases. The authors argued that SLBs were formed all over the substrate, with the edges of the pits, that is to say areas of high curvature increasing the speed of bilayer formation. The interconnectivity of the SLBs was however not discussed in the paper with no strong indication that the SLB was continuous. This indicated that it is very challenging to form a defect-free bilayer on a structured surface.

SLBs on strongly curved surfaces were first investigated in the early 1990s. Bayerl and Bloom (1990) used sub-micron glass microbeads in order to measure the diffusion coefficients of deposited lipids. Whilst their results indicated that they were measuring a single SLB layer it was not until Brisson and coworkers (Mornet *et al.*, 2005) had conducted cryo-TEM measurements on 110 nm silica particles coated with SLBs that the existence of such curved SLBs covering the particles was conclusively proven.

Minko and coworkers (Roiter *et al.*, 2009, 2008) adsorbed SiO_2 nanoparticles of diameters between 1 nm and 140 nm on smooth SiO_2 surfaces in order to create substrates with controlled roughness. In all cases, as investigated by FRAP, an SLB was predominantly present. AFM revealed that pores in the SLBs were formed around particles between 1.2 nm and 22 nm in diameter, while a continuous bilayer was formed over particles with diameter less than 1.2 nm or larger than 22 nm (*Figure 2*). However there was no proof of whether bilayer fragments on top of particles larger than 22 nm were connected with the surrounding SLB. In summary the influence of nanoscale roughness on SLB formation and interconnectedness was found to be large and varied.

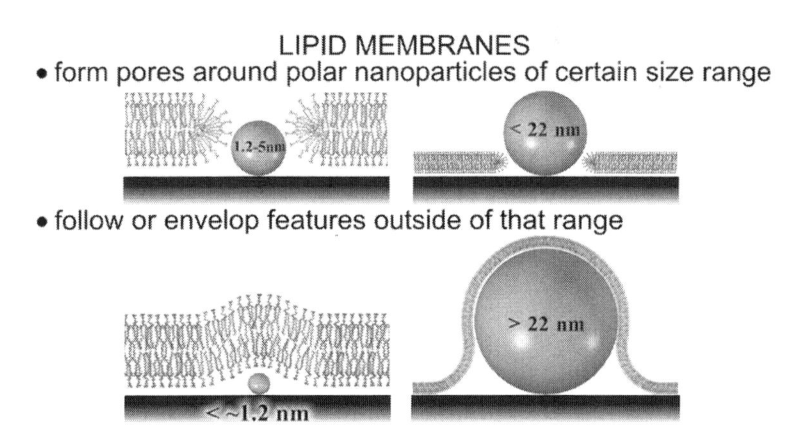

Figure 2. Illustration of the various results presented by Minko et al. The SLB covered particles smaller than 1.2nm completely. For particles between 1.2 nm and 22 nm pore formation in the SLB around the particle was observed . For smooth particles that were larger than 22 nm, defect free SLBs were formed on the particle but it was not clear if the SLB on the particle and around the particle were connected.

Another example of curved surfaces as substrates for studying the formation and behavior of SLBs was the use of polysilicon nanowires, by Noy and coworkers (Misra *et al.*, 2009; Huang *et al.*, 2007). The wires were 5 μm in length, and 20-200 nm in diameter. The lipids were observed to diffuse on the wires. However, due to their small width, diffusion was treated as occurring in one dimension along the wires, neglecting the diffusion across the wires. Mixed DOPC/DOPE vesicles formed continuous SLBs as elucidated by FRAP measurements. The mobile fractions were in the range of 90-100%. The membrane mobility along the nanowire also seemed to increase with increasing curvature. To further quantify the results, the effect of curvature was modeled using the free space diffusion model (Galla *et al.*, 1979). The model provided a good description of the effect of lipid mobility on curved surfaces for wire diameters between 50 nm and 200 nm, but breaks down for a size range of ~20-50 nm.

Groves and coworkers used a photolithographically patterned and etched fused quartz substrate to show that membrane curvature can control lipid phase separation (Parthasarathy *et al.*, 2006). Lines were anisotropically etched into the substrate and smoothened by an HF wet etching step. Phase separation seemed only occur with a curvature higher than 0.8 μm^{-1}. In order to rule out effects of asymmetric direct interaction of lipids and substrate, a dual SLB system was employed. After the formation of the first SLB through small vesicle fusion, a second SLB was formed on top by rupture of giant vesicles. SLBs directly formed on the curved substrates did not display the same phase segregation.

SLBS ON NOVEL NANOSTRUCTURED MEASUREMENT PLATFORMS

When the nanostructure is a sensor element or defines the measurement area it is imperative to know the geometry of the structure. The groups of Craighead (Samiee *et al.*, 2006) and Lenne (Wenger *et al.*, 2006) used nanostructured alumina substrates to demonstrate zero-mode waveguide measurements on SLBs. Sub-wavelength apertures of diameters between 70 nm and 150 nm, and 30 nm and 400 nm respectively with a depth of ~100 nm in alumina sputtered on fused silica were used. Using a TIRF microscope, it was possible to show that only fluorophores within the nanoapertures were excited in SLBs formed by liposome fusion which entered the pores. This enabled a very small volume (attoliter) to be specifically probed. Craighead and coworkers also showed that it was possible to probe binding between a labeled tetanus toxin C fragment and a ganglioside expressing lipid membrane within a small volume.

Similar nanotopography was used by the group of Hook and coworkers for nanoplasmonic membrane sensing (Jonsson *et al.*, 2008, 2007; Dahlin *et al.*, 2005). Plasmon active 140 nm sized wells between 30 and 60 nm in depth were fabricated in a gold coated SiO_2 substrate. SLBs were formed either only within the wells or covering both the surface and going into the wells using different surface chemistries. Correlating LSPR and total-internal reflection microscopy it was shown that it was possible to probe recognition events to adsorbed bilayers in real time on a format that can be easily miniaturized.

The motivation behind much of the extensive studies on SLBs is the potential of these substances as a platform to study ion channels and other charge translocating membrane proteins. Thus integration with nanoscale structures that permits measurement down to the single protein level has been sought after. The nanowire supported SLBs described above from the group of Noy (Misra *et al.*, 2009; Huang *et al.*, 2007) have been used to achieve ionic to electronic signal transduction through voltage-gated or chemically-gated ion transport through membrane pores incorporated into the SLB on the nanowire. Other examples of similar efforts have been provided by the groups of Craighead and McEuen (Zhou *et al.*, 2007), where carbon nanotubes functioning effectively as field effect transistors were used instead of silicon nanowires.

With electrophysiological measurements being the dominant techniques to study ion channels and transporters, aperture spanning SLBs which are ideally suited for these measurements have been pursued for decades (Mueller *et al.*, 1963). The main problems plaguing aperture spanning membranes since their inception have been their lack of stability due to their large area and the common presence of solvents which also alters the properties of membranes and incorporated proteins.

Miniaturization of the spanning membrane to nanostructured pores as those just described improves membrane stability (Han *et al.*, 2007b), but poses the additional problem of ensuring that membranes are formed which span rather than avoid or follow the walls of the aperture. Steinem and coworkers have shown that it is possible to self-assemble nanopore spanning membranes on porous alumina and porous silicon (Boecker *et al.*, 2009; Romer and Steinem, 2004; Drexler and Steinem, 2003; Hennesthal and Steinem, 2000). Pore sizes have been varied from 30 nm to 1000 nm, and the membranes could be electrochemically accessed from both sides including by scanning ion conductance microscopy. The best results were obtained when the top surface was functionalized with cholesteryl tethers (Schmitt *et al.*, 2009, 2008).

This part of the field is currently very active with many research groups testing alternative solutions, many of which exploit capillaries and polymer cushions instead of purely nanostructured chips and was recently reviewed (Reimhult and Kumar, 2008).

Patterning of supported lipid bilayers

Platforms for screening a large library of molecules interacting with membrane components rapidly, fully automated, and, ideally, at low cost would greatly aid biological research and applications such as drug screening. A high-throughput system requires spatially addressable arrays where various functionalities are located at different areas on the biosensor chip. To date, several array technologies have been developed presenting different kinds of libraries like DNA ("genomics"), sugar ("glycomics") or proteins ("proteomics"), among which the latter is considered the most important way to comprehend biological systems (Aebersold and Mann, 2003). However only the first has received considerable industry-wide standardized use and the quest for corresponding solutions to screen for interactions of molecules, e.g. drugs, with proteins is still ongoing. Since membrane proteins require a lipid environment to function, patterning of membrane protein arrays necessitates that methods are developed to pattern lipid bilayers down to at least the micron-level. By patterning the lipid bilayer, spatially addressable arrays can be created where different transmembrane proteins of interest are situated at different locations on the biosensor chip, hence providing a library of various functionalities on a single chip.

Despite the significance of membrane proteins, however, extraction and purification of membrane proteins in sufficient amounts in their pure, natural and functional form is notoriously difficult to achieve (Bayley and Cremer 2001). Additionally, as some transmembrane proteins may require a membrane incorporating several lipid species in order to be functional, different lipids might need to be present at different areas of the array, which is difficult to realize since many of the currently developed techniques only allow for a few lipid species to be varied without being too work-intensive. Another point to be considered is that during the patterning process the patterns might get exposed to air and as the size of the patterns preferably should be very small, the corresponding low volumes of solution evaporate very quickly leading to dehydration and subsequent destruction of the patterns. We provide an overview of the attempts to create membrane arrays in the light of the demands on such arrays for future applications.

Groves *et al.* (1997) first patterned a lipid bilayer in 1997 using photolithographically defined SiO_2 corrals in which supported lipid bilayers were formed by liposome fusion separated by metal diffusion barriers on which lipid bilayers do not form. Since then, many different patterning techniques have been developed based on, for example, spotting (Binkert *et al.*, 2009), electronic control (Kumar *et al.*, 2009) or microfluidics (Taylor *et al.*, 2009).

STAMPING

Lipid bilayer patterns of many shapes have been deposited onto surfaces using stamping, typically on glass by a hydrophilic PDMS mold stamp (Hovis and Boxer, 2000; Kung *et al.*, 2000; Hovis and Boxer, 2001; Sapuri *et al.*, 2003; Tanaka *et al.*, 2004; Jung *et al.*, 2005; Sapuri-Butti *et al.*, 2006). With this technique, simple patterns (Hovis and Boxer, 2000; Sapuri *et al.*, 2003) could be made and by backfilling, a second lipid species introduced (Hovis and Boxer, 2001). In the latter case, using a lipid with a high melting temperature a two-phase lipid bilayer was studied (Jung *et al.*, 2005), which can serve as a model system for lipid rafts (Sapuri-Butti *et al.*, 2006). Lipid bilayer corrals can also be patterned by stamping protein lines as barriers, creating protein-lipid bilayer hybrids (Kung *et al.*, 2000) which could also be created on cellulose substrates (Tanaka *et al.*, 2004). Similarly, Majd and Mayer (2005, 2008) used a hydrogel stamp made of agarose to pattern the surface consisting >96% w/w of water. The hydrated, non-fouling environment of the stamp allowed for up to 30 copies without re-inking the stamp with lipid bilayer forming proteoliposomes while requiring only minimal consumption of stamping material (femtomoles of proteins). Multilayered cell membrane fragments could also be delivered, however, only during the first transfer. This is probably because all the membrane fragments were residing on the stamp surface as they were probably too oversized to diffuse into the hydrogel stamp. Patterns with different lipid mixtures could in addition be stamped on the substrate (Majd and Mayer, 2005) through a work intensive process requiring each stamp pillar to be incubated separately.

Even though microcontact printing is a widespread tool, several problems for membrane patterning remain. These include pattern fidelity and defects in the patterns limited transfer efficiency and surface fouling (Glasmastar *et al.*, 2003), and not least difficulty to impart different molecular functionality to each different spot.

PATTERNING BY MICROFLUIDICS

Many of the techniques for patterning a lipid bilayer include the use of microfluidic technology Taylor *et al.*, 2009; Janshoff and Kunneke, 2000; Kam and Boxer, 2000;

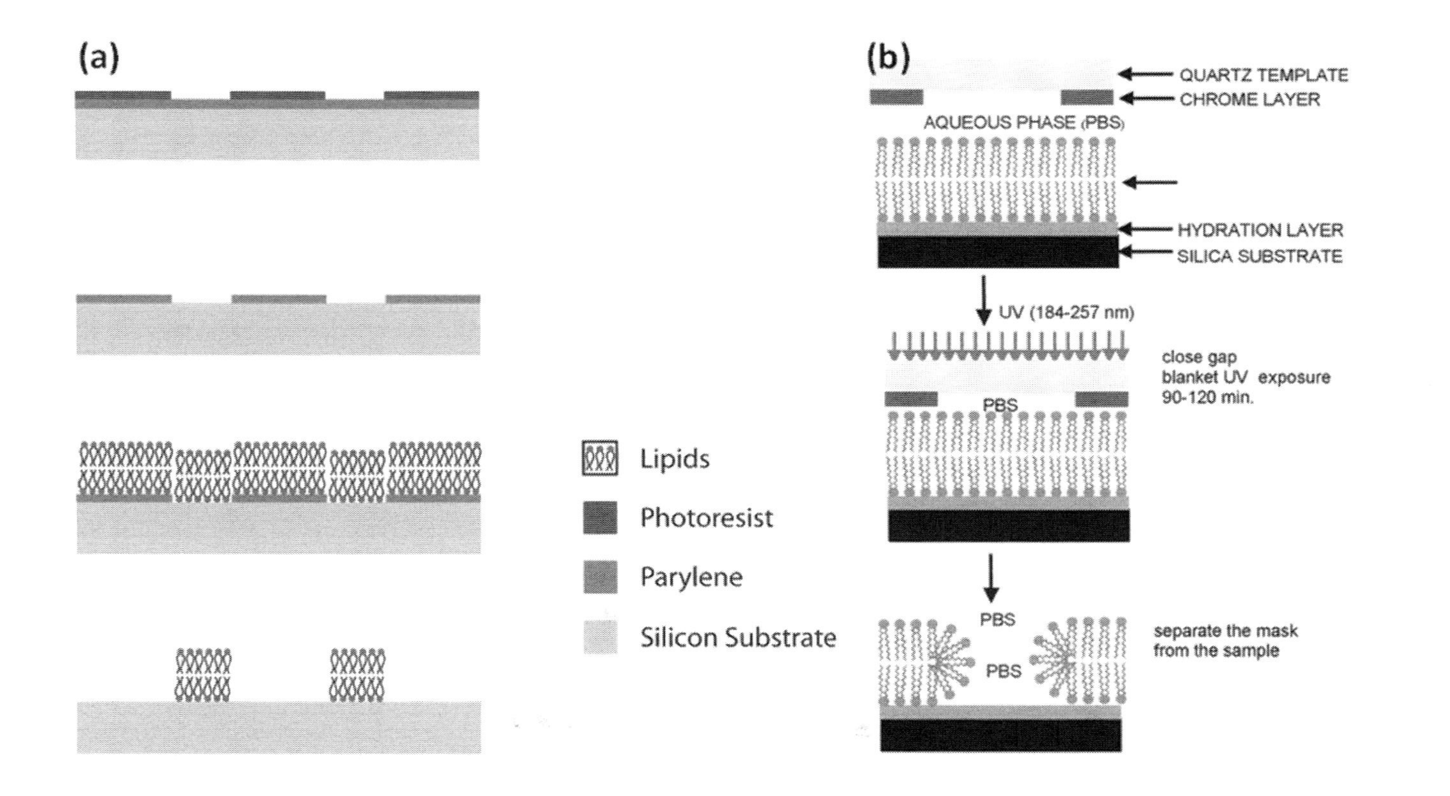

Figure 3. (a) After lipid bilayer formation on the structured polymer film, the polymer was mechanically lifted-off leaving behind a patterned lipid bilayer which could be backfilled with another lipid species. (b) In UV-photolysis, the lipids are decomposed at exposed areas, generating lipid-free areas which can also be backfilled with a second lipid species. Figure (a) adapted from (Orth *et al.*, 2003a) with permission from Elsevier. Figure (b) from (Yee *et al.*, 2004b). Reproduced by permission of The Royal Society of Chemistry.

Yang *et al.*, 2001; Kunneke and Janshoff, 2002; Burridge *et al.*, 2004; Taylor *et al.*, 2007; Shi *et al.*, 2007; Goluch *et al.*, 2008; Shi *et al.*, 2008b; Smith *et al.*, 2008). A substrate suited for analysis with a light microscope or other analytical methods is sealed by a PDMS mold imprinted with microfluidic channels (for review: (Becker and Gaertner 2008)). The vesicle solution is then injected or pumped into the microchannels and supported lipid bilayers are formed localized on the substrate surface exposed in the channels. Patterning of lipid bilayers using parallel single channel patterns (Janshoff and Kunneke, 2000; Yang *et al.*, 2001; Kunneke and Janshoff, 2002; Burridge *et al.*, 2004; Shi *et al.*, 2007; Kim *et al.*, 2006a), crossed channel patterns (Taylor *et al.*, 2009, 2007; Goluch *et al.*, 2008), as well as by other microfluidic geometries and methods (Kam and Boxer, 2000; Shi *et al.*, 2008b; Smith *et al.*, 2008) have been developed. Despite the success of these methods, many of the procedures are work-intensive and require a lot of sample material which might not be readily available in large quantities and hence are cost-intensive. A much discussed obstacle for the implementation of PDMS microfluidics, e.g. for drug screening is the sample and signal loss occurring from non-specific adsorption to the channel walls Although lipid bilayers can form on PDMS (Hovis and Boxer, 2001) and resist non-specific adsorption (Glasmastar *et al.*, 2003), Kim *et al.* (2006a,b) used UV curable low molecular weight PEG polymer such as PEG diacrylate that has been strongly crosslinked to prevent swelling or a combination of PEG and PDMS, instead of pure PDMS as material for the microchannels. As PEG prevents non-specific adsorption of proteins and other molecules (Pasche *et al.*, 2003), microchannels in this material obviate the need for further surface modification, but lead to liquid exchange with the hydrogel type polymer material.

ROBOTIC SPOTTING

Automation of the whole patterning process, starting from the loading of the vesicle solution to delivering it onto the surface whereby different lipid compositions can be brought to designed spots on the surface would greatly simplify and speed up the entire patterning procedure. Transferring the vesicle solution robotically to the surface was first described by Fang *et al.* (2003) and later on by several others using different spotters (Binkert *et al.*, 2009; Yamazaki *et al.*, 2005; Deng *et al.*, 2008). However, due to the small volume that is spotted (~ nL to pL) the spotting solution tends to dry out as soon as it is delivered onto the surface, destroying the vesicles and the lipid bilayers. To ensure hydration and avoid evaporation of the sample solution, the spotting can be carried out at very high humidity as was done by Yamazaki *et al.* (2005). Another method includes the functionalization of the surface prior to vesicle deposition which can be done by cholesteryl-PEG chains still allowing for fully fluid lipid bilayers (Deng *et al.*, 2008). Instead of using high humidity for spotting, cooling of the spotting solution and multiple drop spotting into microwells with glycerol in the buffer, which has a low volatility, was also recently successfully demonstrated (Binkert *et al.*, 2009). With the latter spotting technique array densities of up to 1111 wells cm^{-2} and a lipid bilayer spot size down to 100 μm with a pitch of 300 μm could be reached. Despite the speed and simple handling of this technique, feature sizes smaller than a few μm necessary for sub-cellular patterning or to provide a higher spot density important for library screening might be difficult to achieve. However, the spotting technique allows to address and eventually modify individual spots, therefore making it possible to efficiently spot lipid mixtures of varied composition in a single array.

POLYMER LIFT-OFF

By patterning a thin film of parylene by means of photolithography, lipid bilayer structures down to 1 μm could be achieved (Orth *et al.*, 2003a; Moran-Mirabal *et al.*, 2005; Vats *et al.*, 2008). After exposing the patterns to the vesicle solution, the parylene layer is lifted-off by peeling under water, leaving behind a structured lipid bilayer (*Figure 3(a)*). This patterning method shows a better pattern stability and more uniform patterns over large surface areas compared to the PDMS printing (Orth *et al.*, 2003a). Membranes patterned by this method have been used to study mast cell activation (Orth *et al.*, 2003b), receptor clustering (Wu *et al.*, 2004; Torres *et al.*, 2008), ligand-receptor binding studies (Moran-Mirabal *et al.*, 2005) and two-phase lipid bilayer domains (Vats *et al.*, 2008). Nevertheless, the applications for this methods are limited as only one lipid composition can be used, or in case of backfilling, two.

UV-PHOTOLYSIS

Yee *et al.* (2004a,b) used UV-illumination to pattern a continuous lipid bilayer under aqueous conditions. The UV-irradiation generates activated oxygen which in the exposed pattern decomposes lipids (*Figure 3(b)*). The lipid free parts of the patterns can now be backfilled with a second lipid species, allowing for example the study of metastable lipid bilayer microdomains (Yee *et al.*, 2004a,b). Chemical modification of the surface with OTS (octadecyltrichlorosilane) (Howland *et al.*, 2005; Oliver *et al.*, 2008) or a photocleavable self-assembled monolayer (Han *et al.*, 2007a) creates a mixture of hybrid lipid monolayer and lipid bilayer after exposure to a vesicle solution. Whilst in the latter case well-defined patterns could be achieved (Han *et al.*, 2007a), OTS resulted in a defect region (Howland *et al.*, 2005) between the lipid monolayer and the bilayer of several μm in size. UV-illumination was also used by Tanaka *et al.* (2004) to remove cellulose on glass, followed by BSA adsorption on the ablated glass pattern and vesicle fusion on the remaining cellulose.

PHOTOPOLYMERIZATION

Very stable patterned lipid bilayers can be achieved by polymerizing the lipids themselves using diacetylene phospholipids (DiynePC) and UV-illumination for crosslinking (Morigaki *et al.*, 2001, 2002, 2003, 2004; Morigaki, 2008) or sorbyl-functionalized lipids using UV-irradiation thermal and/or redox radical initiators (Ross *et al.*, 2003a,b, 2005; Mansfield *et al.*, 2007) to crosslink the lipids (*Figure 4(a)*). Whilst for the latter, method vesicle fusion can be used to form a bilayer on the substrate, the Langmuir-Blodgett or Langmuir-Schaefer techniques have to be used for DiynePC since alignment of the reactive moieties in the lipid tails is required for polymerization to occur and has best been achieved by these methods. The polymerized membranes have been shown very stable against surfactants and other treatments that usually destroy lipid bilayers as well as prolonged storage, drying and rehydration, and are resistant against unspecific adsorption of proteins (Ross *et al.*, 2003b; Mansfield *et al.*, 2007). After polymerization of the lipids only in desired areas, unreacted lipids are removed by surfactants leaving a pattern of crosslinked lipids that can be backfilled with a second lipid species by liposome fusion. Vesicle fusion was observed to be enhanced at the edges of the patterns (Okazaki *et al.*, 2006) allowing for facilitated bilayer formation. Also incorporation of biological membrane fractions was found possible with the help of a detergent (Morigaki, 2008).

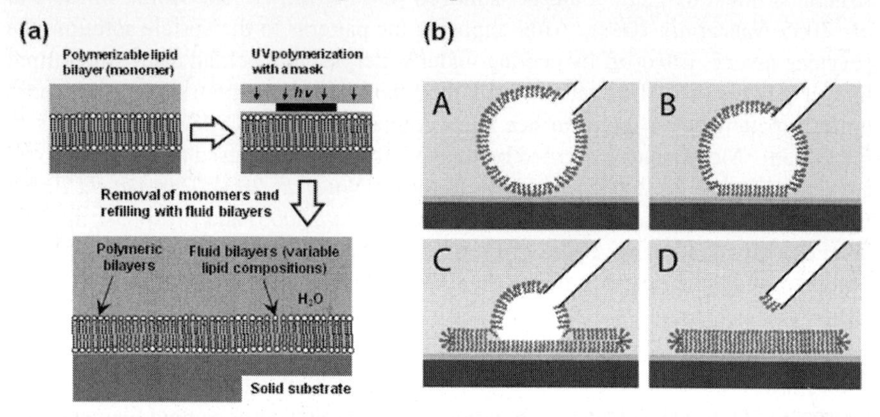

Figure 4. (a) UV light can be used to photopolymerize lipids in a SLB. Unpolymerized lipids are then removed by surfactants. (b) By inking an air bubble with a lipid monolayer, and subsequent removal of the air trapped within the bubble while it is in contact with the surface leading to its collapse, forms a single lipid bilayer patch on the surface. Figure (a) Reprinted with permission from (Morigaki, 2008). Copyright 2008, American Vacuum Society and Figure (b) Reprinted from (Mager and Melosh, 2007). Copyright 2007 American Chemical Society.

PATTERNING BY OTHER METHODS

Lipid bilayers can also be patterned by alternative and less conventional approaches than those described above. Some of these have even reached the sub-micron level (Lenhert *et al.*, 2007; Shi *et al.*, 2008a). In dip-pen nanolithography, an inked atomic force microscope tip directly deposits lipids onto the surface (Lenhert *et al.*, 2007). By careful tuning of the scan speed and relative humidity, feature sizes down to 100 nm as confirmed by AFM could be generated, but fluidity of the bilayers was only shown for larger μm-feature sizes by FRAP. This method was also used to massively deposit lipid bilayers simultaneously by up to 55000 tips per cm^2, however, high humidity is needed. Similar methods also using cantilever tips from atomic force microscopy, include shaving away parts of a continuous lipid bilayer at elevated pH creating a metastable patterned lipid bilayer that can be backfilled with a second lipid species with a resolution of about 1 μm (Jackson and Groves, 2004) or removal of proteins on the surface and subsequent backfilling with vesicles with a resolution down to 100 nm (Shi *et al.*, 2008a).

Air bubble collapse is another method (Mager and Melosh, 2007). Here, an air bubble is inked with a monolayer of lipids and the air within the bubble slowly removed while it is in contact with the substrate surface under water resulting in a fluid lipid bilayer with several hundred μms in size. The size can be tuned by the mechanical force applied on the bubble when it is in contact with the surface and the initial bubble size (*Figure 4(b)*). Manually inking the air bubble also allows for more than one lipid species to be used.

Lipid bilayers can also be patterned by taking advantage of substrate materials contrast, e.g. on patterns of SiO_2 and TiO_2. By first forming an anionic lipid bilayer on SiO_2 without Ca^{2+} in the buffer, the remaining TiO_2 pattern can be backfilled with a second batch of anionic vesicles in the presence of Ca^{2+} whose interaction with TiO_2 is strongly dependent on Ca^{2+} concentration (Rossetti *et al.*, 2005). This method, however, has only been demonstrated for phosphatidyl serine-containing vesicles whose interaction with TiO_2 is strongly dependent on Ca^{2+} concentration. A similar approach but with external control was used by Kumar *et al.* (2009). By using a locally applied voltage to 50 μm indium tin oxide (ITO) electrodes, they could selectively activate electrodes for liposome adsorption by removing coatings of liposome repelling PLL-*g*-PEG. Furthermore, by controlling the voltage bias of the electrodes the formation or desorption of phosphatidyl serine containing supported lipid bilayers could be achieved. Sequentially repeating this process, an array of membranes with different functionalities can be created.

Summary and outlook

Important obstacles to technological implementation of supported lipid bilayers have now begun to be removed and we have described some of the most important recent progress in this area. Detailed investigations of the effect of substrate roughness and texturing have shown that further theoretical and experimental understanding is needed before lipid membranes can be reproducibly self-assembled to either conform to or span engineered nanostructures. While tethered membranes have reached performance levels making them suitable for demanding applications like for example ion channel measurements, this has mostly been achieved at the expense of ease and simplicity of assembly. However, other promising ways of introducing larger polymer spacers have yet to demonstrate the same level of performance. Membrane patterning has taken great strides forward and a large number of approaches have been demonstrated with respective inherent strengths and weaknesses but they are still mostly at the conceptual stage. Perhaps the most obvious area for further exploration is the adaptation of protocols for assembling various model lipid mixture membranes which with greater fidelity capture the diversity of biological membranes in terms of composition and distribution. One can discern two conceptually different approaches for achieving this where the first is to fine-tune liposome-substrate interactions for increasingly complex artificial mixtures, and several research groups have managed to create membrane asymmetry by combining Langmuir-type deposition with liposome fusion. The second approach is to integrate native membrane fragments and liposomes into supported systems. Most likely generic improvements in compositional control have to come in combination with decoupling of the membrane from the underlying substrate.

For applications it is very likely that the separate solutions to the different problems of creating a viable membrane sensor platform have to be combined into a single format. This important and arduous work is likely to result in a new selection of 'best solutions', since several of the currently best methods of solving each separate problem are seemingly incompatible with each other. For example, methods to produce spacers which give suitable wetting conditions of the substrate will be more compatible with for example spotting arrays than other patterning methods. And it is likely that spreading of membranes on polymer cushions opens up new ways of creating artificial replicas of complex natural membranes which have been challenging on solid supports, as it already seems that tethered membranes can help prepare and improve stability of, for example, aperture spanning membranes. The merger of these approaches has only just started.

In summary, SLBs combine added stability, defined localization and application of surface sensitive and on the nanoscale localized measurement techniques, which ensures that they will continue to attract much attention and research effort in the years to come. And while there is no shortage of remaining scientific challenges in the field, there is also no lack of creative solutions being developed to meet these challenges.

References

Aebersold, R. and Mann, M. (2003) Mass spectrometry-based proteomics. *Nature* **422**, 198-207.

Albertorio, F., Diaz, A.J., Yang, T. L. *et al.* (2005) Fluid and air-stable lipopolymer membranes for biosensor applications. *Langmuir* **21**, 7476-7482.

Atanasov, V., Atanasova, P. P. and Vockenroth, I. K. *et al.* (2006) A molecular toolkit for highly insulating tethered bilayer lipid membranes on various substrates. *Bioconjugate Chemistry* **17**, 631-637.

Atanasov, V., Knorr, N., Duran, R. S. *et al.* (2005) Membrane on a chip: a functional tethered lipid bilayer membrane on silicon oxide surfaces. *Biophysical Journal* **89**, 1780-1788.

Barfoot, R. .J., Sheikh, K. H. and Johnson, B. R. G. *et al.* (2008) Minimal F-actin cytoskeletal system for planar supported phospholipid bilayers. *Langmuir* **24**, 6827-6836.

Bayerl, T. and Bloom, M. (1990) Physical properties of single phospholipid bilayers adsorbed to micro glass beads. A new vesicular model system studied by 2H-nuclear magnetic resonance *Biophysical Journal* **58**, 357-362.

Becker, H. and Gaertner, C. (2008) Polymer microfabrication technologies for microfluidic systems. *Analytical and Bioanalytical Chemistry* **390**, 89-111.

Benz, M., Chen, N. and Israelachvili, J. (2004) Lubrication and wear properties of grafted polyelectrolytes, hyaluronan and hylan, measured in the surface forces apparatus. *Journal of Biomedical Materials Research Part A* **71A**, 6-15.

Binkert, A., Studer, P. and Vörös, J. (2009) A microwell array platform for picoliter membrane protein assays. *Small* **5**, 1070-1077.

Boecker, M., Muschter, S., Schmitt, E. K. *et al.* (2009) Imaging and patterning of pore-suspending membranes with scanning ion conductance microscopy. *Langmuir* **25**, 3022-3028.

Burridge, K., Figa, M. and Wong, J. (2004) Patterning adjacent supported lipid bilayers of desired composition to investigate receptor-ligand binding under shear flow. *Langmuir* **20**, 10252-10259.

CRANE, J. M., KIESSLING, V. AND TAMM, L. K. (2005) Measuring lipid asymmetry in planar supported bilayers by fluorescence interference contrast microscopy. *Langmuir* **21**, 1377-1388.

CRANE, J. M. AND TAMM, L. K. (2007) Fluorescence microscopy to study domains in supported lipid bilayers. *Methods in Molecular Biology* **400**, 481-488.

DAHLIN, A., ZACH, M., RINDZEVICIUS, T. *ET AL.* (2005) Localized surface plasmon resonance sensing of lipid-membrane-mediated biorecognition events. *Journal of the American Chemical Society* **127**, 5043-5048.

DEME, B. AND MARCHAL, D. (2005) Polymer-cushioned lipid bilayers in porous alumina. *European Biophysics Journal* **34**, 170-179.

DENG, Y., WANG, Y., HOLTZ, B. *ET AL.* (2008) Fluidic and air-stable supported lipid bilayer and cell-mimicking microarrays. *Journal of the American Chemical Society* **130**, 6267-6271.

DENIAUD, A., ROSSI, C., BERQUAND, A. *ET AL.* (2007) Voltage-dependent anion channel transports calcium ions through biomimetic membranes. *Langmuir* **23**, 3898-3905.

DIAZ, A. .J., ALBERTORIO, F., DANIEL, S. *ET AL.* (2008). Double cushions preserve transmembrane protein mobility in supported bilayer systems. *Langmuir* **24**, 6820-6826.

DREXLER, J. AND STEINEM, C. (2003) Pore-suspending lipid bilayers on porous alumina investigated by electrical impedance spectroscopy. *Journal of Physcal Chemistry B* **107**, 11245-11254.

EGELSEER, E. M., LEITNER, K., .JAROSCH, M. *ET AL.* (1998). The s-layer proteins of two bacillus stearothermophilus wild-type strains are bound via their n-terminal region to a secondary cell wall polymer of identical chemical composition. *Journal of Bacteriology* **180**, 1488-1495.

ELIE-CAILLE, C., FLINIAUX, O., PANTIGNY, J. *ET AL.* (2005) Self-assembly of solid-supported membranes using a triggered fusion of phospholipid-enriched proteoliposomes prepared from the inner mitochondrial membrane. *Langmuir* **21**, 4661-4668.

ERBE, A., BUSHBY, R.J., EVANS, S. D. *ET AL.* (2007) Tethered bilayer lipid membranes studied by simultaneous attenuated total reflectance infrared spectroscopy and electrochemical impedance spectroscopy. *Journal of Physcal Chemistry B* **111**, 3515-3524.

FANG, Y., FRUTOS, A. AND LAHIRI, J. (2003) Ganglioside microarrays for toxin detection. *Langmuir* **19**, 1500-1505.

FISCHLECHNER, M., ZAULIG, M., MEYER, S. *ET AL.* (2008). Lipid layers on polyelectrolyte multilayer supports. *Soft Matter* **4**, 2245-2258.

FRIEDRICH, M. G., PLUM, M. A., SANTONICOLA, M. G. *ET AL.* (2008) In situ monitoring of the catalytic activity of cytochrome c oxidase in a biomimetic architecture. *Biophysical Journal* **95**, 1500-1510.

GALLA, H., HARTMANN, W., THEILEN, U. *ET AL.* (1979) 2-dimensional passive random-walk in lipid bilayers and fluid pathways in biomembranes. *Journal of Membrane Biology* **48**, 215-236.

GLASMASTAR, K., GOLD, J., ANDERSSON, A. *ET AL.* (2003) Silicone transfer during microcontact printing. *Langmuir* **19**, 5475-5483.

GOENNENWEIN, S., TANAKA, M., HU, B. *ET AL.* (2003) Functional incorporation of integrins into solid supported membranes on ultrathin films of cellulose: Impact on adhesion. *Biophysical Journal* **85**, 646-655.

GOLUCH, E. D., SHAW, A. W., SUIGAR, S. G. *ET AL.* (2008) Microfluidic patterning of nanodisc lipid bilayers and multiplexed analysis of protein interaction. *Lab on a Chip*

8, 1723–1728.

GROVES, J., ULMAN, N. AND BOXER, S. (1997). Micropatterning fluid lipid bilayers on solid supports. *Science* **275**, 651-653.

GUFLER, P. C., PUM, D., SLEYTR, U. B. *ET AL.* (2004). Highly robust lipid membranes on crystalline s-layer supports investigated by electrochemical impedance spectroscopy. *Biochimica et Biophysica Acta* **1661**, 154-165.

GYORVARY, E., WETZER, B., SLEYTR, U. B. *ET AL.* (1999). Lateral diffusion of lipids in silane-, dextran, and s-layer-supported mono- and bilayers. *Langmuir* **15,** 1337-1347.

HAN, X., CRITCHLEY, K., ZHANG, L. *ET AL.* (2007a). A novel method to fabricate patterned bilayer lipid membranes. *Langmuir* **23**, 1354-1358.

HAN, X., STUDER, A., SEHR, H. *ET AL.* (2007b). Nanopore arrays for stable and functional free-standing lipid bilayers. *Advanced Materials* **19**, 4466-4470

HENNESTHAL, C. AND STEINEM, C. (2000). Pore-spanning lipid bilayers visualized by scanning force microscopy. *Journal of the American Chemical Society* **122**, 8085- 8086.

HERRIG, A., JANKE, M., AUSTERMANN, J. *ET AL.* (2006). Cooperative adsorption of ezrin on pip2-containing membranes. *Biochemistry* **45**, 13025-13034.

HOVIS, J. AND BOXER, S. (2000) Patterning barriers to lateral diffusion in supported lipid bilayer membranes by blotting and stamping. *Langmuir* **16**, 894-897.

HOVIS, J. AND BOXER, S. (2001) Patterning and composition arrays of supported lipid bilayers by microcontact printing. *Langmuir* **17**, 3400-3405.

HOWLAND, M., SAPURI-BUTTI, A., DIXIT, S. *ET AL.* (2005). Phospholipid morphologies on photochemically patterned silane monolayers. *Journal of the American Chemical Society* **127**, 6752-6765.

HUANG, S.-C.J., ARTYUKHIN, A. B., MARTINEZ, J. A. *ET AL.* (2007) Formation, stability, and mobility of one-dimensional lipid bilayers on polysilicon nanowires. *Nano Letters* **7**, 3355-3359.

HWANG, L. Y., GÖTZ, H., KNOLL, W. *ET AL.* (2008) Preparation and characterization of glycoacrylate-based polymer-tethered lipid bilayers on benzophenone-modified suhstrates. *Langmuir* **24**, 14088-14098.

JACKSON, B. AND GROVES, J. (2004) Scanning prohe lithography on fluid lipid membranes. *Journal of the American Chemical Society*, **126**, 13878-13879.

JANKE, M., HERRIG, A., AUSTERMANN, J. *ET AL.* (2008) Actin binding of ezrin is activated by specific recognition of PIP2-functionalized lipid bilayers. *Biochemistry* **47**, 3762-3769.

JANSHOFF, A. AND KUNNEKE, S. (2000) Micropatterned solid-supported membranes formed by micromolding in capillaries. *European Biophysical Journal* **29**, 549-554.

JANSHOFF, A. AND STEINEM, C. (2006) Transport across artificial membranes - an analytical perspective. *Analytical and Bioanalytical Chemistry* **385**, 433-451.

JEON, T.-.J., MALMSTADT, N. AND SCHMIDT, J. J. (2006) Hydrogel-encapsulated lipid membranes. *Journal of the American Chemical Society* **128**, 42-43.

JEUKEN, L. J. C., BUSHBY, R. J. AND EVANS, S. D. (2007) Proton transport into a tethered bilayer lipid membrane. *Electrochemistry Communications* **9**, 610-614.

JONSSON, M. P., JONSSON, P., DAHLIN, A. B. *ET AL.* (2007). Supported lipid bilayer formation and lipid-membrane-mediated biorecognition reactions studied with a new nanoplasmonic sensor template. *Nano Letters* **7**, 3462-3468.

JONSSON, M. P., JONSSON, P. AND HOOK, F. (2008) Simultaneous nanoplasmonic and quartz crystal microbalance sensing: Analysis of biomolecular conformational changes and quantification of the bound molecular mass. *Analytical Chemistry* **80**, 7988-7995.

JUNG, S., HOLDEN, M., CREMER, P. *ET AL.* (2005) Two-component membrane lithography via lipid backfilling. *ChemPhysChem.* **6**, 423-426.

KALB, E., FREY, S. AND TAMM, L. K. (1992) Formation of supported planar bilayers by fusion of vesicles to supported phospholipid monolayers. *Biochimica et Biophysica Acta* **1103**, 307-316.

KAM, L. AND BOXER, S. (2000) Formation of supported lipid bilayer composition arrays by controlled mixing and surface capture. *Journal of the American Chemical Society* **122**, 12901-12902.

KANG, X. F., CHELEY, S., RICE-FICHT, A. C. *ET AL.* (2007). A storable encapsulated bilayer chip containing a single protein nanopore. *Journal of the Amican Chemical Society* **129**, 4701-4705.

KAUFMANN, S., PAPASTAVROU, G., KUMAR, K. *ET AL.* (2009) A detailed investigation of the formation kinetics and layer structure of poly(ethylene glycol) tether supported lipid bilayers. *Soft Matter* **5**, 2804-2814

KIESSLING, V., CRANE, J. M. AND TAMM, L. K. (2006). Transbilayer effects of raft-like lipid domains in asymmetric planar bilayers measuredby single molecule tracking. *Biophysical Journal* **91**, 3313-3326.

KIESSLING, V., WAN, C. AND TAMM, L. K. (2009) Domain coupling in asymmetric lipid bilayers. *Biochimica et Biophysica Acta* **1788**, 64-71.

KIM, P., JEONG, H. E., KHADEMHOSSEINI, A. *ET AL.* (2006a). Fabrication of non-biofouling polyethylene glycol micro- and nanochannels by ultraviolet-assisted irreversible sealing. *Lab on a Chip* **6**, 1432-1437.

KIM, P., LEE, S., JUNG, H. *ET AL.* (2006b) Soft lithographic patterning of supported lipid bilayers onto a surface and inside microfluidic channels. *Lab on a Chip* **6**, 54-59.

KUMAR, K., TANG, C. S., ROSSETTI, F. F. *ET AL.* (2009) Formation of supported lipid bilayers on indium tin oxide for dynamically- patterned membrane-functionalized microelectrode arrays. *Lab on a Chip* **9**, 718- 725.

KUNDING, A. AND STAMOU, D. (2006) Subnanometer actuation of a tethered lipid bilayer monitored with fluorescence resonance energy transfer. *Journal of the American Chemical Society* **128**, 11328-11329.

KUNG, L., KAM, L., HOVIS, .J. *ET AL.* (2000) Patterning hybrid surfaces of proteins and supported lipid bilayers. *Langmuir* **16**, 6773-6776.

KUNNEKE, S. AND .JANSHOFF, A. (2002) Visualization of molecular recognition events on microstructured lipid-membrane compartments by in situ scanning force microscopy. *Angewandte Chemie International Edition* **41,** 314-316.

LAMBACHER, A., FROMHERZ, P. (1996) Fluorescence interference-contrast microscopy on oxidized silicon using a monomolecular dye layer. *Applied Physics A: Materials Science and Processing* **63**, 297-216.

LAMBACHER, A., FROMHERZ, P. (2002) Luminescence of dye molecules on oxidized silicon and fluorescence interference contrast microscopy of biomembranes. *Journal of the Optical Society of America B* **19**, 1435-1453.

LANG, H., DUSCHL, C. AND VOGEL, H. (1994) A new class of thiolipids for the attachment of lipid bilayers on gold surfaces. *Langmuir* **10**, 197-210.

LEE, S., NA, Y. AND LEE, S. (2009) Nondisruptive micropatterning of fluid membranes through selective vesicular adsorption and rupture. *Langmuir* **25**, 5421-5425.

LENHERT, S., SUN, P., WANG, Y. *ET AL.* (2007) Massively parallel dip-pen nanolithography of heterogeneous supported phospholipid multilayer patterns. *Small* **3**, 71-75.

Mager, M. D. and Melosh, N. A. (2007) Lipid bilayer deposition and patterning via air bubble collapse. *Langmuir* **23**, 9369-9377.

Maid, S. and Mayer, M. (2005) Hydrogel stamping of arrays of supported lipid bilayers with various lipid compositions for the screening of drug-membrane and protein-membrane interactions. *Angewandte Chemie International Edition* **44**, 6697-6700.

Maid, S. and Mayer, M. (2008) Generating arrays with high content and minimal consumption of functional membrane proteins. *Journal of the American Chemical Society* **130**, 16060-16064.

Malmstadt, N., Jeon, L. J. and Schmidt, J. J. (2008) Long-lived planar lipid bilayer membranes anchored to an in situ polymerized hydrogel. *Advanced Materials* **20**, 84-89

Mansfield, E., Ross, E. E., D'Ambruoso, G. D. et al. (2007) Fabrication and characterization of spatially-defined, multiple component, chemically-functionalized domains in enclosed silica channels using cross-linked phospholipid membranes. *Langmuir* 23, 11326-11333.

Melzak, K. and Gizeli, E. (2002) A silicate gel for promoting deposition of lipid bilayers. *Journal of Colloid Interface Science* **246**, 21-28.

Merz, C., Knoll, W., Textor, M. et al. (2008) Formation of supported bacterial lipid membrane mimics. *Biointerphases* **3**, FA41-FA50.

Misra, N., Martinez, J. A., Huang, S.-C. J. et al. (2009) Bioelectronic silicon nanowire devices using functional membrane proteins. *Proceedings of the National Academy of Sciences of the United States of America* **106**, 13780-13784.

Moran-Mirabal, J., Edel, J., Meyer, G. et al. (2005). Micrometer-sized supported lipid bilayer arrays for bacterial toxin binding studies through total internal reflection fluorescence microscopy. *Biophysical Journal* **89**, 296-305.

Morigaki, K. (2008) Micropatterned model biological membranes composed of polymerized and fluid lipid bilayers. *Biointerphases* **3**, FA85-FA89.

Morigaki, K., Baumgart, T., Jonas, U. et al. (2002) Photopolymerization of diacetylene lipid bilayers and its application to the construction of micropatterned biomimetic membranes. *Langmuir* **18**, 4082-4089.

Morigaki, K., Baumgart, T., Offenhausser, A. et al. (2001) Patterning solid-supported lipid bilayer membranes by lithographic polymerization of a diacetylene lipid. *Angewandte Chemie - International Edition* **40**, 172-174.

Morigaki, K., Kiyosue, K. and Taguchi, T. (2004) Micropatterned composite membranes of polymerized and fluid lipid bilayers. *Langmuir* **20**, 7729-7735.

Morigaki, K., Schonherr, H., Frank, C.W. et al. (2003) Photolithographic polymerization of diacetylene-containing phospholipid bilayers studied by multimode atomic force microscopy. *Langmuir* **19**, 6994-7002.

Mornet, S., Lambert, O., Duguet, E. et al. (2005) The formation of supported lipid bilayers on silica nanoparticles revealed by cryoelectron microscopy. *Nano Letters* **5**, 281-285.

Mueller, P., Wescott, W., Rudin, D. et al. (1963) Methods for formation of single bimolecular lipid membranes in aqueous solution. *Journal of Physical Chemistry*, 67, 534-535

Murray, D. H., Tamm, L. K. and Kiessling, V. (2009) Supported double membranes. *Journal of Structural Biology* 168, 183-189

NGUYEN, K. T., LE CLAIR, S. V., YE, S.J. *ET AL.* (2009a) Molecular interactions between magainin 2 and model membranes in situ. *Journal of Physical Chemistry B* **113**, 12358-12363.

NGUYEN, K. T., LE CLAIR, S. V., YE, S. *ET AL.* (2009b) Orientation determination of protein helical secondary structures using linear and nonlinear vibrational spectroscopy. *Journal of Physical Chemistry B* **113**, 12169-12180.

NIKOLOV, V., LIN, .J., MERZLYAKOV, M. *ET AL.* (2007) Electrical measurements of bilayer membranes formed by Langmuir-Blodgett deposition on single-crystal silicon. *Langmuir* **23**, 13040-13045.

OKAZAKI, T., MORIGAKI, K. AND TAGUCHI, T. (2006) Phospholipid vesicle fusion on micropatterned polymeric bilayer substrates. *Biophysical Journal* **91**, 1757-1766.

OLIVER, A. E., KENDALL, E. L., HOWLAND, M. C. *ET AL.* (2008) Protecting, patterning, and scaffolding supported lipid membranes using carbohydrate glasses. *Lab on a Chip* **8**, 892-897.

ORTH, R., KAMEOKA, J., ZIPFEL, W. *ET AL.* (2003a) Creating biological membranes on the micron scale: Forming patterned lipid bilayers using a polymer lift-off technique. *Biophysical Journal* **85**, 3066-3073.

ORTH, R., WU, M., HOLOWKA, D. *ET AL.* (2003b) Mast cell activation on patterned lipid bilayers of subcellular dimensions. *Langmuir* **19**, 1599-1605.

PARTHASARATHY, R. AND GROVES, J. T. (2007) Curvature and spatial organization in biological membranes. *Soft Matter* **3**, 24-33.

PARTHASARATHY, R., YU, C.H. AND GROVES, J. (2006) Curvature-modulated phase separation in lipid bilayer membranes. *Langmuir* **22**, 5095-5099.

PASCHE, S. DE PAUL, S.M., VOROS, J.T. *ET AL.* (2003) Poly(L-lysine)-graft-poly(ethylene glycol) assembled monolayers on niobium oxide surfaces: A quantitative study of the influence of polymer interfacial architecture on resistance to protein adsorption by ToF-SIMS and in situ OWLS. *Langmuir* **19**, 9216-9225.

PEREZ, J. B., MARTINEZ, K. L., SEGURA, J. M. *ET AL.* (2006) Supported cell-membrane sheets for functional fluorescence imaging of membrane proteins. *Advanced Functional Materials* **16**, 306-312.

PFEIFFER, I., SEANTIER, B., PETRONIS, S. *ET AL.* (2008) Influence of nanotopography on phospholipid bilayer formation on silicon dioxide. *Journal of Physical Chemistry B* **112**, 5175-5181.

PUM, D. AND SLEYTR, U. B. (1995) Monomolecular reassembly of a crystalline bacterial cell surface layer (S-layer) on untreated and modified silicon surfaces. *Supramolecular Science* **2**, 193-197.

REIMHULT, E., HOOK, F. AND KASEMO, B. (2003) Intact vesicle adsorption and supported biomembrane formation from vesicles in solution: Influence of surface chemistry, vesicle size temperature and osmotic pressure. *Langmuir* **19**, 1681-1691.

REIMHULT, E. AND KUMAR, K. (2008). Membrane biosensor platforms using nano- and microporous supports. *Trends in Biology* **19**, 82-89.

RENNER, L., OSAKI, T., CHIANTIA, S. *ET AL.* (2008) Supported lipid bilayers on spacious and pH-responsive polymer cushions with varied hydrophilicity. *The Journal of Physical Chemistry B* **112**, 6373-6378.

RICHTER, R. P., BERAT, R. AND BRISSON, A. R. (2006) Formation of solid-supported lipid bilayers: An integrated view. *Langmuir* **22**, 3497-3505.

RICHTER, R. P. AND BRISSON, A. R. (2005) Following the formation of supported lipid bilayers on mica: a study combining AFM, QCM-D, and ellipsometry. *Biophys. J.* **88**, 3422-3433.

RICHTER, R. P., HOCK, K. K., BURKHARTSMEYER, J. *ET AL.* (2007) Membrane-grafted hyaluronan films: A well-defined model system of glycoconjugate cell coats. *Journal of the American Chemical Society* **129**, 5306-5307.

RICHTER, R. P., MAURY, N. AND BRISSON, A. R. (2005) On the effect of the solid support on the interleaflet distribution of lipids in supported lipid bilayers. *Langmuir* **21**, 299-304.

ROITER, Y., ORNATSKA, M., RAMMOHAN, A. R. *ET AL.* (2008). Interaction of nanoparticles with lipid membrane. *Nano Lett.* **8**, 941-944.

ROITER, Y., ORNATSKA, M., RAMMOHAN, A. R. *ET AL.* (2009). Interaction of lipid membrane with nanostructured surfaces. *Langmuir* **25**, 6287-6299.

ROMER, W. AND STEINEM, C. (2004). Impedance analysis and single-channel recordings on nano-black lipid membranes based on porous alumina. *Biophysical Journal* **86**, 955-965.

ROSS, E., MANSFIELD, E., HUANG, Y. *ET AL.* (2005) In situ fabrication of three-dimensional chemical patterns in fused silica separation capillaries with polymerized phospholipids. *Journal of the American Chemical Society* **127**, 16756-16757.

ROSS, E., ROZANSKI, L., SPRATT, T. *ET AL.* (2003a) Planar supported lipid bilayer polymers formed by vesicle fusion. 1. influence of diene monomer structure and polymerization method on film properties. *Langmuir* **19**, 1752—1765.

ROSS, E., SPRATT, T., LIU, S. *ET AL.* (2003b). Planar supported lipid bilayer polymers formed by vesicle fusion 2 adsorption of bovine serum albumin. *Langmuir* **19**, 1766-1774.

ROSSETTI, F., BALLY, M., MICHEL, R. *ET AL.* (2005) Interactions between titanium dioxide and phosphatidyl serine-containing liposomes: Formation and patterning of supported phospholipid bilayers on the surface of a medically relevant material. *Langmuir* **21**, 6443-6450.

ROSSI, C., BRIAND, E., PAROT, P. *ET AL.* (2007) Surface response methodology for the study of supported membrane formation. *Journal of Physical Chemistry B* **111**, 7567-7576.

SAKMANN, B. (1995) *Single-Channel Recording*. New York: Plenum Press, 2nd Edition.

SAMIEE, K., MORAN-MIRABAL, .J, CHEUNG, Y. *ET AL.* (2006) Zero mode waveguides for single-molecule spectroscopy on lipid membranes. *Biophysical Journal* **90**, 3288-3299.

SAPURI, A., BAKSH, M. AND GROVES, J. (2003) Electrostatically targeted intermembrane lipid exchange with micropatterned supported membranes. *Langmuir* **19**, 1606-1610.

SAPURI-BUTTI, A., LI, Q., GROVES, .J. *ET AL.* (2006) Nonequilibrium patterns of cholesterol-rich chemical heterogenieties within single fluid supported phospholipid bilayer membranes. *Langmuir* **22**, 5374-5384.

SCHMITT, E. K., NURNABI, M., BUSHBY, R. J. *ET AL.* (2008) Electrically insulating pore-suspending membranes on highly ordered porous alumina obtained from vesicle spreading. *Soft Matter* **4**, 250-253.

SCHMITT, E. K., WEICHBRODT, C. AND STEINEM, C. (2009) Impedance analysis of gramicidin d in pore-suspending membranes. *Soft Matter* **5**, 3347-3353.

SCHUSTER, B., GUFLER, P. C., PUM, D. *ET AL.* (2004) S-layer proteins as supporting scaffoldings for functional lipid membranes. *IEEE Transactions on Nanobioscience* **3**, 16-21.

SCHUSTER, B., PUM, D., SARA, M. *ET AL.* (2001) S-layer ultrafiltration membranes: A new support for stabilizing functionalized lipid membranes. *Langmuir* **17**, 499-503.

SENGUPTA, K., SCHILLING, J., MARX, S. *ET AL.* (2003). Mimicking tissue surfaces by supported membrane coupled ultrathin layer of hyaluronic acid. *Langmuir* **19**, 1775-1781.

SEU, K. J., PANDEY, A. P., HAQUE, F. *ET AL.* (2007). Effect of surface treatment on diffusion and domain formation in supportedlipid bilayers. *Biophysical Journal* **92**, 2445-2450.

SHAO, Y., JIN, Y. D., WANG, J. L. *ET AL.* (2005) Conducting polymer polypyrrole supported bilayer lipid membranes. *Biosensors and Bioelectronics* **20**, 1373-1379.

SHAW, J. E., EPAND, R. F., HSU, J. C. Y. *ET AL.* (2007) Cationic peptide-induced remodelling of model membranes: Direct visualization by in situ atomic force microscopy. *Journal of Structural Biology* **162**, 121-138.

SHI, J., CHEN, J. AND CREMER, P. S. (2008a) Sub-100 nm patterning of supported bilayers by nanoshaving lithography. *Journal of the American Chemical Society* **130**, 2718-2719.

SHI, J., YANG, T. AND CREMER, P. S. (2008b) Multiplexing ligand-receptor binding measurements by chemically patterning microfluidic channels. *Analytical Chemistry* **80**, 6078-6084.

SHI, J., YANG, T., KATAOKA, S. *ET AL.* (2007) Gm(1) clustering inhibits cholera toxin binding in supported phospholipid membranes. *Journal of the American Chemical Society* **12**, 5954-5961.

SHIM, J. W. AND GU, L. Q. (2007) Stochastic sensing on a modular chip containing a single-ion channel. *Analytical Chemistry* **79(6)**, 2207-2213.

SINNER, E. K. AND KNOLL, W. (2001) Functional tethered membranes. *Current Opinion in Chemical Biology* **5**, 705-711.

SMITH, E. A., COYM, J. W., COWELL, S. M. *ET AL.* (2005) Lipid bilayers on polyacrylamide brushes for inclusion of membrane proteins. *Langmuir* **21**, 9644-9650.

SMITH, K. A., GALE, B. K. AND CONBOY, J. C. (2008) Micropatterned fluid lipid bilayer arrays created using a continuous flow microspotter. *Analytical Chemistry* **80**, 7980-7987.

SÁRA, M., DEKITSCH, C., MAYER., H. F. *ET AL.* (1998) Influence of the secondary cell wall polymer on the reassembly recrystallization, and stability properties of the s-layer protein from Bacillus stearothermophilus pv72/p2. *Journal of Bacteriology* **180**, 4146-4153.

TANAKA, M., KAUFMANN, S., NISSEN, J. *ET AL.* (2001) Orientation selective immobilization of human erythrocyte membranes on ultrathin cellulose films. *Physical Chemistry, Chemical Physics* **3**, 4091-4095.

TANAKA, M. AND SACKMANN, E. (2005) Polymer-supported membranes as models of the cell surface. *Nature* **437**, 656-663.

TANAKA, M., WONG, A. P., REHFELDT, F. *ET AL.* (2004). Selective deposition of native cell membranes on biocompatible micropatterns. *Journal of the American Chemical Society* **126**, 3257-3260.

TAYLOR, .J. D., LINMAN, M. .J., WILKOP, T. *ET AL.* (2009) Regenerable tethered bilayer lipid membrane arrays for multiplexed label-free analysis of lipid-protein interactions on poly(dimethylsiloxane) microchips using spr imaging. *Analytical Chemistry* **81**, 1146—1153.

TAYLOR, J. D., PHILLIPS, K. S. AND CHENG, Q. (2007) Microfluidic fabrication of addressable tethered lipid bilayer arrays and optimization using spr with silane-derivatized nanoglassy substrates. *Lab on a Chip* **7**, 927-930.

THID, D., BENKOSKI, J. J., SVEDHEM, S. *ET AL.* (2007) Dha-induced changes of supported lipid membrane morphology. *Langmuir* **23**, 5878-5881.

TORRES, A. J., VASUDEVAN, L., HOLOWKA, D. *ET AL.* (2008) Focal adhesion proteins connect IgE receptors to the cytoskeleton as revealed by micropatterned ligand arrays. *Proceedings of the National Academy of Sciences of the United States of America*

105, 17238-17244.

Tutus, M., Rossetti, F. F., Schneck, E. et al. (2008) Orientation-selective incorporation of transmembrane f0f1 atp synthase complexfrom *Micrococcus luteus* in polymer-supported membranes. *Macromolecular Bioscience* **8**, 1034-1043.

Van Meer, G. (2005) Cellular lipidomics. *European Molecular Biology Organisation Journal* **24**, 3159-3165.

Vats, K., Kyoung, M. and Sheets, E. D. (2008) Characterizing the chemical complexity of patterned biomimetic membranes. *Biochimica et Biophysica Acta* **1778**, 2461-2468.

Vockenroth, I. K., Ohm, C., Robertson, J. W. F. et al. (2008) Stable insulating tethered bilayer lipid membranes. *Biointerphases* **3**, FA68-FA73.

Wagner, M. L. and Tamm, L. K. (2000) Tethered polymer-supported planar lipid bilayers for reconstitution of integral membrane proteins: silane-polyethyleneglycol-lipid as a cushion and covalentlinker. *Biophysical Journal* **79**, 1400-1414.

Wenger, J., Rigneault, H., Dintinger, J. et al. (2006) Single-fluorophore diffusion in a lipid membrane over a subwavelength aperture. *Journal of Biological Physics* **32**, SN1-SN4.

Werner, J. H., Montano, G. A., Garcia, A. L. et al. (2009) Formation and dynamics of supported phospholipid membranes on a periodic nanotextured substrate. *Langmuir* **25,** 2986-2993.

Weygand, M., Kjaer, K., Howes, P. B. et al. (2002) Structural reorganization of phospholipid headgroups upon recrystallization of an s-layer lattice. *Journal of Physical Chemistry B* **106** 5793-5799.

Wu, M., Holowka, D., Craighead, H. et al. (2004). Visualization of plasma membrane compartmentalization with patternedlipid bilayers. *Proceedings of the National Academy of Sciences of the United States of America* **101**, 13798-13803.

Yamazaki, V., Sirenko, O., Schafer, R. et al. (2005) Cell membrane array fabrication and assay technology. *BMC Biotechnology* **5**, pp. 18.

Yang, T., .Jung, S., Mao, H. et al. (2001). Fabrication of phospholipid bilayer-coated microchannels for on-chip immunoassays. *Analytical Chemistry* **73**, 165-169.

Ye, Q., Konradi, R., Textor, M. et al. (2009) Liposomes tethered to omega-functional peg brushes and induced formation of peg brush supported planar lipid bilayers. *Langmuir,* **25(23)**, 13534-13539.

Yee, C., Amweg, M. and Parikh, A. (2004a) Direct photochemical patterning and refunctionalization of supported phospholipid bilayers. *Journal of the American Chemical Society* **126**, 13962-13972.

Yee, C., Amweg, M. and Parikh, A. (2004b) Membrane photolithography: Direct micropatterning and manipulation of fluid phospholipid membranes in the aqueous phase using deep-uv light. *Advanced Materials* **16**, 1184-1189.

Zhou, X., Moran-Mirabal, J. M, Craighead, H. G. et al. (2007) Supported lipid bilayer/carbon nanotube hybrids. *Nature Nanotechnology* **2**, 185-190.

Biotechnology and Genetic Engineering Reviews - Vol. 27, 217-228 (2010)

Biotechnology developments in the livestock sector in developing countries

SUNEEL K ONTERU, AGATHA AMPAIRE, MAX F ROTHSCHILD*

Department of Animal Science and Center for Integrated Animal Genomics, Iowa State University, Ames, IA 50011, USA

Abstract

Global meat and milk consumption is exponentially increasing due to population growth, urbanization and changes in lifestyle in the developing world. This is an excellent opportunity for developing countries to improve the livestock sector by using technological advances. Biotechnology is one of the avenues for improved production in the "Livestock revolution". Biotechnology developments applied to livestock health, nutrition, breeding and reproduction are improving with a reasonable pace in developing countries. Simple bio-techniques such as artificial insemination have been well implemented in many parts of the developing world. However, advanced technologies including transgenic plant vaccines, marker assisted selection, solid state fermentation for the production of fibrolytic enzymes, transgenic fodders, embryo transfer and animal cloning are confined largely to research organizations. Some developing countries such as Taiwan, China and Brazil have considered the commercialization of biotechnology in the livestock sector. Organized livestock production systems, proper record management, capacity building, objective oriented research to improve farmer's income, collaborations with the developed world, knowledge of the sociology of an area and research on new methods to educate farmers and policy makers need to be improved for the creation and implementation of biotechnology advances in the livestock sector in the developing world.

*To whom correspondence may be addressed (mfrothsc@iastate.edu)

Abbreviations: AI, Artificial insemination; Bt, *Bacillus thurengensis*; ELISA, Enzyme Linked Immuno Sorbent Assay; ESR, Estrogen receptor; FAO, Food and Agriculture Organization; HR, Herbicide Tolerance; GMP, Genetically Modified Plants; ILRI, International Livestock Research Institute; LAMP, Loop-Mediated Isothermal Amplification; LD, Linkage Disequilibrium; LE, Linkage Equilibrium; MAS, Marker Assisted Selection; MC4R, Melanocortin 4 receptor; MOET, Multiple Ovulation and Embryo Transfer; OIE, The World Organization for Animal Health; PABP, Pingtung Agricultural Biotechnology Park; PPR, Peste Des Petits Ruminants.

Introduction

Biotechnology is defined by the United Nations Convention on biological diversity as "any technological application that uses biological systems, living organisms, or derivatives thereof, to make or modify products or processes for specific use". Livestock have played a pivotal role in human development for the past 8,000 to 10,000 years when pigs, chicken, cattle, goats and sheep were first domesticated (Rothschild and Plastow, 2008). Livestock production contributes nearly 43 percent of the gross value for global agricultural production. In developed countries, its contribution is more than half, where as in developing countries; its share is only one third to the total global agricultural production. However, the demand for livestock products is increasing in developing countries (Table 1) because of population growth, urbanization, changes in lifestyles and dietary habits, and increasing the disposable incomes. Meat consumption per capita is relatively stable in the developed world. However, the annual per capita meat consumption doubled in developing countries between 1980 and 2002, and this trend is likely to continue for the foreseeable future. The Food and Agriculture Organization (FAO) estimated that global meat and milk production must double by 2050 to meet population growth. This growth represents both a huge opportunity and challenge for the livestock producers in developing countries (http://www.id21.org/insights/insights72/art05.html). To meet these needs, the next food revolution will be the "Livestock revolution" (Delgado *et al*., 1999). Agricultural biotechnology is one of the major avenues to pave the way for the "Livestock revolution" in developing countries. Many earlier biotechnology reviews have been published with a focus on what might be done using biotechnology techniques rather giving much importance to those initiated or implemented biotechnologies in the livestock sector in developing countries. Therefore, the present review focuses on the biotechnology developments initiated so far in the areas of livestock health, nutrition, breeding and reproduction in developing countries.

Table 1. Increasing demand for meat in developing world (Delgado *et al*, 1999)

	Annual growth of total meat consumption (%)		Total meat consumption (Mt)		
Region	1982–1994	1993–2020	1983	1993	2020
China	8.6	3.0	16	38	85
India	3.6	2.9	3	4	8
Southeast Asia	5.6	3.0	4	7	16
Latin America	3.3	2.3	15	21	39
West Asia/North Africa	2.4	2.8	5	6	15
Sub-Saharan Africa	2.2	3.5	4	5	12
Developing world	5.4	2.8	50	88	188
Developed world	1.0	0.6	88	97	115
Total world	2.9	1.8	139	184	303

Sources: FAO annual data. Total meat consumption for 1983 and 1993 are three-year moving averages. 2020 projections come from IFPRI's global model, IMPACT.
Abbreviation: Mt, million metric tonne.

Biotechnology in livestock health

Sustainable animal productivity depends on sound animal health. The high level of biodiversity and less sophisticated livestock production systems expose livestock to a variety of pathogens in developing countries. The maintenance of wildlife reservoirs for different kinds of infections and infestations, and their proximity to the livestock production systems is an everlasting challenge in the developing world. Therefore, much of the research is focused on the epidemiology and vaccine production to prevent livestock diseases. Using biotechnological approaches, the International Livestock Research Institute (ILRI) is trying to develop new and improved animal vaccines, and diagnostic tools to combat livestock diseases especially the high-priority 'orphan' diseases of the Africa and South Asia. For instance, the major health and management problems of cattle and small ruminants in Africa, Australia, Asia and Latin America are tick borne protozoan diseases such as. Theileriosis, Babesiosis and rickettsial diseases e.g. Heartwater disease or Cowdriosis and Anaplasmosis (Shkap *et al.*, 2007). Presently live attenuated vaccines are being used for East Coast Fever (caused by *Theileria parva*) and Babesiosis (caused by *Babesia bovis*) in cattle. To ensure the production and proper implementation of these vaccines, several technical and cold chain requirements need to be fulfilled. Hence these vaccines are available only in some countries like in Israel, Iran, Morocco, Tunisia, India, China and Uzbekistan (Shkap *et al.*, 2007). The 'muguga cocktail', a live vaccine comprising three Theileria parva stocks has been widely used in central, east and south Africa using the infection and treatment protocol to protect cattle from East Coast Fever (Oura *et al.,* 2004). Preliminary trials with five candidate vaccines are currently underway for a vaccine against East Coast Fever in cattle (Serageldin *et al.,* 2007). A major challenge for vaccine delivery in remote areas in developing countries is maintenance of the cold chain. Research has been focused on production of heat stable vaccines suitable for the use by small scale backyard livestock operations which make up the greater percentage of livestock in these areas. A heat stable poultry vaccine against New Castle disease, which is administered to chickens in feed, was developed in Malaysia by cloning the V4-UPM virus strain and it has been tried in many parts of Asia and Africa (Nwanta *et al.,* 2006). In the future, understanding the biology of parasites by using genome-sequencing data may lead to novel generation of pen side vaccines.

In evolutionary terms, the existence of the domestic livestock is a recent event in comparison to the wild life species. Therefore, domestic livestock have had less time to adapt to certain variable pathogens. The best suited example is the evolution of viruses in different magnitude than that of their hosts. This leads to the development of quasi-species with variant genomes. In order to eradicate such viruses, the viral eradication programs such as the global Rinderpest eradication programs were implemented. Effective strategies of FAO and the OIE (The World Organization for Animal Health) eradicated Rinderpest in many developing countries such as India, China, Sudan, Ghana (http://www.oie.int/eng/Status/Rinderpest/en_RP_free.htm). However, a close form of Rinderpest, Peste Des Petits Ruminants (*PPR*) is causing a lot of economic loss in small ruminants. Using local strains, attenuated vaccines were developed to combat this disease in several developing countries. An effective live attenuated freeze dried vaccine using the sungri strain of PPR virus and an oral vaccine (subunit vaccine) by expressing the Rinderpest hemagglutinin antigen in transgenic peanut plants have been developed (Khandelwal *et al.*, 2004). In addition to attenuated and subunit recombinant vaccines, combined vaccines and naked DNA vaccines have been used to control diseases in the developing world. Recently, a combined DNA vaccine with six immunodominant genes was found to be protective against *Mycobacterium bovis* and *Brucella abortus* in cattle in China (Hu *et al.*, 2009). Under the present globalization of animal vaccine production, the market size of Chinese animal vaccines was about 0.43 billion US Dollars in 2008. This increased Chinese market for animal vaccines is because of the peoples' attention to safety of food and the new law on animal epidemic prevention. Similarly, Taiwan aims to be an animal vaccine hub among developing countries. Recently, it launched the Pingtung Agricultural Biotechnology Park (PABP) with the approval from 59 firms including a German veterinary vaccine company to work in the operating zone (http://www.chinapost.com.tw/taiwan/national/national-news/2009/03/14/200031/Taiwan-aims.html).

Early diagnosis of diseases facilitates early treatment, prevents the loss of animal productivity and trans-boundary transmission of diseases among countries. Traditionally, the diagnosis of diseases involves specific antibody detection. Presently, a specific epitope of an antibody can be detected by the extensive use of ELISA (Enzyme Linked Immuno Sorbent Assay) in which recombinant antigens developed by gene cloning and sequencing are being used (Van Vliet *et al.*, 1995). Advances in biotechnology have revolutionized disease diagnosis through the detection of the nucleic acids, proteins and specific antibodies related to pathogens (http://www.oie.int/eng/Normes/mmanual/2008/pdf/1.1.07_BIOTECHNOLOGY.pdf). However, simpler innovative technologies than PCR were developed for effective diagnosis of diseases in the developing world. One such technique is LAMP (Loop-Mediated Isothermal Amplification), which has been used for the diagnosis of African Trypanosomosis (Kuboki *et al.*, 2003; http://www.finddiagnostics.org/).

Biotechnology in animal breeding

Livestock genetic improvement for production traits has been practiced by herdsmen since the time of animal domestication. Using standard genetic tools such as selection and crossbreeding, many livestock breeds were developed by master breeders in both the developed and developing countries. For example, the growth rate and litter size have been increased considerably in modern breeds with less fat deposition and feed intake than 20 years earlier (Rothschild and Plastow, 2008). This genetic gain was obtained by better genetic evaluations from the development of statistical methods supplementary to the classical genetic tools and increased use of superior sires through artificial insemination (AI). Artificial insemination using superior sires is one of the most effective tools for implementing crossbreeding programs to meet the requirements of milk consumption to the present growing population (9 billion by the year 2040). However, crossbreeds for improved milk yield developed between high yielding (*Bos taurus*) and low yielding (*Bos indicus*) animals are more susceptible to the diseases and environmental conditions in developing countries. Many local breeds have special adaptive traits for disease resistance and the ability to survive in adverse conditions such as high temperature, poor quality feed and water. Therefore, sustainable livestock development likely depends on the preservation of the genetic variation specific to tropical environment (Sugiyama *et al.,* 2003). To preserve local genetic breeds, several policies were developed. Among developing countries, India, the world top milk producer, systemically addressed the pros and cons of the crossbreeding and made a change in a national policy to produce 20 million crossbreeds between *Bos tarus* and *Bos indicus* (http://www.fao.org/docrep/009/t0095e/T0095E01.htm). Simultaneously, a few research institutes were established in India to preserve and characterize the local breeds.

The advancements in molecular genetics and the sequencing technology facilitate opportunities to use the marker assisted selection methods and transgenics over traditional phenotype selection methods. In order to use genetic markers, several genetic maps were developed in 1990s for many species. The sustained efforts on genome projects by developed countries has provided for the release of genetic maps and more recently genome sequences for chicken, cattle, horse and pig among livestock species. To promote genomics research in developing countries, recently the Bovine HapMap analysis included two African and some *Bos indicus* breeds (Boettcher and Hoffman, 2009). The initial maps led to the identification of the role of the individual genes and genome locations controlling some production traits (Anderson and Georges, 2004). Several genetic markers such as causative mutations (e.g., *MC4R* for feed intake and growth rate) (Kim *et al.,* 2000), linkage disequilibrium (LD) markers (e.g., *ESR* for litter size) (Rothschild *et al.,* 1996) and linkage equilibrium (LE) markers (e.g., polled) are in commercial use in developed countries. The commercial use of these markers for Marker Assisted Selection (MAS) in the developed world was extensively reviewed by Dekker's (2004). However, the reports on MAS programs for livestock development in developing countries are very scarce. Many conferences and meetings have been conducted to analyze the feasibility and benefits of MAS implementation in developing countries. For example, an e-mail conference entitled "Molecular marker-assisted selection as a potential tool for genetic improvement of crops, forest trees, livestock and fish in developing countries" was hosted by FAO Biotechnology Forum in 2003 (http://www.fao.org/Biotech/C9doc.html). It was stated that there were no MAS programs delivered to farmers in developing countries; they are still in the evaluation stage at research institutes. The implementation of MAS is difficult in developing countries due to lack of proper infrastructure, recording and management of livestock production systems and equipped laboratories to carry out the genetic testing. Many participants expressed that conventional breeding programs and proper recording systems need to be executed properly prior to the implementation of MAS programs. For instance, MAS programs for Trypanosomosis were not implemented due to the lack of routine recording of production and health traits with limited national molecular research facilities (De Koning, 2003). Therefore, Notter (2003) mentioned that "MAS without recording is unlikely to be very beneficial for most traits".

With the development of genome sequences, further efforts to isolate and identify genes associated with "native" disease resistance would be most useful. Furthermore the right genetic combinations of local and improved breeds could be possibly determined by using large genotyping platforms such as the recently developed cattle, pig, sheep and horse SNP (single nucleotide polymorphism) chips (Illumina, San Diego, CA, USA) . Such evaluations have been suggested for the right crossbred combination to produce milk in Africa but recognizing the most productive gene combinations still will not improve production. Systems of mating to translate the biotechnological advanced genotypes into the needed cattle have not been designed.

A transgenic animal is an animal which carries foreign DNA in its genome. The important field applications of transgenic livestock may be to improve milk production and composition, to increase growth rate, to improve feed efficiency and carcass composition, to increase disease resistance, to enhance reproductive performance and to increase prolificacy. The high cost and limited success involved in the production of transgenic animals has limited the commercial scale of transgenic animal farms even in developed countries. Public perception is also important to implement transgenic animals in the livestock production systems. People favor the transgenic animals for the development of new vaccines and recombinant drugs rather for faster growing animals for food consumption (Wheeler *et al.*, 2007).

Biotechnology in animal nutrition

Animal nutrition is a primary ingredient for animal productivity. Poor nutrient availability and the inefficient utilization of feed by animals are some of the reasons for poor productivity of animals in developing countries. Therefore, biotechnology research in animal nutrition has focused mainly on the important areas of feed base, animal digestive system and animal metabolism.

The feeds available to ruminants in developing countries are high in fiber and lignocellulose content. The digestibility of certain fodder plants (e.g., *Cynodon dactylon*) have been improved by conventional selection methods. Several strategies have been developed to reduce the lignin content in feed (McSweeney, 1999). Some important methods include the use of antisense RNA and ribozymes to develop fodder with low lignin content (e.g., maize, alpha alpha etc) (http://www.i-sis.org.uk/LLGMT.php) and the pretreatment of fodder with lignin digesting enzymes (e.g., lignase) produced by different modern fermentation biotechnologies (e.g., solid-state fermentation) (Graminha *et al.,* 2008). Though research on lignin modification has been done at different research institutes (e.g., Durban university of technology, Durban, South Africa; University of Ibadan, Nigeria etc.,), the application and adaptation of this technology to the field conditions is very scanty in the developing world. Also, grain based feeds for monogastric animals are deficient in several amino acids including lysine, methionine and tryptophan. The deficiency of nutrients in the feed and growing intensification of the pig and poultry production systems has required an extensive use of several feed additives in the developing world. For instance, the amino acid market present in India alone is worth a total of 5 million US dollars. Many of these essential amino acids are being produced from *Escherichia coli* through recombinant DNA technology (http://www.fao.org/biotech/c16doc.html).

Many developed and developing countries adopted the cultivation of the transgenic crops for human and animal nutrition. In 2007, a total of 23 countries cultivated transgenic plants on around 114 million hectares which represented about 5% of arable land worldwide. These areas were concentrated very strongly on five countries in the North and South America in where 88% of the acreage was located (USA: 57.7 million hectares; Argentina: 19.1 million hectares; Brazil 15.0 million hectares; Canada: 7.0 million hectares; Paraguay: 2.6 million hectares), India (6.2 million hectares), China (3.8 million hectares) and South Africa (1.5 million hectares). Even after 12 years of cultivation, only two genetic traits, i.e. herbicide tolerance (HR) and insect resistance to *Bacillus thuringiensis* (Bt), either alone or in combination, account for 99.9% of cultivated genetically modified plants, in only four crop varieties (51.3% soybean, 30.8% maize, 13.1% cotton, and 4.8% rapeseed/canola) (http://www.tab.fzk.de/en/projekt/zusammenfassung/ab128.htm). Many animal studies were conducted to assess the nutritive values of the first generation (feed plants with changed tolerance or resistance and with minor changes in the content of valuable and desirable ingredients) genetically modified plants (GMP) such as Bt-maize, Bt-potatoes, Pat-maize, Pat-sugar beet and roundup ready Soybeans, and the second generation plants. From the results it was concluded that there was no significant difference in the nutritive value between isogenic and first generation transgenic feeds (Flachowsky et al., 2007). Very few companies in developing countries are supplying transgenic soybean protein as an animal feed (e.g., Shandong Rongfeng Biotechnology Development Co., Ltd. China). Further research and extension is required to reduce the cost and to adapt the transgenic crops technology in the field conditions.

Efficient understanding of rumen biology is an imperative task to improve the feed conversion efficiency and to reduce the methane gas production, which is a global concern to prevent the global warming due to ruminants in developing countries. The accumulated knowledge about rumen microorganisms has been based on classical culture based techniques (isolation, enumeration and nutritional characterization). This information can explain the biology of only 10-20% of rumen microorganisms (McCrabb *et al.*, 2005). The use of molecular tools (e.g., rDNA analysis) to understand the biochemistry of rumen microorganisms is scanty in developing countries. Studies on metagenomics of rumen microorganisms have been initiated recently in the developed world (http://www.reeis.usda.gov/web/crisprojectpages/187814.html). The rumen microorganisms vary according to the feed. A good opportunity is for scientists in the developing world to focus on the genomics of rumen microorganisms as sequence information of certain rumen microorganisms can be accessed from developed countries.

Biotechnology in livestock reproduction

Economic stability and sustainable animal productivity of the livestock farmers are also influenced by the efficient reproduction and breeding of farm animals. One of the major obstacles to increased proficiency of livestock production systems is low fertility with less genetic potential of animals in developing countries (Madan, 2005). Artificial insemination is the most commonly adopted technique to improve animal production with animals of high genetic potential in both the developing and developed world. It is the best technique for an effective use of superior male germplasm. In developing countries particularly in Africa, AI has been practiced for over 60 years to improve the production of local hardy breeds (e.g., the east African Zebu cattle) (http://www.agfax.net/radio/detail.php?i=268). In the East African highlands, about 1.8 million small scale farmers have benefited from higher milk yields obtained from the higher genetic potential of cattle resulting from artificial insemination. Around 100 million cattle and pigs have been bred annually by using AI in the developing world (http://siteresources.worldbank.org/INTWDR2008/Resources/WDR_00_book.pdf). Overall, upgrading the genetic potential and the productivity of animals have been improved with the use of AI. However, complete and successful AI programs have not fully been implemented in many regions of developing countries due to the constraints in semen evaluation and technology transfer and problems with electricity and storage issues. For instance, semen evaluation in many semen banks is still based on sperm motility rather with advanced technology and the handling procedures are being inadequately followed in spite of the availability of proper handling guidelines (Madan, 2005). Capable, well trained and dedicated personnel are required to analyze the actual constraints for implementing the AI, to educate farmers about estrus detection and the importance of AI and to implement the technology efficiently in villages. Improvement of such organizational and extension activities according to the regional social status may improve the implementation of the successful AI programs in developing countries.

Multiple ovulation and embryo transfer (MOET) is a biotechnology technique for an efficient use of female germplasm. Many successful embryo transfer events in livestock are confined to research institutes in many developing countries. For example, Srilanka produced two goat kids by embryo transfer recently (http://www-naweb.iaea.org/nafa/aph/stories/2008-calf-embryo2.html). Very few developing countries like Brazil and China adopted embryo transfer through the importation of embryos from the USA and Canada, respectively, on commercial basis for milk and beef purposes. The importance of embryo transfer was recognized in Africa to combat the protozoan diseases such as Trypanosomosis and Theileriosis (http://www.accessmylibrary.com/article-1G1-13065289/embryo-transfer-african-cattle.html).

Another important biotechnology development is cloning of animals. Farm animal cloning might be useful for preserving the elite male and female animals for future breeding schemes, to modify the genetic makeup of domestic animals for the production of human pharmaceutical products ("Pharming") as well as human organs (regenerative medicine). Since Dolly, the legendary sheep was produced by cloning in Rosylin Institute, farm animals including cattle, goats and pigs have been cloned. Among developing countries, China is the big leader in farm animal cloning in the developing world. Following the success in cloning of 45 goats, China established an animal cloning company, Yangling Keyun Cloning Ltd. Recently Indian scientists at the National Dairy Research Institute produced a buffalo calf by using handmade cloning technology.

Conclusions and suggestions

In conclusion, the biotechnological developments in some parts of the developing world are promising (Table 2). However, to improve further, developing countries need to make collaborative ventures, and form multidisciplinary teams within national and regional frameworks in order to implement biotechnology programs successfully. Collaboration would extend resources and lessen the cost. Before implementing any advanced tools for livestock improvement, a clear distinction in the requirements between developed and developing countries needs to be recognized by scientists, policy makers and the public. In developed countries, the requirements have shifted from traditional high quantity of production to high quality and efficiency of production as well as health traits of livestock. In developing countries, the priority requirements are to establish livestock production systems against adverse environmental conditions,

Table 2. Summary of biotechnology developments in developing countries.

Biotechnology area	Implementation in developing countries
Livestock health	Is a very important area in developing countries because of high disease challenge. • Focus is mainly on production of vaccines against 'orphan' diseases of the developing world such as Theileriosis and Babesiosis, and also on early diagnosis of disease. • Live attenuated vaccines, subunit recombinant vaccines, combined vaccines and naked DNA vaccines have been used to control livestock diseases. • Development of heat stable vaccines not requiring refrigeration.
Animal Breeding	One issue which is of importance in developing countries is improvement of animal productivity through crossbreeding while at the same time preserving genotypes of native animals that are adapted to the conditions in developing countries. • There are efforts to characterize local breeds. • Problems associated with inadequate infrastructure hinder the utilization of Marker Assisted Selection and Single Nucleotide Polymorphism technologies in developing countries. • Use of genomics to predict best crossbred performance.
Animal Nutrition	High fiber and lignocelluloses in ruminant feeds lead to poor nutrient availability and inefficient utilization of feed. Grain based feeds for monogastric animals are also deficient in some key nutrients. • Digestibility of certain fodder plants has been improved by conventional selection methods • Research on lignin modification is on-going at several research institutes but the technology has largely not been adopted in the field. • There is extensive use of feed additives to correct nutrient deficiencies in animal feeds. • Research on manipulation of rumen micro-organisms to increase feed efficiency and reduce methane gas production is still limited to developed countries.
Livestock reproduction	A major limitation of livestock production systems in developing countries is low fertility and less genetic potential of animals. • Artificial insemination is the most widely adopted reproduction technology. It has been used to upgrade the genetic potential and increase the productivity of animals. Technical as well as extension constraints limit the complete success of the technology in some areas. • In many developing countries multiple ovulation and embryo transfer is still confined to research centers. There is potential to use the technology to combat disease. A few countries are importing embryos on commercial basis • Some developing countries have successfully cloned animals.

to improve fodder resources and effective storage of fodder in drought situations, to face disease challenge and to analyze the quality of livestock products for effective marketing. To achieve such tasks, scientists need to do research with an objective to increase the income of farmers. They need to collaborate with extension workers to analyze the prior needs in an area and find out the systematic scientific solutions with the consideration of sociological perspectives. In addition, effective methods need to be developed to educate the farmers, policy makers and politicians to understand the international collaborations and the benefit of the scientific methods in field conditions.

References

Andersson, L. and Georges, M. (2004) Domestic animal genomics: Deciphering the genetics of complex traits. *Nature Reviews Genetics* **5**, 202-212.

Boettcher, P.J. and Hoffman, I. (2009) Livestock genomics in developing countries. *Science* **324,** 1515.

De Koning, D.J. (2003) MAS for livestock in developing countries In: *Proceedings of FAO conference, Marker-assisted selection – Current status and future perspectives in crops, livestock, forestry and fish.* Chap. 21, Marker-assisted selection as a potential tool for genetic improvement in developing countries: debating the issues. Ed by Jonathan Robinson and John Ruane, pp. 433.

Dekkers, J.C.M. (2004) Commercial application of marker and gene-assisted selection in livestock: strategies and lessons. *Journal of Animal Science* **82**, E313-328.

Delgado, C., Rosegrant, M., Steinfeld, H., Ehui, S. and Courbois, C. (1999) "Livestock to 2020 – The Next Food Revolution". Food, Agriculture and the Environment Discussion Paper 28. IFPRI/FAO/ILRI.

Flachowsky, G., Aulrich, K., Bohmea, H. and Halle, I. (2007) Studies on feeds from genetically modified plants (GMP) – Contributions to nutritional and safety assessment. *Animal Feed Science and Technology* **133,** 2-30.

Graminha, E.B.N., Goncalves, A.Z.L., Pirota, R.D.P.B., Balsalobre, M.A.A., Da Silvia, R. and Gomes, E. (2008) Enzyme production by solid-state fermentation: application to animal nutrition. *Animal Feed Science Technology* **144**, 1-22.

Hu, X., Yu, D., Chen, S., Li, S. and Cai, H. (2009) A combined DNA vaccine provides protective immunity against *Mycobacterium bovis* and *Brucella abortus* in cattle. *DNA Cell Biology* **28,** 191-199.

Khandelwal, A., Renukaradhya, G.J., Rajasekhar, M., Sita, G. L. and Shaila, M.S. (2004) Systemic and oral immunogenicity of hemagglutinin protein of rinderpest virus expressed by transgenic peanut plants in a mouse model. *Virology* **323,** 284-291.

Kim, K.S., Larsen, N., Short, T., Plastow, G. and Rothschild, M.F. (2000) A missense variant of the porcine melanocortin 4 receptor (MC4R) gene is associated with fatness, growth, and feed intake traits. *Mammalian Genome* **11**, 131–135.

Kuboki, N., Inoue, N., Sakurai, T., Di Cello, F., Grab, D. J., Suzuki, H., Sugimoto, C. and Igarashi, I. (2003) Loop-mediated isothermal amplification for detection of african trypanosomes. *Journal of Clinical Microbiology* **41**, 5517-5524.

Madan, M.L. (2005) Animal biotechnology: applications and economic implications in developing countries. *Revue scientificet technique. Office of International Epizootics.* **24,** 127-128.

McCrabb, G.J., McSweeney, C.S., Denman, S., Mitsumori, M., Fernandez-Rivera, S. and Makkar, H.P.S. (2005) Application of molecular microbial ecology tools to facilitate the development of feeding systems for ruminant livestock that reduce greenhouse gas emissions. In: *Applications of gene based technologies for improving animal production and health in developing countries* (Ed. H.P.S. Makkar and G.J. Viljoen), IAEA, FAO, Springer, The Netherlands, pp. 387-395.

McSweeney, C.S., Dalrymple, B.P., Gobius, K.S., Kennedy, P.M., Krause, D.O., Mackie, R.I. And Xue, G.P. (1999) The application of rumen biotechnology to improve the nutritive value of fibrous feedstuffs: pre- and post-ingestion. *Livestock Production Science* **59**, 265–283.

Notter, D. (2003) Use of MAS and related technologies in livestock improvement In: *Proceedings of FAO conference, Marker-assisted selection – Current status and future perspectives in crops, livestock, forestry and fish.* Chap. 21, Marker-assisted selection as a potential tool for genetic improvement in developing countries: debating the issues. Ed by Jonathan Robinson and John Ruane, pp. 433.

Nwanta, J.A., Umoh, J.U., Adbu, P.A., Ajogi, I. And Adeiza, A.A. (2006) Field trial of Malaysian thermostable Newcastle disease vaccine in village chickens in Kaduma State, Nigeria. *Livestock Research forRural Development* 18, article 60.

Oura, C.A.L., Bishop, R., Wampande, E.M., Lubega, G.W. And Taita, A. (2004) The persistence of component Theileria Parva stocks in cattle immunized with the 'Muguga cocktail' live vaccine against East Coast Fever in Uganda. *International Journal for Parasitology* **33**, 1641-1653.

Rothschild, M. F., Jacobson, C., Vaske, D. A., Tuggle, C., Wang, L., Short, T., Eckardt, G., Sasaki, S., Vincent, A., Mclaren, D. G., Southwood, O., Van Der Steen, H., Mileham, A. And Plastow. G. (1996) The estrogen receptor locus is associated with a major gene influencing litter size in pigs. *Proceedings of the National Academy of Sciences. USA* **93**,201-205.

Rothschild, M.F. And Plastow, G. S. (2008) Impact of genomics on animal agriculture and opportunities for animal health. *Trends in Biotechnology* **26**, 21-25.

Serageldin., Ismail. And Juma C. (2007) Africa Warms Up to Biotechnology. Business Daily (Nairobi), November 15.

Shkap, V., De Vos, A.J., Zweygarth, E. And Jongejan, F. (2007) Attenuated vaccines for tropical theileriosis, babesiosis and heartwater: the continuing necessity. *Trends in Parasitology* **23**, 420-426.

Sugiyama, M., Iddamalgoda, A., Oguri, K. And Kamiya, N. (2003) Development of livestock sector in Asia: an analysis of present situation of livestock sector and its importance for future development. (http://ci.nii.ac.jp/naid/110004436581/en).

Van Vliet, A.H.M., Van Der Zeijst, B.A.M., Camus, E., Mahan, S.M., Martinez, D. And Jongejan, F. (1995) Use of a specific immunogenic region on the *Cowdria ruminantium* MAP1 protein in a serological assay. *Journal of Clinical Microbiology* **33**, 2405–2410.

Wheeler M. B. (2007) Agricultural applications for transgenic livestock. *Trends in Biotechnology* **25**, 204-210.

Websites

http://siteresources.worldbank.org/INTWDR2008/Resources/WDR_00_book.pdf

http://www.accessmylibrary.com/article-1G1-13065289/embryo-transfer-african-cattle.html

http://www.agfax.net/radio/detail.php?i=268

http://www.chinapost.com.tw/taiwan/national/national-news/2009/03/14/200031/Taiwan-aims.html

http://www.fao.org/biotech/c16doc.html

http://www.fao.org/Biotech/C9doc.html

http://www.fao.org/docrep/009/t0095e/T0095E01.html

http://www.finddiagnostics.org

http://www.id21.org/insights/insights72/art05.html

http://www.i-sis.org.uk/LLGMT.php

http://www.oie.int/eng/Normes/mmanual/2008/pdf/1.1.07_BIOTECHNOLOGY.pdf

http://www.oie.int/eng/Status/Rinderpest/en_RP_free.html

http://www.reeis.usda.gov/web/crisprojectpages/187814.html

http://www.tab.fzk.de/en/projekt/zusammenfassung/ab128.html

http://www.naweb.iaea.org/nafa/aph/stories/2008-calf-embryo2.html

Biotechnology and Genetic Engineering Reviews - Vol. 27, 229-256 (2010)

16 years research on lactic acid production with yeast – ready for the market?

MICHAEL SAUER[1,2,*], DANILO PORRO[3], DIETHARD MATTANOVICH[1,2], PAOLA BRANDUARDI[3]

[1] *School of Bioengineering, University of Applied Sciences FH Campus Wien, Muthgasse 18, 1190 Vienna, Austria;* [2] *Department of Biotechnology, BOKU University of Natural Resources and Applied Life Sciences, Muthgasse 18, 1190 Vienna, Austria;* [3] *Dipartimento di Biotecnologie e Bioscienze, Università degli Studi di Milano-Bicocca, P.zza della Scienza 2, 20126 Milano, Italy*

Abstract

The use of plastic produced from non-renewable resources constitutes a major environmental problem of the modern society. Polylactide polymers (PLA) have recently gained enormous attention as one possible substitution of petroleum derived polymers. A prerequisite for high quality PLA production is the provision of optically pure lactic acid, which cannot be obtained by chemical synthesis in an economical way. Microbial fermentation is therefore the commercial option to obtain lactic acid as monomer for PLA production. However, one major economic hurdle for commercial lactic acid production as basis for PLA is the costly separation procedure, which is needed to recover and purify the product from the fermentation broth. Yeasts, such as *Saccharomyces cerevisiae* (bakers yeast) offer themselves as production organisms because they can tolerate low pH and grow on mineral media what eases the purification of the acid. However, naturally yeasts do not produce lactic acid. By metabolic engineering, ethanol was exchanged with lactic acid as end product of fermentation. A vast amount of effort has been invested into the development of yeasts for lactic acid production since the first paper on this topic by Dequin and

*To whom correspondence may be addressed (michael.sauer@fh-campuswien.ac.at)

Abbreviations: ADH, alcohol dehydrogenase; CYB2, cytochrome b2, L-lactate-cytochrome c oxidoreductase; DSP, down stream processing; LAB, lactic acid bacteria; LDH, lactate dehydrogenase; MSW, municipal solid waste, PDC, pyruvate decarboxylase; PLA, poly lactic acid, ROS, reactive oxygen species; SSF, simultaneous saccharification and fermentation.

Barre appeared 1994. Now yeasts are very close to industrial exploitation – here we summarize the developments in this field.

Introduction

Plastic production and the persistence of plastic products after their use in the environment constitute one major environmental problem of the modern society. In 2005 only packaging waste accounted for 56.3 million tons or 25% of the total municipal solid waste (MSW) in Europe (78.81 million tons or 31.6% of the MSW in 2003 in the USA) (Kale et al. 2007). Most of the current packaging polymers are derived from petroleum, thereby relying on a world-wide decreasing resource. Furthermore, production and degradation of plastics contribute to green house gas emissions, which have been identified as one major cause for global warming (http://www.ipcc.ch/pdf/assessment-report/ar4/syr/ar4_syr.pdf). Substitution of non-biodegradable petroleum-derived polymers is consequently one important milestone on the road to a sustainable society.

Polylactide polymers (poly lactic acid, PLA) have recently gained enormous attention for being truly biodegradable and produced from renewable resources.

Initially, PLA has been proposed for use in medical applications because it is biocompatible and bioresorbable (Auras et al. 2004). Nowadays, PLA polymers are becoming a cost-effective alternative to commodity petrochemical-based materials. PLA is particularly suitable as an economically feasible packaging polymer. PLA has been introduced for use all across Europe, the USA and Japan, mainly in the area of fresh products as a food packaging material for short shelf life products such as fruit and vegetables. Package applications include containers, cups, cutlery, plates and saucers, as well as overwrap and lamination films and blister packages (Auras et al. 2004; Datta and Henry 2006).

Traditional processing such as injection molding, sheet extrusion or thermoforming can be applied to PLA. In fact PLA can be processed with conventional injection molding equipment, allowing the substitution of current polymers by PLA without major changes in the manufacturing industry.

Three different routes to obtain high molecular mass PLA have been described: direct condensation, azeotropic dehydrative condensation and polymerization through lactide formation. The latter one is currently used for large-scale production of PLA (patented by Cargill (US 514,20,23). Essentially, D-lactic acid, L-lactic acid or a mixture of the two is converted into lactides. A lactide is the cyclic dimer of lactic acid. High molecular mass PLA is subsequently formed by ring-opening polymerization of the lactides. A prerequisite for pure lactides and therefore high quality PLA is the provision of optically pure lactic acid (D or L), which cannot be obtained by chemical synthesis in an economical way.

Microbial fermentation is therefore the commercial way to obtain lactic acid as monomer for PLA production. Existing industrial production processes rely on homofermentative lactic acid bacteria (Datta and Henry 2006; Wee et al. 2006), which are capable to produce optically pure lactic acid from a variety of carbohydrate sources, such as corn syrup, dextrose, cane or beet sugar.

The major economic hurdle for commercial lactic acid production as basis for PLA is the complex separation procedure, which is needed to recover and purify the product from the fermentation broth.

Improvements of the downstream technologies are one part to improve the economical outcome. Another possible way is to improve the production organism. Lactic acid bacteria have complex nutrient requirements, which have to be addressed by the addition of (costly) supplements such as yeast extract and corn steep liquor, which later on cause problems for the purification of lactic acid. Even if only as little supplement is added for still achieving high lactic acid yield in a reasonable fermentation time, the majority of the supplement is actually not utilised. It just contributes to a major increase in the concentration of impurities at the end of the fermentation with a corresponding increase in separation costs, in addition to raw material costs (Fitzpatrick et al. 2003). Furthermore, the limited pH tolerance of lactic acid bacteria leads to high amounts of gypsum (or other salts) produced when the lactic acid is recovered. Yeasts, such as *Saccharomyces cerevisiae* (bakers yeast) offer themselves as production organisms because they can tolerate low pH and grow on mineral media. However, yeasts do not produce lactic acid naturally. Metabolic engineering allows the exchange of ethanol with lactic acid as end product of fermentation (Fig. 1). A vast amount of effort has been invested into the development of yeasts for lactic acid production since the first paper on this topic by (Dequin and Barre 1994) appeared 16 years ago (Tab 1). Now yeasts are very close to substitute lactic acid bacteria for large-scale industrial exploitation - we feel therefore that this is the right moment for a summary of the ongoings in this field.

Enabling of *S. cerevisiae* to ferment pyruvate to lactate

It is quite evident that lactic acid, like other bulk fermentation products, needs to be produced in large quantities and at very low price to be competitive on the market (Sauer et al. 2008). This is particularly important for the substitution of plastics derived from petroleum.

One of the key points to keep in mind is that the desired end product of the fermentation is free lactic acid. For isolation of free lactic acid the pH of the broth has to be low, significantly lower than the pKa of lactic acid (3,78). However, lactic acid bacteria (LAB), the traditional production organisms, are inhibited by lactic acid and especially by low pH, in addition of displaying complex nutritional requirements. Nevertheless, the fermentative capacity of LAB (for example *Lactobacilli*) is very high, with final product concentrations of over 100 g/L (Benninga 1990) and a prolonged productivity of 23 g/(L·h) (Hofvendahl and Hahn-Hägerdal 2000), that can reach over 100 g/(L·h) in continuous fermentation with cell recycling (González-Vara et al. 2000). The search for an alternative microbial host system had and still has to take these parameters into consideration.

Yeasts were among the first microbes to be studied. They were considered because of their peculiar ability to ferment. In particular their capacity for alcoholic fermentation of sugars led to a long and still lasting story of exploitation of yeasts for biotechnological developments (Barnett 2003): starting from traditional processes such as baking, brewing and wine making to the more recently developed highly efficient bioethanol production (Matsushika et al. 2009). All of them are characterised by the demand for high fermentative capacity and strain robustness. These features, together with a profound knowledge concerning genetics, physiology and biochemistry made *S. cerevisiae* an ideal candidate for different metabolic engineering approaches, resulting

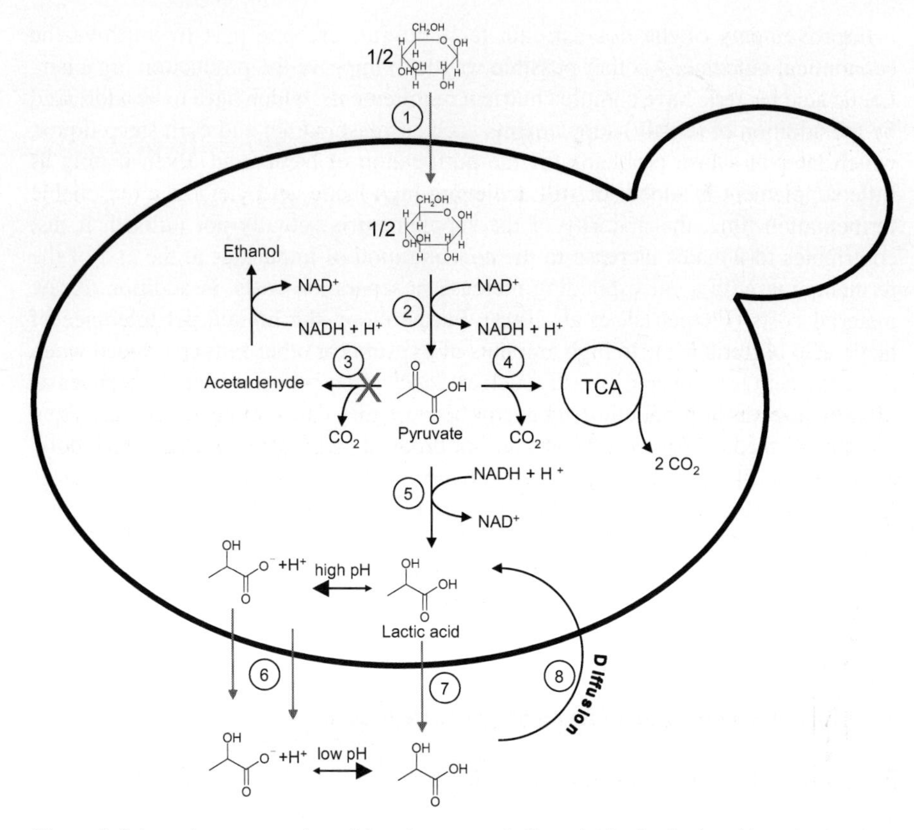

Figure 1. Schematic representation of the relevant metabolic activities for lactic acid production from glucose by yeasts.

1. Glucose uptake by hexose transporters; a rate determining step.
2. Glycolysis, yielding one molecule of ATP per molecule of pyruvate formed (not shown) and one molecule of NADH, which has to be recycled to NAD^+.
3. Pyruvate decarboxylase (PDC), channelling pyruvate to ethanol production, this activity has to be deleted or decreased by genetic engineering for efficient lactic acid production (indicated by **X**).
4. Pyruvate dehydrogenase, feeding pyruvate into the oxidative decarboxylation via the tri-carboxylic-acid cycle (TCA), the relevance of this activity depends on the degree of fermentation vs. respiration.
5. Lactate dehydrogenase (LDH), heterologous activity introduced by genetic engineering, channelling pyruvate to lactic acid.
6. Lactate/H^+ symport. Most probable way of lactic acid export, requires most probably energy; the respective transporters and their properties are currently unknown.
7. Lactic acid export. It is not clear at the time being how exactly lactic acid is exported from the cell.
8. If the extracellular pH is low, lactic acid is present in its protonated form and can therefore enter the cell by diffusion. If this happens, the acid will dissociate in the cytoplasm, and has to be exported actively – this creates an energy requiring circle with reaction 6 (for which the details are still unknown).

in the production of non-natural products (Nevoigt 2008), among them lactic acid. In respect to lactic acid, probably the most attractive physiological trait is the high natural tolerance of *S. cerevisiae* to low pH. This allows the production of organic acids in the desired protonated form, what results in a considerable reduction of down stream processing (DSP) costs.

Table 1. Comparison of final lactic acid concentrations and yields of different yeast strains at high and low pH.

Yeast	L-LDH	pH < pKa			pH > pKa			Culture conditions*	Reference
		Conc. LA	Conc. EtOH	Time	Conc. LA	Conc. EtOH	Time		
S. cerevisiae	*L. casei* (multicopy vector)				12 g/L	17 g/L	66 h	Shake-flask	(Dequin and Barre 1994)
S. cerevisiae	*B. taurus* (integrated, 1 copy)	20 g/L	20 g/L	4 h				Fed-batch, no areation	(Porro et al. 1995)
S. cerevisiae	*B. taurus* (multicopy plasmid)				11.4 g/L	44.6 g/L	38 h	Anaerobic fed-batch	(Adachi et al. 1998)
S. cerevisiae	*B. taurus* (multicopy plasmid)	6.1 g/L	19.0 g/L	26 h				Anaerobic batch	(Adachi et al. 1998)
S. cerevisiae	*L. plantarum* (integrated, 1 copy)	58 g/L (pH3.6)	nd					Batch, production phase with resting cells	(Colombié et al. 2003)
S. cerevisiae	*B. taurus* (integrated, 2 copies)	50.6 g/L	13.8 g/L	72 h	55.6 g/L	16.9 g/L	72 h	Batch growth	(Ishida et al. 2005)
S. cerevisiae	*B. taurus* (integrated, 6 copies)				120 g/L	40 g/L	48 h	Batch fermentation, microaerobic (on juice-based medium)	(Ishida et al. 2006a; Saitoh et al. 2005)
S. cerevisiae	*L. mesenterioides* (D-LDH, integrated, 2 copies)	53.2 g/L	20.7 g/L	72 h	61.5 g/L	17.3 g/L	72 h	Anaerobic batch	(Ishida et al. 2006c)
S. cerevisiae	*B. taurus* (integrated, 2 copies)				82.3 g/L	2.8 g/L	192 h	Microaerobic shake-flask	(Ishida et al. 2006b)
S. cerevisiae	*R. oryzae* (multicopy plasmid)				38 g/L	20 g/L	54 h	Batch aerobic fermentation	(Skory 2003)
S. cerevisiae	*L. plantarum* (multicopy plasmid)	62.4 g/L	-	76 h				Shake-flask	(Liu and Lievense 2005)

Yeast	L-LDH	pH < pKa			pH > pKa			Culture conditions*	Reference
		Conc. LA	Conc. EtOH	Time	Conc. LA	Conc. EtOH	Time		
S. cerevisiae	*L. plantarum* (multicopy plasmid)	70 g/L	-	72 h				Shake-flask	(Valli et al. 2006)
S. cerevisiae	*B. taurus* (integrated, 8 copies)				80 g/L	-	120 h	Buffered microaerobic spherical flat-bottom flask (on cellobiose)	(Tokuhiro et al. 2008)
S. cerevisiae	*B. taurus* (integrated, 2 copies)				74.1 g/L	> 5 g/L	48 h	Shake-flask, microaerobic	(Tokuhiro et al. 2009)
S. cerevisiae	*B. taurus* (integrated, 2 copies)				71.8 g/L	> 5 g/L	63 h	Batch fermentation, microaerobic	(Tokuhiro et al. 2009)
K. lactis	*B. taurus* (low copy number plasmid, 5 copies)	11.3 g/L	-	86 h				Batch, under areation	(Porro et al. 1999)
K. lactis	*B. taurus* (multicopy plasmid)				60 g/L	-	500 h	Fed-batch, under areation	(Bianchi et al. 2001)
K. marxianus	*L. helveticus* (integrated into PDC1 locus)				99 g/L	-	48 h	Buffered shake-flask	(Hause et al. 2009)
K. marxianus	*L. helveticus* (integrated into PDC1 locus)	9.1 g/L	-	72 h		-		Unbuffered shake-flask	(Hause et al. 2009)
I. orientalis	*L. helveticus* (integrated, 1 copy)	66 g/L	-	160 h				Fed-batch	(Miller et al. 2007)
I. orientalis	*L. helveticus* (integrated, 1 copy)	70 g/L	-	77 h				Batch fermentation, microaerobic	(Dundon et al. 2009)

Yeast	L-LDH	pH < pKa			pH > pKa			Culture conditions*	Reference
		Conc. LA	Conc. EtOH	Time	Conc. LA	Conc. EtOH	Time		
S. pombe	*R. oryzae*				80-100 g/L				(Pavlovich et al. 2006)
C. utilis	*B. taurus* - optimised (integrated, 2 copies)				103.3 g/L	< 10 mg/L	33 h	Buffered spherical flat-bottom flask	(Ikushima et al. 2009)
C. boidinii	*B. taurus* - optimised (integrated, 1 copy)				85.9 g/L	1.5 g/L	48 h	Batch, under areation	(Osawa et al. 2009)
P. stipitis	*L. helveticus* (integrated, 1 copy)	15 g/L	3-7 g/L	88 h	58 g/L	4.5 g/L	147 h	Buffered and unbuffered shake-flask (on xylose)	(Ilmén et al. 2007)

* the carbon source is glucose, if not differently specified

As mentioned, lactic acid is not a natural fermentation product of *S. cerevisiae*, despite a cytoplasmic D-LDH was identified whose expression is stimulated by damage to mitochondria (Chelstowska et al. 1999). However, from a biochemical and metabolic point of view the fermentation from pyruvate to ethanol can be replaced by fermentation to lactate, since the driving force, that is the regeneration of NAD^+, is equally satisfied and the reactions are perfectly balanced. In addition, the theoretical yield of lactic acid (g/g) on glucose is 100%, as for lactic acid bacteria, while in the ethanol production it is only 50%, since part of the carbon is lost as CO_2.

The first example of a partial redirection of pyruvate to lactate dates 16 years ago, when the construction of a recombinant *S. cerevisiae* strain expressing the *Lactobacillus casei* L-LDH gene was described (Dequin and Barre 1994). The resulting mixed lactic acid-alcoholic fermentation can be desired in some processes, as for example wine yeast making.

The first paper where the production of lactic acid as bulk chemical from engineered yeasts is stated as the final goal was published one year later (Porro et al. 1995). By the expression of a mammalian L-LDH gene in *S. cerevisiae* mixed lactic acid-alcoholic fermentation was achieved, reaching a production of 20 g/L of lactic acid and, by optimising growth and production parameters, a maximal productivity of 11 g/(L·h), but only for a very short period of time. However, both product concentration and yield were far from being optimal, mainly because of the parallel ethanol fermentation, an undesirable reaction leading to the loss for pyruvate. To completely redirect pyruvate conversion to lactate, competing reactions have to be eliminated.

Towards a homolactic yeast

The very efficient pathway converting pyruvate into ethanol starts with pyruvate decarboxylase (PDC; EC 4.1.1.1). Three different genes (*PDC*1, *PDC*5 and *PDC*6) encode this activity in *S. cerevisiae*. In addition of being induced by glucose (Hohmann and Cederberg 1990; Schmitt et al. 1983), the expression of the *PDC* genes is under autoregulation at the transcriptional level (Hohmann 1991b; Hohmann and Cederberg 1990). Therefore, the deletion of *PDC*1, encoding for the enzyme mainly responsible for acetaldehyde production, does not lead to significant ethanol reduction since it indirectly causes *PCD*5 and to some extent *PDC*6 expression to be enhanced.

Other yeasts posses a different molecular organization at this level of the energy metabolism. The first yeast engineered in this respect for lactic acid production was the Crabtree negative yeast *Kluyveromyces lactis*. Deletion of its sole PDC gene, *KlPDC*1 (Bianchi et al. 1996) together with the expression of a mammalian (bovine) LDH allowed to obtain the first example of the complete redirection of the metabolic flux from pyruvate instead to ethanol to lactic acid. The obtained homolactic fermenting yeast (Porro et al. 1999) was efficient in every scale from shake flask to bioreactor achieving the remarkable yield of 0.58 g/g (grams of lactic acid per grams of glucose consumed).

Another very efficient activity utilizing pyruvate as substrate is the pyruvate dehydrogenase, funnelling the pyruvate into the respiratory metabolism. By additional deletion of this activity, the yield was increased to 0.85 g/g (Bianchi et al. 2001). Another possibility to decrease substrate catabolism by respiration would be the lowering of the aeration rate during the production process. However, this resulted in a sudden as well as almost total reduction of lactate production (while the cells were still able to grow, however). The precise physiological reason for this effect is not understood.

Moreover, the best results were obtained controlling the pH value at 4.5. At lower pH a significant decrease of product accumulation was observed (Porro et al. 1999). In conclusion, the product concentration reached and the productivity were very high, but still not competitive with LAB based productions. Two main properties hindered further research on *K. lactis*: the very low tolerance to the presence of the organic acid at low pH and the absolute requirement of oxygen for sustaining the production.

S. cerevisiae approaching industry

The *S. cerevisiae* triple PDC deleted strain (Δ*pdc1*, Δ*pdc5*, Δ*pdc6*) does not show any detectable PDC activity either in glucose or in ethanol medium but it can grow only very slowly on glucose (Hohmann 1991a). This feature did not seem very attractive for the development of an efficient cell factory. However, the expression of a heterologous LDH led to a significant increase of the growth rate of the mutant strain on glucose. This is very likely due to the complementation of the missing pathway for cytoplasmic NAD^+ regeneration (Porro et al. 1995). Unfortunately, at that time the constructed yeast strain showed only a very low LDH activity and, consequently, only a very low lactic acid production (Porro et al. 1995). Taken together with the impaired growth on glucose this pointed to a complicated process design using such strains.

Based very likely on such considerations research was focused on a sequential but not complete PDC deletion in the following years. Adachi et al. reported about LDH expressing strains with single *PDC* gene deletions (Δ*pdc1* or Δ*pdc5* or Δ*pdc6*) or with a double *PDC5* and *PDC6* inactivation (Adachi et al. 1998). These strains reached a maximum yield of 0.20 g/g at pH values not lower than 3.5. Interestingly, Ishida et al. (2005) reported some years later a production of almost 55 g/L for a strain where the *PDC*1 locus was used as a site of insertion for the bovine LDH, posed under the control of the *PDC*1 promoter, which is strongly induced by glucose. It has to be underlined that the lactate dehydrogenase gene was present in two copies, since the used *S. cerevisiae* strain was diploid.

Amazingly, this strain performed much better (around 3 times better) than the same Δ*pdc1* strain bearing the bovine LDH on a yeast multicopy plasmid. Moreover, this confirmed the stability and the high lactate production (58 g/L) of a diploid *S. cerevisiae* strain bearing an integrated bacterial LDH gene, previously reported by Colombié et al. (2003). They clearly proved that a reduction of the ethanol concentration led to an increasing yield of lactate (0.29 g/g vs 0.62 g/g with *PDC*1 inactivation). Similar yields and productions were found by inserting two copies of a D-LDH from *Leuconostoc mesenterioides* (Ishida et al. 2006c): highly pure D-lactic acid was obtained, confirming the potential of the microbial production strategy in contrast to chemical synthesis. Remarkably, just by increasing the number of integrated LDH copies from 2 to 6 the product concentration reached 120 g/L (Saitoh et al. 2005) (Ishida et al. 2006a). However, still over 40 g/L of ethanol were accumulated which is detrimental for an optimal yield. Consequently, the same authors went back to the Δ*pdc1*, Δ*pdc5* strain inserting only two copies of the bovine LDH, obtaining a product concentration of over 80 g/L with an ethanol accumulation under 5 g/L (Ishida et al. 2006b). The yield was 0.815 g/(L·h).

Tokuhiro et al. report about lactate production with *S. cerevisiae* deleting *PDC1* and *ADH1* (Tokuhiro et al. 2009). Deleting only *PDC1*, 48 g/L of lactate is produced in 24 h with a yield of 0.45 g/g. Deletion of only *ADH1* leads to a severe growth defect and significantly reduced lactic acid yield compared to a wt strain, most probably due to accumulation of the toxic intermediate acetaldehyde (Skory 2003). A combination of *PDC1* and *ADH1* deletion, leads to a strain accumulating 74 g/L in 48 h with a yield of 0.69 g/g (Tokuhiro et al. 2009).

It has to be stressed however, that the highest yields and product concentrations were obtained under neutralising conditions. The same recombinant strains show a significant loss in respect to yield and product concentration as soon as a lower pH is used for the process. So one of the main goals for changing the production hosts was hardly addressed.

This consideration is crucial for the research activities that followed. How the development of the different hosts was addressed depended on which was the main task to be solved (for example pH tolerance, productivity, or viability, just to mention a few). Obviously, all of this included the conceived DSP. Unfortunately, the purification strategy was rarely, if ever, mentioned in detail. The key point is that the very high organic acid concentration at very low pH required for an economical production process is still detrimental for growth and viability even though yeasts are much more tolerant to low pH than lactic acid bacteria. For that reason it became clear that a search for an LDH having a higher turnover rate and/or a higher affinity for pyruvate could lead to a quicker production process, diminishing the time of exposure of the cell factory to the stressing agents and leading therefore to higher product concentrations and/or yield.

The first fungal LDH successfully tested derives from the natural producer *Rhizopus oryzae* (Skory 2000). The advantages regard the optimal pH for activity, which is more similar to the cytoplasmic pH value generally maintained by eukaryotic microorganisms and the independence from fructose 1,6-bisphosphate, required for allosteric activation of most of the bacterial LDHs (Garvie 1980). The expression was tested in different *S. cerevisiae* strains. Comparing haploid and diploid strains, the latter reached the highest product concentration and yield, 38 g/L and 0.44 g/g, respectively. However, also in this case the performance was registered at pH 5. The authors showed that lowering the pH decreases the production (Skory 2003). Interestingly, they also transformed a *S. cerevisiae* strain with a point mutation in the *PDC*1 gene. Said mutation results in a protein with a very low residual enzymatic activity. However, the activity is still detectable for the cell and consequently *PDC*5 expression is not induced (Eberhardt et al. 1999). The rate of glucose utilization was low, though, determining a lower production.

All together, the experiments briefly summarised pointed out that the influence of the yeast background and of the LDH utilised were maybe underestimated. This was clearly shown in (Branduardi et al. 2006) where four different *S. cerevisiae* strains were transformed with six different wild type and one mutagenised LDH genes. The resulting yield values varied from as low as 0.0008 g/g to as high as 0.52 g/g, an astonishing number considering that in all these yeasts the *PDC* genes were entirely functional.

Based on this knowledge a new laboratory strain was developed, starting from one of the better performing strains, *S. cerevisiae* CEN.PK, completely impaired in *PDC* activity and transformed with the very efficient *L. plantarum* LDH (independent of fructose 1,6-bisphosphate), resulting in the strain RWB876 (Liu and Lievense 2005). This strain shows a very high and efficient glucose consumption corresponding to high lactic acid production and yield at very low pH (50.6 g/L, final pH 2.68). Said numbers were obtained dividing the production in two stages, a growth phase, performed on ethanol as carbon source, and a production phase started with high biomass and high glucose concentration. Addition of few calcium carbonate at the beginning of the fermentation prolonged the productive phase. This trick creates an initial buffering while lactate is produced till a critical concentration is reached where most of the product results in the undissociated form. At this point the strain was improved by sequential manipulation, by utilising directed and selection approaches. A step forward was realised with an indirect approach by selection for a more lactic acid and low pH tolerant strain with the re-inoculum technique starting from RWB876. The best performing selected strain m850 reached 62,35 g/L of lactic acid in 76 h reaching a pH of 2.6 (Liu and Lievense 2005). Said strain was further improved with two parallel approaches. In one case, a tuned selection was based on the observation that producing cells have a lower internal pH at the end of the process than at the beginning. Moreover, it was possible to distinguish a sub-population able to maintain a higher intracellular pH within the whole population. By cellular mutagenesis followed by fluorescence activated cell sorting for high intracellular pH the production values were further improved (around 70 g/l in 70 h and with a general quicker production) (Valli et al. 2006). A similar improvement (around 70 g/l in 70 h, with a general improvement of product concentration and productivity of around 15%) was obtained by overexpression of glucose transporters. The inward flux of the carbon source is widely recognised as a rate determining step of glycolysis (Ye et al. 1999). It was shown that an increase of carbon source uptake improves the ethanol production rate (Gutiérrez-Lomelí et al. 2008). This was only recently applied for organic acid production (Branduardi et al. 2008).

Despite of this successful story of improvement, it seems that the numbers are still not sufficiently competitive to substitute the bacterial fermentation on industrial scale.

How to "re-nature" yeast

The aims for strain improvement include high yields (ideally 100%), high product concentrations (aiming at 100 to 150 g/L) in very short process times. Particularly the yield constraint is obviously against the nature of yeast. Microorganisms evolved maximising growth that is converting the carbon source into biomass. Engineering approaches for strain improvement aim consequently against the natural system. This explains why it is so difficult to approach this goal rationally. At the time being we hardly understand how the natural cell is designed to function. Thus, it is much more challenging to engineer a cell to a certain human defined goal. Yet, a lot of interesting work has been devoted to do exactly this. We see three main topics in this context:

- Metabolism: redesigning the metabolic flux to speed up the process and increase yield (that is decrease by-products)

- Transport processes - hardly understood, but central for organic acid productions
- Stress-tolerance: up to now the process ends with dying yeasts

METABOLISM

Interestingly, it has been shown, that for a strain with impaired but not abolished PDC activity, the sugar consumption rate correlates with the lactate production rate (Tokuhiro et al. 2008). However, the sugar consumption rate negatively correlates in these strains with lactate yield. A higher consumption rate leads to a lower yield. This trade-off has to be addressed.

Colombié et al. set out to analyse the impact of process conditions on lactic acid productivity of a heterofermentative wine yeast (Colombié et al. 2003). They showed that increasing the initial glucose concentration leads to an increased lactic acid yield up to an optimal initial glucose concentration of 200 g/L. Exceeding this concentration the lactate yield decreases again. They furthermore show that an increase of the nitrogen source leads to a decreased lactic acid yield, but to a significantly increased productivity. This corresponds to the fact, that the major part of lactic acid is produced in the stationary phase. With increased nitrogen content, the biomass yield is increased, thereby decreasing lactate yield, but increasing lactate productivity due to an increased biocatalyst concentration (Colombié et al. 2003).

PHYSIOLOGY AND METABOLISM AND THEIR LINK TO PRODUCTIVITY

The metabolic engineering approach for construction of a lactic acid producing yeast appears quite straight forward. Glucose conversion to ethanol is equivalent with glucose conversion to lactic acid both in terms of ATP yield and redox balance. Nevertheless, it turned out that the physiologic outcome of the redirection of the carbon flux from ethanol to lactic acid is quite distinct.

Wild-type strains of *S. cerevisiae* grow fast under aerobic as well as anaerobic conditions. With excess of glucose or lack of oxygen ethanol is produced very efficiently. However, homofermentative lactic acid production cannot sustain anaerobic growth (van Maris et al. 2004b) in contrast to *K. lactis*. When the oxygen supply is decreasing the fraction of glucose fermented to ethanol is increasing. Alcoholic fermentation yields ATP in a stoichiometric manner via substrate-level phosphorylation. Consequently, less oxygen is required to form a certain amount of ATP as more glucose is fermented to ethanol. Hence, a decreasing oxygen supply leads to a decrease of the specific oxygen consumption rate and an apparent increase of the biomass yield on oxygen as has been shown already by Fiechter et al. (Fiechter et al. 1981). However, the relationship between oxygen supply and biomass yield found with homolactic *S. cerevisiae* under conditions of limited oxygen supply differs drastically from that of wild-type *S. cerevisiae*. The biomass yield on oxygen is found to be independent of the oxygen supply instead of steadily increasing with decreasing oxygen supply. One possible explanation for this phenomenon is that the export of lactate requires energy in form of ATP. The ATP depletion of the cells by anaerobic homolactate fermentation was confirmed (Abbott et al. 2009b). However, the exact underlying cause could not be elucidated up to now. Further, detailed studies are required to shed light on the relevant physiological processes. As outlined below transport processes are among the very important ones, but maintenance of intracellular pH, uncoupling of the proton motive force, intracellular lactate concentration, just to name a few constitute a network of parameters and processes which are not understood so far.

A similar conclusion namely that cells possess a complex network consisting of gene expression, proteins, and metabolites let Hirasawa et al. set out to evaluate DNA microarray analysis for the identification of genes involved in lactic acid production (Hirasawa et al. 2009). They compared the genome-wide transcription level of cells producing lactic acid with cells not producing lactic acid, both at pH 5, in order to exclude pH/stress reactions. A large number of genes out of a variety of functional categories were identified to be differentially expressed upon lactic acid production. Deletion of a subset of these genes affected lactic acid production more severely than deletion of randomly selected genes, proving that DNA microarray data can be useful for strain optimization. However, no clear picture evolved, which processes are particularly important. As the authors state an integrated analysis of transcriptomics with metabolic and phenotypic data appears to be required for identification of potential targets for genetic manipulation (Hirasawa et al. 2009).

SPECIFIC METABOLIC TRAITS IN THE LIMELIGHT FOR LACTIC ACID PRODUCTION

Colombié et al. showed that the B-vitamin nicotinic acid controls lactate production in *S. cerevisiae* (Colombié and Sablayrolles 2004). Starting point of their study was the observation that lactic acid production by a recombinant *S. cerevisiae* strain is much more efficient on SM medium than on YNB medium. Both are defined media. So a straightforward study, which difference in the medium causes the process improvement was possible. It turned out that the single vitamin nicotinic acid, a precursor for the co-factors NAD and NADP is the improving factor. The lactate production correlates with the nicotinic acid concentration over a wide range in the medium. It remains unclear how exactly the metabolism is influenced.

In order to optimize lactic acid yield it has to be assured that lactic acid is not further metabolized in the cells. L-lactate–cytochrome c oxidoreductase (or cytochrome b2, *CYB2*) is a soluble protein from the inter-membrane space of mitochondria which catalyses the oxidation of L-lactate into pyruvate and the transfer of electrons to cytochrome c. It has been shown, that *CYB2* is up regulated by lactate, but repressed by glucose (Ramil et al. 2000). Ookubo et al. set out to analyse lactic acid production in a *CYB2* deletion strain of *S. cerevisiae* (Ookubo et al. 2008). Interestingly, they showed that *CYB2* deletion has no significant impact on lactate yield at high pH. However, at a pH of 3.5 *CYB2* disruption led to an increase in lactate yield of about 50%. This corresponds to the results from the yeast *Issachenkia orientalis* disclosed by (Miller et al. 2007). While at high pH lactate titer and yield are slightly decreased by *CYB2* deletion, at pH 3 the lactate titer is increased from 56 g/L to 66 g/L. *CYB2* disruption abolishes also the accumulation of acetate and pyruvate as by-products, increases however, the accumulation of glycerol significantly. For *Kluyveromyces marxianus* a deletion of *CYB2* nearly doubles the lactic acid titer and increases the yield by a factor of 1.3. However, the strains produce only about 22 g/L of lactic acid in 92 h under the disclosed conditions.

Summarizing, it can be stated that strains with a *CYB2* disruption have been found to produce lactate at higher productivities and higher lactate titers, especially under low pH conditions, compared to strains having a functional L-lactate–cytochrome c oxidoreductase gene. Such strains cannot grow on a medium containing lactate as its sole carbon source. Strikingly, Miller et al. show, that the deletion of *CYB2* improves the acid resistance of the yeast strain. The resistance against for example glycolic acid is sufficiently pronounced to permit the selection of transformed cells using an acidic medium. This makes it possible to avoid using antibiotic or other resistance gene markers when transforming the strains to delete *CYB2* (Miller et al. 2007).

TRANSPORT

Since the cytoplasmic pH value in yeast cells is much higher than the lactic acid pKa value, almost all of the lactic acid produced is in the dissociated form and, therefore, has to be actively transported outside the cells (Fig. 1). A limitation of lactic acid transport will inevitably lead to an increase in the cytoplasmic lactate concentration, inhibiting the LDH activity (Branduardi et al. 2006) and leading to the reduction or arrest of the lactate production. Like already mentioned above, it has been suggested that the export of lactate from recombinant yeast involves ATP consumption (van Maris et al. 2004a). It is easily comprehensible that the export of lactic acid against the concentration gradient requires energy. However, if the ATP consumption is direct or indirect has not been shown yet and the molar ratio of ATP required for lactate export is not clear at the time being. In the experiments, indicating the so called "absence of a net ATP production" the lactic acid producing yeast cells were cultivated in a chemostat culture, that means they were growing with a constant growth rate. Therefore, ATP was used for biomass formation, lactic acid export and all other growth and maintenance related activities.

Assuming the required energy is available, the overexpression of *JEN*1 has been a tentative to increase lactate export and/or its production in aerobically growing engineered *S. cerevisiae*. Since yeast cells can grow on lactate even at pH values higher than the pKA value of the organic acid (the lactate ion is the predominant form), a specific transporter has to be involved in the uptake of lactate. In *S. cerevisiae*, synthesis of a lactate permease takes place after transcription of the *JEN*1 gene. *JEN*1 encodes for the only known monocarboxylate permease able to actively transport lactate across the *S. cerevisiae* plasma membrane (Casal et al. 1999). Its expression is differently modulated at the transcription, post-transcription, and post-translational levels, in distinct genetic backgrounds and growth conditions (Andrade and Casal 2001). It has been shown that the over-expression of *JEN*1 in both *S. cerevisiae* and *Pichia pastoris* resulted in an increased activity of the lactate (inward) transport, while the deletion of *JEN*1 impaired the growth on both lactate and pyruvate (Soares-Silva et al. 2003).

Unfortunately, the constitutive expression of *JEN*1 in yeast cells producing lactic acid does lead only to a very marginal improvement (Branduardi et al. 2006). In fact, the effect of Jen1 permease overexpression becomes undetectable when the lactate production approaches about 8-10 g/L, suggesting a possible saturation of the Jen1 transport mechanism. Furthermore, by comparing the lactate production from wild type and Jen1 deleted strains transformed with a LDH gene, no differences were observed

(Branduardi et al. 2006). Since lactate cannot diffuse through the membranes, this simple observation indicates that at least another lactate transporter must be operative (Casal et al. 2008).

The monocarboxylate permease family of *S. cerevisiae* comprises five proteins. No experimental evidences that the monocarboxylate transporter-homologous (Mch) proteins of *S. cerevisiae* are involved in the uptake or secretion of monocarboxylates such as lactate, pyruvate or acetate across the plasma membrane have been obtained (Makuc et al. 2001). Skory et al. set out to characterize lactate permeases from lactic acid producing filamentous fungi of the genus *Rhizopus* (Skory et al. 2010). Functionally expression of lacA in *S. cerevisiae* complements the Jen1 deletion phenotype. If this protein is involved in lactate export remains unclear however. Indications for *R. delamar* are that the gene is not expressed, even when lactic acid is produced and therefore exported. The search for lactate exporting proteins is consequently still ongoing (Skory et al. 2010).

Another interesting concept in this context was suggested by Burgstaller (Burgstaller 2006). He points out that the low pH of the culture broth could be in fact the driving force for organic acid export. At low pH the concentration of the deprotonated acid anion is low outside the cell, but high inside the cell owing to the higher intracellular pH. Hence, this ion could leave the cell by (facilitated) passive diffusion. The energy requiring process would consequently be the proton balance, involving mechanisms, which have hardly been studied up to now in context of lactic acid production. At this point transport and stress resistance are strongly interrelated, as elucidated below.

Concluding, besides different studies looking over the transport of lactate across membranes (Merezhinskaya and Fishbein 2009), the lactate outward transport from metabolically engineered yeast host probably remains the most unexplored and understood phenomenon.

Research on export mechanisms and energetics should therefore be an integral part of the development of microbial production processes for this as well as all other compounds requiring an active transport.

STRESS TOLERANCE

According to the presented overview, it is evident that as long as the homolactic fermentation substitutes a typical alcoholic fermentation the engineered yeasts can cope with the unnatural catabolism. The true problems start with high initial glucose concentrations resulting in high lactic acid concentrations, and low pH. As a consequence, the productivity decreases and a typical lactic acid fermentation ends with a very high percentage of dead cells. This limits productivity and prevents the development of a process based on biomass recirculation. A deeper understanding of the toxicity caused by the organic acid at low pH and the defensive mechanisms of the yeast cells could greatly improve the production process. However, and despite significant research effort, this subject is still far from being understood. One reason is inevitably connected to the multiplicity and complexity of the stress response, rendering the analyses and even more the possible interpretations still very dubious. Another reason is related to the experimental settings used for examining this problem. Many studies are in fact planned to examine what happens after the stress is imposed as a pulse, giving indications mainly on immediate responses. The defensive mechanisms switched on during a permanent stressful situation are completely different, and can be examined if the stress persists all over the culturing time. However, this also does not correspond precisely to what happens during the process of production, since the stressful situation is gradually created, giving to the cells the possibility to switch adaptation responses gradually on, as demonstrated for *S. cerevisiae* at low pH or in the presence of acetic acid (Giannattasio et al. 2005).

ANALYSIS OF THE REACTIONS OF YEASTS TO LACTIC ACID

The combination of parameters such as lactic acid concentration and pH value creates a threshold limiting not only the final product concentration but also the product yield. This is clearly evident from the data obtained with the above mentioned *S. cerevisiae* strain m850, which was selected for increased acid resistance, and its improved versions. Only, if the cells can quickly produce a certain amount of lactic acid before the pH of the culture medium drops too low final yield and/or concentration will be sufficiently high, otherwise the same cells will perform much worse. If from one side this observation helps to explain the obtained improvements and could suggest new strategies for future manipulations, on the other side this puts the attention on the still poor robustness of the system.

This can be better reconsidered taking into account that even low concentrations of lactic acid have a detrimental effect on growth and lag phase. 0.2% and 0.4%, respectively, are sufficient to affect these parameters. Low concentration of lactic acid also reduces the glucose consumption rate (Narendranath et al. 2001). Moreover, there is a very significant effect determined by the pH value in association to the presence of the weak acid. When the pH of the culture medium is above the pKa of the acid, said molecule is mainly present in the dissociated form. Because of its hydrophilic nature, the lactate cannot freely permeate into the cells and this allows the cells to tolerate high lactate concentrations. It was shown that buffering the medium at pH 4.5 determines this mechanism of product inhibition in the growth medium. Additionally it creates a buffering system that can retard the normal acidification occurring during yeast growth, having an positive effect on biomass accumulation (Thomas et al. 2002). However, as soon as the pH drops down the undissociated acid freely permeates into the cells. There it dissociates due to a higher pH, causing increased energy consumption for the maintenance of the intracellular pH or finally the drop of the intracellular pH value with all its bad consequences. However, not only the quantity of the undissociated acid is important the detrimental effect on growth, but also the total concentration, as it has been demonstrated for acetic acid and, to some extend, for lactic acid (Thomas et al. 2002).

This might be related to the anion concentration inside of the cells. As reported by Piper et al. the dissociation of the organic acid in the cytoplasm obliges the cells to take care of the released proton but also of the anion, that according to the respective weak acid and to the specific yeast can have weaker or stronger effects on cellular viability (Piper et al. 2001). Independently from the organic acid and from the yeast, the cells have to spend a lot of energy for pumping out the proton and in some cases also the anion. In particular, for the yeast *S. cerevisiae* it was demonstrated that the ATP-binding cassette efflux pump Pdr12 is one of the major effector molecules counteracting intracellular anion accumulation. Pdr12 is in fact induced under stress by the weak acid stress response regulon (Schüller et al. 2004). No evidence for improved lactic acid production of *S. cerevisiae* cells engineered for overexpression of *PDR*12 has been reported so far. Interestingly, *Zygosaccharomyces bailii*, a particularly acid and low pH tolerant yeast does not express a Pdr12 related gene under similar conditions (Piper et al. 2001). This shows that the complex phenotype of "resistance" can be achieved by distinct mechanisms.

Transcriptomic analyses of *S. cerevisiae* show that further pathways are turned on under weak acid stress conditions, but their precise role within the stress response and for stress tolerance remains to be clarified. Among the differentially regulated genes are genes involved in metal metabolism and in cell wall construction. Moreover, it was experimentally shown that in addition to the major plasma membrane proton pump Pma1, the vacuolar class of H^+ proton pump encoded by genes *VMA*2, *VMA*4 and *VMA*6 are also involved in weak organic acid tolerance (Kawahata et al. 2006). It has been demonstrated that transcriptional upregulation and nuclear translocation of the transcription factor Aft1 under organic acid stress, leads to upregulation of genes involved in metal homeostasis and metabolism, particularly in iron metabolism. These data were confirmed and even enlarged by physiological and transcriptional analyses performed on *S. cerevisae* cells exposed to a high lactic acid concentration during anaerobic glucose limited chemostat cultures at different pH values (Abbott et al. 2008). In particular, the importance of iron homeostasis was confirmed but in this case only at higher pH values (pH 5). At lower pH values (pH 3) another regulon was identified including the gene *HAA*1, responsive to high concentrations of lactic acid in the undissociated form and confirmed by the strong growth defect of the Δ*haa1* strain in the presence of lactic acid (Abbott et al. 2008). It has been suggested that the importance of Haa1 could rely on increasing the transcription levels of H^+ antiporters, despite their deletion does not confirm the growth defect of the *HAA*1 null mutant.

APPROACHES TO RENDER YEASTS MORE LACTIC ACID TOLERANT

The presented literature has the merit to show the complexity and the variety of *S. cerevisiae* responses to lactic acid stress, and to point out once more that the process conditions can make a profound difference, rendering the results of rationally designed manipulations not easily predictable.

The complementary aspect of the above presented approaches can be seen in the following examples, where the attention was moved from understanding the cause of the stress and of the stress responses to the alleviation of the most pronounced detrimental effects.

In literature the pro-oxidant effect of organic acids on the cells has been very often mentioned, despite only recently this was at least partially directly demonstrated for formic acid (Du et al. 2008). This implies that, in addition of coping with low intracellular pH, high anion concentration and energy depletion, the lactic acid producing yeasts also have to deal with ROS (reactive oxygen species) accumulation. This was indirectly demonstrated by showing the stronger detrimental effect on growth determined by lactic acid addition to cells growing in liquid medium under vigorous shaking in respect to cells growing without agitation (Porro et al. 2009). Said cells can be "cured" if they are engineered for increasing the endogenous antioxidant levels. This was the case of *S. cerevisiae* cells that were engineered for the non-natural ascorbic acid production (Branduardi et al. 2007). Flow cytometric analyses showed the lower ROS accumulation of vitamin C producing yeasts exposed to oxidative stresses. A further confirmation of this "helping therapy" was accomplished by the overexpression of the gene *CTT*1, encoding for a cytosolic catalase which resulted for reduction of the lactic acid stress through ROS removal (Abbott et al. 2009a).

Another effect to cope is the lowering of internal pH, as also demonstrated by the sorting experiment previously reported (Valli et al. 2005; Valli et al. 2006). Instead of sorting for mutagenised cells able to maintain higher internal pH, *S. cerevisiae* cells were engineered for *PMA*1 overexpression, and this resulted in a higher tolerance to high concentration of lactic acid at low pH (Porro et al. 2009).

Despite helping for better depicting the scenario for further manipulations and of cellular responses, unfortunately these are all examples applied on *S. cerevisiae* cells not engineered for lactic acid production, and so far lacking the definitive confirmation of their effectiveness on improving the process of production.

EXPLOITING BIODIVERSITY

As described above a lot of insight has been gained and research is heading for rationally improved production strains. Yet, a rational approach for optimization of common lab strains is not the only possible solution. In fact it has been shown in the past, that the exploitation of natural properties was very fruitful. Consequently, other yeast hosts than *S. cerevisiae* are still in the focus of attention also for lactic acid production. Avoiding conflicts with external intellectual property, utilization of new carbon sources or improving stress tolerance are among the reasons to explore biodiversity in search of the optimal host organism.

One of the few yeast strains naturally producing lactic acid is *Kluyveromyces thermotolerans* (Rajgarhia et al. 2007; Witte et al. 1989). However, the efficiency is very low. So, not a lot of effort was put into the development of this yeast as production organism. However, it appears worthwhile to explore existing yeast strain collections for the ability to produce lactic acid. Witte et al. provided an agar plate based assay for this purpose (Witte et al. 1989). However, since the suggestion of this assay 20 years ago, no further yeasts producing lactic acid have been identified. Metabolic engineering of host cells with favourable properties was therefore the appropriate strategy chosen in the past.

IMPROVED BIO-REACTION CONDITIONS

Tolerance to weak acids at low pH and availability of genetic tools were the reasons for choosing *S. cerevisiae* as new host for lactic acid production. However, other yeasts are known to be even more acid resistant. These yeasts are on one hand promising candidates as production hosts, but on the other hand natural examples worth to be studied in order to understand complex properties such as stress resistance in more detail. However, the development of genetic tools can be challenging and time-consuming.

Kluyveromyces marxianus is a very promising host for industrial production processes due to a high growth rate at high temperatures (up to 52°C) and utilization of a wide range of inexpensive carbon sources including pentoses (Fonseca et al. 2008). Beyond that *K. marxianus* grows at relatively low pH values. At pH 2.5 this yeast is still able to grow, at pH 2.0 not anymore (Rajgarhia et al. 2007). Taking these facts together this yeast appears as a promising host also for lactic acid production. NatureWorks LLC is investing into the development of this yeast host as various patents and patent applications disclose. Pecota et al. established the genetic tool of sequential gene integration into *K. marxianus* (Pecota et al. 2007). Two copies of the *Bacillus megaterium* LDH led to the accumulation of about 25 g/L of lactic acid, when cells of a batch culture were introduced into glucose medium with 100 g/L of glucose. While the conditions are surely not optimized for production and a true comparison remains difficult, these results underline the potential of this yeast. Hause et al. disclose that a strain expressing two copies of the *L. helveticus* LDH produces 69 g/L of lactic acid in 24 h in a buffered shake flask. This corresponds to a remarkable productivity of 3.6 g/(L·h) (Hause et al. 2009).

PDC deletion lowers growth of the organisms somewhat, but to a much lesser extent than in other yeasts such as *S. cerevisiae*. In a buffered shake flask 99 g/L of lactic acid are produced from 100 g/L of glucose in 48 h, this corresponds to a productivity of 2.1 g/(L·h), and 98% yield (Hause et al. 2009). However, in unbuffered conditions, this strain produces only 9.1 g/L in 72 h, reaching a pH of about 3. Further engineering is consequently required to obtain a suitable production strain.

Issatchenkia orientalis (*Candida krusei*) is another apparently low pH tolerant yeast under investigation by NatureWorks LLC (Dundon et al. 2009). Interestingly, the suggested production mode is a fed-batch, operating at pH 3.0. Approximately, 70 g/L of lactic acid are produced in 77 h. Overall lactate production rate is 1.06 g/(L·h), and overall yield to lactate is 70%.

One example for a remarkably stress resistant yeast is *Zygosaccharomyces bailii* (Branduardi et al. 2004) and references therein). This organism is well-known as a food spoilage agent, often growing in environments with high sugar concentration and/or low pH, such as preserved fruit or pickles (Cole and Keenan 1986).

A first step for metabolic engineering of *Z. bailii* for lactic acid production has been taken (Branduardi et al., 2004). Expressing a bovine lactate dehydrogenase (LDH) the production of lactic acid simultaneously with ethanol could be demonstrated. The product concentration is low at the time being (few g/L), but the potential of this yeast host is high. *Z. bailii* is furthermore an interesting organism to study the molecular basis of "low pH tolerance" and/or "tolerance to weak acid stress". The understanding of these processes could help for improving *S. cerevisiae* as production organism.

For example, Quintas et al. showed that an increase of the weak acid concentration results in decreased growth rate and biomass yield (Quintas et al. 2005). The growth rate was shown to be a linear function of the concentration of the undissociated acid. However, furthermore it was shown the metabolic rate of sugar consumption is not altered, but more and more energy is spent for maintenance under weak acid stress. Interestingly, the net ATP consumption for the increased maintenance under weak acid stress is the same for *Z. bailii* and *S. cerevisiae*, which is much less resistant (Leyva and Peinado 2005), indicating that the improved stress resistance of *Z. bailii* may not be due to a different metabolic response, but due to other physiological properties, *e.g.* membrane composition, just to name one.

Another low pH tolerant yeast is the Crabtree-negative, methylotrophic, haploid *Candida boidinii,* which can easily be genetically modified. Osawa et al. set out to analyse the potential of this yeast as lactic acid producer (Osawa et al. 2009). Deletion of one PDC structural gene and introduction of the bovine LDH led to a strain readily accumulating lactic acid. Unfortunately, in spite of the fact that *C. boidinii* has a striking low pH tolerance (growth at pH 2.0), the production performance at such a low pH was not reported. At neutral pH values the recombinant *C. boidinii* strain produces 85.9 g/L of lactic acid in 48 h (yield: 1.01 g/g) (Osawa et al. 2009).

Pavlovich et al disclosed the production of lactic acid with recombinant *Schizosaccharomyces pombe* carrying the *LDHA* gene from the fungus *Rhizopus oryzae*. Biosynthesis of lactic acid is carried out at a relatively low pH of 4.0 reaching concentrations of lactic acid in the medium of up to 80-100 g/L (Pavlovich et al. 2006).

Candida utilis is another industrially important yeast. It is a Crabtree-negative tetraploid yeast which can grow on a variety of inexpensive carbon sources, such as pulping-waste liquors from the paper industry which most yeasts cannot use. Recent advances with the development of genetic tools allowed constructing a strain lacking PDC activity and expressing a bovine L-LDH. Interestingly, the PDC deletion (4 copies, since the strain is tetraploid) did not abolish ethanol production completely, but the ethanol concentration remained under 10 mg/L. Growth is inhibited by the PDC deletion but not as severely as for *S. cerevisiae*. Consequently, the process employing the recombinant *C. utilis* strain is extremely efficient: 103,3 g/L of lactic acid were produced in only 33 h - yield: 95,1%, maximal production rate: 4.9 g/(L·h) (Ikushima et al. 2009). Finally, the pH reached a value of 4. So, the reaction time can be reduced significantly with this yeast, but no data were published for low pH values.

NEW CARBON SOURCES

Sugar remains the major basis for microbial production commodity chemicals up to now. In fact the sugar industry can be viewed as the cradle of the modern bio-industry (Villadsen 2009). However, recently more efficient exploitation of lignocellulosic plant biomass has received increased attention as carbon source to feed biorefineries (Octave and Thomas 2009; Ohara 2003; Sauer et al. 2008). Such plant-derived substrates contain hexoses such as glucose but also significant amounts of pentose sugars, particularly xylose and arabinose (Saha 2003). The majority of yeasts is not able to convert the wide range of sugars as is required for efficient direct utilization of plant biomass. Therefore, an exploration of biodiversity for organisms naturally suited to ferment these sugars is ongoing. *Pichia stipitis* is a yeast reportedly growing on xylose. Ilmén et al. examined lactic acid production from xylose by *P. stipitis* (Ilmén et al. 2007). Expressing the LDH from *L. helveticus* 58 g/L of lactic acid could be produced from 100 g/L of xylose. When glucose and xylose were present at the same time, both sugars were consumed simultaneously and converted to lactate. PDC activity was not abolished in this study, therefore also ethanol was accumulated. However, lactic acid was the predominant end product of metabolism, indicating once more the suitability of *P. stipitis* for biorefinery approaches.

Another important constituent of plant material is cellulose. Saccharification of cellulose leads to soluble oligosaccharides, which can finally be degraded to glucose and used by conventional cell factories. The disaccharide cellobiose is a potent inhibitor of cellobiohydrolases, which are used for cellulose hydrolysis. A rapid degradation of cellobiose to glucose is therefore required for efficient degradation of cellulose, particularly when saccharification and fermentation occur simultaneously (simultaneous saccharification and fermentation – SSF). A promising attempt for construction of microorganisms for SSF approaches is to display the enzymes for carbon source provision on the cell surface. Tokuhiro et al. report about lactic acid fermentation of cellobiose by a *S. cerevisiae* strain displaying beta-galactosidase on the cell surface (Tokuhiro et al. 2008). The maximum rate of lactate production is with 2.8 g/(L·h) similar to the production rate from glucose, proving the suitability of enzyme surface display. Overall, about 80 g/L of lactic acid are produced from 95 g/L of cellobiose in 120 h.

In this context it is important to keep in mind that for production of lactic acid as basis for PLA purification is crucial. Purification costs are therefore to consider in the same way as are the costs for the raw material (carbon source). A shift to a cheaper carbon source, comprising more impurities might finally increase the costs of the overall process, because the increase in purification costs outweighs the savings in costs for the carbon source. Consequently, a conceptual optimisation is necessary to find a balance between the substitution of expensive nutrients and the limitation of interfering or undesirable components of natural raw materials (Sauer et al. 2008; Venus 2006).

Conclusion

After 15 years of research lactic acid production with yeast is approaching industrial application. Metabolically engineered yeasts will substitute the natural production organisms lactic acid bacteria. Lactic acid producing yeasts are the first example for rationally designed microorganisms where one metabolic end product (ethanol) was entirely substituted by another – "foreign" – product (lactic acid).

We believe that the strategy is right and that the integrated development of bioconversion and purification is the way to go. In this case the change of production organism is dictated by the need for a cheaper purification process.

However, the research of the last 15 years in this area is not only important to establish economic lactic acid production, but to document a paradigmatic development process for the rational design of a microbial cell factory.

We have gained a lot of experience and insight. Some concepts are under development that have already been useful and will be even more used in the future for other production processes – like malic, succinic, 3-hydroxy propionic acid and many more (Abbott et al. 2009c). A balanced exploitation of natural host properties and rational strain design appears as a generally practicable strategy. Stress tolerance dictated by the desired process constraints is a fast moving field, which will have major impact on production costs and thereby on the decision if a given process can be commercially viable.

We also see an important message to learn: One of the basic questions for applied microbiology is: "What are you screening for???" In our opinion, frequently this question has not been considered adequately or questions have not been phrased in the correct way. Examples: If the process is designed such, that the producing cells will not be growing, chemostat cultures might not be the right system to gain process insight. If pH stress is used as basis for DNA microarray analyses, in order to improve the host, what exactly is addressed? Growth? Or productivity? They might be connected, but can be negatively correlated. A better growing strain might not be a better producer. So if the question was growth, the answer might not be what was initially intended (productivity).

A major task for the future is to learn to ask the right questions – a lot of studies intended to lead to better productivity, did lead to interesting results, but NOT to better production strains.

Taking together what we learned from lactic acid production with yeasts, we see a bright future for bulk and fine chemical production with these versatile hosts.

Acknowledgements

The authors thank Minoska Valli for invaluable help for preparation of this manuscript. Research on metabolic engineering and organic acid production by M.S. and D.M. is financially supported by the Translational Research Program of FWF Austria, Project L391 and the program FH*plus* of FFG Austria, Project METORGANIC.

References

Abbott D., Suir E., van Maris A. and Pronk J. (2008) Physiological and transcriptional responses to high concentrations of lactic acid in anaerobic chemostat cultures of *Saccharomyces cerevisiae*. *Appl. Environ. Microbiol.* **74**, 5759-5768.

Abbott D., Suir E., Duong G., de Hulster E., Pronk J. and van Maris A. (2009a) Catalase overexpression reduces lactic acid-induced oxidative stress in Saccharomyces cerevisiae. *Appl. Environ. Microbiol.* **75**, 2320-2325.

Abbott D., van den Brink J., Minneboo I., Pronk J. and van Maris A. (2009b) Anaerobic homolactate fermentation with Saccharomyces cerevisiae results in depletion of ATP and impaired metabolic activity. *FEMS Yeast Res.* **9**, 349-357.

Abbott D., Zelle R., Pronk J. and van Maris A. (2009c) Metabolic engineering of Saccharomyces cerevisiae for production of carboxylic acids: current status and challenges. FEMS Yeast Res. **9**, 1123-36.

Adachi E, Torigoe M., Sugiyama M., Nikawa J.I. and Shimidzu K. (1998) Modification of metabolic pathways of *Saccharomyces cerevisiae* by expression of lactate dehydrogenase and deletion of pyruvate decarboxylase genes for the lactic acid fermentation at low pH. *J. Ferment. Bioeng.* **86**, 284-289.

Andrade R, and Casal M. (2001) Expression of the lactate permease gene JEN1 from the yeast Saccharomyces cerevisiae. *Fungal. Genet. Biol.* **32**, 105-111.

Auras R., Harte B. and Selke S. (2004) An overview of polylactides as packaging materials. *Macromol. Biosci.*, **4**, 835-864.

Barnett J. (2003) Beginnings of microbiology and biochemistry: the contribution of yeast research. *Microbiology* **149**, 557-567.

Benninga H.A. (1990) *A History of Lactic Acid Making*. Dordrecht, The Netherlands: Kluyver Academic Publisher.

Bianchi M., Tizzani L., Destruelle M., Frontali L. and Wésolowski-Louvel, M. (1996) The 'petite-negative' yeast Kluyveromyces lactis has a single gene expressing pyruvate decarboxylase activity. *Mol. Microbiol.* **19**, 27-36.

Bianchi M., Brambilla L., Protani F., Liu C., Lievense J. and Porro D. (2001) Efficient homolactic fermentation by Kluyveromyces lactis strains defective in pyruvate utilization and transformed with the heterologous LDH gene. *Appl. Environ. Microbiol.* **67**, 5621-5625.

Branduardi P., Valli M., Brambilla L., Sauer M., Alberghina L. and Porro D. (2004) The yeast *Zygosaccharomyces bailii*: a new host for heterologous protein production, secretion and for metabolic engineering applications. *FEMS Yeast Res.* **4**, 493-504.

Branduardi P., Sauer M., De Gioia L., Zampella G., Valli M, Mattanovich D. and Porro D. (2006) Lactate production yield from engineered yeasts is dependent from the host background, the lactate dehydrogenase source and the lactate export. *Microb. Cell Fact* **5**, 4.

Branduardi P., Fossati T., Sauer M., Pagani R., Mattanovich D. and Porro D. (2007) Biosynthesis of vitamin C by yeast leads to increased stress resistance. *PLoS ONE* **2**(10):e1092.

Branduardi P., Sauer M. and Porro D. (2008) Improved Yeast Strains for Organic Acid Production. *European Patent Application* EP 08009693.6.

Burgstaller W. (2006) Thermodynamic boundary conditions suggest that a passive transport step suffices for citrate excretion in *Aspergillus* and *Penicillium*. *Microbiology* **152**, 887-893.

Casal M., Paiva S., Andrade R., Gancedo C. and Leão C. (1999) The lactate-proton symport of Saccharomyces cerevisiae is encoded by JEN1. *J. Bacteriol.* **181**,2620-2623.

Casal M., Paiva S., Queirós O. and Soares-Silva I. (2008) Transport of carboxylic acids in yeasts. *FEMS Microbiol. Rev.* **32**, 974-994.

Chelstowska A., Liu Z., Jia Y., Amberg D. and Butow R. (1999) Signalling between mitochondria and the nucleus regulates the expression of a new D-lactate dehydrogenase activity in yeast. *Yeast* **15**, 1377-1391.

Cole M. and Keenan M. (1986) Synergistic effects of weak-acid preservatives and pH on the growth of Zygosaccharomyces bailii. *Yeast* **2,** 93-100.

Colombié S., Dequin S. and Sablayrolles J.M. (2003) Control of lactate production by Saccharomyces cerevisiae expressing a bacterial LDH gene. *Enzyme and Microbial Technology* **33**, 38-46.

Colombié S. and Sablayrolles J. (2004) Nicotinic acid controls lactate production by K1-LDH: a Saccharomyces cerevisiae strain expressing a bacterial LDH gene. *J. Ind. Microbiol. Biotechnol*. **31,** 209-215.

Datta R. and Henry M. (2006) Lactic acid: recent advances in products, processes and technologies - a review. *Journal of Chemical Technology and Biotechnology* **81**, 1119-1129.

Dequin S. and Barre P. (1994) Mixed lactic acid-alcoholic fermentation by Saccharomyces cerevisiae expressing the *Lactobacillus casei* L(+)-LDH. *Biotechnology* (New York) **12**, 173-177.

Du L., Su Y., Sun D., Zhu W., Wang J., Zhuang X., Zhou S. and Lu Y. (2008) Formic acid induces Yca1p-independent apoptosis-like cell death in the yeast *Saccharomyces cerevisiae. FEMS Yeast Res.* **8**, 531-539.

Dundon C.A., Suominen P., Aristidou A., Rush B.J., Koivuranta K., Hause B.M.,

MCMULLIN T.W. AND ROBERG-PEREZ K. (2009) Yeast cells having disrupted pathway from dihydroxyacetone phosphate to glycerol. *United States Patent Application* (20090053782).

EBERHARDT I., CEDERBERG H., LI H., KÖNIG S., JORDAN F. AND HOHMANN, S. (1999) Autoregulation of yeast pyruvate decarboxylase gene expression requires the enzyme but not its catalytic activity. *Eur. J. Biochem.* **262**, 191-201.

FIECHTER A., FUHRMANN G. AND KÄPPELI O. (1981) Regulation of glucose metabolism in growing yeast cells. *Adv. Microb. Physiol.* 22, 123-183.

FITZPATRICK J.J., MURPHY C., MOTA F.M. AND PAULI, P. (2003) Impurity and cost considerations for nutrient supplementation of whey permeate fermentations to produce lactic acid for biodegradable plastics. *International Dairy Journal* **13**, 575–580.

FONSECA G., HEINZLE E., WITTMANN C. AND GOMBERT A. (2008) The yeast *Kluyveromyces marxianus* and its biotechnological potential. *Appl. Microbiol. Biotechnol* . **79**, 339-354.

GARVIE E. (1980) Bacterial lactate dehydrogenases. *Microbiol .Rev*. **44,** 106-139.

GIANNATTASIO S., GUARAGNELLA N., CORTE-REAL M., PASSARELLA S., AND MARRA E. (2005) Acid stress adaptation protects *Saccharomyces cerevisiae* from acetic acid-induced programmed cell death. *Gene* **354**, 93-98.

GONZÁLEZ-VARA, Y.R.A, VACCARI G., DOSI E., TRILLI A. AND ROSSI M, MATTEUZZI D. (2000). Enhanced production of L-(+)-lactic acid in chemostat by *Lactobacillus casei* DSM 20011 using ion-exchange resins and cross-flow filtration in a fully automated pilot plant controlled via NIR. *Biotechnol. Bioeng*. **67**, 147-156.

GUTIÉRREZ-LOMELÍ M., TORRES-GUZMÁN J., GONZÁLEZ-HERNÁNDEZ G., CIRA-CHÁVEZ L., PELAYO-ORTIZ C. AND RAMÍREZ-CÓRDOVA J.J. (2008) Overexpression of ADH1 and HXT1 genes in the yeast Saccharomyces cerevisiae improves the fermentative efficiency during tequila elaboration. *Antonie Van Leeuwenhoek* **93**,363-371.

HAUSE B, RAJGARHIA V. AND SUOMINEN P. (2009) Methods and materials for the production of L-lactic acid in yeast. *US Patent* US000007534597.

HIRASAWA T., OOKUBO A., YOSHIKAWA K., NAGAHISA K., FURUSAWA C., SAWAI H. AND SHIMIZU H. (2009) Investigating the effectiveness of DNA microarray analysis for identifying the genes involved in L: -lactate production by *Saccharomyces cerevisiae*. Appl. Microbiol. Biotechnol. **84**, 1149-59.

HOFVENDAHL K. AND HAHN-HÄGERDAL B. (2000) Factors affecting the fermentative lactic acid production from renewable resources. *Enzyme Microb. Technol.* **26**, 87-107.

HOHMANN S. AND CEDERBERG H. (1990) Autoregulation may control the expression of yeast pyruvate decarboxylase structural genes PDC1 and PDC5. *Eur. J. Biochem.* **188**, 615-621.

HOHMANN S. (1991a) Characterization of PDC6, a third structural gene for pyruvate decarboxylase in Saccharomyces cerevisiae. *J. Bacteriol.* **173**, 7963-7969.

HOHMANN S. (1991b) PDC6, a weakly expressed pyruvate decarboxylase gene from yeast, is activated when fused spontaneously under the control of the PDC1 promoter. Curr. Genet. **20**, 373-378.

IKUSHIMA S., FUJII T., KOBAYASHI O., YOSHIDA S. AND YOSHIDA A. (2009) Genetic engineering of Candida utilis yeast for efficient production of L-lactic acid. *Biosci. Biotechnol . Biochem.* **73**, 1818-1824.

ILMÉN M., KOIVURANTA K., RUOHONEN L., SUOMINEN P. AND PENTTILÄ M. (2007) Efficient production of L-lactic acid from xylose by Pichia stipitis. *Appl. Environ. Microbiol.* **73,** 117-123.

ISHIDA N., SAITOH S., TOKUHIRO K., NAGAMORI E., MATSUYAMA T., KITAMOTO K. AND TAKAHASHI H. (2005) Efficient production of L-Lactic acid by metabolically engineered Saccharomyces cerevisiae with a genome-integrated L-lactate dehydrogenase gene. *Appl. Environ. Microbiol.* **71**, 1964-1970.

ISHIDA N., SAITOH S., OHNISHI T., TOKUHIRO K., NAGAMORI E., KITAMOTO K. AND TAKAHASHI H. (2006a) Metabolic engineering of *Saccharomyces cerevisiae* for efficient production of pure L-(+)-lactic acid. *Appl. Biochem. Biotechnol.* **131**, 795-807.

ISHIDA N., SAITOH S., ONISHI T., TOKUHIRO K., NAGAMORI E., KITAMOTO K., AND TAKAHASHI H. (2006b) The effect of pyruvate decarboxylase gene knockout in Saccharomyces cerevisiae on L-lactic acid production. *Biosci. Biotechnol .Biochem.* **70**, 1148-1153.

ISHIDA N., SUZUKI T., TOKUHIRO K., NAGAMORI E., ONISHI T., SAITOH S., KITAMOTO K. AND TAKAHASHI H. (2006c). D-lactic acid production by metabolically engineered Saccharomyces cerevisiae. *J. Biosci. Bioeng.* **101**, 172-177.

KALE G., KIJCHAVENGKUL T., AURAS R., RUBINO M., SELKE S. AND SINGH S. (2007) Compostability of bioplastic packaging materials: an overview. *Macromol. Biosci.* **7**, 255-277.

KAWAHATA M., MASAKI K., FUJII T. AND IEFUJI H. (2006) Yeast genes involved in response to lactic acid and acetic acid: acidic conditions caused by the organic acids in Saccharomyces cerevisiae cultures induce expression of intracellular metal metabolism genes regulated by Aft1p. *FEMS Yeast Res.* **6**, 924-936.

LEYVA J. AND PEINADO J. (2005) ATP requirements for benzoic acid tolerance in *Zygosaccharomyces bailii. J . Appl. Microbiol.* **98**, 121-126.

LIU CL. AND LIEVENSE J. (2005) Lactic acid producing yeast. *US Patent application* 20050112737.

MAKUC J., PAIVA S., SCHAUEN M., KRÄMER R., ANDRÉ B., CASAL M., LEÃO C. AND BOLES E. (2001) The putative monocarboxylate permeases of the yeast *Saccharomyces cerevisiae* do not transport monocarboxylic acids across the plasma membrane. *Yeast* **18**, 1131-1143.

MATSUSHIKA A., INOUE H., KODAKI T. AND SAWAYAMA S. (2009) Ethanol production from xylose in engineered *Saccharomyces cerevisiae* strains: current state and perspectives. Appl. Microbiol. Biotechnol. **84**, 37-53.

MEREZHINSKAYA N., FISHBEIN W. (2009) Monocarboxylate transporters: past, present, and future. *Histol. Histopathol.* **24**, 243-264.

MILLER M., SUOMINEN P., ARISTIDOU A., HAUSE B.M., VAN HOEK P. AND DUNDON C.A. (2007) Lactic acid-producing yeast cells having nonfunctional L-or D-lactate:ferricytochrome C oxidoreductase gene. International Patent Application WO2007117282.

NARENDRANATH N, THOMAS K, INGLEDEW W. (2001. Effects of acetic acid and lactic acid on the growth of Saccharomyces cerevisiae in a minimal medium. *J. Ind. Microbiol. Biotechnol.* **26**, 171-177.

NEVOIGT E. (2008) Progress in metabolic engineering of *Saccharomyces cerevisiae. Microbiol Mol. Biol. Rev.* **72**, 379-412.

OCTAVE, S. AND THOMAS, D. (2009) Biorefinery: Toward an industrial metabolism. *Biochimie* **91**, 659-664.

OHARA H. (2003) Biorefinery. *Appl. Microbiol. Biotechnol.* **62**, 474-477.

OOKUBO A., HIRASAWA T., YOSHIKAWA K., NAGAHISA K., FURUSAWA C., AND SHIMIZU H. (2008) Improvement of L-lactate production by CYB2 gene disruption in a recombinant *Saccharomyces cerevisiae* strain under low pH condition. *Biosci. Biotechnol. Biochem.* **72**, 3063-3066.

OSAWA F., FUJII T., NISHIDA T., TADA N., OHNISHI T., KOBAYASHI O., KOMEDA T. AND YOSHIDA S. (2009) Efficient production of L-lactic acid by Crabtree-negative yeast *Candida boidinii*. Yeast **26**,485-496.

PAVLOVICH S.S., MIKHAJLOVIC V.M., VLADIMIROVICH J.T., ALEKSANDROVICH R.J., ISAAKOVNA R.E., GEORGIEVNA T.N., ALEKSANDROVNA V.M., MIKHAJLOVNA A.A. AND GEORGIEVICH D.V. (2006). Method for microbiological synthesis of lactic acid and recombinant strain of yeast Schizosaccharomyces pombe for its realization. *Russian Patent Application* RU000002268304.

PECOTA D., RAJGARHIA V. AND DA SILVA N. (2007) Sequential gene integration for the engineering of Kluyveromyces marxianus. *J. Biotechnol* . **127**, 408-416.

PIPER P., CALDERON C., HATZIXANTHIS K., AND MOLLAPOUR M. (2001) Weak acid adaptation: the stress response that confers yeasts with resistance to organic acid food preservatives. *Microbiology* **147**, 2635-2642.

PORRO D., BRAMBILLA L., RANZI B., MARTEGANI E. AND ALBERGHINA L. (1995) Development of metabolically engineered Saccharomyces cerevisiae cells for the production of lactic acid. *Biotechnol. Prog.* **11**, 294-298.

PORRO D., BIANCHI M., BRAMBILLA L., MENGHINI R., BOLZANI D., CARRERA V., LIEVENSE J., LIU C., RANZI B., FRONTALI L. AND OTHERS. (1999) Replacement of a metabolic pathway for large-scale production of lactic acid from engineered yeasts. Appl. Environ. Microbiol. **65**, 4211-4215.

PORRO D., DATO L. AND BRANDUARDI P. (2009) Method for improving acid and low pH tolerance in yeast. *International Patent Application* WO2008153890.

QUINTAS C., LEYVA J., SOTOCA R., LOUREIRO-DIAS M. AND PEINADO J. (2005) A model of the specific growth rate inhibition by weak acids in yeasts based on energy requirements. *Int. J. Food Microbiol.* **100**, 125-130.

RAJGARHIA V., HATZIMANIKATIS V., OLSON S., CARLSON T., STARR J.N., KOLSTAD J.J. AND EYAL A. (2007) Methods for the synthesis of lactic acid using crabtree-negative yeast transformed with the lactate dehydrogenase gene. *United States Patent* 7,229,805.

RAMIL E., AGRIMONTI C., SHECHTER E., GERVAIS M. AND GUIARD B. (2000) Regulation of the CYB2 gene expression: transcriptional co-ordination by the Hap1p, Hap2/3/4/5p and Adr1p transcription factors. *Mol. Microbiol.* **37**, 1116-1132.

SAHA B. (2003) Hemicellulose bioconversion. *J. Ind. Microbiol. Biotechnol.* **30**, 279-291.

SAITOH S., ISHIDA N., ONISHI T., TOKUHIRO K., NAGAMORI E., KITAMOTO K. AND TAKAHASHI H. (2005) Genetically engineered wine yeast produces a high concentration of L-lactic acid of extremely high optical purity. *Appl. Environ. Microbiol.* **71**, 2789-2992.

SAUER M., PORRO D., MATTANOVICH D. AND BRANDUARDI P. (2008). Microbial production of organic acids: expanding the markets. *Trends Biotechnol.* **26**, 100-108.

SCHMITT H., CIRIACY M. AND ZIMMERMANN F. (1983) The synthesis of yeast pyruvate decarboxylase is regulated by large variations in the messenger RNA level. *Mol. Gen. Genet.* **192,** 247-252.

SCHÜLLER C., MAMNUN Y., MOLLAPOUR M., KRAPF G., SCHUSTER M., BAUER B., PIPER P., KUCHLER K. (2004) Global phenotypic analysis and transcriptional profiling defines the weak acid stress response regulon in *Saccharomyces cerevisiae. Mol .Biol. Cell.* **15**, 706-720.

SKORY C. (2000) Isolation and expression of lactate dehydrogenase genes from *Rhizopus oryzae. Appl. Environ. Microbiol.* **66**, 2343-2348.

SKORY C. (2003) Lactic acid production by Saccharomyces cerevisiae expressing a Rhizopus oryzae lactate dehydrogenase gene. *J. Ind. Microbiol. Biotechnol.* **30**, 22-27.

SKORY C.D., HECTOR R.E., GORSICH S.W., RICH J.O. (2010) Analysis of a functional lactate permease in the fungus Rhizopus. *Enzyme and Microbial Technology* **46**, 43–50.

SOARES-SILVA I., SCHULLER D., ANDRADE R., BALTAZAR F. AND CÁSSIO F, CASAL M. (2003). Functional expression of the lactate permease Jen1p of Saccharomyces cerevisiae in Pichia pastoris. *Biochem. J.* **376**, 781-787.

THOMAS K, HYNES S., INGLEDEW W. (2002) Influence of medium buffering capacity on inhibition of Saccharomyces cerevisiae growth by acetic and lactic acids. Appl *Environ. Microbiol.* **68**, 1616-1623.

TOKUHIRO K, ISHIDA N, KONDO A, TAKAHASHI H. (2008) Lactic fermentation of cellobiose by a yeast strain displaying beta-glucosidase on the cell surface. *Appl. Microbiol. Biotechnol..* **79**, 481-488.

TOKUHIRO K., ISHIDA N., NAGAMORI E., SAITOH S., ONISHI T., KONDO A., TAKAHASHI H. (2009) Double mutation of the PDC1 and ADH1 genes improves lactate production in the yeast Saccharomyces cerevisiae expressing the bovine lactate dehydrogenase gene. *Appl. Microbiol. Biotechnol* . **82**, 883-890.

VALLI M., SAUER M., BRANDUARDI P., BORTH N., PORRO D., MATTANOVICH D. (2005) Intracellular pH distribution in *Saccharomyces cerevisiae* cell populations, analyzed by flow cytometry. *Appl. Environ. Microbiol.* **71**, 1515-1521.

VALLI M, SAUER M., BRANDUARDI P., BORTH N., PORRO D. AND MATTANOVICH D. (2006) Improvement of lactic acid production in *Saccharomyces cerevisiae* by cell sorting for high intracellular pH. *Appl. Environ. Microbiol* .**72**, 5492-5499.

VAN MARIS A., KONINGS W., VAN DIJKEN J. AND PRONK J. (2004a) Microbial export of lactic and 3-hydroxypropanoic acid: implications for industrial fermentation processes. *Metab. Eng.* **6**, 245-255.

VAN MARIS A., WINKLER A., PORRO D., VAN DIJKEN J., PRONK J. (2004b) Homofermentative lactate production cannot sustain anaerobic growth of engineered Saccharomyces cerevisiae: possible consequence of energy-dependent lactate export. *Appl .Environ. Microbiol.* **70**, 2898-2905.

VENUS J. (2006) Utilization of renewables for lactic acid fermentation. *Biotechnol. J.* **1**, 1428-1432.

VILLADSEN J. (2009 The sugar industry - the cradle of modern bio-industry. *Biotechnol . J.* **4**, 620-631.

WEE Y.-J., KIM J.-N. AND RYU H.-W. (2006) Biotechnological Production of Lactic Acid and Its Recent Applications. *Food Technology and Biotechnology* **44**, 163-172.

WITTE V., KROHN U. AND EMEIS C. (1989) Characterization of yeasts with high L[+]-lactic acid production: lactic acid specific soft-agar overlay (LASSO) and TAFE-patterns. *J. Basic Microbiol.* **29**, 707-716.

YE L., KRUCKEBERG A., BERDEN J., AND VAN DAM, K. (1999) Growth and glucose repression are controlled by glucose transport in *Saccharomyces cerevisiae* cells containing only one glucose transporter. *J. Bacteriol.* **181**, 4673-4675.

Biotechnology and Genetic Engineering Reviews - Vol. 27, 257-284 (2010)

Polysaccharide drug delivery systems based on pectin and chitosan

GORDON A. MORRIS[1,*], M. SAMIL KÖK[2], STEPHEN E. HARDING[1] AND GARY G. ADAMS[1,3]

[1]NCMH Laboratory, School of Biosciences, University of Nottingham, Sutton Bonington LE12 5RD, UK; [2]Department of Food Engineering, Abant Izzet Baysal University, 14280 Bolu, Turkey; [3]Insulin and Diabetes Experimental Research (IDER) Group, Faculty of Medicine and Health Science, University of Nottingham, Clifton Boulevard, Nottingham NG7 2RD, UK

Abstract

Chitosans and pectins are natural polysaccharides which show great potential in drug delivery systems.

Chitosans are a family of strongly polycationic derivatives of poly-N-acetyl-D-glucosamine. This positive charge is very important in chitosan drug delivery systems as it plays a very important role in mucoadhesion (adhesion to the mucosal surface). Other chitosan based drug delivery systems involve complexation with ligands to form chitosan nanoparticles with can be used to encapsulate active compounds.

Pectins are made of several structural elements the most important of which are the homogalacturonan (HG) and type I rhamnogalacturonan (RG-I) regions often described in simplified terms as the "smooth" and "hairy" regions respectively. Pectin HG regions consist of poly-glacturonic acid residues which can be partially methyl esterified. Pectins with a degree of methyl esterification (DM) > 50% are known as high methoxyl (HM) pectins and consequently low methoxyl (LM) pectins have a DM < 50%. Low methoxyl pectins are of particular interest in drug delivery as they can form gels with calcium ion (Ca^{2+}) which has potential applications especially in nasal formulations.

*To whom correspondence may be addressed (gordon.morris@nottingham.ac.uk)

Abbreviations: GlcN: D-glucosamine; GlcNAc: N-acetyl-D-glucosamine; DA/ DAc: degree of acetylation; TPP: tripolyphosphate; HG: homogalacturonan; RG-I: type I rhamnogalacturonan; DM: degree of methylation; [h]: intrinsic viscosity; $s^0_{20,w}$: sedimentation coefficient; M_w: weight average molecular weight; r_g: radius of gyration; f/f_0: translational frictional ratio; L_p: persistence length

In this chapter we will discuss the physicochemical properties of both chitosans and pectins and how these translate to current and potential drug delivery systems.

Introduction

The route for the delivery of drugs that is still the most popular with medical staff and patients alike is through the mouth and down the alimentary tract: the oral route. The major site for drug absorption by this route is the small intestine which offers ≈ 100 m^2 of surface epithelia across which transfer can at least in principle take place. If the drug is poorly soluble, or is in the form of a controlled release dosage form, significant absorption of the drug may also occur in the large intestine (Davis, 1989). However, the clearance time through the whole alimentary tract is generally too short (4–12 h), rendering oral drug administration a very inefficient process, with much of the drug unabsorbed. More recently interest has focused on drug absorption through nasal epithelia, where again clearance problems are an issue. Consideration is also given to other delivery routes (*e.g.* vaginal and ocular). Other important issues are the degradation of peptide-based drugs in the gastrointestinal tract and low trans-mucosal permeability. Macromolecular based carrier and mucoadhesive systems have been considered for several years and two polysaccharide based systems have emerged as particularly promising: chitosans (cationic) and low methoxy pectins (anionic).

Chitosan

CHEMICAL STRUCTURE

Chitosan is the generic name for a family of strongly polycationic derivatives of poly-N-acetyl-D-glucosamine (chitin) it is found in the exoskeletons of crustaceans such as crabs and shrimps, but can also be found in the cell wall of fungi and bacteria (Tombs and Harding, 1998; Rinaudo, 2006; Yen and Mau, 2007). In chitosan (*Figure 1*) the N-acetyl group is replaced either fully or partially by NH_2 therefore the degree of acetylation can vary from DA = 0 (fully deacteylated) to DA = 1 (fully acetylated *i.e.* chitin). Acetylated monomers (GlcNAc; A-unit) and deacteylated monomers (GlcN; D-unit) have been shown to be distributed randomly or block wise (Vårum, *et al.*, 1991a, 1991b).

Chitosan is biodegradable, non-toxic, non-immunogenic and biocompatible (Terbojevich and Muzzarelli, 2000) and as the only naturally occurring polycationic polymer chitosan and its derivatives have received a great deal of attention from, for example, the food, cosmetic and pharmaceutical industries. Important applications include water and waste treatment, antitumor, antibacterial and anticoagulant properties (Illum, 1998; Rinaudo, 2006; Muzzarelli, 2009).

PHYSICAL PROPERTIES

Chitosan is a semi-crystalline polymer (solid), which exhibits a degree of polymorphism (Ogawa and Yui, 1994). In an aqueous acidic environment, chitosan is promptly solubilised, as a result of the removal of the acetyl moieties present in the amine

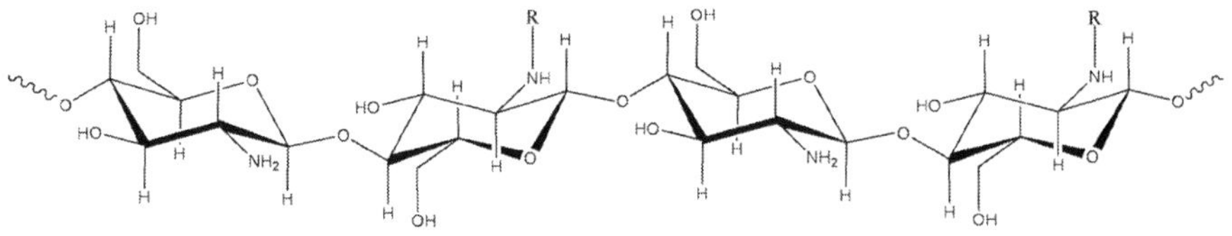

Figure 1. Schematic representation of the structure repeat units of chitosan, where R = Ac or H depending on the degree of acetylation.

functional groups. This solubility is limited, however, in inorganic acids compared to its solubility in organic acids. Solubilisation occurs as a consequence of the protonation of $-NH_2$ functional groups on the C-2 position of D-glucosamine residues. Chitosan is a weak base with pKa values ranging from 6.2 to 7 and at physiological pH 7.4 or higher, low solubility is shown (Park, *et al.,* 1983). However, chitosan's solution properties are dependent on the distribution of its acetyl groups and the molecular weight of the polymer (Kubota and Eguchi, 1997). The solubility of chitosan in water increases with increasing DA (Vårum *et al.*, 1994). With the addition of electrolytes to the solution, the aqueous solubility of chitosan is affected and salting out of chitosan can be seen as in the case of excessive hydrochloric acid use, and the resulting formation of chitosan chlorhydrate (Rinaudo, 2006). Salting out can also be used to recover chitosan from solution and salting out efficiency of anions follows the Hofmeister series $SO_4^{2-} > H_2PO_4 \approx HPO_4^{2-} > NO_3^-$ (LeHoux and Depuis, 2007). With an extended chitosan conformation, due to the repelling effect of each positively charged deacetylated unit, the addition of electrolytes reduces the inter-chain repulsion and induces a more random coil-like conformation in the molecule (Terbojevich and Muzzarelli, 2000).

Molecular weight, pH, ionic strength, and temperature are all factors which affect the viscosity of chitosan. Hydrodynamic studies based on intrinsic viscosity ($[\eta]$), sedimentation coefficient ($s^0_{20,w}$), radius of gyration (r_g) and weight average molecular weight (M_w) have focussed on qualitative/ semi-quantitative methods of estimating the conformation based around "power law" Mark-Houwink-Kuhn-Sakurada relations (Tombs and Harding, 1998) which link intrinsic viscosity, sedimentation coefficient and radius of gyration with molar mass $[\eta] \sim M^a$, $s^0_{20,w} \sim M^b$ and $r_g \sim M^c$, where a, b and c have defined values for specific conformation types (*Table 1*). The translational frictional ratio, f/f_o (Tanford, 1961), sedimentation conformation zoning (Pavlov, *et al.*, 1997; 1999), the "Wales-van Holde" ratio, $k_s/[\eta]$ (Wales and van Holde, 1954) and the persistence length, L_p (Kratky and Porod, 1949) have also been used to estimate dilute solution conformation (*Table 2*).

This has resulted in chitosan being reported to have either a rigid rod-type structure (Terbojevich, *et al.*, 1991; Errington, *et al.*, 1993; Cölfen, *et al.*, 2001; Fee, *et al.*, 2003; Kasaai, 2006; Morris, *et al.*, 2009a) or a semi-flexible-coil (Rinaudo, *et al.*, 1993; Berth, *et al.*, 1998; Brugnerotto, *et al.*, 2001; Schatz, *et al.*, 2003; Mazeau and Rinaudo, 2004; Vold, 2004; Lamarque, *et al.*, 2005; Velásquez, *et al.*, 2008). It has also been shown that flexibility (in terms of persistence length) is moderately influenced by DA (Terbojevich, *et al.*, 1991; Mazeau and Rinaudo, 2004).

Table 1. The Mark-Houwink-Kuhn-Sakurada (MHKS) power law exponents (a, b and c), and the Wales – van Holde ($k_s/[\eta]$) for the conformations described by sedimentation conformation zoning.

	Zone A Extra-rigid rod	*Zone B* Rigid rod	*Zone C* Semi-flexible coil	*Zone D* Random coil	*Zone E* Spherical
a	> 1.4	0.8 – 1.4	0.5 – 0.8	0.2 – 0.5	0.0
b	< 0.2	0.2 – 0.4	0.4 – 0.5	0.5 – 0.6	0.67
c	> 0.8	0.6 – 0.8	0.5 – 0.6	0.4 – 0.5	0.33
$k_s/[\eta]$	< 0.2	0.2 – 0.4	0.4 – 1.0	1.0 - 1.4	1.6

Table 2. Estimations of the dilute solution conformation for pectin and chitosan.

	Pectin	*Chitosan*
a	0.62 – 0.94	0.77 – 1.1
b	0.17	0.24 – 0.25
c	0.57	0.55 – 0.56
$k_s/[\eta]$	0.10 – 0.85	0.16 – 0.73
f/f_0	7 – 10	11 - 16
L_p *(nm)*	10 - 15	4 - 35
Zone	A/B/C	B/C
References	Anger and Berth, 1985; Axelos, *et al.*, 1987; Axelos and Thibault, 1991, Berth, *et al.*, 1977; Harding, *et al.*, 1991; Garnier, *et al.*, 1993; Malovikova, *et al.*, 1993; Tombs and Harding, 1998; Morris, *et al.*, 2000, 2002, 2008; Fishman, *et al.*, 2001, 2006	Terbojevich, *et al.*, 1991; Errington, *et al.*, 1993; Ottøy, *et al.*, 1996; Berth, *et al.*, 1998; Cölfen, *et al*, 2001; Brugnerotto, *et al.*, 2001; Fee, *et al.*, 2003; Schatz, *et al.*, 2003; Mazeau and Rinaudo, 2004; Vold, 2004; Lamarque, *et al.*, 2005; Rinaudo, 2006; Kasaai, 2006; Velásquez, *et al.*, 2008; Morris, *et al.*, 2009

CHITOSAN COMPLEXATION

The ability of chitosan to complex with other ligands, metals for example, is well known (Rhazi, *et al.*, 2002a,b). The proclivity for chelation is dependent on physical state, $-NH_2$ content and distribution of chitosan, degree of polymerization, pH and cation content. It has been shown that following a higher degree of deacetylation, there is a characteristic increase in the degree of chelation. The chelation takes place in chitosan with a degree of polymerization greater than 6 monomeric residues (Rhazi, *et al.*, 2002b). In addition, the intensity of chitosan chelation is governed by nature of cation in solution. Studies have shown that the affinity of chitosan for divalent and trivalent cations of chloride salts shows selectivity in the following order: $Cu^{2+} >> Hg^{2+} > Zn^{2+} > Cd^{2+} > Ni^{2+} > Co^{2+} > Ca^{2+}$, $Eur^{3+} > Nd^{3+} > Cr^{3+} > Pr^{3+}$ (Rhazi, *et al.*, 2002b).

USAGE IN DRUG DELIVERY

Chitosan is of great interest to the pharmaceutical industry in drug delivery and the number of publications on this subject has increased by almost an order of magnitude in the last decade (*Figure 2*). Many aspects including biodegradation, biodistribution and toxicity (Kean and Thanou, 2010); formulations for delivery of DNA and siRNA (Mao, *et al.*, 2010); delivery systems for protein therapeutics (Amidi, *et al.*, 2010); hydrogels for controlled, localized drug delivery (Bhattarai, *et al.*, 2010); nanostructures for delivery of ocular therapeutics (de la Fuente, *et al.*, 2010) and the targeted delivery of low molecular drugs (Park, *et. al.*, 2010) have been reviewed in the most recent volume of Advanced Drug Delivery Reviews (Volume 62).

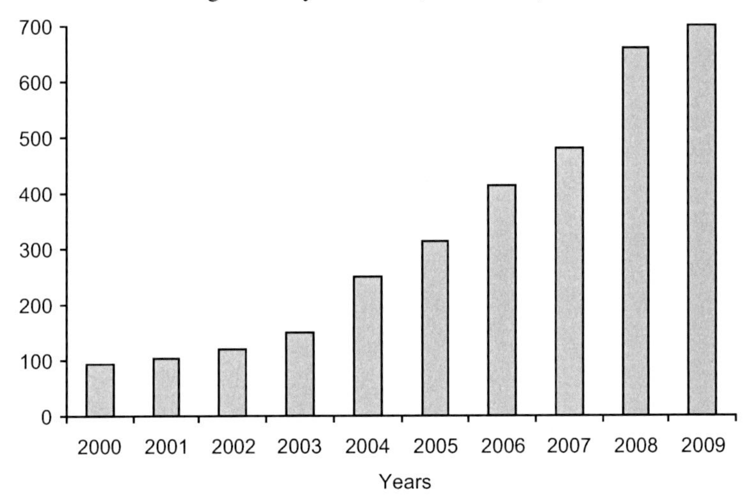

Figure 2. Number of publications on chitosan in drug delivery over the last 10 years (adapted from **Figure 1** in Amidi and Hennink (2010)). Reproduced with the permission of Elsevier.

The mucoadhesive properties of chitosan play an important role in its usage in oral, nasal and ocular drug delivery (Harding, *et al.*, 1999; Illum, 2002; Harding, 2006).

MUCOADHESION

Mucoadhesion is the specific term for adhesion when one of the surfaces is mucus (Harding, *et al.*, 1999). Mucus consists largely of water (> 95 %) and the high molecular weight glycoprotein mucin (Harding, *et al.*, 1999; Harding, 2003; Harding, 2006). The key sugar residues for mucoadhesive interaction are the acidic ones (N-acetyl neuraminic acid or "sialic acid", and some sulphated galactose) and the hydrophobic methyl containing fucose. Despite the polydispersity of these molecules compared to unglycosylated proteins, their structural hierarchy is also well understood. They consist of M ~ 500 000 g/mol basic units linked linearly into "subunits" of M ~ 2 500 000 g/mol. These subunits are further linearly arrayed into macroscopic structures (M between 5 and 50 000 000 g/mol) seen under the electron microscope (Harding, *et al.*, 1983) or using atomic force microscopy (Deacon, *et al.*, 2000). Chitosan interacts strongly with the negatively sialic acid residues (Fiebrig, *et al.*, 1994a,b; 1995a,b; Anderson, *et al.*, 1989; Deacon, *et al.*, 1999; Rossi, *et al.*, 2000; 2001; Dodou, *et al.*, 2005) although hydrogen bonding and hydrophobic interactions are also important (Deacon, *et al.*, 1999; Qaqish and Amiji, 1999; Dodou, *et al.*, 2005; Sogias, *et al.*, 2008). The different theories explaining mucoadhesion and the properties of mucoadhesives are shown in *Figure 3* (Dodou, *et al.*, 2005 and references therein). The chitosan/ mucin interaction depends on the zeta potential of the mucin (*Figure 4)* (Takeuchi, *et al.* 2005) and this change in zeta potential is related to the concentration, molecular weight and charge of the chitosan (*Figure 5*) and to the pH (Takeuchi, *et al.* 2005; Sogias, *et al.*, 2008). This change in zeta potential is associated with a change particle size (Fiebrig, *et al.*, 1994a,b; 1995a,b; Anderson, *et al.*, 1989; Takeuchi, *et al.* 2005; Sogias, *et al.*, 2008) (**Table 3**). As a consequence the degree of chitosan/ mucin interaction is also dependent on the biological source of mucin (*Figure 6*). Drug delivery systems involving chitosan, therefore, show great potential and the use of encapsulation technology involving chitosan nano- or microparticles are increasing in populararity.

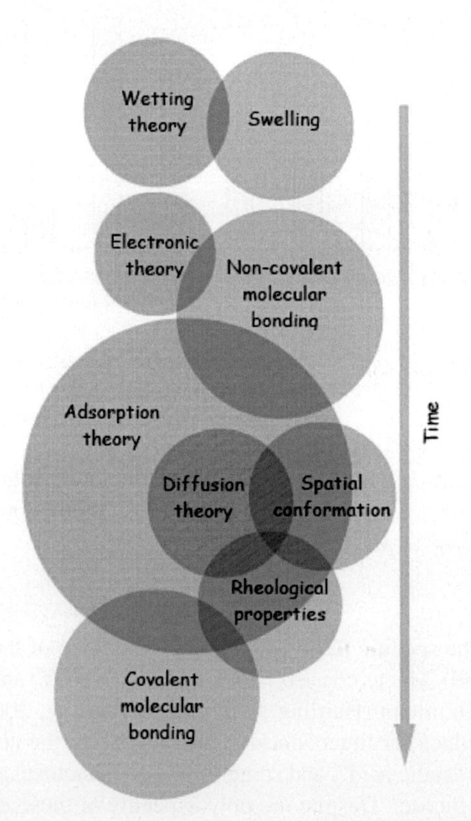

Figure 3. Theories of mucoadhesion (red circles) and material properties of mucoadhesives (blue circles). The overlapping areas between the circles of the material properties and the mucoadhesive theories indicate how and to what extent the former are connected to the latter (adapted from **Figure 2** in Dodou, *et al.* (2005)). Reproduced with the permission of Elsevier.

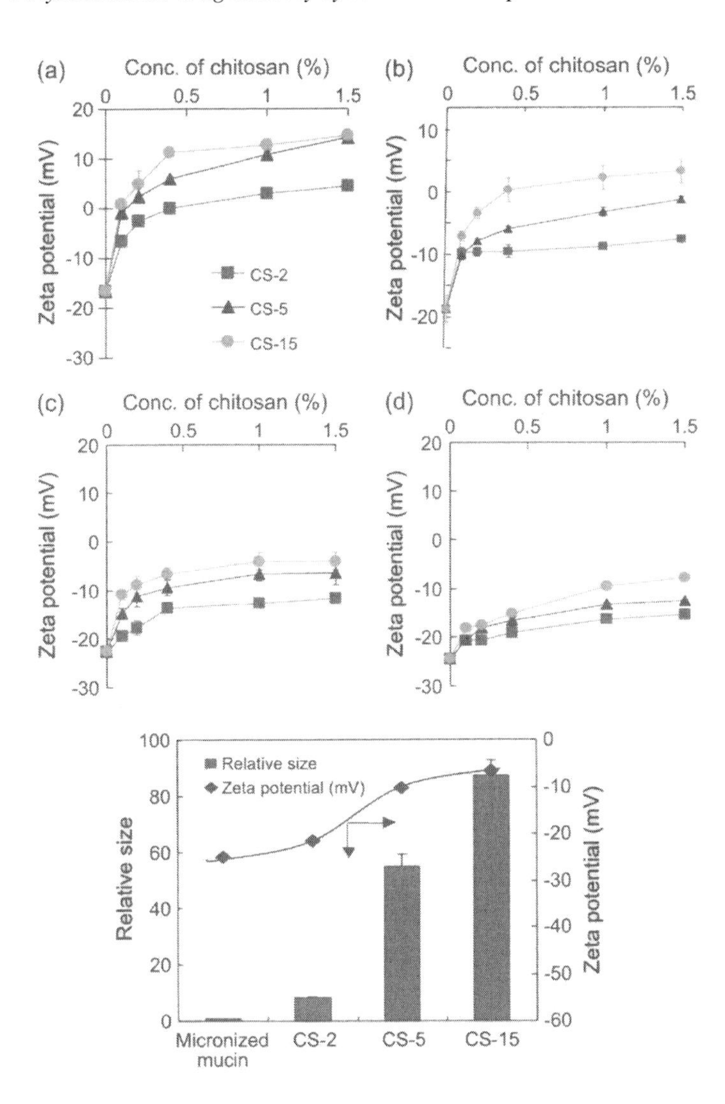

Figure 4. Zeta potential of coarse mucin particles in the solutions of chitosan having different molecular weight with various concentrations and different pH. (a) pH 5.0, (b) pH 6.8, (c) pH 7.4, (d) pH 9.0, (e) Change in observed particle size of micronized mucin particles when mixed with the chitosan solutions. Concentration of chitosan solution: 1.5% w/v. pH of solution: 6.8. Molecular weight of chitosan: CS-2 = 20000; CS-5 = 50000 and CS-15 = 150000 g/mol respectively (adapted from **Figures 1** and **3** in Takeuchi, *et al.* (2005)). Reproduced with the permission of Elsevier.

NANOPARTICLES

Chitosan has been widely used in the preparation of nanoparticles for drug delivery (Dyer *et al.*, 2002; Fernández-Urrasuno, *et al.*, 1999; Gan and Wang, 2007; Gan, *et al.*, 2005; Luangtana-anan, *et al.*, 2005; Shu and Zhu, 2000; Tsai, *et al.*, 2008; Xu and Du 2005). Chitosan nanoparticles can be prepared by at least three different methods (Kumari, *et. al.*, 2010):

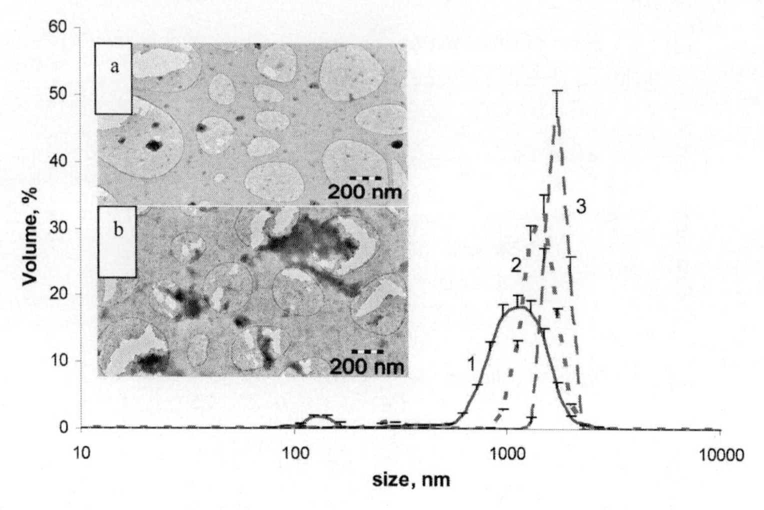

Figure 5. Dynamic light scattering size measurements of pig gastric mucin mixed with chitosan at pH 2.0 (1), half acetylated chitosan at pH 7.0 (2), and half acetylated chitosan at pH 2.0 (3) at [polymer]/[mucin] weight ratio = 0.05. Insets: pig gastric mucin at pH 2.0 before (a) and after (b) addition of chitosan (adapted from **Figure 3** in Sogias, *et al.* (2008)). Reproduced with the permission of American Chemical Society Publications.

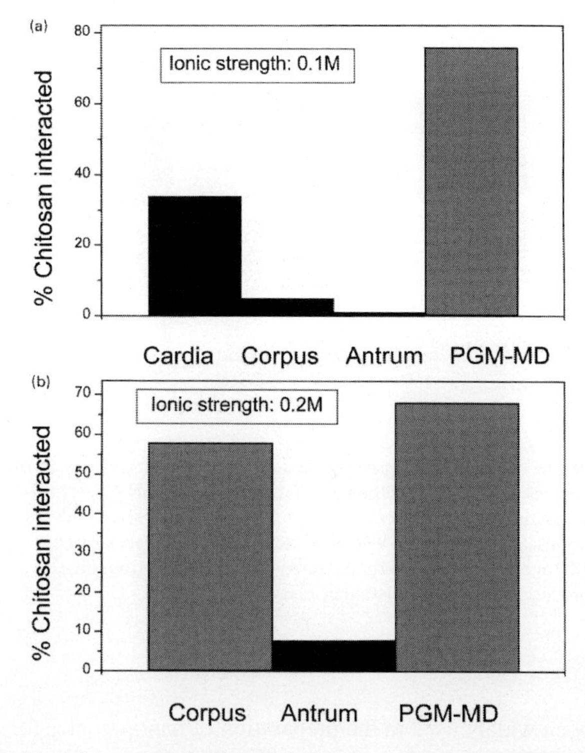

Figure 6. Comparison of the interaction between "SC210+" chitosan with three mucin populations purified from different regions of the porcine stomach (cardia, corpus-LD and antrum-LD) and one mucin population purified from the whole porcine stomach (PGM-MD). (a) I = 0.1 M, (b) I = 0.2 M (adapted from **Figure 2** in Deacon, *et al.* (1999)).

Table 3. Mucoadhesive analysis. The sedimentation coefficient ratio ($s_{complex}/s_{mucin}$) as an index of (muco)adhesiveness (from Fiebrig, *et al.*, 1994a,b; 1995a,b; Anderson, *et al.*, 1989) (adapted from **Table 1** in Harding (2003)).

Mucoadhesive	$s_{complex}/s_{mucin}$	Conditions
DEAE-dextran	1.1–1.9[a]	pH 6.8, 20 °C
	1.2–1.4[a]	pH 6.8, 37 °C
Chitosan (FA ≈ 0.11)	48	pH 6.5, 20 °C
	15	pH 4.5, 20 °C
	22	pH 2.0, 20 °C
	12	pH 2.0, 37 °C
	26	pH 4.5, 20 °C + 3 mM bile salt
	35	pH 4.5, 37 °C + 3 mM bile salt
	18	pH 4.5, 20 °C + 6 mM bile salt
	14	pH 4.5, 37 °C + 6 mM bile salt
Chitosan (FA ≈ 0.42)	31	pH 4.5, 20 °C
	44	pH 4.5, 37 °C

[a]Depends on the mixing ratio

Electrostatic interaction and resultant ionotropic gelation between chitosan and the for example tripolyphosphate (TPP) polyanion (He, *et al.*, 1998, 1999; Dyer, *et al.*, 2002; Luangtana-anan, *et al.*, 2005; Janes, *et al.*, 2001; Shu and Zhu, 2000; Gan, *et al.*, 2005; Morris, *et al.*, 2010a) (*Figure 7*).

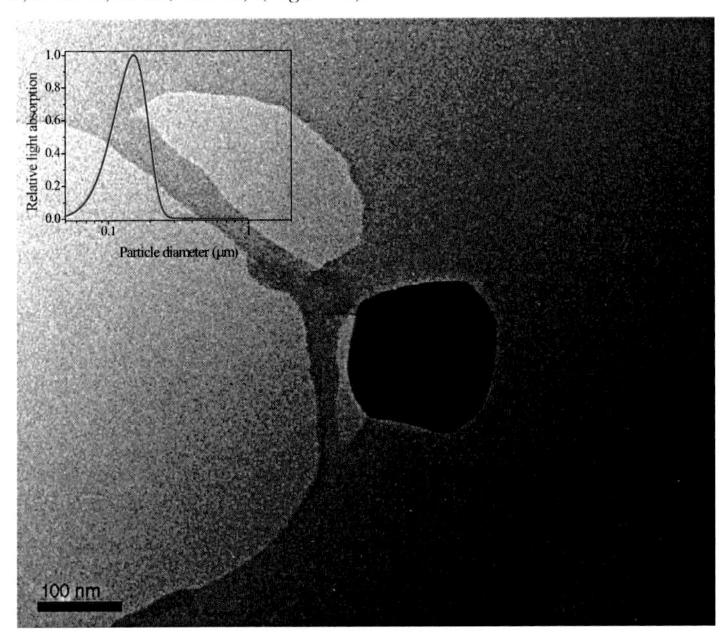

Figure 7. Transmission electron microscopy image of a chitosan–TPP nanoparticle of diameter 140 – 250 nm (adapted from **Figure 10** in Gan, *et al.* (2005)). Inset: particle size distribution measured by differential sedimentation for TPP-chitosan nanoparticles with a mean particle size of 141 nm (adapted from **Figure**

3 in Morris, *et al.* (2010a)). Reproduced with the permission of Elsevier.

Micro-emulsion for preparation of chitosan – glutaraldehyde complexes for example (Genta, *et. al.*, 1998; Dhawan, *et al.*, 2004).

Polyelectrolyte complex (PEC) formation with for example pectin (MacLeod, *et al.*, 1999; Ofori-Kwakye and Fell, 2001) or hyaluronic acid (Lim, *et al.*, 2000; Kim, *et al.*, 2004; Kujawa, *et al.*, 2007). This is of particular importance when a constant drug release profile is not desired (MacLeod, *et al.*, 1999; Ofori-Kwakye and Fell, 2001).

The size of the nanoparticles depends on the molecular weight of the chitosan polymer and higher molecular weight chitosans produce larger nanoparticles (Luangtana-anan *et al.*, 2005; Morris, *et al.*, 2010a). The method of cross-linking affects the mucoadhesive strength and stability of the nanoparticles (Genta, *et. al.*, 1998; Dhawan, *et al.*, 2004).

STABILITY

The stability (shelf-life) of chitosan in terms of molar mass, viscosity and conformation is very important to pharmaceutical industry as these properties play an important role in the function of chitosan in formulations (Skaugrud, *et al.*, 1999; Terbojevich and Muzzarelli, 2000). Chitosan storage conditions and particularly temperature may be important but whether or not chitosan depolymerisation will be detrimental to its intended application will depend on the functional significance of the changes that occur. Depolymerisation of chitosan in both the polymeric and nanoparticle form is temperature dependent (Nguyen, *et al.*, 2007; Morris, *et al.*, 2009b; Morris, *et al.*, 2010a). For example it has been reported that low molar mass chitosans can cause more cell damage (Aspden, *et al.*, 1996), although they may also prevent diabetes mellitus progression in mice to a greater extent than high molar chitosans (Kondo, *et al.*, 2000), show greater antibacterial activity compared with high molar mass chitosans (Lui, *et al.*, 2001) and whilst the high viscosities of high molar mass chitosans limit its biological usefulness, low molar mass chitosan is more soluble at neutral pH and therefore potentially more available *in vivo* (Harish Prashanth and Tharanathan, 2007). However, it has also been reported that high molar mass chitosans show greater antibacterial activity compared with low molar mass chitosans (No, *et al.*, 2006), that nasal insulin delivery (Aspden, *et al.*, 1997; Davis and Illum, 2000) is more effective with chitosan of molar mass greater than 100000 g/mol and the reversibility of transepithelial chemical resistance (TEER) values decrease with decreased chitosan molar mass (Holme, *et al.*, 2000).

Pectin

CHEMICAL STRUCTURE

Pectins are a complex family of heteropolysaccharides that constitute a large proportion of the primary cell walls of dicotyledons and play important roles in growth, development and senescence (van Buren, 1991; Tombs and Harding, 1998; Ridley, *et. al.*, 2001; Willats, *et. al.*, 2001). Pectic polysaccharides are made of several structural elements the most important of which are the homogalacturonan (HG) and type I rhamnogalacturonan (RG-I) regions often described in simplified terms as the "smooth" and "hairy" regions respectively (*Figure 8*). The HG region is composed of (1→4) linked α-D-Gal*p*A residues that can be partially methylated at *C*-6 (Pilnik and Voragen, 1970) and possibly partially acetyl-esterified at *O*-2 and/ or *O*-3 (Rombouts and Thibault, 1986). The degree of methylation (DM) and the degree of acetylation (DAc) are defined as the number of moles of methanol or acetic acid per 100 moles of GalA. The degree of methylation in native pectins is generally in the order of DM ≈ 70-80; whereas degree of acetylation is generally much lower *e.g.* DAc ≈ 35 for sugar beet pectins (Rombouts and Thibault, 1986). Theoretically the degree of methoxyl esterification (DM) can range from 0-100 %. Pectins with a degree of esterification (DM) > 50% are known as high methoxyl (HM) pectins and consequently low methoxyl (LM) pectins have a DM < 50% (Walter, 1991). The RG-I region consists of disaccharide repeating unit $[\rightarrow 4)$-α-D-Gal*p*A-$(1\rightarrow 2)$-α-L-Rha*p*-$(1\rightarrow]_n$ with a variety side chains consisting of L-arbinosyl and D-galactosyl residues (Voragen, *et. al.*, 1995). It has been reported that GalA residues in the RG-I region are partially acetylated (Ishii, 1997; Perrone, *et. al.*, 2002) but not methylated (Komalavilas and Mort, 1989; Perrone, *et. al.*, 2002). In the case of sugar beet pectin the neutral side chain sugars are substituted with ferulic acid (Fry, 1982; Rombouts and Thibault, 1986) and there is evidence indicating that pectin chains can be dimerised via diferulic bridges (Levigne, *et. al.*, 2004a,b). There are a number of different ways in which ferulic acid can dimerise the most common being: 5-5'; 8-O-4'; 8-5' cyclic and 8-5' non-cyclic dimers (Micard, *et. al.*, 1997).

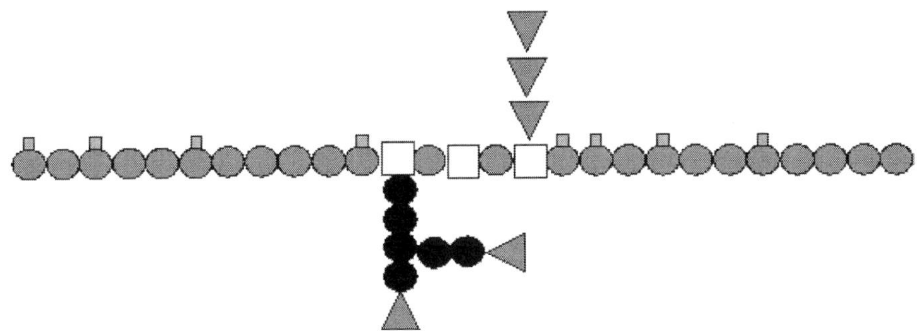

Figure 8. Schematic structure for pectin: galacturonic acid (●); galactose (●); arabinose (▽); rhamnose (□) and methyl groups (■).

PHYSICAL PROPERTIES

The degree of esterification and therefore the charge on a pectin molecule is important to the functional properties in the plant cell wall. It also significantly affects their commercial use as gelling and thickening agents (Lapasin and Pricl, 1995; Tombs and Harding, 1998). HM pectins (low charge) form gels at low pH (< 4.0) and in the presence of a high amount (> 55 %) of soluble solids, usually sucrose (Oakenfull, 1991). HM pectin gels are stabilised by hydrogen-bonding and hydrophobic interactions of individually weak but cumulatively strong junction zones (*Figure 9*) (Oakenfull, 1991; Lopes da Silva and Gonçalves 1994; Pilnik, 1990; Morris, 1979).

Conversely, LM pectins (high charge) form electrostatically stabilised gel networks with/ or without sugar and with divalent metal cations, usually calcium in the so-called "egg-box" model (*Figure 10*) (Morris, *et. al.*, 1982; Pilnik, 1990; Morris, 1980; Oakenfull and Scott, 1998; Axelos and Thibault, 1991), which also depends on the distribution of negative carboxylate groups and structure breaking rhamnose side chains (Powell, *et. al.*, 1982; Axelos and Thibault, 1991). A similar "egg-box" model has been proposed for alginate gels (Wang, *et. al.*, 1994; Morris, 1980) from the results of circular dichroism (CD), small angle X-ray scattering (SAXS) and X-ray fibre diffraction respectively, it is thought that in both pectin and alginate the "egg-box" is formed in a two-step process – dimerisation followed by aggregation of the preformed "egg-boxes" (Thibault and Rinaudo, 1986).

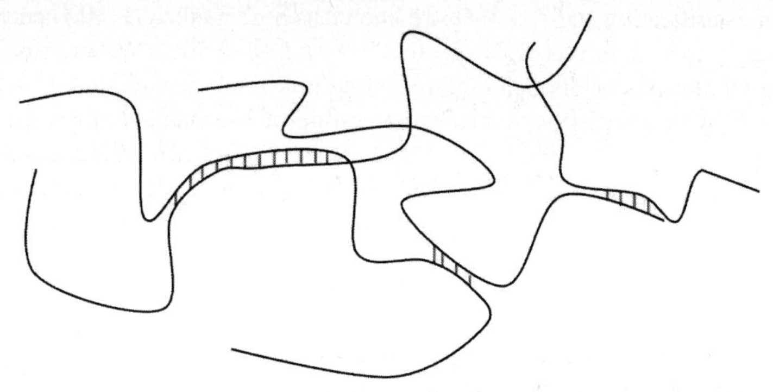

Figure 9. Representation of gelation mechanism in high methoxyl (HM) pectin gels, where the junction zones are indicated with shading.

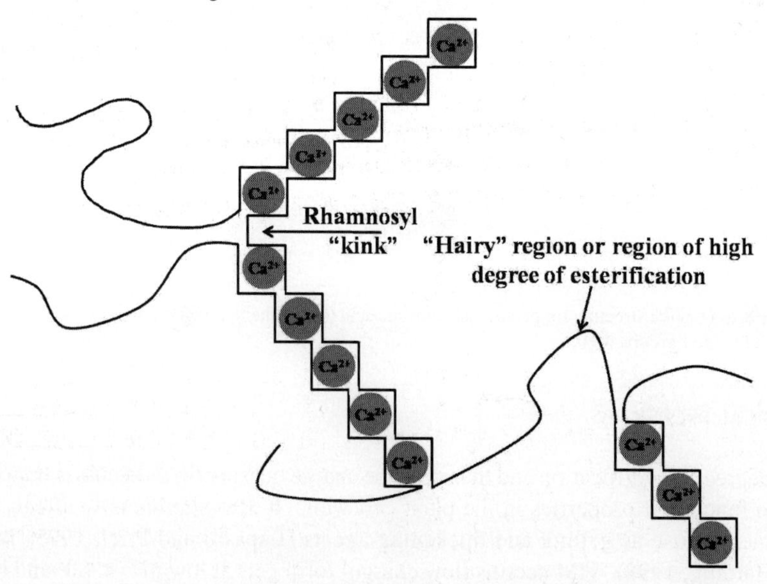

Figure 10. "Egg box" model for the gelation mechanism for low methoxyl (LM) pectin gels.

Solution properties such as viscosity also depend on degree of esterification, solvent environment (*i.e.* salt concentration, sugar concentration and pH) together with temperature (Oakenfull, 1991). Hydrodynamic studies based on intrinsic viscosity ([η]), sedimentation coefficient ($s^0_{20,w}$), radius of gyration (r_g) and weight average molecular weight (M_w) have focussed on qualitative/ semi-quantitative methods of estimating the conformation based around "power law" Mark-Houwink-Kuhn-Sakurada relations (Tombs and Harding, 1998) which link intrinsic viscosity, sedimentation coefficient and radius of gyration with molar mass $[\eta] \propto M^a$, $s^0_{20,w} \propto M^b$ and $r_g \propto M^c$, where a, b and c have defined values for specific conformation types (*Table 1*). The translational frictional ratio, f/f_o (Tanford, 1961), sedimentation conformation zoning (Pavlov, *et al.*, 1997; 1999), the "Wales-van Holde" ratio, $k_s/[\eta]$ (Wales and van Holde, 1954) and the persistence length, L_p (Kratky and Porod, 1949) have also been used to estimate dilute solution conformation (*Table 2*). A picture of a semi-flexible conformation for pectins irrespective of degree of esterification (and charge) has emerged from these studies (Anger and Berth, 1985; Axelos, *et al.*, 1987; Axelos and Thibault, 1991, Berth, *et al.*, 1977; Harding, *et al.*, 1991; Garnier, *et al.*, 1993; Malovikova, *et al.*, 1993; Cros, *et al.*, 1996; Tombs and Harding, 1998; Braccini, *et al.*, 1999; Morris, *et al.*, 2000, 2002, 2008; Fishman, *et al.*, 2001, 2006; Noto, *et al*, 2005). Pectin molecular weight and chain flexibility is important in mucoadhesive interactions (Nafee, *et al.*, 2007).

USAGE IN DRUG DELIVERY

Pectins have been used is a gelling agent for a large number of years, however there has been recent interest in the use of pectin gels in controlled drug delivery (Sungthongjeen, *et al.*, 2004; Lui, *et al.*, 2003; Lui, *et al.*, 2006). This is in part due their long standing reputation of being non-toxic (GRAS – generally regarded as safe) (Lui, *et al.*, 2003; Lui, *et al.*, 2007; Watts and Smith, 2009), their relatively low production costs (Sungthongjeen, *et al.*, 2004) and high availability (Beneke *et al.*, 2009). It is proposed that pectin could be used to deliver drugs orally, nasally and vaginally (*Figure 11*) (Peppas, et al., 2000; Sinha and Kumria, 2001; Lui, *et al.*, 2003; Nafee, et al., 2004; Valenta, 2005; Lui, *et al.*, 2007; Chelladurai, *et al.*, 2008; Thirawong, *et al.*, 2008), which are generally well accepted by patients (Lui, *et al.*, 2003; Lui, *et al.*, 2007; Yadav, *et al.*, 2009).

ORAL DELIVERY

The oral route is of particular interest as in general oral drug administration results in less pain, greater convenience, higher compliance and reduced infection risk as compared to subcutaneous injections (Chen and Langer, 1998, Yadav, et al., 2009). However, there are disadvantages associated with this route of administration such as low bioavailability due to relatively low passage of active agents across the mucosal epithelium, rapid polypeptide degradation due to action of digestive enzymes in the GI tract, enzymatic proteolysis and acidic degradation of orally administered drugs in the stomach (Lui, *et al.*, 2003; Lin, *et al.*, 2007). Various approaches have been made to increase the buccal penetration using permeation enhancers (Mesiha, *et al.*, 1994; Carino, *et al.*, 2000), protease inhibitors (Yamamoto, *et al.*, 1994), enteric coatings (Morishita, *et. al.*, 1993) and (bio)polymer micro-/ nano-sphere formulations

(Sarmento, *et al.*, 2007; Jain, *et al.*, 2005). However, protein drugs are essentially free

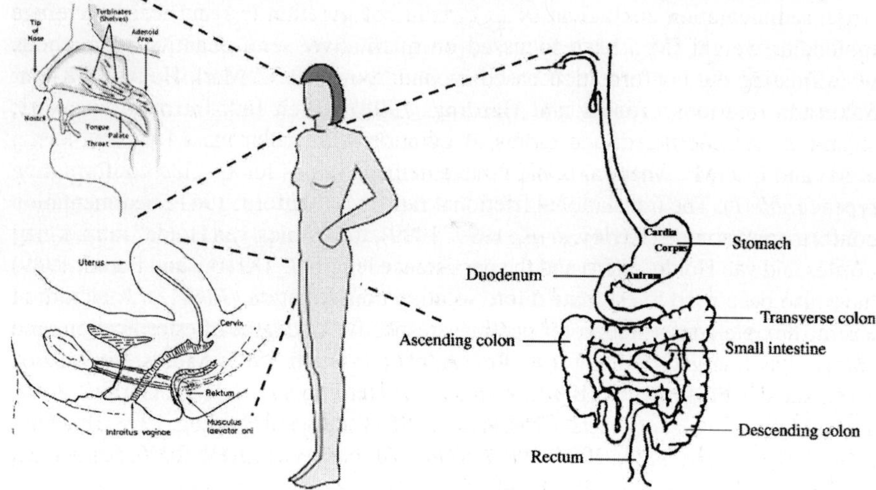

Figure 11. Major sites for pectin drug delivery systems (adapted from Peppas, *et al.*, 2000; Harding, 2003; Lui, *et al.*, 2003; Valenta, 2005) with insets showing the nasal cavity, vagina and GI tract. Reproduced with the permission of Elsevier.

from enzymatic proteolysis and acidic degradation in the colon which has resulted in a concentrated effort to target their delivery to this organ (Sinha and Kumria, 2001; Lui, *et al.*, 2003; Chambin, *et al.*, 2006). Therefore a number of different polymers including pectin have been identified as protective agents against enzymatic proteolysis (Lui, *et al.*, 2003; Sriamornsak, 2003; Pourjavadi and Barzegar, 2009). The pectin-stabilised polypeptide drug therefore mains intact in the stomach and small intestine prior to pectin digestion by the colonic microflora resulting in the release of drug molecule (Sinha and Kumria, 2001; Vandamme, *et al.*, 2002). The susceptibility of pectin to enzymatic attack is increased in the presence of calcium ions (Miler and MacMilan, 1970) and decreased by methyl esterification (Ashford, *et al.*, 1993). One problem with pectin formulations is that they can swell under physiological conditions which may result in premature drug release (Semdé, *et al.*, 2000; Lui, *et al.*, 2006), the effect can be minimised by the use of pectin in combination with other polymers: cellulosic or acrylic polymers (Semdé, *et al.*, 2000); chitosan (Macleod, *et al.*, 1999); hydroxypropylmethyl cellulose (Ofori-Kwakye and Fell, 2001) and zein (Lui, *et al.*, 2005; 2007).

NASAL DELIVERY

However, the clearance time through the whole alimentary tract is generally too short (4–12 h), rendering oral drug administration a very inefficient process, with much of the drug unabsorbed. More recently interest has focused on drug absorption through nasal epithelia, which results in very rapid absorption ~ 15 minutes (Jabbal-Gill, *et al.*, 1998). Other important issues are the degradation of peptide-based drugs in the gastrointestinal tract and low trans-mucosal permeability. Macromolecular based carrier and mucoadhesive systems have been considered for several years and pectin based systems have emerged as particularly promising. Low methoxyl pectins are strongly polyanionic polyuronides from fruit used traditionally in jams and jellies (Rolin, 1993) and can also form weak gels in the presence of Ca^{2+} ions (8 meq/L), which occur naturally in nasal secretions (Chang and Su, 1989; Illum, 2000), and their texture makes them patient friendly in nasal delivery formulations (Dale, *et al.*, 2002; Yadav, *et al.*, 2009). These gels are pseudoplastic (Sriamornsak, 2004; Thirawong, *et al.*, 2008; Chelladurai, *et al.*, 2008) and drug release is diffusion controlled at low pectin concentrations (Lui, *et al.*, 2007; Chelladurai, *et al.*, 2008) and determined by gel dissolution at higher pectin concentration (Lui, *et al.*, 2007). In addition, this may hold an incorporated drug substance in the nasal cavity for a prolonged period and thereby modulate its rate of systemic absorption. Pectins do not act as an absorption enhancer, however they cause tight junctions to open and therefore alter drug release characteristics due to the chelation of calcium (Charlton, *et al.*, 2007; McConaughy, *et al.*, 2009). They are also highly mucoadhesive (Nafee, *et al.*, 2004; Lui, *et al.*, 2007; Thirawong, *et al.*, 2008), although less mucoadhesive than chitosan (Nafee, *et al.*, 2004). Their mucoadhesive power depends on molecular weight, viscosity, the local pH and pectin functional groups (Lui, *et al.*, 2007; Thirawong, *et al.*, 2008).

Nasal drug delivery is limited by the small sample volume that can be delivered ~ 150 µl, which is important in drug formulations especially if the drug is sparingly soluble or if a drug has to be delivered over prolonged period (Lui, *et al.*, 2007).

VAGINAL DELIVERY

Like the nose the vagina is another potential site for drug delivery due to its rich blood supply, large surface area (Vermani and Garg, 2000) and well understood microflora (Valenta, 2005). Drug delivery release rates may vary during the menstrual cycle and this is especially important at the menopause (Valenta, 2005). Drug delivery systems are based on mucoadhesion (Harding, *et al.*, 1999; Harding, 2003; 2006). The vaginal route has been demonstrated to be favourable in the delivery of many drugs *e.g.* propranolol, human growth hormone, etc. (see Valenta, 2005 and references therein). Furthermore it might be expected the vaginal delivery of hormonal contraception may be more efficient than the oral route (Valenta, 2005). The vaginal route offers many of the advantages of the nasal route with main disadvantage being it is only available to females. Pectin-based formulations have demonstrated highest mucoadhesive strength, highest swelling volume and lowest pH reduction in a trial (Balo lu, *et al.*, 2003; 2006).

STABILITY

However, as yet pectin has not fulfilled its potential as drug delivery system this is due variability in pectin formulations and question marks over formulation stability (Lui, *et al.*, 2003; Lui, *et al.*, 2006). According to Morris, *et al.* (2010b) the viscosity of pectin solutions decrease significantly after 6 months storage at 25 °C and 40 °C respectively, and this is reflected by a decrease in gel strength upon addition of calcium ions. This is explained by a depolymerisation of pectin over time (*Figure 12*). However, it has been shown that decreases in viscosity of this magnitude do not significantly change the drug release rates from pectin gels *in vitro* (Nessa, 2003; Chelladurai, *et al.*, 2008). In calcium pectate based tablet formulations drug release time is increased with lower degree of methyl esterification, but higher levels of calcium ions can lead to disintegration of the tablet and increased drug release (Sungthongjeen, *et al.*, 2004).

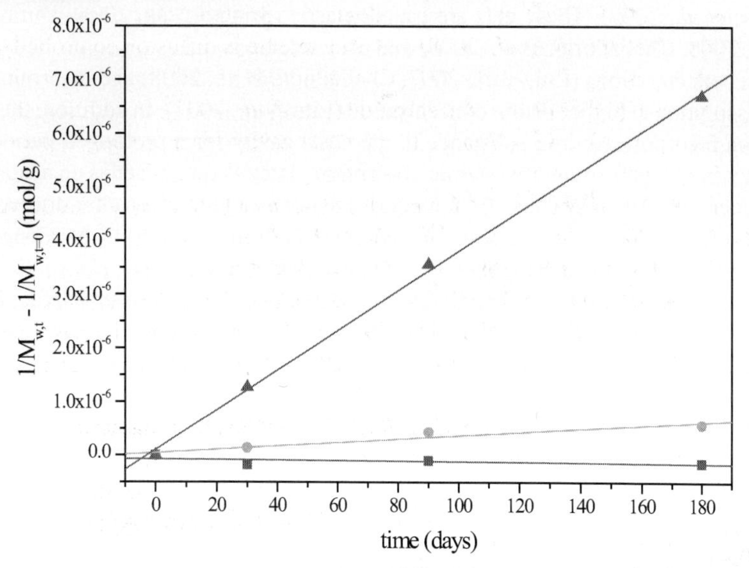

Figure 12. 1st order kinetic plots of (mol/g) vs. time (days) for pectin of DM ~ 19 %, where closed symbols represent molar masses estimated from viscometry at 4 °C (■), 25 °C (●) and 40 °C (▲) (adapted from **Figure 3** in Morris, *et al.* (2010b)). The kinetic rate constants (day^{-1}) are $(-0.8 \pm 1.1) \times 10^{-7}$, $(5.7 \pm 1.1) \times 10^{-7}$ and $(6.7 \pm 0.2) \times 10^{-6}$ at 4 °C, 25 °C and 40 °C, respectively.

Conclusions

In the last decade there has been a great deal of interest in the use of polysaccharides and particularly chitosan and pectin in drug delivery systems. It is clear that both the polysaccharides either individually or together show great potential, however, many important issues still remain to be resolved fully. With chitosans, these include (i) their stability, with the important constraint that they are soluble only at $pH < 6$. (ii) their construction into microparticles capable of surviving the large environmental variation between mouth and intestine for oral drug delivery. In addressing (i) and (ii) issues concerning the optimal degree of acetylation and molecular weight of the chitosan need to be addressed. With low methoxy-pectin systems issues include: (i) optimal molecular weight and degree of esterification (ii) drug diffusivity (iii) interactions with mucosal tissues, (iv) stability (molecular weight/viscosity/gelation).

References

AMIDI, M. AND HENNINK, W. E. (2010). Preface: Chitosan-based formulations of drugs, imaging agents and biotherapeutics. *Advanced Drug Delivery Reviews*, **62**, 1–2.

AMIDI, M., MASTROBATTISTA, E., JISKOOT, W. AND HENNINK, W. E. (2010). ChitosaN based delivery systems for protein therapeutics and antigens. *Advanced Drug Delivery Reviews*, **62**, 59-82.

ANDERSON, M. T., HARDING, S. E. AND DAVIS, S. S. (1989). On the interaction in solution of a candidate mucoadhesive polymer, diethylaminoethyl-dextran, with pig gastric mucus glycoprotein. *Biochemical Society Transactions*, **17**, 1101–1102.

ANGER, H. AND BERTH, G. (1985). Gel-permeation chromatography of sunflower pectin. *Carbohydrate Polymers*, **5**, 241-250.

ASHFORD, M., FELL, J. J., ATTWOOD, D., SHARMA, H. L. AND WOODHEAD, P. J. (1993). An evaluation of pectin as a carrier for drug targeting to the colon. *Journal of Controlled Release*, **26**, 213–220.

ASPDEN, T. J., ILLUM, L. AND SKAUGRUD, Ø. (1996). Chitosan as a nasal delivery system: evaluation of insulin absorption enhancement and effect on nasal membrane integrity using rat models. *European Journal of Pharmaceutical Sciences*, **4**, 23-31.

ASPDEN, T. J., MASON, J. D. T., JONES, N. S., LOWE, J., SKAUGRUD, Ø. AND ILLUM, I. (1997). Chitosan as a nasal delivery system: the effect of chitosan solutions on in vitro and in vivo mucociliary transport rates in human turbinates and volunteers. *Journal of Pharmaceutical Sciences*, **86**, 509-513.

AXELOS, M. A. V. LEFEBVRE, J. AND THIBAULT, J-F. (1987). Conformation of a low methoxyl citrus pectin in aqueous solution, *Food Hydrocolloids*, **1**, 569-570.

AXELOS, M. A. V. AND THIBAULT, J-F. (1991). The chemistry of low-methoxyl pectin gelation. In: Walter R. H. (Ed.) *The Chemistry and Technology of Pectin* (pp 109-118). San Diego: Academic Press.

BALOĜLU, E., ÖZYAZICI, M., HIZARCIOĜLU, S. Y. AND KARAVANA. H. A. (2003). An *in vitro* investigation for vaginal bioadhesive formulations: bioadhesive properties and swelling states of polymer mixtures. *Il Farmaco*, **58**, 391-396.

BALOĜLU, E., ÖZYAZICI, M., HIZARCIOĜLU, S. Y., SENYIĜIT, T., ÖZYURT, D. AND PEKÇETIN, C. (2006). Bioadhesive controlled release systems of ornidazole for vaginal delivery. *Pharmaceutical Development and Technology*, **11**, 477-484.

BENEKE, C. M., VILJOEN, A. M. AND HAMMAN, J. H. (2009). Polymeric plant-derived excipients in drug delivery. *Molecules*, **14**, 2602-2620.

BERTH, G., ANGER, H. AND LINOW, F. (1977). Scattered-light photometric and viscosimetric studies of molecular mass determination of pectins in aqueous-solutions. *Nahrung-Food*, **31**, 939-950.

BERTH, G., DAUTZENBERG, H. AND PETER, M. G. (1998). Physico-chemical characterization of chitosans varying in degree of acetylation. *Carbohydrate Polymers*, **36**, 205-216.

BHATTARAI, N., GUNN, J. AND ZHANG, M. (2010). Chitosan-based hydrogels for controlled, localized drug delivery. *Advanced Drug Delivery Reviews*, **62**, 83-99.

BRACCINI, I., GRASSO, R. P. AND PEREZ, S. (1999). Conformational and configurational features of acidic polysaccharides and their interactions with calcium ions: a molecular modeling investigation *Carbohydrate Research*, **317**, 199-130.

BRUGNEROTTO J., DESBRIÈRES J., ROBERTS G. AND RINAUDO M. (2001). Characterization of chitosan by steric exclusion chromatography. *Polymer*, **42**, 9921–9927.

CHAMBIN, O., DUPUIS, G, CHAMPION, D., VOILLEY, A. AND POURCELOT, Y. (2006). Colon-specific drug delivery: Influence of solution reticulation properties upon pectin beads performance. *International Journal of Pharmaceutics*, **321**, 86-93.

Carino, G. P., Jacob, J. S. and Mathiowitz, E. (2000). Nanosphere based oral insulin delivery. *Journal of Controlled Release*, **65**, 261–269.

Charlton, S. T., Davis, S. S. and Illum, L. (2007). Evaluation of bioadhesive polymers as delivery systems for nose to brain delivery: *In vitro* characterisation studies. *Journal of Controlled Release*, **118**, 225–234.

Chelladurai, S., Mishra, M. and Mishra, B. (2008). Design and Evaluation of Bioadhesive in-Situ Nasal Gel of Ketorolac Tromethamine. *Chemical and Pharmaceutical Bulletin*, **56**, 1596-1599.

Chen, H. and Langer, L. (1998). Oral particulate delivery: status and future trends. *Advanced Drug Delivery Reviews*, **34**, 339–350.

Chien, Y. W., Su, K. S. E. and Chang, S. F. (1989). *Nasal Systemic Drug Delivery* (pp 1-26). New York: Marcel Dekker Inc.

Cros, S. C., Garnier, C., Axelos, M. A. V., Imbery, A. and Perez, S. (1996). Solution conformations of pectin polysaccharides: determination of chain characteristics by small angle neutron scattering, viscometry and molecular modeling. *Biopolymers*, **39**, 339-352.

Cölfen H., Berth, G. and Dautzenberg, H. (2001). Hydrodynamic studies on chitosans in aqueous solution. *Carbohydrate Polymers*, **45**, 373-383.

Dale, O., Hjortkjaer, R. and Kharasch, E. D. (2002). Nasal administration of opioids for pain management in adults. *Acta Anaesthesiologica Scandinavica*, **46**, 759 –770.

Davis, S. S. (1989). Small intestine transit. In: Hardy, J. G., Davis, S. S. and Wilson, C. G. (Eds.) *Drug Delivery to the Gastrointestinal Tract* (pp. 49–61). Chichester: Ellis Horwood.

Davis, S. S. and Illum, L. (2000). Chitosan for oral delivery of drugs. In: Muzzarelli, R. A. A. (Ed.). *Chitosan per os: from Dietary Supplement to Drug Carrier* (pp 137-164). Grottammare: Atec.

Deacon, M. P., Davis, S. S., White, R. J., Nordman, H., Carlstedt, I., N. Errington, N., Rowe, A. J. and Harding, S. E. (1999). Are chitosan–mucin interactions specific to different regions of the stomach? Velocity ultracentrifugation offers a clue. *Carbohydrate Polymers*, **38**, 235–238.

Deacon, M. P., McGurk, S., Roberts, C. J., Williams, P. M., Tendler, S. J. B., Davies, M. C., Davis, S. S. and Harding, S. E. (2000). Atomic force microscopy of gastric mucin and chitosan mucoadhesive systems, *Biochemical Journal*, **348**, 557–563.

de la Fuente, M., Raviña, M., Paolicelli, P., Sanchez, A., Seijo, B. and Alonso, M. J. (2010). Chitosan-based nanostructures: A delivery platform for ocular therapeutics. *Advanced Drug Delivery Reviews*, **62**, 100-117.

Dhawan, S., Singla, A. K. and Sinha, V. R. (2004). Evaluation of mucoadhesive properties of chitosan microspheres prepared by different methods. *AAPS PharmSciTech*, **5**, 1-7.

Dodou, D., Breedveld, P. and Wieringa, P. A. (2005). Mucoadhesives in the gastrointestinal tract: revisiting the literature for novel applications. *European Journal of Pharmaceutics and Biopharmaceutics*, **60**, 1–16.

Dyer, A. M., Hinchcliffe, M., Watts, P. Castile, J., Jabbal-Gill, Nankervis, I. R., Smith, A. and Illum, L. (2002). Nasal delivery of insulin using novel chitosan based formulations: a comparative study in two animal models between simple chitosan formulations and chitosan nanoparticles. *Pharmaceutical Research*, **19**, 998-1008.

Errington, N., Harding, S. E., Vårum, K. M. and Illum, L. (1993). Hydrodynamic characterisation of chitosans varying in degree of acetylation. *International Journal of Biological Macromolecules*, **15**, 113-117.

Fee, M., Errington, N., Jumel, K., Illum, L, Smith, A. and Harding, S. E. (2003). Correlation of SEC/MALLS with ultracentrifuge and viscometric data for chitosans. *European Biophysical Journal*, **32**, 457-464.

Fernández-Urrasuno, R., Calvo, P., Rumuñán-Lopez, C., Vila-Jato, J. L. and Alonso, M. J. (1999). Enhancement of nasal absorption of insulin using chitosan nanoparticles. *Pharmaceutical Research*, **16**, 1576-1581.

Fishman, M. L., Chau, H. K., Hoagland, P. D., and Hotchkiss, A. T. (2006). Microwave-assisted extraction of lime pectin. *Food Hydrocolloids*, **20**, 1170-1177.

Fishman, M. L., Chau, H. K., Kolpak, F. and Brady, J. (2001). Solvent effects on the molecular properties of pectins. *Journal of Agricultural and Food Chemistry*, **49**, 4494-4501.

Fiebrig, I., Davis, S. S. and Harding, S. E. (1995a). Methods used to develop mucoadhesive drug delivery systems: bioadhesion in the gastrointestinal tract. In: Harding, S. E., Hill, S. E. and Mitchell, J. R. (Eds.) *Biopolymer Mixtures* (pp 373–419). Nottingham: Nottingham University Press.

Fiebrig, I., Harding, S. E., & Davis, S. S. (1994a). Sedimentation analysis of the potential interactions between mucins and putative bioadhesive polymer. *Progress in Colloid and Polymer Science*, **93**, 66–73.

Fiebrig, I., Harding, S. E., Rowe, A. J., Hyman, S. C. and Davis, S. S. (1995b). Transmission electron microscopy studies on pig gastric mucin and its interactions with chitosan. *Carbohydrate Polymers*, **28**, 239–244.

Fiebrig, I., Harding, S. E., Stokke, B. T., Vårum, K. M., Jordan, D. and Davis, S. S. (1994b). The potential for chitosan as mucoadhesive drug carrier: studies on its interaction with pig gastric mucin on a molecular level. *European Journal of Pharmaceutical Sciences*, **2**, 185.

Fry, S. C. (1982). Phenolic compounds of the primary cell wall: feruloylated disaccharides of D-galactose and L-arabinose from spinach polysaccharide. *Biochemical Journal*, **203**, 493-502.

Gan, Q, and Wang, T. (2007). Chitosan nanoparticle as protein delivery carrier Systematic examination of fabrication conditions for efficient loading and release. *Colloids and Surfaces B: Biointerfaces*, **59**, 24–34.

Gan, Q, Wang, T., Cochrane, C. and McCarron, P. (2005). Modulation of surface charge, particle size and morphological properties of chitosan–TPP nanoparticles intended for gene delivery. *Colloids and Surfaces B: Biointerfaces*, **44**, 65–73.

Garnier, C., Axelos M. A. V. and Thibault, J-F. (1993). Phase-diagrams of pectin-calcium systems - influence of pH, ionic-strength, and temperature on the gelation of pectins with different degrees of methylation. *Carbohydrate Research*, **240**, 219-232.

Genta, I., Costantini, M., Asti, A., Conti, B. and Montanari, L. (1998). Influence of glutaraldehyde on drug release and mucoadhesive properties of chitosan microspheres. *Carbohydrate Polymers*, **36**, 81-88.

Harding, S. E. (2003). Mucoadhesive interactions. *Biochemical Society Transactions*, **31**, 1036-1041.

Harding, S. E. (2006). Trends in mucoadhesive analysis. *Trends in Food Science and*

Technology, **17**, 255-262.

HARDING, S. E., BERTH, G., BALL, A., MITCHELL, J. R. AND GARCÌA DE LA TORRE, J. (1991). The molecular weight distribution and conformation of citrus pectins in solution studied by hydrodynamics. *Carbohydrate Polymers*, **16**, 1-15.

HARDING, S. E., DAVIS, S. S., DEACON, M. P. AND FIEBRIG, I. (1999). *Biopolymer mucoadhensives*. In: Harding, S. E. (Ed.) *Biotechnology and Genetic Engineering Reviews* Vol. 16 (pp. 41-86). Andover: Intercept.

HARDING, S. E., ROWE, A. J. AND CREETH, J. M. (1983). Further evidence for a flexible and highly expanded spheroidal model for mucus glycoproteins in solution, *Biochemical Journal*, **209**, 893-896.

HARISH PRASHANTH, K. V. AND THARANATHAN R. N. (2007). Chitin/ chitosan: modifications and their unlimited potential – an overview. *Trends in Food Science and Technology*, **18**, 117-131.

HE, P., DAVIS, S. S. AND ILLUM, L. (1998). In-vitro evaluation of the properties of chitosan microspheres *International Journal of Pharmaceutics,* **166**, 75–68.

HE, P., DAVIS, S.S. AND ILLUM, L. (1999). Chitosan microspheres prepared by novel modified spray drying methods. *Journal of Microencapsulation*, **16**, 343-355.

HOLME, H. K., DAVIDSEN, L., KRISTIANSEN, A. AND SMIDSRØD, O. (2008). Kinetics and mechanisms of the depolymerization of alginate and chitosan in aqueous solution. *Carbohydrate Polymers*, **73**, 656-664.

ILLUM, L. (1998). Chitosan and its use as a pharmaceutical excipient. *Pharmaceutical Research*, **15**, 1326-1331.

ILLUM, L. (2000). Improved delivery of drugs to mucosal services. European patent EP0975367B1.

ILLUM, L. (2002) Nasal drug delivery: new. developments and strategies. *Drug Discovery Today*, **7**, 1184–1189.

ISHII, T. (1997). O-acetylated oligosaccharides from pectins of potato tuber cell walls. *Plant Physiology*, **113**, 1265-1272.

JABBAL-GILL, I., FISCHER, A. N., RAPPUOLI, R., DAVIS, S. S. AND ILLUM, L. (1998). Stimulation of mucosal and systemic responses against *Bordetella pertussis* filamentous haemagglutinin and recombinant pertussis toxin after nasal administration with chitosan in mice. *Vaccine*, **16**, 2039 –2046.

JAIN, D., PANDA, A. K. AND MAJUMDAR, D. K. (2005). Eudragit S100 Entrapped Insulin Microspheres for Oral Delivery. *AAPS PharmSciTech*, **6**, E100-E107.

JANES, K. A., CALVO, P. AND ALONSO, M. J. (2001). Polysaccharide colloidal particles as delivery systems for macromolecules. *Advanced Drug Delivery Reviews*, **47**, 83–97.

KASAAI, M. R. (2006). Calculation of Mark–Houwink–Sakurada (MHS) equation viscometric constants for chitosan in any solvent–temperature system using experimental reported viscometric constants data. *Carbohydrate Polymers*, **68**, 477-488.

KEAN, T. AND THANOU, M. (2010). Biodegradation, biodistribution and toxicity of chitosan. *Advanced Drug Delivery Reviews*, **62**, 3-11.

KIM, S. J., SHIN, S. R., LEE, K. B., PARK, Y. D. AND KIM, S. I. (2004). Synthesis and characteristics of polyelectrolyte complexes composed of chitosan and hyaluronic acid. *Journal of Applied Polymer Science*, **91**, 2908–2913.

KOMALAVILAS, P. AND MORT, A. J. (1989). The acetylation of O-3 of galacturonic acid in the rhamnose-rich portion of pectin. *Carbohydrate Research*, **189**, 261-272.

KONDO, Y., NAKATANI, A., HAYASHI, K. AND ITO, M. (2000). Low molecular weight chitosan prevents the progression of low dose streptozotocin induced slowly progressive diabetes mellitus in mice. *Biological and Pharmacological Bulletin*, **23**, 1458-1464.

KRATKY, O. AND POROD, G. (1949). Röntgenungtersuchung gelöster fadenmoleküle. *Recueil Des Travaux Chimiques Des Pays-Bas*, **68**, 1106-1109.

KUBOTA, N. AND EGUCHI, Y. (1997). Facile preparation of water-soluble n-acetylated chitosan and molecular weight dependence of its water-solubility. *Polymer Journal*, **29**, 123-127.

KUMARI, A., YADAV, S. K. AND YADAV, S. C. (2010). Biodegradable polymeric nanoparticles based drug delivery systems. *Colloids and Surfaces B: Biointerfaces*, **75**, 1-18.

KUJAWA, P., SCHMAUCH, G., VIITALA, T., BADIA, A. AND WINNIK, F. M. (2007). Construction of viscoelastic biocompatible films via the layer-by-layer assembly of hyaluronan and phosphorylcholine-modified chitosan. *Biomacromolecules*, **8**, 3169-3176.

LAMARQUE, G., LUCAS, J-M., VITON, C. AND DOMARD, A. (2005). Physicochemical behavior of homogeneous series of acetylated chitosans in aqueous solution: role of various structural parameters. *Biomacromolecules*, **6**, 131-142.

LAPASIN, R. AND PRICL. S. (1995). *Rheology of Industrial Polysaccharides, Theory and Applications*. Blackie, London, UK.

LEHOUX, J-G. AND DEPUIS, G. (2007). Recovery of chitosan from aqueous acidic solutions by salting-out: Part 1. Use of inorganic salts. *Carbohydrate Polymers*, **68**, 295-304.

LEVIGNE, S., RALET, M-C., QUÉMÉNER, B. AND THIBAULT, J-F. (2004a). Isolation of diferulic bridges ester-linked to arabinan in sugar beet cell walls. *Carbohydrate Research*, **339**, 2315-2319.

LEVIGNE, S., RALET, M-C., QUÉMÉNER, B. C., POLLET, B. N-L., LAPIERRE, C. AND THIBAULT, J-F. (2004b). Isolation from sugar beet cell walls of arabinan oligosaccharides esterified by two ferulic acid monomers. *Plant Physiology*, **134**, 1173-1180.

LIM, S. T., MARTIN, G. P., BERRY, D. J. AND BROWN, M. B. (2007). Preparation and evaluation of the in vitro drug release properties and mucoadhesion of novel microspheres of hyaluronic acid and chitosan. *Journal of Controlled Release*, **66**, 281–292.

LIN, Y. H., CHEN, C. H., LIANG, H. F., KULKARNI, A. R., LEE, P. W., CHEN, C. H. AND SUNG, H. W. (2007). Novel nanoparticles for oral insulin delivery via the paracellular pathway. *Nanotechnology*, **18**, 1-11.

LOPES DA SILVA, J. A. AND GONÇALVES, M. P. (1994). Rheological study into the ageing process of high methoxyl pectin/ sucrose gels. *Carbohydrate Polymers*, **24**, 235-245.

LUANGTANA-ANAN, M., OPANASOPIT, P., NGAWHIRUNPAT, T., NUNTHANID, J., SRIAMORNSAK, P., LIMMATVAPIRAT, S. AND LIM, L. Y. (2005). Effect of chitosan salts and molecular weight on a nanoparticulate carrier for therapeutic protein. *Pharmaceutical Development and Technology*, **10**, 189–196.

LUI, L., FISHMAN, M. L. AND HICKS, K. B. (2007). Pectin in controlled drug delivery – a review. *Cellulose*, **14**, 15-24.

LUI, L., FISHMAN, M. L., HICKS, K. B. AND KENDE, M. (2005). Interaction of various pectin formulations with porcine colonic tissues. *Biomaterials*, **26**, 5907–5916.

Lui, L., Fishman, M. L., Kost, J. and Hicks, K. B. (2003). Pectin-based systems for colon-specific drug delivery via oral route. *Biomaterials*, **24**, 3333-3343.

Liu, X. F., Guan, Y. L., Yang, D. Z., Li, Z. and De Yao, K. (2001). Antibacterial action of chitosan and carboxymethylated chitosan. *Journal of Applied Polymer Science*, **79**, 1324-1335.

Macleod, G. S., Collett, J. H. and Fell, J. T. (1999). The potential use of mixed films of pectin, chitosan and HPMC for bimodal drug release. *Journal of Controlled Release*, **58**, 303–310.

Malovikova, A., Rinaudo, M. and Milas, M. (1993). On the characterization of polygalacturonate salts in dilute-solution. *Carbohydrate Polymers*, **22**, 87-92.

Mao, S., Sun, W. and Kissel, T. (2010). Chitosan-based formulations for delivery of DNA and siRNA. *Advanced Drug Delivery Reviews*, **62**, 12-27.

Mazeau K. and Rinaudo M. (2004). The prediction of the characteristics of some polysaccharides from molecular modelling. Comparison with effective behaviour. *Food Hydrocolloids*, **18**, 885–898.

McConaughy, S. D., Kirkland, S. E., Treat, N. J. Stroud, P. A. and McCormick, C. L. (2009). Tailoring the network properties of ca2+ crosslinked *aloe vera* polysaccharide hydrogels for in situ release of therapeutic agents. *Biomacromolecules*, **9**, 3277-3287.

Mesiha, M., Plakogiannis, F. and Vejosoth, S. (1994). Enhanced oral absorption of insulin from desolvated fatty acid-sodium glycocholate emulsions. *International Journal of Pharmaceutics*, **111**, 213–216.

Micard, V., Grabber, J. H., Ralph, J., Renard, C. M. C. G. and Thibault, J-F. (1997). Dehydrodiferulic acids from sugar beet pulp. *Phytochemistry*, **44**, 1365-1368.

Miller, L. and MacMilan, J. D. (1970). Mode of action of pectic enzymes. 2. Further purification of exopolygalacturonate lyase and pectinesterases from *Clostridium multifermerans*. *Journal of Bacteriology*, **102**, 72–78.

Morishita, I., Morishita, M., Takayama, K., Machida, Y. and Nagai, T. (1993). Enteral insulin delivery by microspheres in 3 different formulations using Eudragit L100 and S100. *International Journal of Pharmaceutics*, **91**, 29-37.

Morris, E. R. (1980). Physical probes of polysaccharide conformation and interactions. *Food Chemistry*, **6**, 15-39.

Morris, E. R., Powell, D. A., Gidley, M. J. and Rees, D. A. (1982.) Conformation and interactions of pectins I. Polymorphism between gel and solid states of calcium polygalacturonate. *Journal of Molecular Biology*, **155**, 507-516.

Morris, G. A., Foster, T. J. and Harding, S. E. (2000). The effect of degree of esterification on the hydrodynamic properties of citrus pectin. *Food Hydrocolloids*, **14**, 227-235.

Morris, G. A., Foster, T. J. and Harding, S. E. (2002). A hydrodynamic study of the depolymerisation of a high methoxy pectin at elevated temperatures. *Carbohydrate Polymers*, **48**, 361-367.

Morris, G. A., García De La Torre, J., Ortega, A., Castile, J., Smith, A., and Harding, S. E. (2008). Molecular flexibility of citrus pectins by combined sedimentation and viscosity analysis. *Food Hydrocolloids*, **22**, 1435-1442.

Morris, G. A., Castile, J., Smith, A., Adams, G. G., and Harding, S. E. (2009a). Macromolecular conformation of chitosan in dilute solution: a new global hydrodynamic approach. *Carbohydrate Polymers*, **76**, 616-621.

MORRIS, G. A., CASTILE, J., SMITH, A., ADAMS, G. G., AND HARDING, S. E. (2009b). The kinetics of chitosan depolymerisation at different temperatures. *Polymer Degradation and Stability*, **94**, 1344-1348, 2009.

MORRIS, G. A., CASTILE, J., SMITH, A., ADAMS, G. G., AND HARDING, S. E. (2010a). The effect of different storage temperatures on the stability of tripolyphosphate (TPP) – chitosan nanoparticles (submitted).

MORRIS, G. A., CASTILE, J., SMITH, A., ADAMS, G. G. AND HARDING, S. E. (2010b). The effect of different storage temperatures on the physical properties of pectin solutions and gels (submitted).

MUZZARELLI, R. A. A. (2009). Genipin-crosslinked chitosan hydrogels as biomedical and pharmaceutical aids. *Carbohydrate Polymers*, **77**, 1-9.

NAFEE, N. A., ISMAIL, F. A., BORAIE, N. A. AND MORTADA, L. M. (2004). Mucoadhesive delivery systems. I. Evaluation of mucoadhesive polymers for buccal tablet formulation. *Drug Development and Industrial Pharmacy*, **30**, 985-993.

NESSA, M. U. (2003). Physiochemical characterisation of pectin solution as a vehicle for nasal drug delivery. MSc Dissertation, University of Nottingham, U.K.

NGUYEN, T. T. B., HEIN, S., NG, C-H. AND STEVENS, W. F. (2007). Molecular stability of chitosan in acid solutions stored at various conditions. *Journal of Applied Polymer Science*, **107**, 2588–2593.

NO, H. K., KIM, S. H., LEE, S. H., PARK, N. Y. AND PRINYAWIWATKUL, W. (2006). Stability and antibacterial activity of chitosan solutions affected by storage temperature and time. *Carbohydrate Polymers*, **65**, 174-178.

NOTO, R., MARTORANA, V., BULONE, D. AND SAN BIAGIO, B. L. (2005). Role of charges and solvent on the conformational properties of poly(galacturonic acid) chains: a molecular dynamics study. *Biomacromolecules*, **6**, 2555-2562.

OAKENFULL, D. G. (1991). The Chemistry of High-Methoxyl Pectins. In: Walter RH (Ed.) *The Chemistry and Technology of Pectin*, pp 87-108. Academic Press, San Diego.

OAKENFULL, D. G. AND SCOTT, A. (1998). Milk gels with low methoxy pectins. In: Phillips, G. O., Williams, P. A. and Wedlock, D. J. (Eds.) *Gums and Stabilisers for the Food Industry 9* (pp 212-221). Oxford: Oxford University Press.

OFORI-KWAKYE, K. AND FELL, J. T. (2001). Biphasic drug release: the permeability of films containing pectin, chitosan and HPMC. *International Journal of Pharmaceutics*, **226**, 139–145.

OGAWA, K. AND YUI, T. (1994). Effect of explosion on the crystalline polymorphism of chitin and chitosan. *Bioscience Biotechnology and Biochemistry*, **58**. 968-969.

PARK, J. H., SARAVANAKUMAR, G., KIM, K. AND KWON, I. C. (2010). Targeted delivery of low molecular drugs using chitosan and its derivatives. *Advanced Drug Delivery Reviews*, **62**, 28-41.

PARK, J. W., CHOI, K-H. AND PARK, K. K. (1983). Acid-base equilibria and related properties of chitosan. *Bulletin of the Korean Chemical Society*, **4**, 68-72.

PAVLOV, G. M., ROWE, A. J. AND HARDING, S. E. (1997). Conformation zoning of large molecules using the analytical ultracentrifuge. *Trends in Analytical Chemistry*, **16**, 401-405.

PAVLOV, G. M. HARDING, S. E. AND ROWE, A. J. (1999). Normalized scaling relations as a natural classification of linear macromolecules according to size. *Progress in Colloid and Polymer Science*, **113**, 76-80.

PEPPAS, N. A., BURES, P., LEOBANDUNG, W. AND ICHIKAWA, H. (2000). Hydrogels in pharmaceutical formulations. *European Journal of Pharmaceutics and Biopharmaceutics*, **50**, 27-46.

Perez, S., Rodríguez-Carvajal, M. A. and Doco, T. (2003). A complex plant cell wall polysaccharide: rhamnogalacturonan II. A structure in quest of a function. *Biochimie*, **85**, 109-121.

Perrone, P., Hewage, H. C., Thomson, H. R., Bailey, K., Sadler, I. H. and Fry, S. C. (2002). Patterns of methyl and O-acetyl esterification in spinach pectins: a new complexity. *Phytochemistry*, **60**, 67-77.

Pilnik, W. (1990). Pectin - a many splendored thing. In: Phillips, G. O., Williams, P. A. and Wedlock, D. J. (Eds.) *Gums and Stabilisers for the Food Industry 5* (pp. 209-222). Oxford: IRL Press.

Pilnik, W. and Voragen A. G. J. (1970). Pectin substances and their uronides. In: Hulme, A. C. (Ed.) *The Biochemistry of Fruits and their Products* (pp 53-87). New York: Academic Press.

Pourjavadi A. and Barzegar, S. (2009). Smart pectin-based superabsorbent hydrogel as a matrix for ibuprofen as an oral non-steroidal anti-inflammatory drug delivery. *Starch/ Stärke*, **61**, 173-187.

Powell, D. A., Morris, E. R., Gidley, M. J. and Rees, D. A. (1982). Conformation and interactions of pectins II. Influence of residue sequence on chain association in calcium pectate gels. *Journal of Molecular Biology*, **155**, 317-331.

Qaqish, R. B. and Amiji, M. M. (1999). Synthesis of a fluorescent chitosan derivative and its application for the study of chitosan-mucin interactions. *Carbohydrate Polymers*, **38**, 99-107.

Rhazi, M., Desbriéres J., Tolaimate, A., Rinaudo, M., Vottero, P. and Alagui, A. (2002a). Contribution to the study of the complexation of copper by chitosan and oligomers. *Polymer*, **43**, 1267–76.

Rhazi, M., Desbriéres J., Tolaimate, A., Rinaudo, M., Vottero, P., Alagui, A. and El Meray, M. (2002b). Influence of the nature of the metal ions on the complexation with chitosan.: Application to the treatment of liquid waste. *European Polymer Journal*, **38**, 1523-1530.

Ridley, B. L., O'Neil, M. A. and Mohnen, D. (2001). Pectins: structure, biosynthesis and oligogalacturonide-related signalling. *Phytochemistry*, **57**, 929-967.

Rinaudo, M. (2006). Chitin and chitosan: properties and applications. *Progress in Polymer Science*, **31**, 603-632.

Rolin, R. (1993). Pectin. In: Whistler, R. L. and BeMiller, J. N. (Eds.) *Industrial Gums* (pp 257-293). New York: Academic Press.

Rombouts, F. M. and Thibault, J-F. (1986). Sugar beet pectins: chemical structure and gelation through oxidative coupling. In: Fishman, M. L. and Jen, J. J. (Eds.) *The Chemistry and Function of Pectins* (pp 49-60). Washington DC: American Chemical Society.

Rossi, S., Ferrari, F., Bonferoni, M. C. and Caramella, C. (2000). Characterization of chitosan hydrochloride–mucin interaction by means of viscosimetric and turbidimetric measurements. *European Journal of Pharmaceutical Sciences*, **10**, 251–257.

Rossi, S., Ferrari, F., Bonferoni, M. C. and Caramella, C. (2001). Characterization of chitosan hydrochloride–mucin rheological interaction: influence of polymer concentration and polymer: mucin weight ratio. *European Journal of Pharmaceutical*

Sciences, **12**, 479–485.

Sarmento, B., Ribeiro, A., Veiga, F., Ferreira, D. and Neufeld, R. (2007). Oral bioavailability of insulin contained in polysaccharide nanoparticles. *Biomacromolecules*, **8**, 3054-3060

Schatz, S. Viton, C., Delair, T. Pichot, C. and Domard, A. (2003). Typical physicochemical behaviors of chitosan in aqueous solution. *Biomacromolecules*, **4**, 641-648.

Semdé, R., Amighi, K., Devleeschouwer, M. J. and Moës, A. J. (2000). Studies of pectin HM/Eudragit® RL/Eudragit® NE film-coating formulations intended for colonic drug delivery. *International Journal of Pharmaceutics*, **197**, 181–192.

Shu, X. Z., and Zhu, K. J. (2000). A novel approach to prepare tripolyphosphate: chitosan complex beads for controlled release drug delivery. *International Journal of Pharmaceutics*, **201**, 51–58.

Sinha, V. R. and Kumria, R. (2001). Polysaccharides in colon-specific drug delivery. *International Journal of Pharmaceutics*, **224**, 19–38.

Skaugrud, Ø., Hagen, A., Borgersen, B. and Dornish, M. (1999). Biomedical and pharmaceutical applications of alginate and chitosan. In: Harding, S. E. (Ed.) *Biotechnology and Genetic Engineering Reviews Vol. 16* (pp 23-40). Andover: Intercept.

Sogias, I. A., Williams, A. C. and Khutoryanskiy, V. V. (2008). Why is chitosan mucoadhesive? *Biomacromolecules*, **9**, 1837–1842.

Sriamornsak, P. (2003). Chemistry of pectin and its pharmaceutical uses: a review. *Silpakorn University International Journal*, **3**, 206–228.

Sungthongjeen, S., Sriamornsak, P., Pitaksuteepong, T., Somsiri, A. and Puttipipatkhachorn, S. (2004). Effect of degree of esterification of pectin and calcium amount on drug release from pectin-based matrix tablets. *AAPS PharmSciTech*, **5**, 1-9.

Takeuchi, H., Thongborisute, J., Matsui, Y., Sugihara, H., Yamamoto, H. and Kawashima, Y. (2005). Novel mucoadhesion tests for polymers and polymer-coated particles to design optimal mucoadhesive drug delivery systems. *Advanced Drug Delivery Reviews*, **57**, 1583– 1594.

Tanford, C. (1961). *Physical Chemistry of Macromolecules*. New York: John Wiley & Sons.

Terbojevich, M. Cosani, A., Conio, G., Marsano, E. and Bianchi, E. (1991). Chitosan: chain rigidity and mesophase formation. *Carbohydrate Research*, **209**, 251-260.

Terbojevich, M. and Muzzarelli, R. A. A. (2000). Chitosan. In: Phillips, G. O. and Williams, P. A. (Eds.). *Handbook of Hydrocolloids* (pp 367-378). Cambridge: Woodhead Publishing Ltd.

Thibault, J-F. and Rinaudo, M. (1986). Chain association of pectic molecules during calcium-induced gelation. *Biopolymers*, **25**, 455-468.

Thirawong, N., Kennedy, R. A. and Sriamornsak, P. (2008). Viscometric study of pectin–mucin interaction and its mucoadhesive bond strength. *Carbohydrate Polymers*, **71**, 170-179.

Tombs M. P., and Harding, S. E. (1998). *An Introduction to Polysaccharide Biotechnology*. London: Taylor and Francis.

Tsai, M. L., Bai, S. W., and Chen, R. H. (2008). Cavitation effects versus stretch effects resulted in different size and polydispersity of ionotropic gelation chitosan–sodium tripolyphosphate nanoparticle. *Carbohydrate Polymers*, **71**, 448-457.

Valenta, C. (2005). The use of mucoadhesive polymers in vaginal delivery. *Advanced Drug Delivery Reviews*, **57**, 1692– 1712.

van Buren, J. P. (1991). Function of pectin in plant tissue structure and firmness. In: Walter, R. H. (Ed). *The Chemistry and Technology of Pectin* (pp 1-22). San Diego: Academic Press.

Vandamme, T. F., Lenourry, A., Charrueau, C. and Chaumeil, J-C. (2002). The use of polysaccharides to target drugs to the colon. *Carbohydrate Polymers*, **48**, 219-231.

Vårum, K. M., Anthonsen, M. W., Grasdalen, H. and Smidsrød, O. (1991a). Determination of the degree of N-acetylation and distribution of N-acetyl groups in partially N-deacteylated chitins (chitosans) by high-field NMR spectroscopy. *Carbohydrate Research*, **211**, 17-23.

Vårum, K. M., Anthonsen, M. W., Grasdalen, H. and Smidsrød, O. (1991b). 13C NMR studies of the acetylation sequences in partially N-deacteylated chitins (chitosans). *Carbohydrate Research*, **217**, 19-27.

Vårum, K. M., Ottøy, M. H. and Smidsrød, O. (1994). Water-solubility of partially N-acetylated chitosans as a function of pH: effect of chemical composition and depolymerisation. *Carbohydrate Polymers*, **25**, 65-70.

Velásquez, C. L., Albornoz, J. S. and Barrios, E. M. (2008). Viscosimetric studies of chitosan nitrate and chitosan chlorhydrate in acid free NaCl aqueous solution. *E-Polymers*, 014.

Vermani, K. and Garg, S. (2000). The scope and potential of vaginal drug delivery, *Pharmaceutical Science & Technology Today*, **3**, 359– 364.

Vold, I. M. N. (2004). *Periodate Oxidised Chitosans: Structure and Solution Properties*. PhD Dissertation, Norwegian University of Science and Technology, Trondheim, Norway.

Voragen, A. G. J., Pilnik, W., Thibault, J-F., Axelos, M. A. V. and Renard, C. M. G. C. (1995). Pectins. In: Stephen, A. M. (Ed.) *Food polysaccharides and their interactions* (pp 287-340). New York: Marcel Dekker.

Wales, M. and van Holde, K. E. (1954). The concentration dependence of the sedimentation constants of flexible macromolecules. *Journal of Polymer Science*, **14**, 81-86.

Wang, Z-Y., White, J. W., Konno, M., Saito, S. and Nozawa, T. (1994). A small-angle X-ray scattering study of alginate solution and its sol-gel transition by addition of divalent cations. *Biopolymers*, **35**, 227-238.

Watts, P. and Smith, A. (2009). PecSys: in situ gelling system for optimised nasal drug delivery. *Expert Opinion on Drug Delivery*, **6**, 543-552.

Willats, W. G. T., McCartney, L., Mackie, W. and Knox J. P. (2001). Pectin: cell biology and prospects for functional analysis. *Plant Molecular Biology*, **47**, 9-27.

Xu, Y. and Du, Y. (2005). Effect of molecular structure of chitosan on protein delivery properties of chitosan nanoparticles. *International Journal of Pharmaceutics*, **250**, 215-226.

Yadav, N., Morris, G. A., Harding, S. E., Ang, S. and Adams, G. G. (2009). Various non-injectable delivery systems for the treatment of diabetes mellitus. *Endocrine, Metabolic & Immune Disorders - Drug Targets*, **9**, 1-13.

Yamamoto, A., Taniguchi, T., Rikyuu, K., Tsuji, T., Fujita, T., Murakami, M. and

MURANISHI, S. (1994). Effects of various protease inhibitors on the intestinal absorption and degradation of insulin in rats. *Pharmaceutical Research*, **11**, 1496–1500.

YEN, M-T. AND MAU, J-L. (2007). Physico-chemical characterization of fungal chitosan from shiitake stipes. *LWT - Food Science and Technology*, **40**, 472-479.

Biotechnology and Genetic Engineering Reviews - Vol. 27, 285-304 (2010)

Stem cells: The therapeutic role in the treatment of diabetes mellitus

GARY G. ADAMS,[1/3*], LEE BUTTERY,[2], SNOW STOLNIK[2], GORDON MORRIS[3], STEPHEN HARDING[3] AND NAN WANG [1]

[1/3]University of Nottingham, Faculty of Medicine and Health Sciences, Insulin Diabetes Experimental Research Group, Clifton Boulevard, Nottingham NG7 2RD UK; [2]University of Nottingham, School of Pharmacy, Faculty of Science, University Park, Nottingham, NG7 2RD, UK; [3]University of Nottingham, National Centre for Macromolecular Hydrodynamics, School of Biosciences, University of Nottingham, Sutton Bonington, LE12 5RD, U.K.

Abstract

The unlimited proliferative ability and plasticity to generate other cell types ensures that stem cells represent a dynamic system apposite for the identification of new molecular targets and the production and development of novel drugs. These cell lines derived from embryos could be used as a model for the study of basic and applied aspects in medical therapeutics, environmental mutagenesis and disease management. As a consequence, these can be tested for safety or to predict or anticipate potential toxicity in humans. Human ES cell lines may, therefore, prove clinically relevant to the development of safer and more effective drugs for patients presenting with diabetes mellitus.

*To whom correspondence may be addressed (Gary.Adams@nottingham.ac.uk)

Abbreviations: EC: embryonic carcinoma cells; EG: embryonic germ cell; PGC: primordial germ cells; ES:Embryonic stem cells; MSC: mesenchymal stem cells; ICM: inner cell mass; LIF: Leukaemia inhibitory factor; IL-6: interleukin-6, IL-11: interleukin-11, OSM: oncostatin M; CNTF: ciliary neurotrophic factor; CT-1: cardiotrophin-1; STAT: signal transduction and activation of transcription; SSEA-1: specific cell surface antigens; BIO: 6-bromoindirubin-3'-oxime; GSK-3: glycogen synthase kinase-3; DM: diabetes mellitus

Introduction

Stem cells are cells with the capacity for unlimited or prolonged self-renewal that can produce at least one type of highly differentiated descendant (Watt and Hogan 2000). Under the right conditions, or given the right signals, stem cells can give rise (differentiate) to the many different cell types that make up the organism. Stem cells have the potential to develop into mature cells with characteristic shapes and specialized functions, such as heart cells, skin cells, or nerve cells.

The first entity of life, the fertilized egg, has the ability to generate an entire organism, a capacity, defined as totipotency, which is retained by early progeny of the zygote up to the eight-cell stage of the morula. Hence, totipotent stem cells have the ability to form any embryonic or extra-embryonic cell type, including germ cells. Subsequently, cell differentiation results in the formation of a blastocyst composed of an outer trophoblast and undifferentiated inner cells, which are no longer totipotent, but retain the ability to develop into all cell types of the embryo proper. These pluripotent stem cells are capable of differentiating into cells from each of the three embryonic germ layers. Multipotent stem cells, which are able to form multiple organ specific cell types, are progressively restricted in their potential to differentiate (Figure 1).

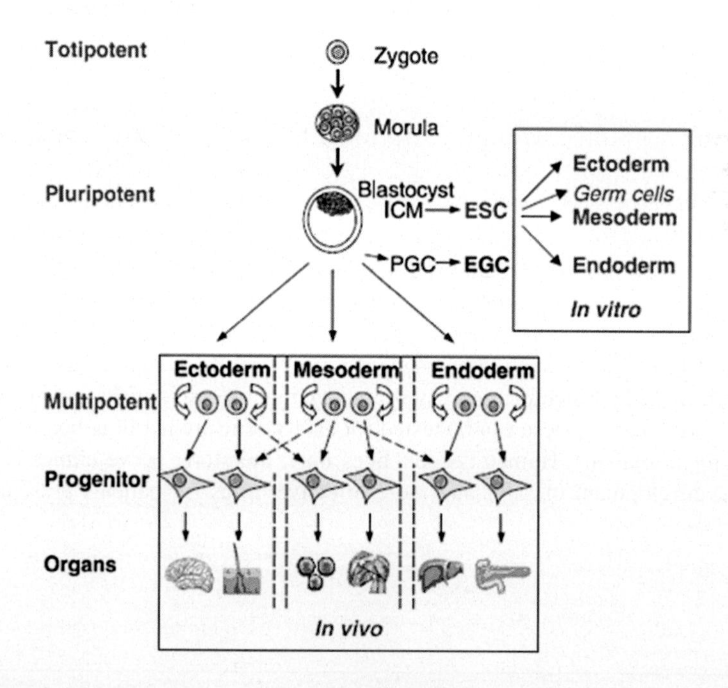

Figure 1. Stem cell and their hierarchy (Wobus, and Boheler 2005).

Stem cells research dates back to the early 1970s, when embryonic carcinoma (EC) cells, the stem cells of germ line tumours called teratocarcinomas (Smith, Heath *et al.* 1988), were established as cell lines (Jakob, Boon *et al.* 1973; Gearhart and Mintz 1974). Embryonic stem (ES) cells, and their tumorigenic counterparts, embryonal carcinoma (EC) cells, were among the first stem cells to be isolated and characterized extensively.

However, it has been suggested that the EC cells did not retain the pluripotent capacities of early embryonic cells and had undergone cellular changes during the transient tumorigenic state *in vivo* (Andrews 2002). In 1981, two groups succeeded in cultivating pluripotent cell lines from mouse blastocysts (Evans and Kaufman 1981; Martin 1981). These cells, originate from the inner cells mass of epiblast, termed as ES cells, can be maintained *in vitro* without any apparent loss of differentiation potential.

It has been confirmed that mouse ES cells showed their capacity to reproduce the various somatic cell types (Evans and Kaufman 1981; Wobus, Holzhausen *et al.* 1984; Doetschman, Eistetter *et al.* 1985), and moreover they were found to develop into cells of the germ line (Hubner, Fuhrmann *et al.* 2003; Geijsen, Horoschak *et al.* 2004). The other pluripotent embryonic cells type is primordial germ cells (PG cells), which form normally within the developing genital ridges. Isolation and cultivation of mouse PG cells on feeder cells led to the establishment of mouse embryonic germ (EG) cell lines (Resnick, Bixler *et al.* 1992; Labosky, Barlow *et al.* 1994; Stewart, Gadi *et al.* 1994). ES cells and EG cells differ in the conditions required for their isolation, culture, life span *in vitro*, and differentiation ability. ES cells can proliferate for as long as 300 population doublings and can be passaged for over a year in culture; EG cells can proliferate for only 80 population doublings (Thomson, Itskovitz-Eldor *et al.* 1998; Odorico, Kaufman *et al.* 2001). EG cells hold high ability to proliferate and differentiate *in vitro,* however, once transferred into blastocysts, EG cells retain the capacity to erase gene imprints. Human EG cell lines (Matsui, Zsebo *et al.* 1992) showed multilineage development *in vitro* but have a limited proliferation capacity and currently can only be propagated as embryoid body derivatives (Shamblott, Axelman *et al.* 2001). Over the past 4 decades, stem cells have been isolated from a wide range of sources, including mesenchymal stem cells (Prockop 1997) neural stem cells (Reynolds and Weiss 1992), and amniotic fluid stem cells (De Coppi, Bartsch *et al.* 2007) which were isolated from adult tissues (Oktem 2009).

Types of stem cells and their properties

Zygote and early cell division stages (blastomeres) to the ovular stage are defined as totipotent, because they can generate a complex organism. At the blastocyst stage, only the cells of the inner cell mass (ICM) retain the capacity to build up all three primary germ layers, the endoderm, mesoderm, and ectoderm as well as the primordial germ cells (PGC), the founder cells of male and female gametes. In adult tissues, multipotent stem and progenitor cells exist in tissues and organs to replace lost or injured cells. At present, it is not known to what extent adult stem cells may also develop (transdifferentiate) into cells of other lineages or what factors could enhance their differentiation capability (dashed lines). Embryonic stem (ES) cells, derived from the ICM, have the developmental capacity to differentiate *in vitro* into cells of all somatic cell lineages as well as into male and female germ cells (Wobus and Boheler 2005).

Embryonic stem cell (ES cell)

The embryonic origin of mouse and human ES cells is the major reason that research in this field is a topic of great scientific interest and vigorous public debate, influenced by both ethical and legal positions (Wobus and Boheler 2005; Bukovsky 2009). Undifferentiated ES cells share two unique properties, which are the unlimited self-renewal capacity and the ability to differentiate via precursor cells into terminally differentiated somatic cells.

ES cell is derived from the blastocyst stage of the embryo, a stage of embryonic development prior to implantation in the uterine wall. At this stage, the pre-implantation embryo is made up of 150 cells and consists of a sphere made up of an outer layer of cells (the trophectoderm), a fluid-filled cavity (the blastocoel), and a cluster of cells on the interior (the inner cell mass). Cells of the ICM are no longer totipotent but retain the ability to develop into all cell types of the embryo. Studies of ES cells derived from mouse blastocysts became possible 20 years ago with the discovery of techniques that allowed the cells to be grown in the laboratory. One of the current perceived advantages of using ESCs rather than adult stem cells is that ESCs have the ability to proliferate for long periods *in vitro* and can be directed to differentiate into a broad range of cell types.

Mouse ES cells

Mouse ES cell lines were first established in the early 1980s (Evans and Kaufman 1981; Axelrod 1984; Doetschman, Eistetter *et al.* 1985). The pre-implantation embryos (blastocysts) was isolated and cultivated on mouse embryonic fibroblasts followed by the expansion of primary ES cells outgrowths through careful trypsinization (Robertson 1987). The mouse ES cell lines displayed their unlimited *in vitro* proliferation ability (Smith 2001) and maintenance capacity to differentiate into all cell lineages (Ezhkova 2009; Teramura 2009).

The generation of mouse ES cell lines required the inactivated feeder layer cells; hence the property of feeder layer cells may be crucial to affect ES cells' proliferation and differentiation. It has been identified that Leukaemia inhibitory factor (LIF) is an important factor which responsible for controlling ES cells activities (Williams, Hilton *et al.* 1988). It has been demonstrated that cytokines, including LIF, IL-6, IL-11, OSM (oncostatin M), CNTF (ciliary neurotrophic factor), cardiotrophin-1 (CT-1), can be applied to maintain the pluripotency of mouse ES cells(Smith, Heath *et al.* 1988). They act via a membrane-bound gp130 signalling complex to regulate a variety of cell functions through signal transduction and activation of transcription (STAT) signalling (Burdon, Stracey *et al.* 1999). Quantitative measurements of ES cell phenotypic markers demonstrated a superior ability of LIF to maintain ES cell pluripotentiality at higher concentrations (>500 pM) which indicated a ligand /receptor signalling threshold model of ES cell fate modulation that requires appropriate types and levels of cytokine stimulation to maintain self-renewal (Viswanathan, Benatar *et al.* 2002).

It is now well established that undifferentiated mouse ES cells express specific cell surface antigens (SSEA-1) (Solter and Knowles 1978) and membrane-bound receptors (gp130) (Niwa, Burdon *et al.* 1998; Burdon, Chambers *et al.* 1999) and possess enzyme activities for alkaline phosphatase (ALP) (Wobus, Holzhausen *et al.* 1984) and telomerase (Prelle, Vassiliev *et al.* 1999; Armstrong, Lako *et al.* 2000). ES cells also contain the epiblast/germ cell-restricted transcription factor Oct-3/4 (Scholer, Hatzopoulos *et al.* 1989; Pesce, Anastassiadis *et al.* 1999). In ES cells, continuous Oct-3/4 function at appropriate levels is necessary to maintain pluripotency. A less than two-fold increase in expression causes differentiation into primitive endoderm and mesoderm, whereas loss of Oct-3/4 induces the formation of trophectoderm concomitant with a loss of pluripotency (Niwa, Miyazaki *et al.* 2000).

Recently, two groups identified the homeodomain protein Nanog as another key regulator of pluripotentiality (Chambers, Colby *et al.* 2003; Mitsui, Tokuzawa *et al.* 2003). It has been demonstrated that the dosage of Nanog is a critical determinant of cytokine-independent colony formation, and forced expression of this protein confers constitutive self-renewal. Nanog may (Vallier 2009) therefore act to restrict the differentiation-inducing potential of Oct-3/4. Study also implicated that Wnt pathway activation by a specific pharmacological inhibitor (BIO; 6-bromoindirubin-3'-oxime) of glycogen synthase kinase-3 (GSK-3) maintains the undifferentiated phenotype in both mouse and human ES cells and sustains expression of the pluripotent stage (Sato, Meijer *et al.* 2004). Hence, the ES cell property of self-renewal, therefore, depends on a stoichiometric balance among various signalling molecules, and an imbalance in any one can cause ES cell identity to be lost (Figure 2).

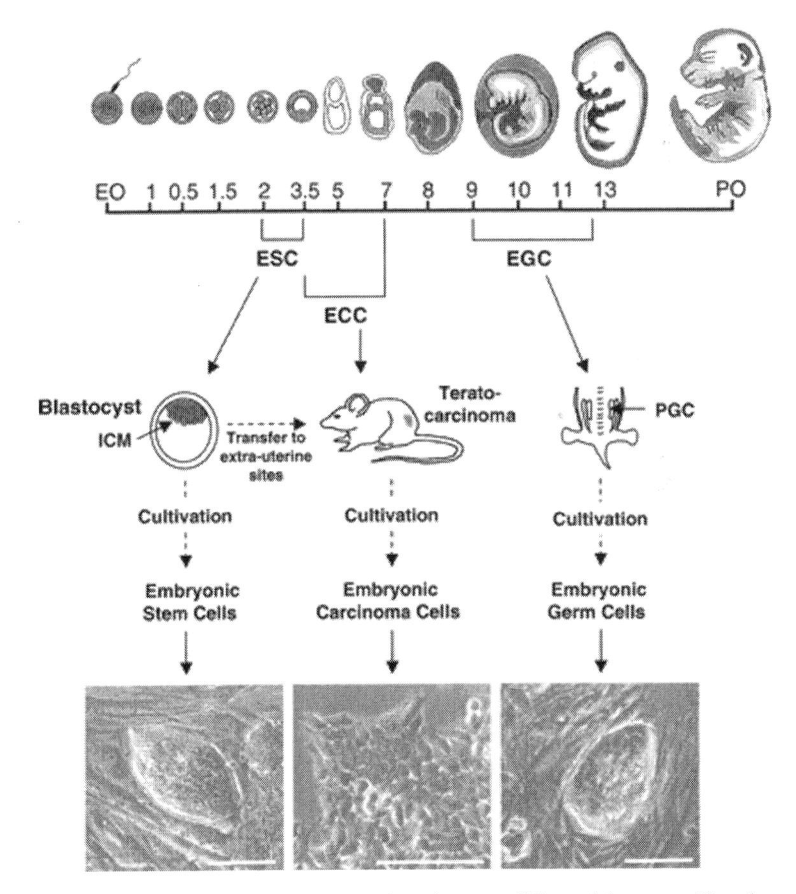

Figure 2. Developmental origin of pluripotent embryonic stem cell lines of the mouse. The scheme demonstrates the derivation of embryonic stem cells (ESC), embryonic carcinoma cells (ECC), and embryonic germ cells (EGC) from different embryonic stages of the mouse. ECC are derived from malignant teratocarcinomas that originate from embryos (blastocysts or egg cylinder stages) transplanted to extrauterine sites. EGC are cultured from primordial germ cells (PGC) isolated from the genital ridges between embryonic day *9* to *12.5*. Bar = 100 μm (Boheler, Czyz ***et al.*** 2002).

Human ES cells

Pluripotent human embryonic stem (Doss MX) cell lines were successfully derived from the inner cell mass (ICM) of human blastocysts in the late 1990s. hES cells have similar characters with mES cells such as the Oct3/4 expression, telomerase activity, and the formation of teratomas containing derivatives of all three primary germ layers in immunodeficient mice (Thomson, Itskovitz-Eldor *et al.* 1998; Richards, Fong *et al.* 2002). Compared to mouse ES cells, hES cells present a longer average population doubling time (30–35 h vs. 12–15 h) (Amit, Carpenter *et al.* 2000). It has been demonstrated that LIF is insufficient to inhibit the differentiation of hES cells (Thomson, Itskovitz-Eldor *et al.* 1998; Reubinoff, Pera *et al.* 2000), which continue to be cultured routinely on feeder layers of MEFs or feeder cells from human tissues. LIF is unable to maintain the pluripotent state of hES cells (Daheron, Opitz *et al.* 2004), hence it has been demonstrated that the application of extracellular matrix-associated factors (such as Matrigel and laminin with MEF-conditioned media) need to be employed to improve the culture and maintenance of pluripotent hES cells (Xu, Inokuma *et al.* 2001; Kalkunte 2009).

Although the successful establishment of hES lines has been reported, these cells suffer from significant limitations including possible murine retrovirus infections (from the feeder cells) that have rendered them inappropriate for therapeutic applications in the previous studies (Wobus and Boheler 2005). By the end of 2001 about 70 hES lines had been established using feeder layers of mouse embryonic fibroblasts. As of December 2004, only 22 of the cell lines listed in the NIH register had been successfully propagated *in vitro* (Wobus and Boheler 2005). Importantly, hES cell lines have now been cultivated both on human feeder cells to avoid xenogenic contamination (Richards, Fong *et al.* 2002; Amit, Margulets *et al.* 2003; Amit and Itskovitz-Eldor 2006) and in the absence of feeder cells under serum-free conditions (Lee, Lee *et al.* 2005).

ES cells of other species

Pluripotent stem cell lines have been generated from other species, including chicken (Chang, Jeong *et al.* 1997), hamster (Doetschman, Williams *et al.* 1988), rabbit (Graves and Moreadith 1993) (Schoonjans, Albright *et al.* 1996), and rat (Brenin, Look *et al.* 1997; Vassilieva, Guan *et al.* 2000). Among all of these, the establishment of monkey ES cells represented special importance due to the possible application for human stem cell research, including Rhesus monkey (Lester LB 2004; Pau and Wolf 2004), common marmoset (Sasaki, Hanazawa *et al.* 2005) and cynomolgus monkey (Suemori, Tada *et al.* 2001). Monkey ES cells, characterized by typical markers of human ES (Oct-4, SSEA-4, TRA-1–60, TRA-1–81), have a high differentiation capacity *in vitro* (Thomson, Kalishman *et al.* 1995). These properties may qualify these cell lines as alternative and substitute model systems for hES cell lines. Moreover, after *in vivo* parthenogenetic development of *Macaca fascicularis* eggs to blastocyst-stage embryos, a pluripotent monkey stem cell line (Cyno-1) has been established that showed all the properties of hES cells, such as high telomerase and ALP activity; expression of Oct-3/4, SSEA-4, TRA 1–60, and TRA 1–81; and the ability to differentiate into various cell lineages (Vrana, Hipp *et al.* 2003).

Adult stem cell

It is already known that fetal organs as well as adult tissues, with high self-renewal capacity, contain regenerative stem cell populations : blood, skin and gut. Adult stem cells are also known as fetal/somatic stem cells or tissue-derived stem cells which include both multi-potentiality stem cells, such as adult mesenchymal stem cells (hMSC) (Meirelles 2009; Chugh 2009), and the uni-potentiality cell, such as epidermal stem cells. Adult stem cells are present in most tissues and are responsible for the replenishment of those tissues throughout life. Adult stem cells differ from embryonic stem cells not only in their different lineage potentials, but also in the mechanism by which they proliferate. Adult stem cells divide asymmetrically to maintain their number in the tissue, while at the same time giving rise to cells committed to becoming differentiated tissues and organs; however, embryonic stem cells normally divide symmetrically (Serakinci and Keith 2006).

Adult stem cells are rare but the list of adult tissues reported to contain stem cells is growing and includes bone marrow, peripheral blood, brain, spinal cord, dental pulp, blood vessels, skeletal muscle, epithelia of the skin and digestive system, cornea, retina, liver, and pancreas (2001) and their plasticity restricted. A number of recent investigations have suggested that they may be more plastic than previously thought. Haematopoietic stem cells (Ainsworth 2009; Parrish 2009) from the bone marrow had the capacity to develop into neural, myogenic and hepatic cell types and neuronal or muscle stem cells developed into the haematopoietic lineage (Bjornson, Rietze *et al.* 1999). Moreover only an estimated 1 in 10,000 to 15,000 cells in the bone marrow is a hematopoietic (blood forming) stem cell (HSC) (Weissman 2000).

Their primary functions are to maintain the steady state functioning of a cell (homeostasis), with limitations, to replace cells that die because of injury or disease (Holtzer 1978). These observations follow that somatic cells of the adult organism may yet have a high plasticity, and their developmental potential may not be restricted to one lineage, but could be determined by the tissue environment in the body (Watt and Hogan 2000). Furthermore, adult stem cells are dispersed in tissues throughout the mature animal and behave very differently, depending on their local environment.

Hence, adult stem cells offer new potential for the biomedical therapy, although the cells proliferation and differentiation mechanisms remains unclear. The limitation of adult stem cell include: 1) the proliferation and differentiation capacity may be lower than ES cells; 2) adult stem cells may contain DNA abnormalities caused by exposure to environmental factors, and stem cells of certain patients may contain genetic defects and therefore would be inappropriate for transplantation.

Stem cells: Therapeutic interventions in the treatment of Diabetes mellitus

Due to their unlimited unique self-renewal and potential development properties, ES cells can be derived into multi-lineage cells and used as alternative resource of transplantation therapy. To fulfill the therapeutic application, efficiency of directed differentiation into specific somatic cell populations need to be improved to satisfy clinic requirements. Stem cells represent the nature of embryonic development and tissue regeneration. The established permanent ES cell lines can be regarded as a versatile biological model system that leads to major advances in cell and developmental biology and this includes the pancreas.

The mature pancreas plays a central role in glucose homeostasis and metabolism, has two morphologically and physiologically distinct components, the endocrine pancreas and the exocrine pancreas (White 1973). The exocrine pancreas comprises the bulk of the tissue and is made up of acinar cells that specialise in the secretion of digestive enzymes via the epithelial pancreatic ducts. The ductal system transports the digestive enzymes to the intestine where they ensure nutrient digestion and absorption. The endocrine part of pancreas is scattered irregularly throughout the exocrine pancreas, with the majority identified as small spheroid clusters of cells, called the islets of Langerhans (Soria, Roche *et al.* 2000). The islets are distributed throughout the whole pancreas and vary in size with a higher density in the tail of the organ. Apart from the typical islets, small endocrine clusters and single pancreatic endocrine cells can also be found occasionally in the pancreas. The islets are a heterogeneous population of 4 main types in the following proportions: 65-90% insulin-producing -cells, 15-20% glucagon-producing α-cells, 3-10% somatostatin-producing δ-cells and 1% pancreatic polypeptide producing PP-cells (Rahier 1988) and under normal physiological conditions glucose homeostasis and metabolism is maintained; in the patient presenting with DM, however, imbalance can lead to micro-and-macrovascular disorders.

During gastrulation, the epiblast, consisting of multi-potential cells, generates three embryo layers, including endoderm, mesoderm and ectoderm (Figure 3). The gastrointestinal organs, including the pancreas, are derived from the endodermal layer. The induction of the endoderm appears to be governed by nodal/transforming growth factor (TGF-β) signalling from the adjacent ectoderm and mesoderm within the primitive streak and the node (Wells and Melton 1999; Lewis and Tam 2006). After gastrulation, the endoderm is patterned in an anterior-posterior fashion in response to fibroblast growth factor (FGF-4) signals from adjacent mesoderm (Dessimoz, Opoka *et al.* 2006). It has long been recognized that mesenchymal-epithelial signalling is important during subsequent growth and differentiation of the pancreatic epithelium (Golosow and Grobstein 1962). Epidermal growth factor is another mesenchymal factor that has been shown to regulate proliferation of the developing pancreas both *in vitro* and *in vivo* (Miettinen, Huotari *et al.* 2000). Other mesenchymal factors, such as follistatin, regulate cell-type specific differentiation within the pancreatic epithelium. In the absence of mesenchyme, pancreatic epithelium gives rise to endocrine cells, but not exocrine cells (Gittes, Galante *et al.* 1996; Miralles, Czernichow *et al.* 1998). Notch signalling is another key regulator of pancreatic cell growth and differentiation. During the early stages following the formation of pancreatic primordial, Notch signalling is essential for the expansion of undifferentiated progenitor cells (Hart, Papadopoulou *et al.* 2003; Norgaard, Jensen *et al.* 2003). During subsequent stages, Notch is frequently used to control the sequential generation of different cell types from a common progenitor cell and consequently helps to specify endocrine cell differentiation (Esni, Ghosh *et al.* 2004). It has been reported recently that the Wnt signalling pathway is also crucial in pancreatic organogenesis (Murtaugh, Law *et al.* 2005) (Papadopoulou and Edlund 2005) and that activation of Wnt/β-catenin signalling at a later time point in pancreas development causes enhanced proliferation of acinar cells (Heiser, Lau *et al.* 2006).

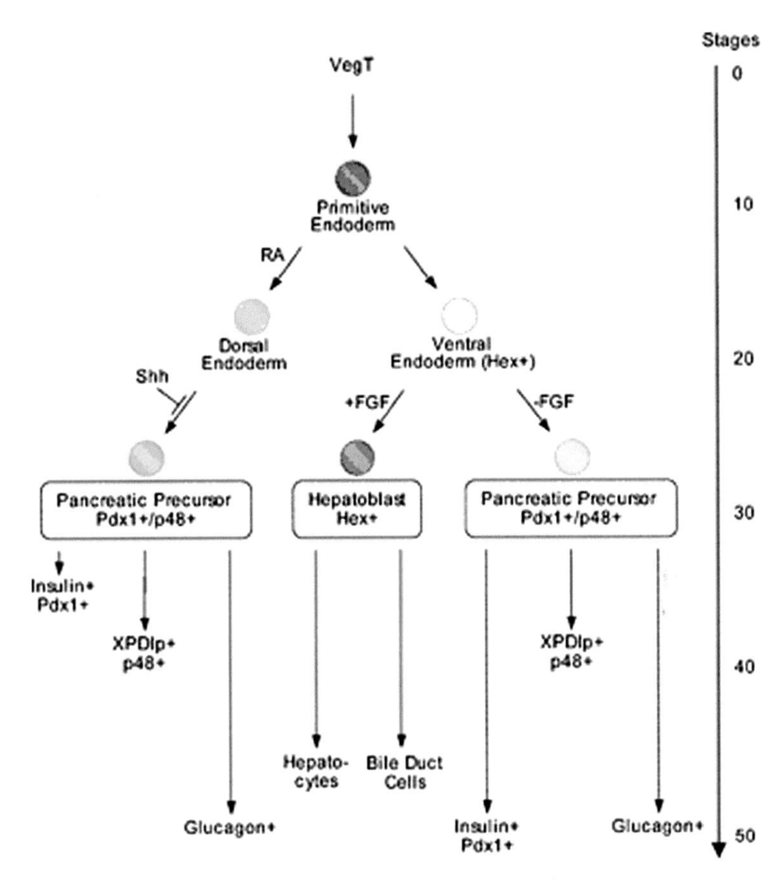

Figure 3. Forgotten and novel aspects in pancreas development (Pieler and Chen 2006).

Differentiation of ES cells is a heterogeneous process, and desired lineages can be enriched using various strategies (Zhao 2009), which includes the induction using chemokines, co-culture with differentiated cell types, and genetic manipulations on the ES cells. Differentiation is normally induced by culturing ES cells as aggregates (EBs) in the absence of the self-renewal signals provided by feeder layers or LIF, either in hanging drops (Yamada, Yoshikawa *et al.* 2002; Kurosawa 2007), in liquid "mass culture" (Doetschman, Eistetter *et al.* 1985), or in methylcellulose (Wiles and Keller 1991; Dang, Kyba *et al.* 2002). The adherent monolayer cultures in the absence of LIF (Ying, Stavridis *et al.* 2003) have also been used to differentiate mouse ES cells *in vitro*. Scalable production of ES-derived cells could furthermore be achieved through the use of stirred suspension bioreactors with encapsulation techniques (Dang and Zandstra 2005; Cameron, Hu *et al.* 2006).

During the differentiation process, ES cells spontaneously develop to form cells within all three germ layers (Qu 2008), endoderm, ectoderm and mesoderm (Figure 4).

Endodermal differentiation

In mammalian development, both the pancreas and liver originate from the definitive endoderm (Shiraki 2009). The pancreas develops from dorsal and ventral regions of the foregut, and the liver from the foregut adjacent to the ventral pancreas compartment (Slack 1995). Pancreatic and hepatic cells are of special therapeutic interest for the treatment of hepatic failure (Gupta and Chowdhury 2002) and diabetes mellitus (Blyszczuk and Wobus 2006), and both pancreatic endocrine and hepatic cells could be developed successfully *in vitro* from ES cells.

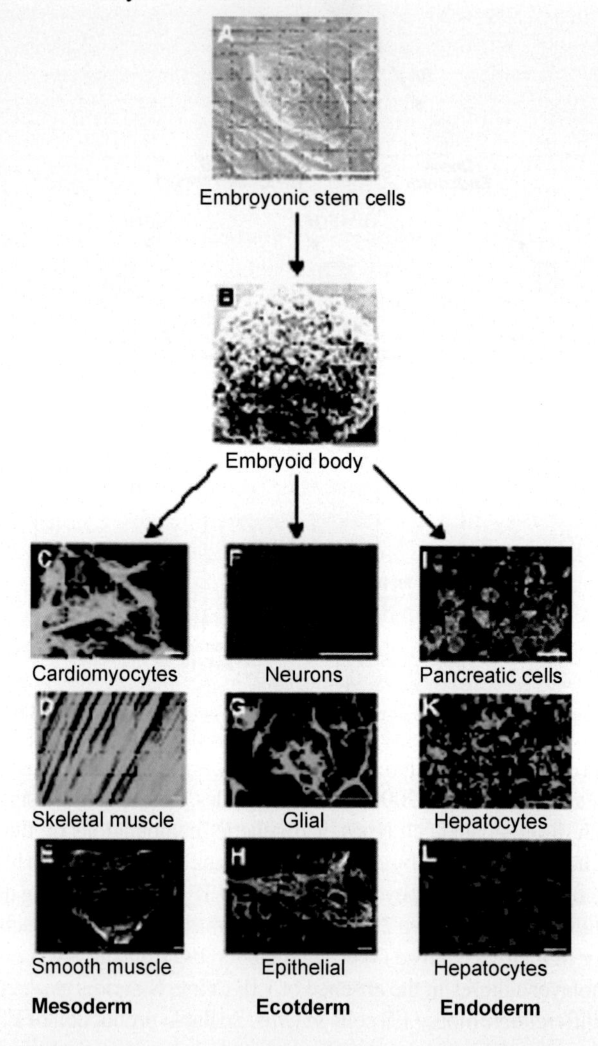

Figure 4. *In vitro* differentiation of ES cells. Undifferentiated mouse ES cells (***A***) develop *in vitro* via three-dimensional aggregates (embryoid body, ***B***) into differentiated cell types of all three primary germ layers. Shown are differentiated cell types labeled by tissue-specific antibodies (in parentheses). ***C***: cardiomyocytes (titin Z-band epitope). ***D***: skeletal muscle (titin Z-band epitope). ***E***: smooth muscle (smooth muscle α-actin). ***F***: neuronal (ßIII tubulin). ***G***: glial (glial fibrillary acidic protein, GFAP). ***H***: epithelial cells (cytokeratin 8). ***I***: pancreatic endocrine cells [insulin (red), C-peptide (Sato, Meijer ***et al.***), insulin and C-peptide colabeling (yellow)]. ***K*** and ***L***: hepatocytes (***K***, albumin; ***L***, α 1-antitrypsin). Bars = 0.5 μm (***H***), 20 μm (***I***), 25 μm (***C***, ***D***, ***E***), 30 μm (***K***, ***L***), 50 μm (***B***, ***G***), and 100 μm (***A***, ***F***). (Wobus and Boheler 2005)

The generation of ES derived insulin-producing cells holds great potential as an alternative treatment for T1DM. The first successful induction of pancreatic differentiation from ES cells was obtained by stable transfection with a vector containing a neomycin-resistance gene under the control of the insulin promoter (Soria, Roche *et al.* 2000). In contrast, the spontaneous differentiation of mouse ES cells *in vitro* generated only a small fraction of insulin-producing cells (0.1%) (Shiroi, Yoshikawa *et al.* 2002). By modifying the differentiation protocols and using genetically modified mouse ES cells, the differentiation efficiency has been improved greatly (Lumelsky, Blondel *et al.* 2001; Blyszczuk P 2004; Miyazaki, Yamato *et al.* 2004). Although the modification of differentiation protocol allowed the generation of insulin-producing cells from mouse and human ES cells, further improvements are necessary to generate functional pancreatic islet like clusters.

Although ES cells could be used as an available source to obtain large amounts of transplantable cells as regenerative medicine as in the case of T1DMs, the major obstacle to the successful and safe clinical use of differentiated cells is the possibility of immune rejection and teratomas or teratocarcinoma formation in the recipients (Mimeault and Batra 2006; Trounson 2006).

More recently, Kroon *et al* showed that pancreatic endoderm derived from human embryonic stem (Doss MX) cells efficiently generates glucose-responsive endocrine cells after implantation into mice. Upon glucose stimulation of the implanted mice, human insulin and C-peptide are detected in sera at levels similar to those of mice transplanted with ~3,000 human islets. Moreover, the insulin-expressing cells generated after engraftment exhibit many properties of functional ß-cells, including expression of critical ß-cell transcription factors, appropriate processing of proinsulin and the presence of mature endocrine secretory granules. They also showed that implantation of hES cell–derived pancreatic endoderm protects against streptozotocin-induced hyperglycemia (Kroon 2008).

Encapsulation of embryonic stem cells

A way of ensuring that immune rejection does not occur is to establish an appropriate encapsulation method that releases insulin in a glucose-responsive manner and that the diffusivity of insulin being produced by the stem cells post-encapsulation can be assessed. In our systems, alginate was used in various concentrations: 0.5, 1.0 and 2% w/v in order to examine diffusion parameters. A key concern with the use of alginate is the long-term stability, and too low an alginate concentration may easily cause a collapse of the gel structure and cell release. 0.5% w/v alginate gel can only form relatively weak structure and cannot maintain cells long term in culture system and 2% w/v gels inhibit the influx of nutrients and the efflux of cellular wastes. As a result, only the 1% w/v alginate-based gel was used. An essential requirement for alginate micro-beads to serve as 3D environments for stem cells differentiation is the permeability towards diffusion of the growth factors and supplements required for cells' endocrinal differentiation into the micro-beads. Diffusion studies were conducted by monitoring penetration of fluorescently-labelled model molecules in a range of molecular weights (FITC of 327 Da, FITC-IgG of 150kDa and FITC-dextran of 500kDa) into alginate beads (Figure 5). We assessed the viability of encapsulated ES cells such that a scalable tissue culture environment using alginate encapsulation could be utilised to promote embryonic stem cell differentiation. The cells were released immediately post-encapsulation (within 12 hours) and cell viability analysis carried out. 95.2% of the original 10.3 million confluent mouse ES cells harvested were live cells and following encapsulation and immediate disintegration of the beads, 9.6 million cells were collected (data not shown). There was no significant difference between cell viability before and after encapsulation (P=0.1205) and no indication that encapsulation significantly affects ES cell viability. This shows that in the initial stages, the cells may continue to follow a pre-programmed route until they become acclimatised to the local surrounding environment (Maguire 2006). Once acclimatised the viability begins to minimally diminish over time but this diminution is not statistically significant.

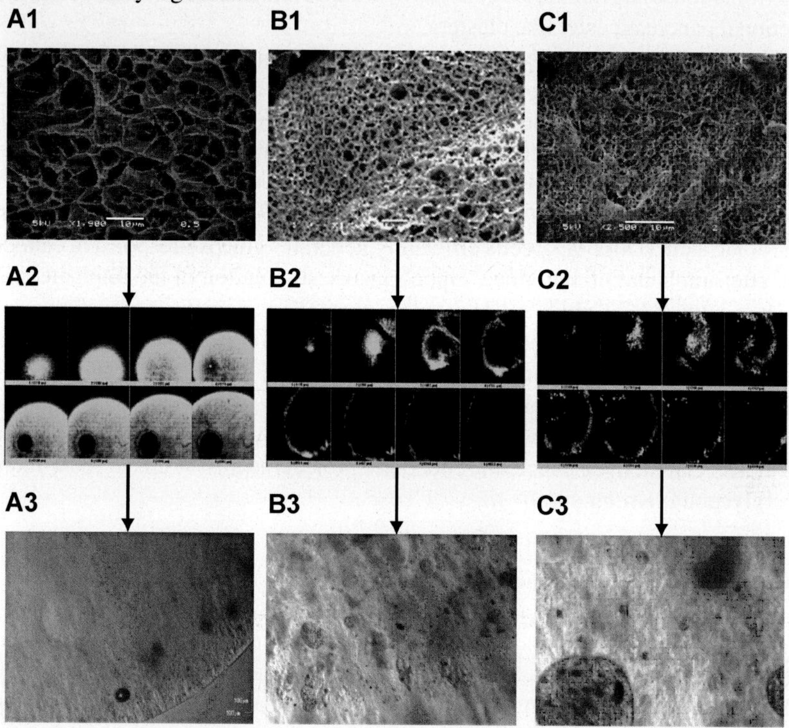

Figure 5. Cryo-SEM images of various concentrations alginates A1 (0.5% w/v), B1 (1% w/v), and C1 (2% w/v). Study of FITC conjugated Dextran (Mw 500KDa) diffusion into alginate micro-beads at 1 and 24 hours time points. A2) Confocal microscopy section images of 0.5% alginate micro-beads incubated with FITC- Dextran after 24 hours; B2) Confocal microscopy section images of 1% alginate micro-beads incubated with FITC- Dextran after 24 hours; C2) Confocal microscopy section images of 2% alginate micro-beads incubated with FITC- Dextran after 24 hours. Alginate beads with encapsulated stem cells A3) 0.5% alginate beads; B3) 1% alginate beads; C3) 2% alginate beads.

The use of alginate may promote entrapment either by cell–cell contact or decrease intercellular spacing, both of which could be conducive to intercellular signalling, which is mediated either by direct contact or soluble mediators. Any additional entrapment by alginate that followed this could further promote this interaction; the inclusion of hyaluronic acid as a constituent of the encapsulate does not, however, further significantly affect the viability of the cells.

These findings substantiate our method of encapsulation as opposed to the traditional formation strategies, for example those described by Cameron *et al*, where embryoid body formation is required (Hu 2003; Cameron 2006). These systems are probably used because they are known to generate three germ layers characteristic of embryo development, but EB formation is not an essential requirement in the production of insulin.

Conclusion

Due to their unique characters, the unlimited proliferative ability and plasticity to generate other cell types, stem cells represent a dynamic system suitable for the identification of new molecular targets and the development of novel drugs, which can be tested *in vitro* for safety or to predict or anticipate potential toxicity in humans (Ahuja, Vijayalakshmi *et al*. 2007). These cell lines derived from embryo could be used as a model for the study of basic and applied aspects in medical therapeutics, environmental mutagenesis and disease management. Moreover Human ES cell lines may, therefore, prove clinically relevant to the development of safer and more effective drugs for human diseases (Davila, Cezar *et al*. 2004) especially with those patients presenting with diabetes mellitus.

Acknowledgements

The authors would like to thank Pfizer Global and The University of Nottingham for funding.

References

Ahuja, Y. R., Vijayalakshmi, V. (2007). Stem cell test: A practical tool in toxicogenomics. *Toxicology* **231**(1), 1-10.

Ainsworth, C. D., Crowther, M.A., Treleaven, D., Evanovitch, D., Webert, K.E., Blajchman, M.A. (2009). Severe hemolytic anemia post-renal transplantation produced by donor anti-D passenger lymphocytes, case report and literature review. *Transfusion Medical Review* **23**(2), 155-9.

Amit, M., Carpenter, M. K. (2000). Clonally derived human embryonic stem cell lines maintain pluripotency and proliferative potential for prolonged periods of culture. *Developmental Biology* **227**(2), 271-278.

Amit, M. and Itskovitz-Eldor, J. (2006). Maintenance of human embryonic stem cells in animal serum- and feeder layer-free culture conditions. *Methods Molecular Biology* **331**, 105-13.

Amit, M., Margulets, V. (2003). Human feeder layers for human embryonic stem cells. *Biological Reproduction* **68**(6), 2150-6.

Andrews, P. W. (2002). From teratocarcinomas to embryonic stem cells. *Philosophical Transactions of the Royal Society B* **357**(1420), 405-17.

Armstrong, L., Lako, M. (2000). mTert expression correlates with telomerase activity during the differentiation of murine embryonic stem cells. *Mechanisms of Development* **97**(1-2), 109-16.

Axelrod, H. R. (1984). Embryonic stem cell lines derived from blastocysts by a simplified technique. *Developmental Biology* **101**(1), 225-8.

Bjornson, C. R., Rietze, R. L. (1999). Turning brain into blood, a hematopoietic fate adopted by adult neural stem cells in vivo. *Science* **283**(5401), 534-7.

Blyszczuk P, A. C., Rozzo A, Kania G, St-Onge L, Rupnik M, Wobus Am. (2004). Embryonic stem cells differentiate into insulin-producing cells without selection of nestin-expressing cells. *International Journal of Developmental Biology* **48**(10), 1095-104.

Blyszczuk, P. and. Wobus A. M (2006). In vitro differentiation of embryonic stem cells into the pancreatic lineage. *Methods Molecular Biology* **330**, 373-85.

Boheler, K. R., Czyz, J. (2002). Differentiation of pluripotent embryonic stem cells into cardiomyocytes. *Circulation Research* **91**(3), 189-201.

Brenin, D., Look, J. (1997). Rat embryonic stem cells, a progress report. *Transplantion Proceedings* **29**(3), 1761-5.

Bukovsky, A., Caudle, M.R., Virant-Klun, I., Gupta, S.K., Dominguez, R., Svetlikova, M., XU, F. (2009). Immune physiology and oogenesis in fetal and adult humans, ovarian infertility, and totipotency of adult ovarian stem cells. *Birth Defects Research C Embryo Today* **87**(1), 64-89.

Burdon, T., Chambers, I. (1999). Signaling mechanisms regulating self-renewal and differentiation of pluripotent embryonic stem cells. *Cells Tissues Organs* **165**(3-4), 131-43.

Burdon, T., Stracey, C.(1999). Suppression of SHP-2 and ERK signalling promotes self-renewal of mouse embryonic stem cells. *Developmental Biology* **210**(1), 30-43.

Cameron, C. M., Hu, W. S.(2006). Improved development of human embryonic stem cell-derived embryoid bodies by stirred vessel cultivation. *Biotechnology Bioengineering* **94**(5), 938-48.

Cameron, C. M., Hu, W.S., Kaufman, D.S. (2006). Improved development of human embryonic stem cell-derived embryoid bodies by stirred vessel cultivation. *Biotechnology and Bioengineering* **94**(5), 938-948.

Chambers, I., Colby, D. (2003). Functional expression cloning of Nanog, a pluripotency sustaining factor in embryonic stem cells. *Cell* **113**(5), 643-55.

Chang, I. K., Jeong, D. K.(1997). Production of germline chimeric chickens by transfer of cultured primordial germ cells. Cell Biology International 21(8), 495-9.

Chugh, A. R., Zuba-Surma, E.K., Dawn, B. (2009). Bone marrow-derived mesenchymal stems cells and cardiac repair. *Minerva Cardioangiologica* **57**(2), 185-202.

Daheron, L., Opitz, S. L.(2004). LIF/STAT3 signaling fails to maintain self-renewal of human embryonic stem cells. *Stem Cells* **22**(5), 770-8.

Dang, S. M., Kyba, M. (2002). Efficiency of embryoid body formation and hematopoietic development from embryonic stem cells in different culture systems. *Biotechnology Bioengineering* **78**(4), 442-53.

Dang, S. M. and Zandstra P. W. (2005). Scalable production of embryonic stem cell-derived cells. *Methods Molecular Biology* **290**, 353-64.

Davila, J. C., Cezar, G. G. (2004). Use and application of stem cells in toxicology. *Toxicology Science* **79**(2), 214-23.

De Coppi, P., Bartsch, Jr., G. (2007). Isolation of amniotic stem cell lines with potential for therapy. *Nature Biotechnology* **25**(1), 100-6.

Dessimoz, J., Opoka, R. (2006). FGF signaling is necessary for establishing gut tube domains along the anterior-posterior axis in vivo. *Mechanisms of Development* **123**(1), 42-55.

Doetschman, T., Williams, P. (1988). Establishment of hamster blastocyst-derived embryonic stem (ES) cells. *Developmental Biology* **127**(1), 224-7.

Doetschman, T. C., Eistetter, H. (1985). The in vitro development of blastocyst-derived embryonic stem cell lines, formation of visceral yolk sac, blood islands and myocardium. *Journal of Embryological Experimental Morphology* **87**, 27-45.

Doss Mx, K. C., Gissel C, Hescheler J, Sachinidis A. (2004). Embryonic stem cells, a promising tool for cell replacement therapy. *Journal of Cell Molecular Medicine* **8**(4),465-73.

Esni, F., Ghosh, B. (2004). Notch inhibits Ptf1 function and acinar cell differentiation in developing mouse and zebrafish pancreas. *Development* **131**(17), 4213-4224.

Evans, M. J. and Kaufman M. H. (1981). Establishment in Culture of Pluripotential Cells from Mouse Embryos. *Nature* **292**(5819), 154-156.

Ezhkova, E., Pasolli, H.A., Parker, J.S., Stokes, N., Su, I.H., Hannon, G., Tarakhovsky, A., Fuchs, E. (2009). Ezh2 orchestrates gene expression for the stepwise differentiation of tissue-specific stem cells. *Cell* **136**(6), 1122-35.

Gearhart, J. D. and Mintz B. (1974). Contact-mediated myogenesis and increased acetylcholinesterase activity in primary cultures of mouse teratocarcinoma cells. *Proceedings National Academy of Sciences U S A* **71**(5), 1734-8.

Geijsen, N., Horoschak, M. (2004). Derivation of embryonic germ cells and male gametes from embryonic stem cells. *Nature* **427**(6970), 148-54.

Gittes, G. K., Galante, P. E. (1996). Lineage-specific morphogenesis in the developing pancreas, Role of mesenchymal factors. *Development* **122**(2), 439-447.

Golosow, N. and Grobstein C. (1962). Epitheliomesenchymal Interaction in Pancreatic Morphogenesis. *Developmental Biology* **4**(2), 242-&.

Graves, K. H. and Moreadith R. W. (1993). Derivation and characterization of putative pluripotential embryonic stem cells from preimplantation rabbit embryos. *Molecular Reproductive Development* **36**(4), 424-33.

Gupta, S. and Chowdhury J. R. (2002). Therapeutic potential of hepatocyte transplantation. *Seminars in Cell & Developmental Biology* **13**(6), 439-446.

Hart, A., Papadopoulou, S. (2003). Fgf10 maintains notch activation, stimulates proliferation, and blocks differentiation of pancreatic epithelial cells. *Developmental Dynamics* **228**(2), 185-193.

Heiser, P. W., J. Lau, (2006). Stabilization of beta-catenin impacts pancreas growth. *Development* **133**(10), 2023-2032.

Holtzer, H. (1978). Cell lineages, stem cells and the quantal cell cycle concept. In, Stem cells and tissue homeostasis, Cambridge University Press.

Hu, A., Cai, J., Zheng, Q., He, X., Pan, Y., Li, L. (2003). Hepatic differentiation from embryonic stem cells in vitro. *Chinese Medicine Journal* (Engl) **116**(12), 1893-1897.

Hubner, K., Fuhrmann, G. (2003). Derivation of oocytes from mouse embryonic stem cells. *Science* **300**(5623), 1251-1256.

Jakob, H., Boon, T.et al. (1973). [Teratocarcinoma of the mouse, isolation, culture and properties of pluripotential cells. *Annals of Microbiology* (Paris) **124**(3), 269-82.

Kalkunte, S. S., Mselle, T.F., Norris, W.E., Wira, C.R., Sentman, C.L., Sharma, S. (2009). Vascular endothelial growth factor C facilitates immune tolerance and endovascular activity of human uterine NK cells at the maternal-fetal interface. Journal of *Immunology* **182**(7), 4085-92.

KROON, E., MARTINSON, L.A., KADOYA, K., BANG, A.G., KELLY, O.G., ELIAZER, S., YOUNG, H., RICHARDSON, M., SMART, N.G., AND CUNNINGHAM J. (2008). Pancreatic endoderm derived from human embryonic stem cells generates glucose-responsive insulin-secreting cells *in vivo. Nature Biotechnology* **26** 443-452

KUROSAWA, H. (2007). Methods for inducing embryoid body formation, *in vitro* differentiation system of embryonic stem cells. *Journal of Bioscience and Bioengineering* **103**(5), 389-98.

LABOSKY, P. A., BARLOW, D. P. (1994). Mouse Embryonic Germ (Eg) Cell-Lines - Transmission through the Germline and Differences in the Methylation Imprint of Insulin-Like Growth-Factor 2 Receptor (Igf2R) Gene Compared with Embryonic Stem (Es) Cell-Lines. *Development* **120**(11), 3197-3204.

LEE, J. B., LEE, J. E. (2005). Establishment and maintenance of human embryonic stem cell lines on human feeder cells derived from uterine endometrium under serum-free condition. *Biological Reproduction* **72**(1), 42-9.

LESTER LB, K. H., ANDREWS L, NAUERT B, WOLF DP. (2004). Directed differentiation of rhesus monkey ES cells into pancreatic cell phenotypes. *Reproductive Biological Endocrinology*. **16**, 42.

LEWIS, S. L. AND TAM, P. P. (2006). Definitive endoderm of the mouse embryo, formation, cell fates, and morphogenetic function. *Developmental Dynamics* **235**(9), 2315-29.

LUMELSKY, N., BLONDEL, O. (2001). Differentiation of embryonic stem cells to insulin-secreting structures similar to pancreatic islets. *Science* **292**(5520), 1389-94.

MAGUIRE, T., NOVIK, E., SCHLOSS, R., YARMUSH, M. (2006). Alginate-PLL microencapsulation, effect on the differentiation of embryonic stem cells into hepatocytes. *Biotechnology Bioengineering* **93**(3), 581-91.

MARTIN, G. R. (1981). Isolation of a pluripotent cell line from early mouse embryos cultured in medium conditioned by teratocarcinoma stem cells. *Proceedings National Academy of Sciences U S A* **78**(12), 7634-8.

MATSUI, Y., ZSEBO, K. (1992). Derivation of pluripotential embryonic stem cells from murine primordial germ cells in culture. *Cell* **70**(5), 841-7.

MEIRELLES, L., NARDI, N.B. (2009). Methodology, biology and clinical applications of mesenchymal stem cells *Frontiers in Bioscience* **14**, 4281-98.

MIETTINEN, P. J., HUOTARI, M. A. (2000). Impaired migration and delayed differentiation of pancreatic islet cells in mice lacking EGF-receptors. *Development* **127**(12), 2617-2627.

MIMEAULT, M. AND BATRA S. K. (2006). Concise review, recent advances on the significance of stem cells in tissue regeneration and cancer therapies. *Stem Cells* **24**(11), 2319-45.

MIRALLES, F., CZERNICHOW, P. (1998). Follistatin regulates the relative proportions of endocrine versus exocrine tissue during pancreatic development. *Development* **125**(6), 1017-1024.

MITSUI, K., TOKUZAWA, Y. (2003). The homeoprotein Nanog is required for maintenance of pluripotency in mouse epiblast and ES cells. *Cell* **113**(5), 631-42.

MIYAZAKI, S., YAMATO, E. (2004). Regulated expression of pdx-1 promotes in vitro differentiation of insulin-producing cells from embryonic stem cells. *Diabetes*

53(4), 1030-7.

Murtaugh, L. C., Law, A. C. (2005). beta-Catenin is essential for pancreatic acinar but not islet development. *Development* **132**(21), 4663-4674.

Niwa, H., Burdon, T. (1998). Self-renewal of pluripotent embryonic stem cells is mediated via activation of STAT3. *Genes Development* **12**(13), 2048-60.

Niwa, H., Miyazaki, J. (2000). Quantitative expression of Oct-3/4 defines differentiation, dedifferentiation or self-renewal of ES cells. *Nature Genetics* **24**(4), 372-6.

Norgaard, G. A., Jensen, J. N. (2003). FGF10 signaling maintains the pancreatic progenitor cell state revealing a novel role of Notch in organ development. *Developmental Biology* **264**(2), 323-338.

Odorico, J. S., Kaufman, D. S. (2001). Multilineage differentiation from human embryonic stem cell lines. *Stem Cells* **19**(3), 193-204.

Oktem, O., Oktay, K. (2009). Current knowledge in the renewal capability of germ cells in the adult ovary. *Birth Defects Research C Embryo Today* **87**(1), 90-95.

Papadopoulou, S. and Edlund H. (2005). Attenuated Wnt signaling perturbs pancreatic growth but not pancreatic function. *Diabetes* **54**(10), 2844-2851.

Parrish, Y. K., Baez, I., Milford, T.A., Benitez, A., Galloway, N., Rogerio, J.W., Sahakian, E., Kagoda, M., Huang, G., Hao, Q.L., Sevilla, Y., Barsky, L.W., Zielinska, E., Price, M.A., Wall, N.R., Dovat, S., Payne, K.J. (2009). IL-7 Dependence in human B lymphopoiesis increases during progression of ontogeny from cord blood to bone marrow. *Journal of Immunology* **182**(7), 4255-66.

Pau, K. Y. And Wolf D. P. (2004). Derivation and characterization of monkey embryonic stem cells. *Reproductive Biological Endocrinology* **2**, 41.

Pesce, M., Anastassiadis, K. (1999). Oct-4, lessons of totipotency from embryonic stem cells. *Cells Tissues Organs* **165**(3-4), 144-52.

Pieler, T. and Chen Y. (2006). Forgotten and novel aspects in pancreas development. *Biol Cell* **98**(2), 79-88.

Prelle, K., Vassiliev, I. M. (1999). Establishment of pluripotent cell lines from vertebrate species--present status and future prospects. *Cells Tissues Organs* **165**(3-4), 220-36.

Prockop, D. J. (1997). Marrow stromal cells as stem cells for nonhematopoietic tissues. *Science* **276**(5309), 71-4.

Qu, X. B., Pan, J., Zhang, C., Huang, S.Y. (2008). Sox17 facilitates the differentiation of mouse embryonic stem cells into primitive and definitive endoderm in vitro. *Developmental Growth and Differentiation* **50**(7), 585-93.

Rahier, J. (1988). The diabetic pancreas, a pathologist's view. New York, Springer-Verlag.

Resnick, J. L., Bixler, L. S. (1992). Long-Term Proliferation of Mouse Primordial Germ-Cells in Culture. *Nature* **359**(6395), 550-551.

Reubinoff, B. E., Pera, M. F. (2000). Embryonic stem cell lines from human blastocysts, somatic differentiation in vitro. *Nature Biotechnology* **18**(4), 399-404.

Reynolds, B. A. and Weiss S. (1992). Generation of neurons and astrocytes from isolated cells of the adult mammalian central nervous system. *Science* **255**(5052), 1707-10.

Richards, M., Fong, C. Y. (2002). Human feeders support prolonged undifferentiated growth of human inner cell masses and embryonic stem cells. *Nature Biotechnology* **20**(9), 933-936.

ROBERTSON, E. (1987). Embryo-derived stem cell lines. In, Teratocarcinoma and embryonic stem cells, a practical approach. Oxford, UK, IRL.

SASAKI, E., HANAZAWA, K. (2005). Establishment of novel embryonic stem cell lines derived from the common marmoset (*Callithrix jacchus*). *Stem Cells* **23**(9), 1304-13.

SATO, N., MEIJER, L. (2004). Maintenance of pluripotency in human and mouse embryonic stem cells through activation of Wnt signaling by a pharmacological GSK-3-specific inhibitor. *Nature Medicine* **10**(1), 55-63.

SCHOLER, H. R., HATZOPOULOS, A. K. (1989). A family of octamer-specific proteins present during mouse embryogenesis, evidence for germline-specific expression of an Oct factor. *Embo Journal* **8**(9), 2543-50.

SCHOONJANS, L., ALBRIGHT, G. M. (1996). Pluripotential rabbit embryonic stem (ES) cells are capable of forming overt coat color chimeras following injection into blastocysts. *Molecular Reproductive Development* **45**(4), 439-43.

SERAKINCI, N. AND KEITH W. N. (2006). Therapeutic potential of adult stem cells. European *Journal of Cancer* **42**(9), 1243-6.

SHAMBLOTT, M. J., AXELMAN, J. (2001). Human embryonic germ cell derivatives express a broad range of developmentally distinct markers and proliferate extensively in vitro. *Proceedings of the National Academy of Sciences U S A* **98**(1), 113-118.

SHIRAKI, N., HIGUCHI, Y., HARADA, S., UMEDA, K., ISAGAWA, T., ABURATANI, H., KUME, K., KUME, S. (2009). Differentiation and characterization of embryonic stem cells into three germ layers *Biochemical and Biophysical Research Communications* 278-283

SHIROI, A., YOSHIKAWA, M. (2002). Identification of insulin-producing cells derived from embryonic stem cells by zinc-chelating dithizone. *Stem Cells* **20**(4), 284-92.

SLACK, J. M. W. (1995). Developmental Biology of the Pancreas. *Development* **121**(6), 1569-1580.

SMITH, A. G. (2001). Embryo-derived stem cells, of mice and men. *Annual Review Cell Developmental Biology* **17**, 435-62.

SMITH, A. G., HEATH, J. K. (1988). Inhibition of pluripotential embryonic stem cell differentiation by purified polypeptides. *Nature* **336**(6200), 688-90.

SOLTER, D. AND KNOWLES B. B. (1978). Monoclonal antibody defining a stage-specific mouse embryonic antigen (SSEA-1). *Proceedings of the National Academy of Sciences, USA* **75**(11), 5565-9.

SORIA, B., ROCHE, E. (2000). Insulin-secreting cells derived from embryonic stem cells normalize glycemia in streptozotocin-induced diabetic mice. *Diabetes* **49**(2), 157-62.

STEWART, C. L., GADI, I. (1994). Stem-Cells from Primordial Germ-Cells Can Reenter the Germ-Line. *Developmental Biology* **161**(2), 626-628.

SUEMORI, H., TADA, T. (2001). Establishment of embryonic stem cell lines from cynomolgus monkey blastocysts produced by IVF or ICSI. *Developmental Dynamics* **222**(2), 273-9.

TERAMURA, T., ONODERA, Y., MURAKAMI, H., ITO, S., MIHARA, T., TAKEHARA, T., KATO, H., MITANI, T., ANZAI, M., MATSUMOTO, K., SAEKI, K., FUKUDA, K., SAGAWA, N., HOSOI, Y. (2009). Mouse Androgenetic Embryonic Stem Cells Differentiated to Multiple

Cell Lineages in Three Embryonic Germ Layers *In Vitro*. *Journal of Reproductive Development*. **55**(3),283-92.

THOMSON, J. A., ITSKOVITZ-ELDOR, J. (1998). Embryonic stem cell lines derived from human blastocysts. Science 282(5391), 1145-1147.

THOMSON, J. A., KALISHMAN, J. (1995). Isolation of a primate embryonic stem cell line. *Proceedings of the National Academy of Sciences, USA* **92**(17), 7844-8.

TROUNSON, A. (2006). The production and directed differentiation of human embryonic stem cells. *Endocrine Review* **27**(2), 208-19.

VALLIER, L., MENDJAN, S., BROWN, S., CHNG, Z., TEO, A., SMITHERS, L.E., TROTTER, M.W., CHO, C.H., MARTINEZ, A., RUGG-GUNN, P., BRONS, G., PEDERSEN, R.A. (2009). Activin/Nodal signalling maintains pluripotency by controlling Nanog expression. *Development* **136**(8), 1339-49.

VASSILIEVA, S., GUAN, K. (2000). Establishment of SSEA-1- and Oct-4-expressing rat embryonic stem-like cell lines and effects of cytokines of the IL-6 family on clonal growth. *Experimental Cell Research* **258**(2), 361-73.

VISWANATHAN, S., BENATAR, T. (2002). Ligand/receptor signaling threshold (LIST) model accounts for gp130-mediated embryonic stem cell self-renewal responses to LIF and HIL-6. *Stem Cells* **20**(2), 119-38.

VRANA, K. E., HIPP, J. D. (2003). Nonhuman primate parthenogenetic stem cells. *Proceedings of the National Academy of Sciences, USA* **100** Suppl 1, 11911-6.

WATT, F. M. AND HOGAN B. L. (2000). Out of Eden, stem cells and their niches. *Science* **287**(5457), 1427-30.

WEISSMAN, I. L. (2000). Stem cells, units of development, units of regeneration, and units in evolution. *Cell* **100**(1), 157-68.

WELLS, J. M. AND D. A. MELTON (1999). Vertebrate endoderm development. *Annual Review Cell Developmental Biology* **15**, 393-410.

WHITE, T. (1973). Surgical anatomy of the pancreas. Saint Louis, The CV Mosby.

WILES, M. V. AND KELLER G. (1991). Multiple hematopoietic lineages develop from embryonic stem (ES) cells in culture. *Development* **111**(2), 259-67.

WILLIAMS, R. L., HILTON, D. J. (1988). Myeloid-Leukemia Inhibitory Factor Maintains the Developmental Potential of Embryonic Stem-Cells. *Nature* **336**(6200), 684-687.

WOBUS, A. M. AND BOHELER K. R. (2005). Embryonic stem cells, prospects for developmental biology and cell therapy. *Physiological Review* **85**(2), 635-78.

WOBUS, A. M., HOLZHAUSEN, H. (1984). Characterization of a Pluripotent Stem-Cell Line Derived from a Mouse Embryo. *Experimental Cell Research* **152**(1), 212-219.

XU, C., INOKUMA, M. S. (2001). Feeder-free growth of undifferentiated human embryonic stem cells. *Nature Biotechnology* **19**(10), 971-4.

YAMADA, T., YOSHIKAWA, M. (2002). In vitro differentiation of embryonic stem cells into hepatocyte-like cells identified by cellular uptake of indocyanine green. *Stem Cells* **20**(2), 146-54.

YING, Q. L., STAVRIDIS, M. (2003). Conversion of embryonic stem cells into neuroectodermal precursors in adherent monoculture. *Nature Biotechnology* **21**(2), 183-6.

ZHAO, Y., LIN, B., DARFLINGER, R., ZHANG, Y., HOLTERMAN, M.J., SKIDGEL, R.A. (2009). Human cord blood stem cell-modulated regulatory T lymphocytes reverse the autoimmune-caused type 1 diabetes in nonobese diabetic (NOD) mice. *PLoS ONE* **4**(1), e4226.

Biotechnology and Genetic Engineering Reviews - Vol. 27, 305-330 (2010)

Mechanistic insights into laccase-mediated functionalisation of lignocellulose material

GIBSON S. NYANHONGO*, TUKAYI KUDANGA, ENDRY NUGROHO PRASETYO AND GEORG M. GUEBITZ

Graz University of Technology, Institute of Environmental Biotechnology, Petersgasse 12/1, A-8010, Graz, Austria

Abstract

Recent emerging studies on the grafting mechanisms of functional molecules onto complex lignocellulose moieties have shown useful insights and possibilities in opening new frontiers in the enzymatic development of multifunctional polymers. Thanks to these studies which have demonstrated in principle the ability of laccases to mediate the coupling of antimicrobial compounds, hydrophobic molecules, including application processes for the development of fibreboards, particle boards, laminates etc. Further, laccase mediated grafting strategies developed using small reactive molecules e.g. phenolic amines which impart reactive properties to an inert polymer demonstrates the remarkable opportunities of enzyme meditated functionalization of polymers. Therefore recent studies focusing on understanding the mechanistic basis of the coupling mechanisms in order to make meaningful contribution to the development of new processes and products are a welcome development.

*To whom correspondence may be addressed (gnyanhongo@yahoo.com; g.nyanhongo@tugraz.at)

Abbreviations: H: *p*-hydroxyphenyl units; G: guaiacyl units; S: syringyl; Py-GC/MS: Pyrolysis Gas Chromatography –Mass Spectroscopy; HSQC: Heteronuclear Single Quantum Coherence; HSQC–NMR: Heteronuclear Single Quantum Coherence- Nuclear magnetic resonance; FTIR: Fourier transform infrared spectroscopy; HBT: 1-hydroxybenzotriazole; ^{13}C-NMR: ^{13}Carbon Nuclear magnetic resonance; ^{31}P NMR: ^{31}Phosphate Nuclear magnetic resonance; TMP: thermomechanical pulps; PF: Phenol formaldehyde; MDF: medium density fiberboards

Introduction

Enzyme technology entered its most fascinating and exciting phase since the beginning of the last decade, as biomaterials engineers, confronted with restrictive governmental regulations owing to global climate change, became aware of the great potential of biological systems. Driving this new thrust are three major new goals, that is, maximizing the exploitation of renewable resources as sources of raw materials for the production of multifunctional polymers, development of environmentally friendly processes, and development of biodegradable products. Among the natural polymers highly sought after are lignocellulose polymers, which are the most abundant available renewable resource consisting mainly of cellulose (40 - 50 %), lignin (18 - 35 %) and hemicellulose (15 - 25 %) (Friedman, 1998; Sjostrom, 1993). Lignocellulose materials touch all aspects of our lives; uses range from sources of food, construction material, and furniture to raw materials used for the production of fine chemicals with many industrial and medical applications. In the construction, textile and furniture industries lignocellulose materials occupy a special place because of their impressive range of attractive properties, including low thermal extension, low density and high mechanical strength. Despite these attractive properties, lignocellulose materials are highly hygroscopic and susceptible to biodeterioration, factors which are partly responsible for the continued search for better lignocellulose processing technologies in the construction, furniture, textile, and other industries (Nugroho-Prasetyo *et al.* 2009). Among the enzymes, laccase mediated modification of lignocellulose materials has recently attracted a lot of scientific interest and is emerging as one of the most attractive options to impart a variety of functional groups resulting in tailor-made functional polymers (polymers modified to contain chemical groups that serve a specific function, whether biological, chemical or physical) for targeted applications. In recent years, there has been a lot of research output describing novel processes and products derived from enzymatic lignocellulose functionalisation. This review provides a general overview of the recent advances in laccase-mediated grafting of functional molecules onto lignocellulose materials. In addition fundamental aspects of the reaction mechanisms are summarized.

Laccase: distribution and brief characterization

Laccases were first discovered in plants (Japanese lacquer tree - *Rhus vernicifera*) by Yoshida in 1883. Since then laccases have been reported in insects and are increasingly being reported in bacterial systems (Claus, 2003; Claus, 2004). Nevertheless, fungi, among them the white-rots, are the best producers of laccases although some Ascomycetes and Deuteromycetes have been shown to produce significant amounts of laccases too. In organisms, laccases have been reported to participate in insect sclerotisation and bacterial melanisation (Allexandre and Zhullin, 2000), spore resistance (Aramayo and Timberlake, 1993), lignification of plant cell walls (O' Malley *et al.* 1993), biodegradation of lignin as well as detoxification processes (Hermann *et al.*, 1983) and as virulence factors in pathogenic organisms (Williamson *et al.* 1998). Laccase distribution, biochemical characterization and other general applications have been subject of many elegant reviews (Yaroplov *et al.* 1994; Claus, 2004; Baldrian,

2006; Minussi *et al.* 2002; Kobayashia and Higashimura, 2003; Mayer and Staples, 2002; Riva, 2006; Rodriguez Cuoto and Toca, 2006; Kunamneni *et al.* 2009) and thus will not be addressed here.

Laccases (benzenediol: oxygen oxidoreductase, EC 1.10.3.2) are glycoproteins (520 - 550 amino acids including a N-terminal secretion peptide) with a carbohydrate content of 10 - 30%, four copper atoms bound to 3 redox sites (T1, T2 and T3) at its active center and average molecular weight between 50 and 80 kDa (Kim *et al.* 2002). They catalyze the oxidation of various phenolic substrates (phenols, polyphenols, anilines, aryl diamines, methoxy-substituted phenols, hydroxyindols, benzenethiols and many others) simultaneously reducing molecular oxygen to water (Nyanhongo *et al.* 2005). At the active centre, the trinuclear cluster (containing Cu T2 and two Cu T3) is located approximately 12 Å away from the T1 site and it is the place where molecular oxygen is reduced to water (Alcalde, 2007). Since the enzyme is able to abstract one-electron at a time, the enzyme reaction mechanism has been speculated to operate as a battery by storing electrons from the four individual oxidation reactions of four molecules of substrate required to reduce molecular oxygen to two molecules of water. In the active center, the oxidation of substrate at the T1 site and the electrons flow towards the trinuclear cluster where molecular oxygen is reduced to two water molecules (*Figure 1*).

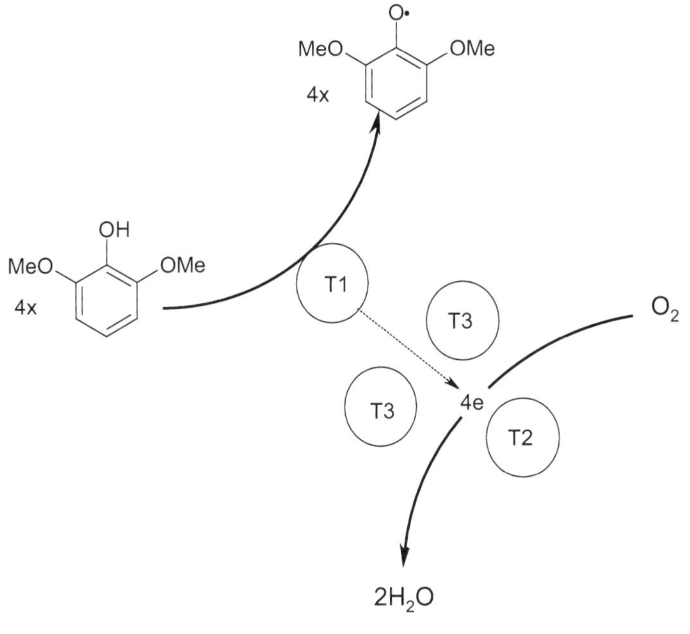

Figure 1. Laccase reaction mechanism.

The ability of laccase to abstract an electron and simultaneously oxidize a variety of phenolic molecules thereby generating reactive species forms the basis of their application in polymer functionalisation. The generated reactive species provide ideal sites for cross-linking (coupling) desired functional molecules leading to polymerization reactions and consequently formation of new materials. Although the redox potential

(E°) of laccases is between +450 and +790 mV (Yaropolov *et al.* 1994, Rebrikov *et al.*, 2006, Tadesse *et al.*, 2008, Melo *et al.*, 2007), laccase activity can still be extended to molecules with higher E° values by using small molecules (redox mediators) which when oxidized by laccases form highly reactive species able to attack and/or even diffuse inside complex polymers where laccase alone have no access. By combining laccase and redox mediators the potential application of laccase in wastewater treatment, detoxification, degradation of pollutants, bioremediation, dye decolourization, delignification, construction of biosensors, pulp bleaching and modification of natural and synthetic polymers, removal of phenolic compounds in beverages and biofuel cells has been made possible (Nyanhongo *et al.* 2007).

LIGNIN AS THE PRIMARY TARGET POLYMER FOR FUNCTIONALISATION OF LIGNOCELLULOSE MATERIALS

Lignin is the primary target for lignocellulose functionalization processes because unlike cellulose and hemicellulose which are attacked by hydrolytic enzymes (cellulases and hemicellulases), enzymes attacking lignin (peroxidases and laccases) restrict their activities to modifying the functional groups thereby leaving the aromatic backbone largely unaltered. Lignin is structurally and strategically positioned as it forms an encrusting material on and around the carbohydrate fraction which makes it easily accessible for modification. Structurally, lignin is a three dimensional optically inactive polymer formed by the dehydrogenative polymerization of three *p*-hydroxycinnamyl alcohol precursor molecules coumaryl alcohol, coniferyl alcohol and sinapyl alcohol (*Figure* 2) linked together in an irregular manner (Sjöström, 1993; Boudet, 2000; Martínez *et al.*, 2008). ρ-Coumaryl alcohol is a minor component of grass and forage type lignins and is also a minor precursor of softwood and hardwood lignins.

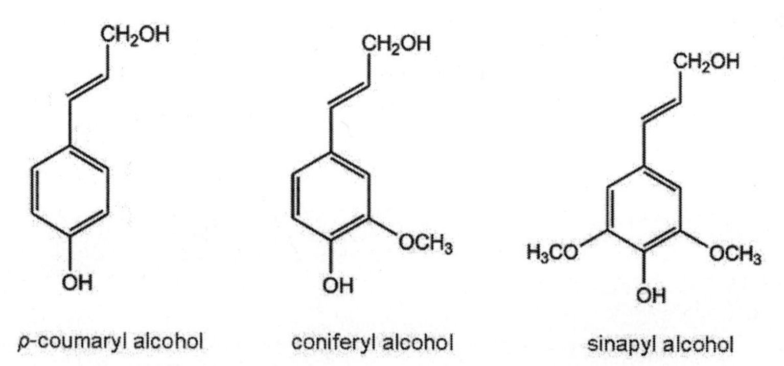

Figure 2. Major lignin precursor molecules.

Coniferyl alcohol is the predominant lignin monomer found in softwoods. Both coniferyl and sinapyl alcohols are the building blocks of hardwood lignin. However, acetylated (e.g. acetylated, *p*-coumaroylated, *p*-hydroxybenzoylated) lignins (at the γ-carbon) have also been observed (Martínez *et al.*, 2008). Due to plasticity in lignin structure, its composition is generally characterized by the relative abundance of *p*-hydroxyphenyl (H), guaiacyl (G) and syringyl (S) units (derived from each of the

3 primary monolignols, respectively) and by the distribution of interunit linkages in the polymer (Ralph *et al.*, 2004). The pathway for lignin biosynthesis is relatively well characterized (Li *et al.*, 2000; Dixon *et al.*, 2001; Li *et al.*, 2001; Humphreys and Chapple, 2002; Baucher *et al.*, 2003; Boerjan *et al.*, 2003) and interestingly involving the participation of peroxidases or laccases in radical formation. Peroxidase or laccase oxidation of lignin building blocks result in a series of reactive species which are in equilibrium with one another through resonance stabilization (*Figure 3*).

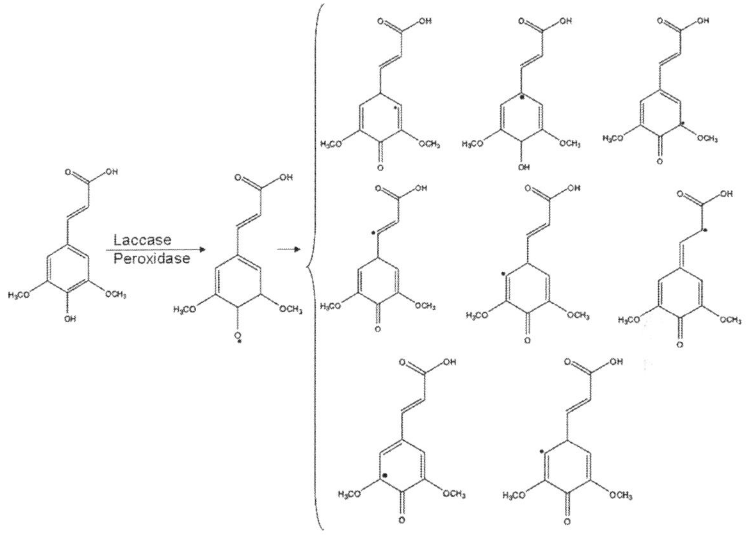

Figure 3. The major resonance stabilized free radicals of sinapyl alcohol.

This gives rise to the multitude of coupling possibilities depending on reacting molecule and/or position of the radical thereby generating a series of inter-unit linkages. Characteristic linkages such as β-*O*-4, β-5, 5-5, β-β, 5-*O*-4 and the more recently discovered dibenzodioxocin 5-5-*O*-4 (Karhunen *et al.*, 1995a;b) (*Figure 4*) are very common. These inter-unit linkages are globally classified as non-condensed (mainly aryl-alkyl-β-*O*-4 ether bonds) and condensed lignin (mainly C-C bonds) (Kukkola *et al.* 2003). The percentage abundance of the linkages has been determined and it has been shown that the β-*O*-4 linkage is the predominant linkage estimated to be as high as 50 % in softwood and 60 % in hardwood (Kukkola *et al.*, 2003; Adler, 1977; Sakakibara, 1980).

Enzymatic lignocellulose polymer functionalization processes mimics the same natural processes, involving oxidation of lignin moieties of the lignocellulose polymer resulting in a radical-rich reactive polymer to which molecules of interest can be grafted (Kudanga *et al.*, 2008). Logically the oxidative enzymes intensely being investigated for lignocellulose functionalization are the same enzymes involved in the biosynthesis of lignin although enzymes of fungal origin are preferred partly due to their easy of production. Among these enzymes, laccases are gaining favor instead of peroxidases since the later require an expensive cofactor (hydrogen peroxide). To date a number of strategies for lignocellulose functionalisation have been described catalysed by laccases.

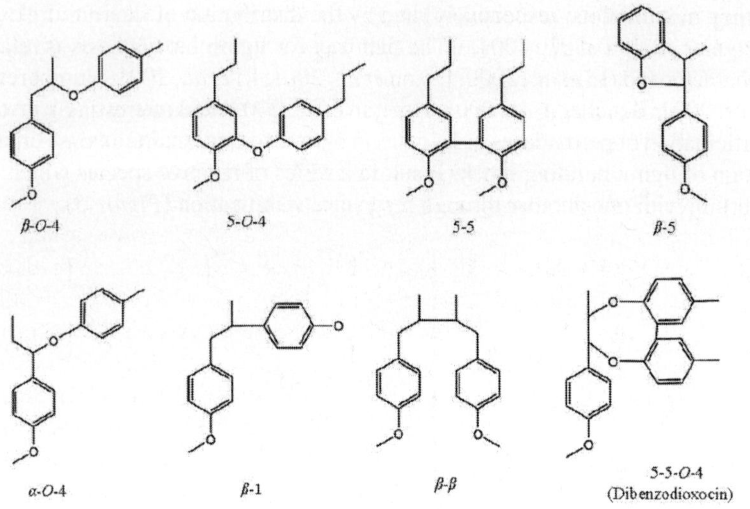

Figure 4. Common linkages in lignin.

Creating reactive polymers

FUNCTIONAL POLYMERS DERIVED FROM LIGNIN

Lignin is the major by product of the pulp and paper industry with only 2 % exploited commercially (1 million tons/year of lignosulphonates and less than 100,000 tons/year of kraft lignins) (Gargulak and Lebo, 2000; Gosselink *et al.* 2004; Lora and Glasser, 2002). Massive exploitation of lignin is hampered by its huge physico-chemical heterogeneity owing to its inherent variety of aromatic units, inter-unit linkages, functional groups, and molecular size (Nugroho Prasetyo *et al.* 2010). This is a result of both the heterogeneity of this plant polymer and its random degradation during the pulping process resulting in a heterogenous polymer lacking interfacial adhesion properties, making it immiscible in polymer blends. For example, increasing lignin concentration in thermoplastics and rubber blends negatively affected the tensile force and melt flow index of the product (Alexy *et al.* 2000), while its poor adhesion and dispersion properties limits its addition to between 5 and 10 % of the level of the resin weight in adhesive synthesis (Mansouri and Salvado, 2007; Turunen *et al.*, 2003). It is not surprising therefore that, grafting of small reactive molecules onto lignin in an effort to produce phenolic resins was perhaps the first demonstration of the potential industrial importance of laccases and heme peroxidases. Phenolic resins are widely used in surface coatings, laminates, molding, friction materials, abrasives, flame retardants, carbon membranes and as adhesives among other uses (Eker *et al.* 2009). Dordick and coworkers, using horse raddish peroxidase, were the first to demonstrate the incorporation (coupling) of phenolics onto lignin leading to the formation of reactive polymers with great potential as phenolic resins. Nevertheless, Milstein *et al.* 1994 subsequently used laccase to graft a variety of compounds onto organosolv lignin in dioxane-H_2O. Laccase mediated grafting of guaiacol sulfonate and vanillylamine onto Indulin (kraft lignin) formed a precipitate containing nitrogen indicating co-polymerization, while the reaction with guaiacol sulfonate did not form a precipitate due to the presence of the sulphur groups. Acrylamide was also successfully grafted onto lignin resulting in increased Mw (Mai *et al.* 1999, 2000). This observation has great industrial implications since acrylamide is used as a precursor in the synthesis of water-soluble thickeners with applications in gel electrophoresis, papermaking, synthesis of dyes and press fabrics. Recently, a highly thermostable polymer with good thermosetting properties in ionic liquids was produced (Eker *et al.* 2009). In another very interesting development, treatment of brown-rot-fungus-decayed wood with sodium borohydride followed by mixing with polyethylenimine resulted in a formaldehyde free, strong and water-resistant wood adhesive (Li and Geng, 2005). The pre-treatment of the wood with brown-rot fungi enabled preferential degradation of cellulose and hemicellulose in the wood while simultaneously the oxidoreductases (laccases, peroxidases etc) oxidized and demethylated lignin (Lin and Geng, 2005) resulting in an oxidized ortho-quinone reactive polymer. The reactivity of lignin depends on nature of lignins, position of the radical after laccase oxidation or introduced functional molecules. For example highly methoxylated lignin are less reactive as the methoxyl groups occupy potential coupling sites. Recently, Nugroho Prasetyo *et al.* 2010 provided a detailed insight into laccase modifications of lignins which may partly explain the observed modifications novel products with excellent properties described above. For example, fluorescence spectroscopy showed the ability of *Trametes villosa* and *Trametes hirsuta* laccase-HBT (hydroxybenzotriazole) systems to modify conjugated carbonyls, biphenyls, phenylcoumarins and stilbene groups of the lignosulphonates resulting in a gradual loss of fluorescence.

The polymerizations leading to the formation of new ether and C-C aryl-aryl (characteristic of condensed lignin) or aryl-alkyl linkages were accompanied by a decrease in phenolic hydroxyl groups and carboxylic groups. The polymerization was so intense that HSQC–NMR could not detect the aromatic signal (*Figure 5*).

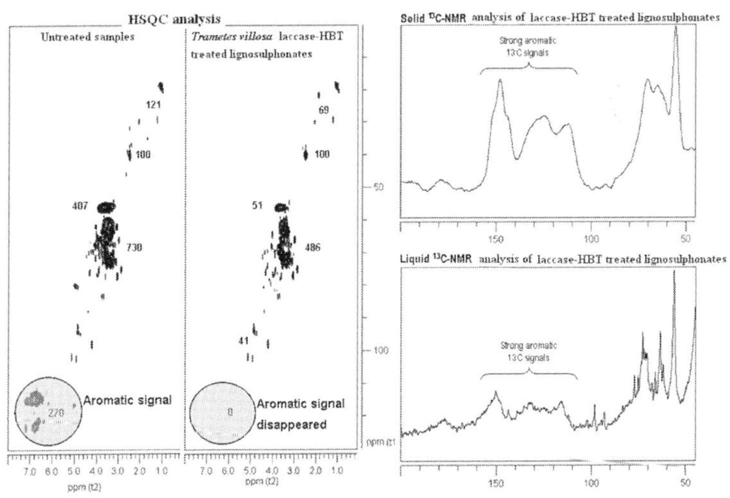

Figure 5. HSQC, Solid ^{13}C- and lquid ^{13}C-NMR analysis of lignosulphates modified by *Trametes villosa* laccase-HBT system (Nugroho Prasetyo *et al.*, 2010).

Nevertheless, FTIR spectroscopy, ^{13}C-NMR and Py-GC/MS analysis suggested no changes in the aromatic backbone, a good indication of the laccases-HBT system to limit its activities to the functional groups. The modifications resulted in an increase in the dispersion properties of the lignosluphonates, a positive development for improving lignin polymer blends. Py-GC/MS analysis of the same samples showed a decrease in the intensity of 4-methylguaiacol, 4-ethylguaiacol, guaiacylacetone, 4-vinylguaiacol, homovanillylalcohol, eugenol, *cis-* and *trans*-isoeugenol, dihydroconiferyl alcohol and *trans*-coniferaldehyde.

CREATING A STABLE REACTIVE SURFACE

Once an electron is abstracted, the substrate forms a radical which usually tries to stabilize itself through resonance or by reacting with any other molecule in its environment (*Figure 3*). Therefore it is not surprising that up to 90 % of radicals formed on wood surfaces were reported to be quenched within a few hours in laccase-bleached thermo-mechanical pulp (Grönqvist *et al.* 2006). In addition, the reactivity of the resonance structures depends on the position of the radical. If the radical is shielded by substituent molecules like the methoxyl groups as shown in *Figure 3,* it becomes unreactive. Recently, in an effort to create a stable reactive surface, Kudanga *et al.* 2008 developed a novel strategy based on grafting various structurally different phenolic amines onto laccase-oxidized beech veneers to act as anchor groups. Grafting of these reactive aromatic amines onto beech veneers resulted in increased coupling of antifungal molecules onto the reactive surface as compared to the control. The authors also provided the first mechanistic evidence for the covalent binding of aromatic amines onto different lignin model compounds like syringylglycerol- guaiacylether and dibenzodioxocin (Kudanga *et al.* 2009, 2010a) demonstrating that the $-NH_2$ group was indeed free for further coupling reactions as shown in *Figure 6*. Regardless of the chemical structure, all the phenolic amines preferred to couple onto dibenzodioxocin through their position 5 resulting in C-C (5-5) coupling (typical of condensed lignin) (*Figure 6*).

However, in the presence of syringylglycerol- guaiacylether, a 4-*O*-5 coupling was preferred. From the coupling products obtained with syringylglycerol- guaiacylether it is evident that the presence of methoxyl groups blocks the 5 position promoting binding at the 4th position (*Figure 7*). Nevertheless, regardless of the nature of coupling, $-NH_2$ group was free for further grafting reactions thereby acting as anchor groups and providing a stable reactive surface.

Improving hydrophobicity

As already mentioned elsewhere in this review, lignocellulose materials are hydrophilic leading to swelling thereby increasing volume and reducing the mechanical strength of lignocellulose polymers. Improving hydrophobicity is one way of reducing biodeterioration as this will discourage the growth of microorganisms. Traditionally physico-chemical methods are used to improve hydrophobicity. These include application of water repellents like waxes, oils, natural or synthetic resins (Borgin and Corbett,1970; Hyvönen *et al.*, 2006), use of silicon compounds (Mai and Militz, 2004), impregnation with hydrophobic oil (Ulvcrona *et al.*, 2006), using binders containing hydrophobic diluents and thermal wood processing technologies. However, the water repellent substances are mainly bonded to the cell wall mainly by relatively weak Van der Wall forces and over a long time they are washed off (Razzaque 1982).

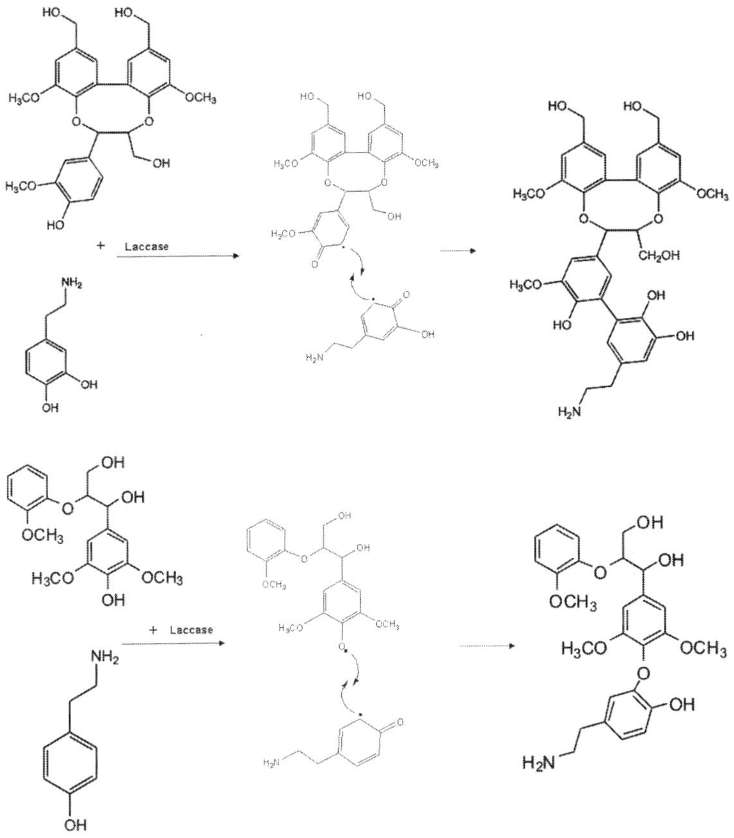

Figure 6. *In vitro* grafting studies of different phenolic amines (anchor groups) onto dibenzodioxocin (a complex lignin model)

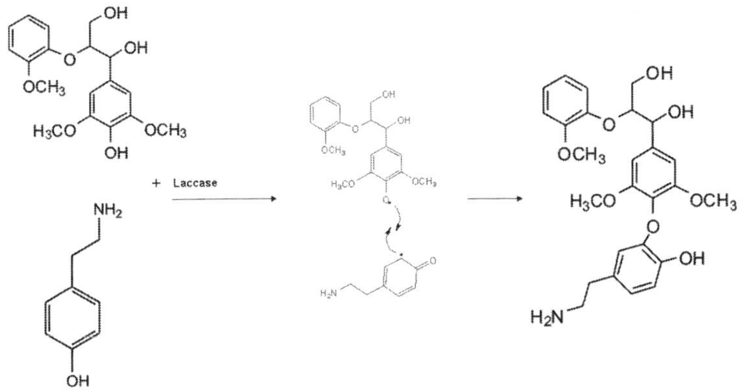

Figure 7. Grafting of tyramine onto syringylglycerol- guaiacylether.

Modifications with silicon compounds usually involves complicated and expensive multi-step processes while due to high chemical and weathering stability, such treatments are usually recommended for wood exposed to conditions of hazard class III

(EN 335, outside exposure without soil contact) (Mai and Militz, 2004). Alternative technologies developed to improve dimensional stability such as chemical modification of wood through for example acetylation (Obataya *et al.*, 2002) and thermal degradation of hemicellulose components (Viitaniemi and Jämsä, 1996) have been shown to reduce mechanical properties such as tensile strength (Vick and Rowell, 1990; Ramsden and Blake, 1997).

Enzymatic modification has therefore emerged as an attractive alternative for increasing lignocellulose hydrophobicity. Although no mechanistic evidence was provided, laccase mediated grafting of laurylgallate onto pulps increased their hydrophobicity (Suurnäkki *et al.*, 2006). Recently, laccase-catalysed coupling of fluorophenols was shown to increase hydrophobicity of wood veneers (Kudanga *et al.*, 2010a, 2010b). The authors also provided the first *in vitro* evidence for the possibility of establishing covalent bonds between complex lignin models [syringylglycerol β-guaiacylether (G-S-β- ether), guaiacylglycerol β-guaiacyl (erol) and dibenzodioxocin] (Kudanga *et al.*, 2010a, 2010b). The proposed mechanism showed that the fluorophenols were bonded to sinapyl units via 4-*O*-5 linkages while coupling to guaicyl units occurred through 5-5 linkages (*Figure 8*).

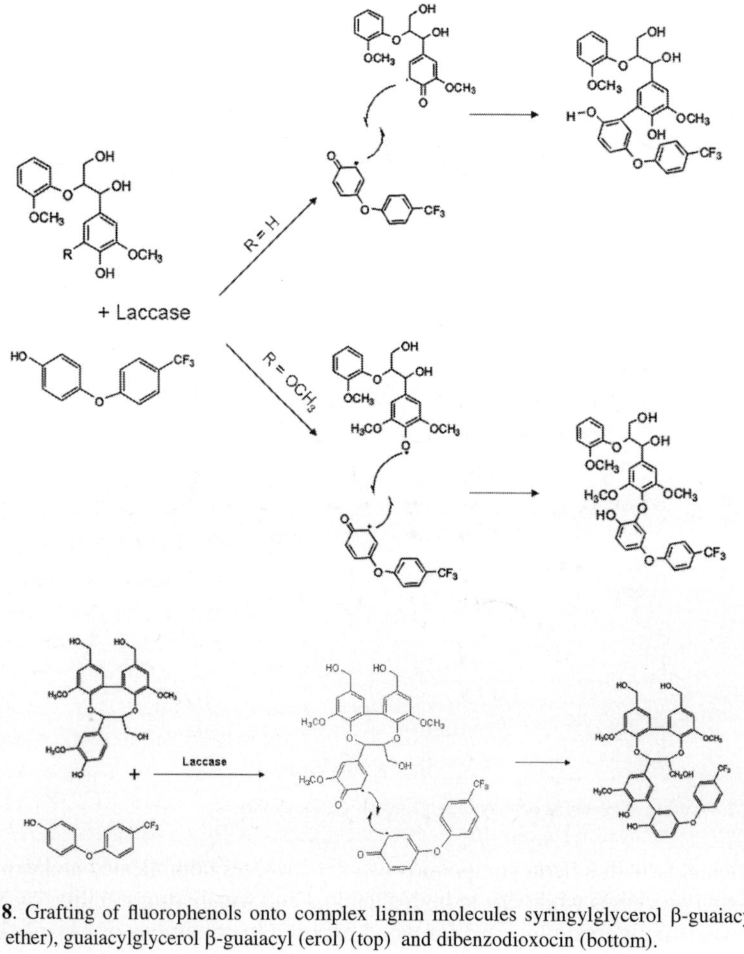

Figure 8. Grafting of fluorophenols onto complex lignin molecules syringylglycerol β-guaiacylether (G-S-β- ether), guaiacylglycerol β-guaiacyl (erol) (top) and dibenzodioxocin (bottom).

The same authors also elegantly demonstrated the ability of laccase to couple long chain alkylamines onto wood veneers resulting in an increase in hydrophobicity (Kudanga *et al.*, 2010c). The authors further provided evidence for the possibility of grafting alkyl amines onto diverse phenolics and complex lignin models as structural components of lignocellulose materials. Mechanistic evidence showed that both simple phenolics and lignin models bonded with the amines mainly through –C-N coupling (*Figure 9*).

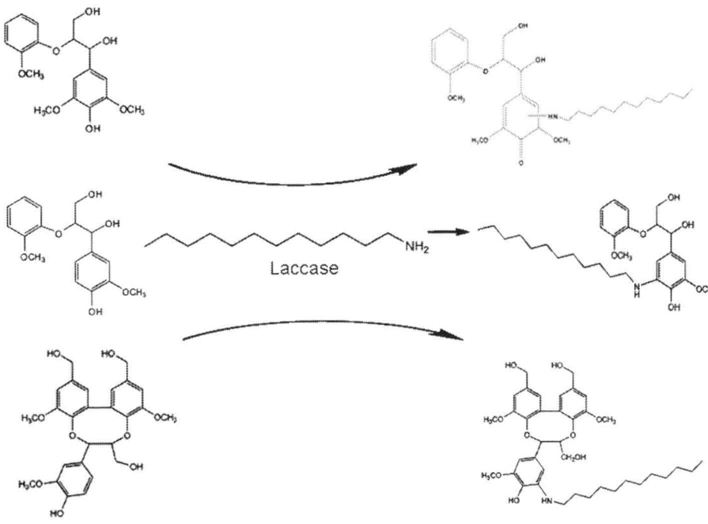

Figure 9. Grafting of an alkylamine onto different lignin molecules.

Antimicrobial properties

Plants produce a remarkable diverse array (>100,000) secondary metabolites mostly derived from the isoprenoid, phenylpropanoid, alkaloid or fatty acid/polyketide pathways (Dixon, 2001). This rich diversity results in part from an evolutionary process driven by selection for acquisition of improved defense against microbial attack or insect/animal predation (Dixon, 2001). This has recently become evident as more bioactive compounds are being discovered in plants that act as effective antimicrobials. For example, antimicrobial properties of several phenolic compounds against a wide range of microorganisms are increasingly being reported (Nohynek *et al.*, 2006; Banes-Marshall, 2001; Burt, 2004; Kubo *et al.*, 2003; Almeida *et al.*, 2006; Rodríguez Vaquero *et al.*, 2007 a,b; Knobloch *et al.*, 1989; Ultee *et al.*, 2002; Harkental *et al.*, 1999). Accompanying this progress a new dimension has emerged where material scientists have seen the possibility of developing green technologies and products by incorporating plant derived antimicrobial compounds into polymers using enzymatic approaches. Consequently, a number of studies have reported laccase-mediated grafting of natural antimicrobial agents to lignocellulose materials (Schroeder *et al.*2007; Elegir *et al.*, 2008; Widstein *et al.*, 2010). Schröder *et al.* (2007) showed that laccase-assisted modification of flax fibres with ferulic acid, hydroquinone, methyl-3-hydroxy-4-methoxybenzoate, or 2-methoxy-5-nitrophenol improved their antibacterial activity toward the gram-positive *Bacillus subtilis* and gram-negative *Escherichia coli* while 2-methoxy-5-nitrophenol was also effective against *Klebsiella pneumoniae*. Elegir *et al.* (2008) successfully developed antimicrobial lignocellulose packaging material through laccase-mediated grafting of phenolic compounds. Handsheets treated with phenolic acids (caffeic acid and *p*-hydroxybenzoic acid) showed strong antibactericidal effect on *Staphylococcus aureus* and *Escherichia coli* while isoeugenol was also shown to have bacteriostatic effect against *Bacillus subtilis*. The antimicrobial activities were attributed to the presence of hydroxyls and delocalisation of the electrons on their structure (Knobloch *et al.*, 1989; Ultee *et al.*, 2002). The molecular structures of the caffeic acid and isoeugenol oligomers formed during grafting reactions were elaborated by FTIR and ^{13}C NMR/^{31}P NMR studies. The caffeic acid oligomer showed formation of lignin-like ether bonds while isoeugenol oligomers showed a prevalence of β–*O*–4 linkages with a significant amount of intermonomeric β–5 bonds also detected. Recently, our group (Widstein *et al.*, 2010) reported antimicrobial activities of tannins against *S. aureus* and *E. coli*. Laccase assisted treatments with hydrolysable tannins (tannic acid and chestnut tannin) significantly improved the antibacterial resistance of veneers and paper made from tannin-treated pulp against *Staphylococcus aureus* while a more modest protective effect was observed against *Escherichia coli*. Condensed tannin (Mimosa tannin and Quebracho tannin) improved the antibacterial resistance against *S. aureus*, albeit less than hydrolysable tannin, but had little effect on *E. coli*. A cationic condensed tannin derivative bearing a quaternary amino group provided far better protection against *S. aureus* and *E. coli* than the corresponding unmodified condensed tannin. Widsteins *et al.* (2010), provided the mechanistic evidence using putative lignin monomers and catechin as a model of flavonoid-based tannins and gallic acid as model of hydrolysable tannins to elucidate the coupling products. Catechin was bonded to the phenolic molecules mainly through 5-5 linkages while bonding to monomers with sinapyl units was mainly through 4-*O*-5 linkages (*Figure 10*). Although these substructures are known to be the most abundant in lignin (Kandanarachchi *et al.*, 2002a, b), they were not observed with catechin. This is probably because the position 5 is free in catechin molecules and 5-5 and 4- *O*-5 bonds are favored as they have a lower heat of formation than the β-*O*-4 bonds, although β-*O*-4 linkages are more abundant in lignin, dimerization and lignin synthesis are substantially different processes (Adler, 1977).

In contrast, β-*O*-4 coupling products were observed when gallic acid was coupled onto the unsaturated phenolic molecules caffeic acid, ferulic acid and sinapic acid respectively (*Figure 11*)

Preformed lignin oligomers usually couple through 5-5 and 4-*O*-5 linkages (Boerjan *et al.*, 2003), while β-O-4 bonds are favoured *in vivo*, which probably explains the different frequencies of lignin substructures obtained *in vitro* from those obtained *in vivo*.

Improving aesthetic appearance of lignocellulose polymers

The more than 6 000 years old Chinese lacquer artwork demonstrates the remarkable ability of laccase in producing colourful artwork products. However, it was not until recently that this principle has intensively being pursued for *in situ* dyeing of textile

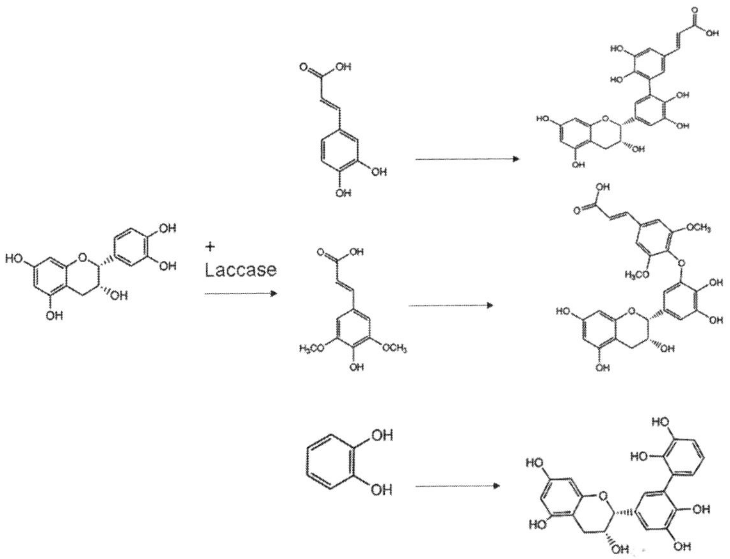

Figure 10. Grafting of catechin onto different lignin monomers

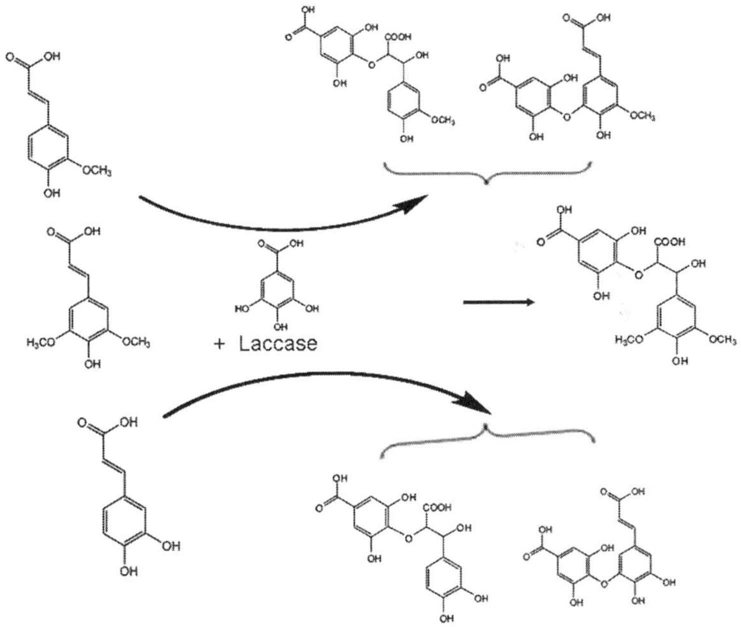

Figure 11. Coupling of gallic acid to different lignin moieties.

fibres. Laccases have the added advantage of being able to oxidise various phenolic and amine substituted molecules into coloured oligomeric forms (Li *et al.* 1999; Robles *et al.* 2000). Such a development is welcome in the textile industry since a number of auxiliary chemicals, some even toxic, drastic pH conditions and elevated temperatures are currently employed during the dyeing process (Hadzhiyska *et al.*, 2006). Laccase mediated oxidation of some colourless phenolic (hydroquinones, dopamine, guaiacol, catechol and ferulic acid) in the presence of wool resulted in fibers with different colour shades (Shin *et al.* 2001). Building on the success of this approach, Shin *et al.* 2001; Tzanov *et al.* 2003; Barfoed *et al.* 2004; Sørensen. 1999; Calafell *et al.* 2007; Barfoed *et al.* 2001, 2003, 2004; Kim *et al.* 2007, and Kim *et al.* 2008 repeatedly demonstrated the feasibility of using laccase-assisted colouring of wool, lignocellulose and cellulosic materials. Using laccase, Schroeder *et al.* (2007) achieved different colour shades of flax fibres depending on the phenol used, for example yellow with 2-methoxy-5-nitrophenol, orange with ferulic acid and red with guaiacol. However, color fastness properties were low in all the cases. Recently using the anchor group concept (creating reactive surface approach), tysolated cotton cellulose was dyed by coupling catechol resulting in coloured product covalently fixed on the fabric (Blanco *et al.*, 2009). Similarly, amine-functionalised cellulose was coated *"in situ"* with enzymatically-synthesized poly(catechol) in the presence of *Trametes villosa* laccase as a way of colouring the cellulose fibres (Kim *et al.*, 2007a). The authors also demonstrated the molecular mechanisms of the coupling (*Figure 12*) where the polycatechol covalently bind to cellulose through the $-NH_2$ group. Laccase has also been used to catalyze the oxidation of flavonoids in solution producing quinones that can be further polymerised and grafted onto natural flavonoids present on the surface of the cotton providing colour which depended on the flavonoids used (Kim *et al.*, 2007b). The natural flavonoids present on unbleached cotton acted as anchor groups, facilitating the grafting reaction, improving dyeing and colour fastness. Patents describing enzymatic (including laccases) processes for colouring of a wide range of natural and synthetic materials including lignocellulose materials have been filed (Barfoed and Kirk, 1999; Barfoed *et al.* 2004).

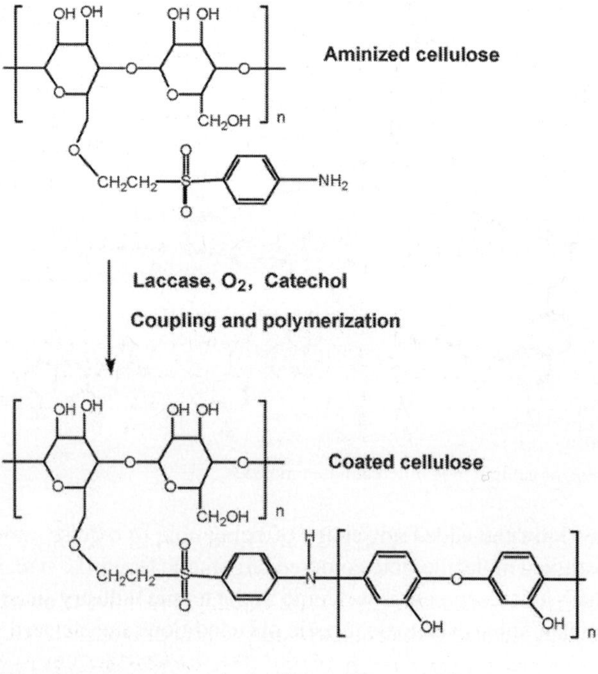

Figure 12. Grafting of catechol onto aminized cellulose (Kim *et al.* 2007a).

IMPROVING INTERNAL BOND STRENGTH OF THERMOMECHANICAL PULPS

Application of laccases has been intensively investigated in the field of pulp and paper for biopulping, biobleaching, deinking, mill process water and effluent treatment. However, recently research has also shifted toward functionalization of pulps in order to produce novel paper products. The pioneering works of Yamaguchi and co-authors (1991, 1992, 1994) demonstrated the ability of laccase to polymerize various phenolic compounds to form dehydrogenative polymers which were subsequently coupled to thermomechanical pulps (TMP) in the presence of peroxidase. Incorporation of phenolics (vanillic acid, catechol, mimosa tannin and tannic acid) in the presence of laccase enabled production of paper with increased tensile strength and water resistant properties attributed to the coupling of free phenolic groups on the fiber surface with the added dehydrogenative polymers. Huttermann *et al.* (1998), combined laccase pretreated kraft lignin with spruce sulfite pulp resulting in handsheets with higher tear strength and higher wet strength than untreated controls. The lignin was shown to irreversibly bind onto the fibers as evidenced by subsequent X-ray microanalysis in combination with transmission electron microscopy revealing a high proportion of lignin in contact areas between the fibers. In another attempt, Felby *et al.* 1997a,b observed an increase in the wet strength and decrease in bulk thickness after treatment of beech TMP fibers with laccase and a ferulic acid-arabinoxylan dimer. They noted an increase in dry tensile strength from 8.1 Nm/g to 47.0 N.m/g and attributed it to the coupling of ferulic acid to the pulp fiber leading to enhanced fiber bonding by the arabinoxylan portion of the dimer. Using a low dose of laccase/HBT, Wong *et al.* (2000, 2003) improved the strength of paper made from high-yield kraft pulp by 5-10 %. Further, Lund and Felby 2001 reported that the addition of lignin-rich extractives to laccase treated high-yield kraft pulp improved wet strength of the paper. The potential of laccase-assisted biografting of low Mw phenolics was also extensively studied for improving strength properties of paper made from high-kappa pulps by Chandra and co-authors 2001, 2002a,b, 2003. The observed improvements in paper tensile and burst strength were attributed to the capacity of carboxyl groups to promote hydrogen bonding and to the crosslinking of phenoxy radicals in the paper sheet. Surface-grafting of the cationic dye phenol celestine blue also increased the strength of high-kappa kraft paper (Chandra *et al.* 2003). Recently Witayakran and Ragauskas, 2009, successfully demonstrated the ability of laccase to graft various amino acids onto high-lignin content pulps (*Figure 13*). Laccase-histidine treated pulps showed an increase in strength properties of the resulting paper.

The addition of methyl syringate to unbleached kraft pulp in the presence of laccase resulted in doubling of its wet tensile strength (Liu *et al.* 2009). Elegir *et al.* 2007 reported that treatment of kraft liner pulp with laccase in the presence of a low-molecular-weight ultra-filtered lignin fraction leads to a two-fold increase in the wet strength of the produced kraft liner pulp hand sheets.

INCREASING INTERNAL BONDING STRENGTH OF LIGNOCELLULOSE BOARDS

Attempts to enhance lignin reactivity by use of enzymes in wood started in the mid 1960s. Empirical evidence for the ability of peroxidase or laccase modified lignosulphonates to act as binders for the production of particle boards were subsequently

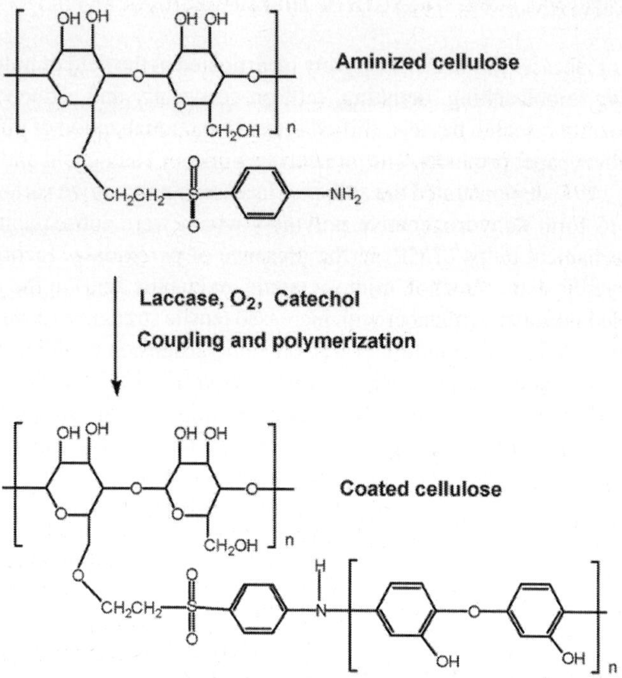

Figure 13. Propose mechanism for the grafting treatment of linerboard pulp with laccase and amino acids

reported by Nimz *et al.* 1972; 1976. They successfully produced particle boards that surpassed the requirements of transverse tensile strength (DIN 52365 test) of 0.35 MPa specified by European standard EN 312-4. Unfortunately the particle boards swelled in water due to the presence of the sulphonate groups, making them suitable for indoor use only. However, using brown-rotted lignin isolated from decayed wood instead of lignosulphonates resulted in a 35 % substitution of phenol and boards with strength comparable with that of boards glued by conventional PF resins (Jin *et al.*1990a,b) but still the wood laminates swelled in water (Jin *et al.* 1991). However, Huttermann and Kharazipour 1996, and Huttermann *et al.* 2001 reduced swelling by incorporating 1 % methylene diphenyl diisocyanate. Fungal pre-treatment of wood chips with the white-rot fungus *Trametes hirsuta* and the brown-rots *Gloeophyllum trabeum* had an implication on energy consumption, reducing it by 40 %, compared with untreated wood chips (Unbehaun *et al.* 1999, 2000). The fungi pretreated wood chips were able to bond without extra added adhesives under pressing, conditions usually applied for conventional gluing. In another different attempt, forced-air-drying of enzyme pretreated MDF increased internal bond strength and reduced swelling (Felby *et al.* 1997, 1998, 2002 and Unbehaun *et al.* 2000). Both laccase and peroxidase bonded MDF boards achieved the European standard CIN DIN 622-5. Kharazipour *et al.* (1997, 1998) used laccase to treat mechanical pulp fibres that were later used for the formation of MDF resulting in increased internal bond strength from 0.1 to 0.52 N/mm^2. Although peroxidases gave the same results as boards treated by laccase, the requirement of hydrogen peroxide and its rapid degradation constituted its main drawback (Kharazipour *et al.* 1998 a,b). MDFs produced from rape-straw fibres incubated with whole culture fermentation broth were shown to produce MDF with twice the strength compared to those obtained by purified *T. versicolor* laccase (Kuhne and Dittler 1999, Unbehaun *et al.* 2000). Laccase-catalyzed functionalization of spruce wood particles with 4-hydroxy-3-methoxybenzylurea significantly increased the internal bond strength of fibre boards in the subsequent processing steps (Fackler *et al.* 2008). This demonstrates that the phenolic molecules in the whole cultures acted as cross-linkers. Recently, Widsten *et al.* (2009) successfully produced superior fiberboards in terms of mechanical strength by incorporating tannins in the presence of laccase while further addition of wax to the tannic acid-laccase formulation improved the dimensional stability of the boards.

Concluding remarks

Important progress has been made with respect to the development of environmentally friendly functional polymers using laccases. The examples presented here show the great potential for further exploitation of laccases. Although many researchers have reported successful functionalization of lignocellulose polymers, studies on insights into reaction mechanisms are just emerging. It is absolutely important and imperative that research focuses on understanding reaction mechanisms as this will help to clear bottlenecks and pave the way for the development of optimized processes. Already, the lignin monomers and complex lignin model compounds used reveal interesting information that must be considered when working with the real polymer. Also, such model molecules will be of great help in enzyme engineering where they would allow higher throughput analysis in contrast to polymeric molecules where sophisticated analysis is required. Studies in lignin biosynthesis, chemical and structural composition of lignocellulose has already provided important background on the lignin building blocks, types of lignin available and other information useful for modeling *in vitro* experiments which mimic the lignocellulose polymer. Interestingly, available mechanistic evidence summarized in this review and other previous relevant reviews (Kobayashi and Higashimur, 2003) show that functionalization of lignocellulose materials depends on the nature and type of the lignin polymer (condensed or hydrolysable lignin) which in turn determines its reactivity. For example the reactivity is influenced by the nature (electron-withdrawing or electron-donating), position and number of substituents on the benzene ring (*p*- and *m*-substituted phenolics have been widely reported as good substrates for oxidative polymerization) while the presence of aliphatic side-chains block reaction sites. The resonance structures arising from enzymatic oxidation also determine whether successful bonding is possible or not. However, as demonstrated by some authors it is possible to overcome the loss of generated radicals by coupling anchor groups e.g. $-NH_2$ groups in order to create a stable reactive surface. In the case of commercial lignin as a raw material, the method used to obtain the lignin determines the reactivity properties. It is evident that, as our understanding of enzymatic reaction mechanisms increases and new laccases with higher redox potentials are discovered/and or engineered, cost-effective technologies will be developed which will facilitate the use of renewable resources for a cleaner and sustainable environment.

References

ADLER E. (1977) Lignin chemistry—past, present and future. *Wood Science and Technology* **11**, 69-218

ALCALDE, M. (2007) Laccase: biological functions, molecular structure and industrial applications. In Industrial Enzymes: structure, function and applications. Ed: J Polaina J, MacCabe AP. 459-474: New York, Springer.

ALEXANDRE, G. AND ZHULLIN, I.B. (2000) Laccases are widespread in bacteria. *Trends in Biotechnology*, **18**, 41-42

ALEXY, P., KOŠÍKOVÁ, B. AND PODSTRÁNSKA, G. (2000) The effect of blending lignin with polyethylene and polypropylene on physical properties. *Polymer* **41**, 4901-4908.

Almeida, G., Gibbs, P., Hogg, T. and Teixeira, P. (2006) Listeriosis in Portugal: an existing but unreported infection. *BMC Infectious Diseases* **6**, 1471-2334

ARAMAYO, R. AND TIMBERLAKE, W. E. (1993) The *Aspergillus nidulans* yA gene is regulated by abaA. *EMBO Journal* **12**, 2039-2048.

BALDRIAN, P. (2006) Fungal laccases – occurrence and properties. *FEMS Microbiology Reveviews,* **30**:215-242.

BANES-MARSHALL, L., CAWLEY, P. AND PHILLIPS, C. A. (2001). *In vitro* activity of Melaleuca alternifolia (tea tree) oil against bacterial and *Candida* sp. isolates from clinical specimens. *British Journal of Biomedical Sciences* **58**, 139–145.

BAUCHER, M., HALPIN, C., PETIT-CONIL, M., BOERJAN, W. (2003) Lignin: genetic engineering and impact on pulping. *Critical Review Biochemistry and Molecular Biology* **38**, 305–350

BARFOED, M. AND KIRK, O. (1999) Method for dyeing a material with a dyeing system which contains an enzymatic oxidizing agent *US Patent 5972042*

BARFOED, M., KIRK, O., SALMON, S. (2004) - Enzymatic method for textile dyeing. *US Patent* 6805718 US

BARFOED, M., KIRK, O. AND SALMON, S. (2001) Enzymatic method for textile dyeing. US2001037532, Novozymes A/S (DK)

BARFOED, M., KIRK, O. AND SALMON, S. (2003) Enzymatic method for textile dyeing. US2003226215, Novozymes North America Inc. (US)

BARFOED, M., KIRK, O. AND SALMON, S. (2005) Enzymatic method for textile dyeing. PL365996 Novozymes North America Inc. (US)

BOERJAN, W., RALPH, J. AND BAUCHER, M. (2003) Lignin biosynthesis. *Annual Review in Plant Biology* **54**, 519–46

BORGIN, K., AND CORBETT, K. (1970) The stability and weathering properties of wood treated with various oils. *Plastics, Paint and Rubber* **14**, 69 – 72.

BOUDET, A.M. (2000) Lignins and lignification: Selected issues. *Plant Physiology and Biochemistry* **38**, 81-96.

BURT, S. (2004) Essential oils: their antibacterial properties and potential applications in foods—a review, International Journal of Food Microbiology **94** , 223–253.

CALAFELL, M., DÍAZ, C., HADZHIYSKA, H., GIBERT J. M., DAGÀ, J.M., TZANOV T. (2007) Bio-catalyzed coloration of cellulose fibers. *Biocatalysis and Biotransformations* **25,** 336 - 340

CHANDRA, R.P. AND RAGAUSKAS, A.J. (2001) Laccase: the renegade of fiber modification, *Tappi Pulping Conference*, 1041–1051.

CHANDRA, R.P. AND RAGAUSKAS, A. J. (2002) Evaluating laccase-facilitated coupling of phenolic acids to high-yield kraft pulps. *Enzyme and Microbial Technology* **30**, 855-861.

CHANDRA R.P. AND RAGUASKAS, A.J. (2002) Elucidating the effects of laccase on the physical properties of high-kappa kraft pulps, *Progress in Biotechnology* **21**, 165–172

CHANDRA, R.P., WOLFAARDT, F. AND RAGAUSKAS, A. J. (2003) Biografting of Celestine Blue onto a High Kappa Kraft Pulp. In: Mansfield, S.D. andSaddler, J.N.: Application of enzymes to lignocellulose. *ACS Sym Ser* 66–80

CHANDRA, R.P., LEHTONEN L.K. AND RAGAUSKAS, A.J. (2004) Modification of high lignin content kraft pulps with laccase to improve paper strength properties 1. Laccase treatment in the presence of gallic acid. *Biotechnology Progress* **20**, 255–261

CLAUS, H. (2003) Laccases and their occurrence in prokaryotes. *Archive Microbiology* **179,** 145–150

CLAUS, H. (2004) Laccases: structure, reactions, distribution. *Micron* **35**, 93–96

DÍAZ BLANCO, C., DÍAZ GONZÁLEZ, M., DAGÁ MONMANY, M. AND TZANOV, T. (2009) Dyeing properties, synthesis, isolation and characterization of an in situ generated phenolic pigment, covalently bound to cotton. *Enzyme and Microbial Technology* **44**, 380 -385

DIXON, R. A. (2000) Natural products and plant disease resistance. *Nature* **411,** 843-847

DIXON, R.A., CHEN, F., GUO, D. AND PARVATHI, K. (2001). The biosynthesis of monolignols: A "metabolic grid", or independent pathways to guaiacyl and syringyl units? *Phytochemistry* **57**, 1069–1084

ELEGIR, G., KINDL, A., SADOCCO, P. AND ORLANDI, M. (2008) Development of antimicrobial cellulose packaging through laccase-mediated grafting of phenolic compounds, *Enzyme and Microbial Technology* **43,** 84–92

EKER, B., ZAGOREVSKI, D., ZHU, G., LINHARDT, R. J. AND DORDICK, J.S. (2009) Enzymatic polymerization of phenols in room-temperature ionic liquids. *Journal of Molecular Catalysis B: Enzymatic* **59**, 177–184

FACKLER, K., KUNCINGER, T., TERS, T. AND SREBOTNIK, E. (2008) Laccase-catalyzed functionalization with 4-hydroxy-3-methoxybenzylurea significantly improves internal bond of particle boards. *Holzforschung* **62**, 223–229

FELBY, C. AND HASSINGBOE, J. (1996) The influence of the chemical structure and physical state of native lignin upon the bonding strength of enzymatic bonded dry-process fiberboards. In:Kyoto University (ed) *Third Pacific Rim Bio-based Composites Symposium.* Kyoto University, Kyoto, pp. 283–291

FELBY, C., NIELSEN, B.R., OLESEN, P.O. AND SKIBSTED, L. H. (1997a) Identification and quantification of radical reaction intermediates by electron spin resonance spectrometry of laccase-catalyzed oxidation of wood fibers from beech (*Fagus sylvatica*). *Applied Microbiology and Biotechnology* **48**, 459–464

FELBY, C., PEDERSEN, L. S., NIELSEN, B. R. (1997b) Enhanced autoadhesion of wood fibers using phenol oxidases. *Holzforschung* **51,** 281–286

FELBY, C., OLESEN, P.O. AND HANSEN, T. T. (1998) Laccase catalyzed bonding of wood fibers. In: Eriksson K-EL, Cavaco-Paulo A (eds) Enzyme applications in fiber processing. (ACS symposium series 687) 88–98: American Chemical Society, Washington, D.C.

FELBY, C., HASSINGBOE, J. AND LUND, M. (2002) Pilot-scale production of fiberboards made by laccase oxidized wood fibers: board properties and evidence for cross-linking of lignin. *Enzyme and Microbial Technology* **31**, 736–741

Gargulak, J.D. and Lebo, S. E. (2000) Commercial use of lignin-based materials, In: Glasser, W.G., Northey, R.A. and Schultz, T.P. *Lignin: Historial, Biological, and Material Prospectives*. ACS Sym Ser **742:** 304–320

Gosselink, R.J.A., de Jong, E., Guran, B., Abächerli, A. (2004) Co-ordination network for lignin—standardisation, production and applications adapted to market requirements (EUROLIGNIN). *Industrial Crops and Products* **20**, 121–129.

Grönqvist, S., Rantanen, K., Alén, R., Mattinen, M. L., Buchert, J. and Viikari, L. (2006) Laccase-catalysed functionalisation of TMP with tyramine, *Holzforschung* **60**, 503–508

Friedman, R. (1998) Principles of Fire Protection Chemistry and Physics. 3rd Edition. Sudbury, USA. Jones and Bartlett Publishers.

Harkental, M,, Reichling, J., Geiss, H.K.and Saller, R. (1999) Comparative study on the in vitro antibacterial activity of Australian tea tree oil, cajeput oil, niaouli oil, manuka oil, kanuka oil, and eucalyptus oil. *Pharmazie* **54**, 460–463.

Hadzhiyska H, Calafell, M Gibert, J. M., Daga, J.M., Tzanov T. (2006) Laccase-assisted dyeing of cotton. *Biotechnol Letters* **28**, 755–759.

Hermann, T. E., Kurtz, M. B. and Champe, S. P. (1983) Laccase localised in hülle cells and cleistothecial primordia of *Aspergillus nidulans*. *Journal of Bacteriology* **154**, 955-964

Humphreys, J.M. and Chapple, C. (2002). Rewriting the lignin roadmap. *Current Opinion in Plant Biology* **5**, 224–229

Hüttermann, A. and Kharazipour, A. (1996) Enzymes as polymerization catalysts. In: Maijanen A, Hase A (eds) New catalysts for a clean environment. VTT Symposium 163, 143–148: Technical Research Centre of Finland (VTT), Espoo, Finland

Hüttermann, A., Mai C. and Kharazipour, A. (2001) Modification of lignin for the production of new compounded materials. *Applied Microbiology and Biotechnology* **55,** 387–394

Hyvönen, A., Piltonen P. and Niinimäki, J. (2006). Tall oil/water – emulsions as water repellents for Scots pine sapwood. *European Journal of Wood and Wood products* **64**, 68–73.

Jin, I., Nicholas, D. D. And Kirk, T. K. (1990a) Mineralization of the methoxyl carbon of isolated lignin by brown-rot fungi under solid substrate conditions. *Wood Science and Technology* **24**, 263–276

Jin, L., Schultz, T.P. and Nichols, D.D. (1990b) Structural characterization of brown-rotted lignin. *Holzforschung* **44**, 133–138

Kandanarachchi, P., Guo, A., Demydov, D., Petrovic, Z. (2002a) Kinetics of the hydroformylation of soybean oil by ligand-modified homogeneous rhodium catalysis. *Journal of the American Oil Chemistry Society* **79**, 1221–1225

Kandanarachchi, P., Guo, A. and Petrovic, Z. (2002b) The hydroformylation of vegetable oils and model compounds by ligand modified rhodium catalysis. *Journal of Molecular Catalysis A –Chemical* **184**, 65–71

Karhunen, P., Rummakko, P., Sipilä, J., Brunow, G. and Kilpelinen, I. (1995a). Dibenzodioxocins; a novel type of linkage in softwood lignins. *Tetrahedron Letters* **36**, 169-170.

Karhunen, P., Rummakko, P., Sipil , J., Brunow, G. and Kilpelinen, I. (1995b) The formation of dibenzodioxocin structures by oxidative coupling. A model reaction for lignin biosynthesis. *Tetrahedron Letters* **36**, 4501-4504.

KHARAZIPOUR, A. AND HÜTTERMANN, A. (1998a) Biotechnological production of wood composites. In: Bruce A, Palfreyman JW (eds) *Forest Products Biotechnology*.141–150: Taylor and Francis, London

KHARAZIPOUR, A., BERGMANN, K., NONNINGER, K. AND HÜTTERMANN, A. (1998b) Properties of fibre boards obtained by activation of the middle lamella lignin of wood fibres with peroxidase and H_2O_2 before conventional pressing. *Journal of Adhesion Science and Technology* **12**, 1045–1053

KIM, S.Y., ZILLE, A., MURKOVIC, M., GUEBITZ, G. AND CAVACO-PAULO, A. (2007a) Enzymatic polymerization on the surface of functionalized cellulose fibers. *Enzyme and Microbial Technology* **40**, 1782–1787.

KIM, S., MOLDES, D. AND CAVACO-PAULO, A. (2007b) Laccases for enzymatic colouration of unbleached cotton. *Enzyme and Microbial Technology* **40**, 1788–1793.

KIM, S., LOPEZ, C., GÜEBITZ, G. AND CAVACO-PAULO, A. (2008) Biological coloration of flax fabrics with flavonoids using laccase from *Trametes hirsuta*. *Engineering Life Sciences* **8**, 324–330

KOBAYASHIA, S. AND HIGASHIMURA, H. (2003) Oxidative polymerization of phenols revisited. *Progress in Polymer Science* **28**, 1015–1048

KUBO, 1., FUJITA K. AND NIHEI, K. (2003) Molecular design of multifunctional antibacterial agents against methicillin resistant *Staphylococcus aureus* (MRSA). *Bioorganic and Medicinal Chemistry* **11**, 4255–4262

KUDANGA,T., PRASETYO, E. N., SIPILÄ, J., NOUSIAINEN, P., WIDSTEN, P., KANDELBAUER, A., NYANHONGO, G.S. AND GUEBITZ, G. (2008) Laccase-mediated wood surface functionalization. *Engineering Life Sciences* **8**, 297 -302

KUDANGA, T., NUGROHO PRASETYO, E., SIPILÄ, J., NYANHONGO, G.S. AND GUEBITZ, G. M. (2010) Enzymatic grafting of functional molecules to the lignin model dibenzodioxocin and lignocellulosic material. *Enzyme and Microbial Technology* **46,** 272–280

KUDANGA, T., NUGROHO PRASETYO, E., SIPILA, J., EBERL, A., NYANHONGO, G.S. AND GUEBITZ, G.M. (2009) Coupling of aromatic amines onto syringylglycerol –guaiacylether using *Bacillus* SF spore laccase: A model for functionalization of lignin-based materials, *Journal Molecular Catalysis: Enzymatic* B **61**, 143–149

KUDANGA, T., NUGROHO PRASETYO, E., SIPILA, J., EBERL, A., NYANHONGO, G.S., GUEBITZ, G.M. (2010) Reactivity of long chain alkylamines to lignin moieties: implications on wood hydrophobicity – *Process Biochemistry* (in press)

KUKKOLA, E.M., KOUTANIEMI, S., GUSTAFSSON, M., KARHUNEN, P., RUEL, K., LUNDELL, T.K., SARANPÄ, P., BRUNOW, G., TEERI, T. H. AND FAGERSTEDT, K.V. (2003) Localization of dibenzodioxocin substructures in lignifying Norway spruce xylem by transmission electron microscopy-immunogold labeling. *Planta* **217**, 229-237.

KUNAMNENI, A., CAMARERO, S., GARCÍA-BURGOS, C., PLOU, F.J AND BALLESTEROS, A. AND ALCALDE, M. (2008) Engineering and applications of fungal laccases for organic synthesis. *Microbial Cell Factories* doi:10.1186/1475-2859-7-32

KÜHNE, G., AND DITTLER, B. (1999) Enzymatische Modifizierung nachwachsender Rohstoffe für die Herstellung bindemittelfreier Faserwerkstoffe. *Holz Roh Werkst* **57**, 264- 268

KNOBLOCH, K., PAULI, A., IBERL, N., WEIGAND, N. AND WEIS, H. M. (1989) Antibacterial and antifungal properties of essential oil components. *Journal of Essential Oil Research* **1**, 119-128.

LI, L., CHENG, X.F., LESHKEVICH, J., UMEZAWA, T., HARDING, S.A. AND CHIANG, V. L. (2001) The last step of syringyl monolignol biosynthesis in angiosperms is regulated by a novel gene encoding sinapyl alcohol dehydrogenase. *Plant Cell* **13**, 1567–1585

LI, L., POPKO, J. L., UMEZAWA, T. AND CHIANG, V. L. (2000) 5-Hydroxyconiferyl aldehyde modulates enzymatic methylation for syringyl monolignol formation, a new view of monolignol biosynthesis in angiosperms. *Journal of Biological Chemistry* **275**, 6537–6545

LI, K., AND GENG X. (2005) Formaldehyde-free wood adhesives from decayed wood *Macromolecular Rapid Communications* **26**, 529–532

LI, K., XU, F. AND ERIKSSON, K.E.L. (1999) Comparison of fungal laccases and redox mediators in oxidation of a nonphenol lignin model compound. *Applied Environmental Microbiology* **65**, 2654–2660.

LIU, N., SHI, S., GAO, Y. AND QIN, M. (2009) Fiber modification of kraft pulp with laccase in the presence of methyl syringate. *Enzyme and Microbial Technology* **44**, 89-95

LORA, J.H. AND GLASSER, W.G. (2002) Recent industrial applications of lignins; A sustainable alternative to non-renewable materials. *Journal of Polymers and Environment* **10**, 39-48.

LUND, M. AND FELBY, C. (2001) Wet strength improvement of unbleached kraft pulp through laccase catalyzed oxidation. *Enzyme and Microbial Technology* **28,** 760–765

MAI, C., MILSTEIN, O. AND HÜTTERMANN, A. (2000) Chemoenzymatical grafting of acrylamide onto lignin, *Journal of Biotechnology* **79**, 173–183.

MAI, C., MILSTEIN, O. AND HÜTTERMANN, A. (1999) Fungal laccase grafts acrylamide onto lignin in presence of peroxides. *Applied Microbiology and Biotechnology* **51**, 527–531.

MAI, C. AND MILITZ, H. (2004) Modification of wood with silicon compounds. Treatment systems based on organic silicon compounds - a review. *Wood Science and Technology* **37,** 453 - 461.

MARTÍNEZ, Á. T., RENCORET, J., MARQUES, G., GUTIÉRREZ, A, IBARRA, D., JIMÉNEZ-BARBERO J AND DEL RÍO, J.C. (2008) Monolignol acylation and lignin structure in some nonwoody plants: A 2D NMR study. *Phytochemistry* **69**, 2831-2843

MANSOURI, N., PIZZI A., SALVADO, J (2007) Lignin-basedwood panel adhesives without formaldehyde. *European Journal of Wood and Wood products* **65,** 65–70

MAYER, A. M. AND STAPLES, R.C. (2002) Laccase: new functions for an old enzyme. *Phytochemistry* **60**, 551–565

MELO, E.P., FERNANDES, A.T., DURÃO, P., MARTINS, L.O. (2007) Insight into stability of CotA laccase from the spore coat of *Bacillus subtilis*. *Biochemical Society Transactions* **35**, 1579-1582.

MILSTEIN, O., HUTTERMANN, A., FRUND R. AND LUDEMANN, H.-D. (1994) Enzymatic co-polymerisation of lignin with low molecular mass compounds, *Applied Microbiology and Biotechnology* **40**, 760–767

MINUSSI, R.C., PASTORE, G. M. AND DURAN, N. (2002) Potential applications of laccase in the food industry. *Trends in Food Science and Technology* **13**, 205–216

NIMZ, H., RAZVI, A., MOGHARAB I. AND CLAD, W. (1972) Bindemittel bzw. Klebemittel zur Herstellung von Holzwerkstoffen, sowie zur Verklebung von Werkstoffen verschiedener art, *German Patent* (03.01.1972) DE2221353.

NIMZ, H. (1974) Beech lignin-proposal of a constitutional scheme. Agnew Chew Int Ed 13, 313-321

NOHYNEK, L. J., ALAKOMI, H. L., KÄHKÖNEN, M. P., HEINONEN, M., HELANDER, I. M., OKSMAN-CALDENTEY, K. M. AND PUUPPONEN-PIMIÄ, R. (2006) Berry phenolics: Antimicrobial properties and mechanisms of action against severe human pathogens. *Nutrition and Cancer* **54**, 18-32.

NUGROHO PRASETYO, E., KUDANGA, T., EICHENGER R., NYANHONGO, G.S. AND GUEBITZ, G.M. (2009) Laccase catalysed coupling of functional molecules to lignocellulose: Polymers derived from coupling of lignin onto silanes. *Proceedings of the 6th International Conference on textile and Polymer Biotechnology*, pp76 – 82: Gent Belgium

NUGROHO PRASETYO, E., KUDANGA, T., RENCORET, J., GUTIÉRREZ, A., DEL RÍO, J. C., SANTOS, J. I., NIETO, L., JIMÉNEZ-BARBERO, J., MARTÍNEZ, A. T., LI, J., GELLERSTEDT, G., LEPFIRE, S., SILVA, C., KIM, S.Y., CAVACO-PAULO, A., SELJEBAKKEN KLAUSEN, B., FRODE LUTNAES, B., NYANHONGO, G.S., AND GUEBITZ, G.M. (2010) Polymerisation of lignosulfonates by the laccase-HBT (1-hydroxybenzotriazole) system improves dispersibility. *Bioresource Technology* **101**, 5054-5062

NYANHONGO, G.S., SCHROEDER, M., STEINER W. AND GÜBITZ, G.M. (2005) Biodegradation of 2, 4, 6-trinitrotoluene (TNT):An enzymatic perspective. *Biocataysis and Biotransformation* **23,** 1-17

NYANHONGO, G.S., GUEBITZ, G., SUKYAI, P., LEITNER, C., HALTRICH, D. AND LUDWIG, R. (2007) Oxidoreductases from *Trametes* spp. in Biotechnology:A Wealth of Catalytic Activity. *Food Technology and Biotechnology* **45,** 248–266

OBATAYA, E., TANAKA, F., NORIMOTO, M. AND TOMITA, B. (2000) Hygroscopicity of heat-treated wood: Effects of after-treatments on the hygroscopicity of heat-treated wood. *Mokuzai Gakkaishi* **46**, 77–87

O'MALLEY, D.M., WHETTEN, R., BAO, W., CHEN, C. L., SEEDORF, R. R. (1993) The role of laccase in lignification. *Plant Journal* **4**, 751-757.

RALPH, J., LUNDQUIST, K., BRUNOW, G., LU, F., KIM, H., SCHATZ, P., MARITA, J.M., HATFIELD, R.D., RALPH, S.A., CHRISTENSEN, J.H. AND BOERJAN W. (2004) Lignins: natural polymers from oxidative coupling of 4-hydroxyphenyl-propanoids, *Phytochemistry Reviews* **3**, 29–60

RAMSDEN, M. J. AND BLAKE, F. S. R. (1997) A kinetic study of the acetylation of cellulose, hemicellulose and lignincomponents in wood. *Wood Science Technology* **31**, 45–50

RAZZAQUE, M.A. (1982) The effect of concentration and distribution on the performance of water repellents applied to wood. University of Wales, Ph.D. Thesis.

REBRIKOV, D. N., STEPANOVA, E., KOROLEVA, V.O.V., BUDARINA, Z.I., ZAKHAROVA, M.V., YURKOVA, T.V., SOLONIN, A.S., BELOVA, O.V., POZHIDAEVA, Z.A. AND LEONT'EVSKY, A.A. (2006) Laccase of the lignolytic fungus Trametes hirsuta : Purification and characterization of the enzyme, and cloning and primary structure of the gene. *Applied Biochemical and Biotechnology* **42**, 564-572.

RIVA, S. (2006) Laccases: blue enzymes for green chemistry. *Trends in Biotechnology* **24,** 219 -226

ROBLES, A., LUCAS, R., CIENFUEGOS, G. A. AND GÁLVEZ, A. (2000) Phenol-oxidase (laccase) activity in strains of the hyphomycete *Chalara paradoxa* isolated from olive mill wastewater disposal ponds. *Enzyme Microbial Technology* **26**, 484-490.

Rodríguez Couto, S. and Toca Herrera, J. L. (2006) Industrial and biotechnological applications of laccases: A review. *Biotechnology Advances* **24,** 500–513

Rodríguez-Vaquero, M.J., Alberto, M.R. and Manca de Nadra M.C. (2007) Antibacterial effect of phenolic compounds from different wines. *Food Control* **18**, 93–101.

Rodríguez-Vaquero, M.J., Alberto, M.R. and Manca de Nadra, M.C. (2007) Influence of phenolic compounds from wines on the growth of *Listeria monocytogenes*, *Food Control* **18** 587–593.

Sjostrom, E. (1993) *Wood Chemistry. Fundamentals and Applications*. Second edition ed., p292. San Diego: Academic Press.

Sakakibara, A. (1980) A Structural Model of Softwood Lignin. *Wood Science and Technology* **14**, 89-100.

Schröder, M., Aichernig, N., Gübitz, G.M. and Kokol V. (2007) Enzymatic coating of lignocellulosic surfaces with polyphenols. *Biotechnol. Journal* **2** , 334–341.

Shin H, Gübitz G, Cavaco-Paulo, A. (2001). In situ enzymatically prepared polymers for pool coloration. *Macromolecular Material and Engineering* **286**, 691-694.

Sørensen, N. H. (1999) Enzymatic foam compositions for dyeing of keratinous fibres. WO9915137, Novo Nordisk A/S (DK)

Suurnakki, A., Buchert, J., Gronqvist, S., Mikkonen, H., Peltonen, S. and Viikari, L. (2006) Bringing new properties to lignin rich fiber materials. *VTT Symposium* **244**, 61-70

Tadesse, M.A., D'Annibale, A., Galli, C., Gentilia, P. And Sergi, F. (2008) An assessment of the relative contributions of redox and steric issues tolaccase specificity towards putative substrates. *Organic and Biomolecular Chemistry* **6**, 868-878.

Turunen, M., Alvila, L., Pakkanen, T.T. and Rainio, J. (2003) Modification of phenol–formaldehyde resol resins by lignin, starch, and urea. *Journal of Applied Polymer Science* **88**, 582–588

Tzanov, T., Silva, C., Zille, A., Oliveira, J. and Cavaco-Paulo, A. (2003) Effect of some process parameters in enzymatic dyeing of wool. *Applied Biochemistry and Biotechnology* **111**, 1–14

Ultee, A., Slump, R. A., Steging G., and Smid, E.J. (2000) Antimicrobial activity of carvacrol toward *Bacillus cereus* on rice, *Journal of Food Protection* **63,** 620–624.

Ultee, A., Bennik M.H.J. and Moezelaar, R. (2002) The phenolic hydroxyl group of carvacol is essential for action against the food-borne pathogen *Bacillis cereus*. *Applied Environmental Microbiology* **68**, 1561–1568.

Ulvcrona, T., Lindberg, H. and Bergsten, U. (2006) Impregnation of Norway spruce (*Picea abies* L. Karst.) wood by hydrophobic oil and dispersion patterns in different tissues. *Forestry* **79**, 123-134.

Unbehaun, H., Wolff, M., Kühne, G., Schindel, K., Hüttermann, A., Cohen, R. and Chet, I. (1999) Mechanismen der mykologischen Transformation von Holz für die Holzwerkstoffherstellung. *Holz Roh Werkst* **57**, 92

Unbehaun, H., Dittler, B., Kühne, G. and Wagenführ, A. (2000) Investigation into the biotechnological modification of wood and its application in the wood-based material industry. *Acta Biotechnologica* **20**, 305–312

Vick, C.B., and Rowell, R.M. (1990) Adhesive bonding of acetylated wood, International *Journal of Adhesion and Adhesives* **10**, 263-272

VIITANIEMI, P. AND JÄMSÄ, S. (1996) Modification of wood by heat treatment, 1- 57, VTT publications 814.Espoo, Finland

WIDSTEN, P., HEATHCOTE, C., KANDELBAUER, A., GUEBITZ, G., NUGROHO PRASETYO E., NYANHONGO, G.S. AND KUDANGA, T. (2010) Enzymatic surface functionalisation of lignocellulosic materials with tannins for enhancing antibacterial properties. *Process Biochemistry* doi:10.1016/j.procbio.2010.03.022

WIDSTEN, P. AND KANDELBAUER, A. (2008) Laccase applications in the forest products industry: A review. *Enzyme and Microbial Technology* **42**, 293-307.

WILLIAMSON, P. R., WAKAMATSU, K. AND ITO, S. (1998) Melanin biosynthesis in *Cryptococcus neoformans*. *Journal of Bacteriology* **180**, 1570-1572.

WINANDY, J. E. AND ROWELL, R. M. (2005) Chemistry of wood strength. In: Rowell, R. M. ed. Handbook of wood chemistry and wood composites. 303-347, Boca Raton FL, CRC Press LLC.

WITAYAKRAN, S. AND RAGAUSKAS, A.J. (2009) Modification of high-lignin softwood kraft pulp with laccase and amino acids. *Enzyme and Microbial Technology* **44** , 176-181.

WONG, K. K. Y., RICHARDSON, J. D. AND MANSFIELD, S. D. (2000) Enzymatic treatment of mechanical pulp for improving papermaking properties. *Biotechnology Progress* **16**, 1025–1029

WONG, K.Y., SIGNAL F.A. AND CAMPION, S.H. (2003) Improving linerboard properties with enzymatic treatment of the kraft component of the base sheet. *Appita Journal* **56**, 308–311.

YAMAGUCHI, H., NAGAMORI, N. AND SAKATA, I. (1991) Application of the dehydrogenative polymerization of vanillic acid to bonding of woody fibers. *Mokuzai Gakkaishi* **37**, 220–226

YAMAGUCHI, H., MAEDA, Y., SAKATA, I. (1992) Application of phenol dehydrogenative polymerization by laccase to bonding among woody fibers. *Mokuzai Gakkaishi* **38**, 931–937

YAMAGUCHI, H., MAEDA, Y. AND SAKATA, I. (1994) Bonding among woody fibers by use of enzymatic phenol dehydrogenative polymerization. *Mokuzai Gakkaishi* **40**, 185–190

YAROPOLOV, A.I., SKOROBOGAT'KO, O.V., VARTANOV, S. S. AND VARFOLOMEYEV, S. D. (1994) Laccse: Properties, catalytic mechanism, and applicability. *Appl Biochem and Biotechnology*, **49**, 257-280.

Biotechnology and Genetic Engineering Reviews - Vol. 27, 331-366 (2010)

Structure and function of enzymes acting on chitin and chitosan

INGUNN A. HOELL[1], GUSTAV VAAJE-KOLSTAD, AND VINCENT G.H. EIJSINK*

Department of Chemistry, Biotechnology and Food Science, Norwegian University of Life Sciences, P. O. Box 5003, N-1432 Ås, Norway, [1]and Stord/Haugesund University College, Bjoernsonsgate 45, N-5528 Haugesund, Norway

Abstract

Enzymatic conversions of chitin and its soluble, partially deacetylated derivative chitosan are of great interest. Firstly, chitin metabolism is an important process in fungi, insects and crustaceans. Secondly, such enzymatic conversions may be used to transform an abundant biomass to useful products such as bioactive chito-oligosaccharides. Enzymes acting on chitin and chitosan are abundant in nature. Here we review current knowledge on the structure and function of enzymes involved in the conversion of these polymeric substrates: chitinases (glycoside hydrolase families 18 & 19), chitosanases (glycoside hydrolase families 8, 46, 75 & 80) and chitin deacetylases (carbohydrate esterase family 4).

Introduction

Chitin is a linear insoluble homo-polymer of β-1,4 linked *N*-acetylglucosamine (GlcNAc) (*Figure 1*). In terms of annual production, chitin is often considered the second most abundant polymer in nature with only cellulose being more widespread. In nature chitin predominantly exists as a composite material consisting of ordered crystalline microfibrils embedded in a matrix of protein and minerals. Chitin is an important structural component of cell walls in fungi and yeast, and in the exoskeletons

*To whom correspondence may be addressed (Vincent.eijsink@umb.no)

Abbreviations: CAZy, Carbohydrate-Active Enzymes database; CBM, Carbohydrate-Binding Module; CHOS, chito-oligosaccharides; GH, glycoside hydrolase; GlcN or D, glucosamine; GlcNAc or A, *N*-acetylglucosamine.

of insects and crustaceans (Adams, 2004; Rinaudo, 2006; Gooday, 1990; Raabe *et al.*, 2007). The repeating sugar units in the chitin chain are disaccharides where the monomers are rotated 180° relative to each other (*Figure 1*).

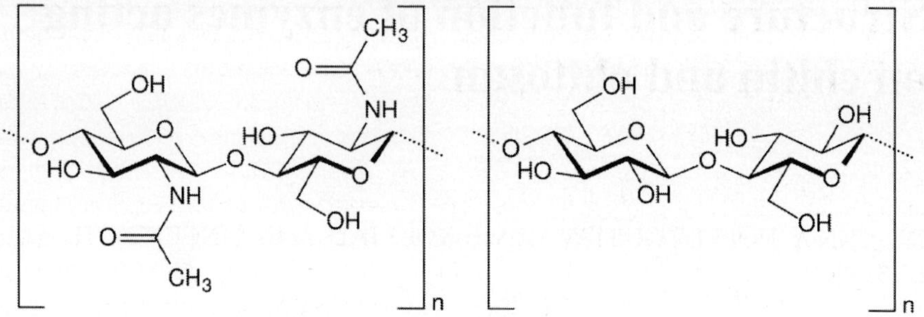

Figure 1. Repeating disaccharide unit in chitin and cellulose. Chitin consists of β-1,4 linked *N*-acetylglucosamine (GlcNAc), where the monosaccharides are rotated 180° to each other. For comparison, the repeating disaccharide in cellulose is shown too, to the right. Cellulose consists of β-1,4 linked D-glucose.

Chitin can be found in three different crystalline polymorphs in nature. α-chitin consists of anti-parallel *N*-acetylglucosamine chains (Minke and Blackwell, 1978), whereas β-chitin contains parallel chains of *N*-acetylglucosamine (Gardner and Blackwell, 1975). In the γ-form sets of two parallel strands alternate with a single anti-parallel strand. The α- and the β-form are the most abundant in nature. The anti-parallel packing in α-chitin is stabilized by a high number of hydrogen bonds, allowing for tight packing of the *N*-acetylglucosamine chains. This arrangement may contribute considerably to mechanical strength and stability (Giraud-Guille, 1986). Both packing tightness and the number of intra-chain hydrogen bonds are reduced in β-chitin, resulting in an increased number of hydrogen bonds with water. This results in a more flexible and softer chitinous structure (Merzendorfer and Zimoch, 2003). For comparison, cellulose is also found in two main crystalline forms, where cellulose I is similar to β-chitin, and cellulose II has anti-parallel chains as in α-chitin (Eijsink *et al.*, 2008, Nishiyama *et al.*, 2002). Chitin and cellulose (*Figure 1*) are both materials that contribute to structure and strength, and as a consequence they play protective roles in organisms.

Deacetylation of chitin yields chitosan, a water-soluble heteropolymer of β(1,4)-linked GlcNAc (A-units) and D-glucosamine (GlcN or D-unit). Chitosans differ in their degree of acetylation, from close to zero to about 65 % (Anthonsen *et al.*, 1993; Varum *et al.*, 1994). The solution properties of chitosans depend on the degree of acetylation, the distribution of acetyl groups along the main chain, and the chain length (Rinaudo and Domard, 1989; Aiba, 1991; Kubota and Eguchi, 1997). Industrial production of chitosan is one of the most important applications of chitin (see below). Deacetylation of chitin does occur in nature and it has been suggested that this process may play an important role in modulating fungal cell walls and in plant-pathogen interactions (Tsigos *et al.*, 2000).

In principle, considerable amounts of chitin, e.g. from shrimp and crab shell waste, are available for exploitation but, currently, only a small part of this biopolymer is utilized. The most dominant application of chitin occurs after its chemical deacetylation to chitosan. Before deacetylation (by concentrated NaOH or by enzymatic hydrolysis) chitin is extracted from the shells by acid treatment (to dissolve calcium carbonate) followed by alkaline extraction (to solubilize proteins) (Rinaudo, 2006). Chitosan has been used in a wide range of applications, not all of which are equally well documented. Possible applications are in wastewater treatment (for removal of heavy metal ions, flocculation/coagulation of dyes and proteins, membrane purification processes), in the food industry (anti-cholesterol and fat binding, preservative, packing material, animal feed additive), in agriculture (seed and fertilizer coating, controlled agrochemical release), in the paper industry (surface treatment, photographic paper) and in personal-care products (moisturizer, body lotion, shampoo) (Felse and Panda, 1999; Shahidi *et al.*, 1999; No and Meyers, 2000; Kurita, 2001; Dutta *et al.*, 2002; Tharanathan and Kittur, 2003; Krajewska, 2004; Guan *et al.*, 2009). In the biomedical field, chitin/chitosan-based materials can be used in antimicrobial and antifungal agents, drug delivery vehicles, gene delivery vehicles, drug controlled release systems, artificial cells, wound healing lotions, haemodialysis membranes, contact lenses, artificial skins and tissue engineering (Felt *et al.*, 1998; Paul and Sharma, 2000; Singla and Chawla, 2001; Khor, 2002; Tharanathan and Kittur, 2003; Krajewska, 2004; Ladet *et al.*, 2008; Strand *et al.*, 2008).

Even though chitosan has several important applications, its low solubility and high viscosity can cause problems. The hydrolyzed products of chitin and chitosan, chitooligosaccharides (CHOS), are soluble in water and of great interest in several applications. Especially, CHOS have attracted much attention in the food, biomedical, and agricultural fields, (reviewed in Tharanathan and Kittur, 2003; Kim and Rajapakse, 2005; Rinaudo, 2006). This is due to their potential antibacterial activities (Jeon *et al.*, 2001; Fernandes *et al.*, 2008), antifungal activity (Xu *et al.*, 2007), immunoenhancing effect (Ngo *et al.*, 2008; Wu *et al.*, 2008; Liu *et al.*, 2009), and antitumor properties (Prashanth and Tharanathan, 2005). CHOS may be obtained by alkaline *N*-deacetylation of chitin and subsequent hydrolysis of the obtained chitosan, but enzymatic processes are to be preferred due to their higher precision and more environmentaly friendly process conditions. In principle, it should be possible to develop enzymes and enzymatic procedures that enable production of CHOS with controlled length, degree of acetylation and acetylation pattern.

Despite an estimated annual production of 10^{10}-10^{11} tons per year (Gooday, 1990), chitin does not accumulate in the environment, indicating that natural chitinolytic machineries are sufficiently efficient to handle all produced chitin. Indeed, it is now well established that many organisms produce chitin-converting enzymes. Chitin-containing organisms need chitinolytic enzymes for growth and proper development, microorganisms produce these enzymes to exploit chitin as a source of energy and nitrogen, while plants produce chitinolytic enzymes as part of their defence against chitin-containing pathogens. *Plasmodium falciparum*, the malaria-causing agent, is an example of an organism that is dependent on chitinases during its life-cycle (Langer and Vinetz, 2001). Interestingly, mammals produce two chitinases, chitotriosidase and AMCase, that possibly are parts of a pathogen combating system (Boot *et al.*, 1995; Renkema *et al.*, 1995; Boot *et al.*, 2001; Bussink *et al.*, 2007).

From a biotechnological point of view, enzymes acting on chitin are important for two major reasons. Firstly, because humans do not contain chitin, chitin metabolism is an interesting target area for the development of compounds to combat chitin-containing plague organisms such as fungi, insects and nematodes. Secondly, chitin-converting enzymes may be exploited as industrial biocatalysts to produce CHOS and potentially other interesting chemicals out of chitin. In the remainder of this report, we will review the classification and occurrence of relevant enzymes and we will summarize available structural and functional information for the most important enzyme classes. We will focus on microbial enzymes since these have the largest potential as biocatalysts. The structure and function of human chitinases has recently been reviewed by others (Guan *et al.*, 2009). Recent literature also contains several reviews on the occurrence and phylogenetic analysis of chitinolytic enzymes in all kingdoms of life (Karlsson and Stenlid, 2009).

Enzymes acting on chitin and chitosan – general description and classification

Chitin can de degraded via two major pathways. When the pathway involves initial hydrolysis of the (1→4)-β-glycosidic bond, the process is termed chitinolytic. Chitinases are found in glycoside hydrolase families 18 & 19 (see below) and hydrolyse chitin into oligosaccharides (mainly dimers) of GlcNAc. Subsequently, β-*N*-acetyl hexosaminidases (family 20 glycoside hydrolases; sometimes referred to as chitobiases; see below) further degrade the oligomers, producing GlcNAc monomers. The second pathway for chitin degradation involves deacetylation of chitin to chitosan. Enzymes capable of deacetylating chitin are called chitin deacetylases, and are found in carbohydrate esterase family 4 (see below). Hydrolysis of the (1→4)-β-glycosidic bond in chitosan is accomplished by chitosanases, which occur in families 5, 7, 8, 46, 75 and 80 of the glycoside hydrolases (see below).

There are several ways to classify enzymes. One of the oldest and most common classifications uses EC-numbers according to the IUBMB nomenclature (International Union of Biochemistry and Molecular Biology). This classification is based on substrate specificity; it is simple and commonly used. Glycoside hydrolases are classified as EC 3.2.1.x, where x depends on the substrate specificity. However, this classification becomes problematic when enzymes are active against several substrates, as is the case for many carbohydrate-active enzymes. Other problems with this classification concern the fact that it does not provide information about the 3D-structure of the enzymes, and the fact that structurally and evolutionary unrelated enzymes can have the same substrate specificities. Furthermore, today's abundant genome sequencing projects provide information on sequence, but not about substrate specificity (Cantarel *et al.*, 2009). Enzymes can also be classified on the basis of their catalytic mechanism. For example, glycoside hydrolases employ two different mechanisms which lead to either retention or inversion of the anomeric configuration (Koshland, 1953). Knowledge of the stereochemistry of the reaction is an important part of the information on mechanism, and the IUBMB classification does not provide information on this aspect.

Enzymes acting on polysaccharides could also be classified by their mode of action. Exo-enzymes attack the polysaccharide chain from the reducing or non-reducing end of the sugar chain, whereas endo-enzymes attack from random points along the polysaccharide chain. Both the endo- and exo- mode of action can occur in combination with processivity (also called "multiple attack"). In processive enzyme action the polymeric substrate is not released after a successful cleavage, but slides through the active site so a new cleavage can take place (Robyt and French, 1967; Robyt and French, 1970). It is not possible to distinguish between these modes of action by looking at sequences alone. However, the structure of the active site region can provide hints as to the mode of action. A pocket active site is often seen for exo-enzymes, a substrate-binding deep cleft or tunnel is often related to an exo- and/or a processive mode of action, whereas non-processive endo-enzymes tend to have shallow substrate-binding clefts (Henrissat and Davies, 1997).

It is important to note that it is not easy to study how chitinases act on chitin because it is difficult to analyze changes in the insoluble substrate with common biochemical techniques. Furthermore intermediate products emerging during reactions, in particular soluble oligomers, are usually much better substrates than the insoluble polymer, which makes them difficult to observe. Electron microscopy of chitin shows that although chitin is a crystalline substrate, it also consists of amorphous regions (Blackwell, 1988). These amorphous regions of the substrate may also be better substrates for enzymes than the crystalline regions. This implies that initial rate measurements on the enzymatic degradation of chitin in fact may not relate the enzyme's ability to attack a truly crystalline substrate. It is difficult to confirm experimentally if enzymes bind the substrate in an exo- or endo-fashion, in particular when inappropriate substrates are used, or when the enzyme is processive (Horn *et al.*, 2006a,b; Eijsink *et al.*, 2008). Literature data on this issue should be used with caution.

In 1991, glycoside hydrolases (GHs) were classified in families based on their amino acid sequences. The Carbohydrate-Active Enzymes database (CAZy) provides a continuously updated list of the GH families and, since a few years, also other families of carbohydrate-active enzymes such as glycosyl transferases and carbohydrate esterases (www.cazy.org). By January 2010 115 different families of glycoside hydrolases had been identified. The CAZy classification is based on amino acid sequence, which gives very useful information since sequence and structure and hence mechanism are related. In other words, a great advantage of this classification is that it groups enzymes with similar folds and catalytic mechanism. This gives the opportunity to predict function and mechanism by sequence only, and to use homology modelling in structure solving (Davies and Henrissat, 1995; Henrissat and Davies, 1997). Since 3D structures are more conserved than sequences, GH families may have similar folds and are therefore further classified into clans. So far, 14 such clans have been described, named GHA → GHN (www.cazy.org).

In addition to a catalytic domain, glycoside hydrolases may contain additional domains, many of which are thought to be involved in interacting with the substrate. These domains, or carbohydrate-binding modules (CBMs), are also classified in the CAZy database. Within the GH families, members can show differences in their domain structure, which may affect their substrate preference (in the case of chitinases e.g. the preferred degree of crystallinity, the degree of acetylation, or the acetylation pattern) and their mode of action (exo-, endo-, processivity and directionality).

In the past, several alternative nomenclatures and classification systems have been proposed for chitinases and several of these are briefly mentioned below. One key problem is how to name enzymes belonging to the same GH family within one organism. Within the CAZy nomenclature, the current practice is to base the enzyme designation on the preferred substrate, the CAZy family and an additional letter that identifies the order in which the enzymes from the specific organism were reported in the literature (e.g. the three family 18 chitinases from *Serratia marcescens* are designated Chi18A, Chi18B and Chi18C). For additional clarification, two letters indicating the organism of origin can be inserted in front of the enzyme name (e.g. Chi18A from *S. marcescens* can be written SmChi18A). A more detailed description of the CAZy nomenclature is provided in Henrissat *et al.* (1998).

One major discriminator between glycoside hydrolase families is their catalytic mechanism. Enzymatic hydrolysis of glycosidic bonds usually takes place via general acid catalysis. This requires two residues, a proton donor, the catalytic acid, and a nucleophile / base (Koshland, 1953; Sinnott, 1990). In most glycoside hydrolases both the proton donor and the nucleophile are aspartate (D/Asp) or glutamate (E/Glu) (Davies and Henrissat, 1995). Hydrolysis can occur via two different main mechanisms which lead to either retention or inversion of the anomeric configuration (Koshland, 1953; Rye and Withers, 2000). In both mechanisms, the position of the proton donor is such that one of its carboxylic oxygens is at hydrogen-binding distance to the glycosidic oxygen. In the retaining mechanism (also called the double displacement mechanism), the first step involves protonation of the glycosidic oxygen (by the catalytic acid) and concomitant nucleophilic attack on the anomeric carbon by the nucleophile. This leads to breakage of the scissile bond and the formation of a covalent bond between the anomeric carbon and the catalytic nucleophile (Vocadlo *et al.*, 2001). In the second step, the intermediate is hydrolyzed by a water molecule that approaches the anomeric carbon from a position close to that of the original glycosidic oxygen. This leads to retention of the anomeric configuration. In inverting enzymes the catalytic nucleophile is a water molecule which is activated by a carboxylic group that acts as a base. This water is located between this carboxylic group and the anomeric carbon. This so-called direct displacement mechanism leads to inversion of the anomeric configuration, because the water molecule approaches the anomeric carbon from the other side of the sugar plane.

Chitinases belonging to class GH18 and *N*-acetylglucosaminidases belonging to class GH20 employ a special variant of the retaining double displacement mechanism, often called the substrate-assisted double displacement mechanism (Tews *et al.*, 1997; Mark *et al.*, 2001; Van Aalten *et al.*, 2001; Vocadlo and Withers, 2005; See *Figure 2*). In these enzymes the *N*-acetylgroup of the sugar bound in the -1 subsite acts as nucleophile, leading to the formation of an oxazolinium ion intermediate. This *N*-acetylgroup is activated by specific interactions with conserved enzyme residues.

In addition to catalytic residues that are directly necessary for hydrolytic activity, glycoside hydrolases contain several residues involved in substrate binding. A sugar-binding subsite is a cluster of amino acids involved in binding of one sugar monomer of a longer sugar chain. Subsites for sugar-binding in glycoside hydrolases are labeled from –n to +n, where –n denotes the non-reducing end and +n at the reducing end. Cleavage occurs between sugars bound in the -1 and +1 subsites (Biely *et al.*, 1981; Davies *et al.*, 1997). Since optimal positioning of the substrate is important for catalytic efficiency, residues that primarily seem to generate substrate-affinity may in fact also be important for catalytic turnover.

Structural and mutagenesis studies of glycoside hydrolases have shown that aromatic amino acids are important for protein-carbohydrate interactions. The side chains of aromatic amino acids can be involved in various interactions with the pyranose rings of the carbohydrate, such a stacking, and hydrophobic and hydrogen bonding

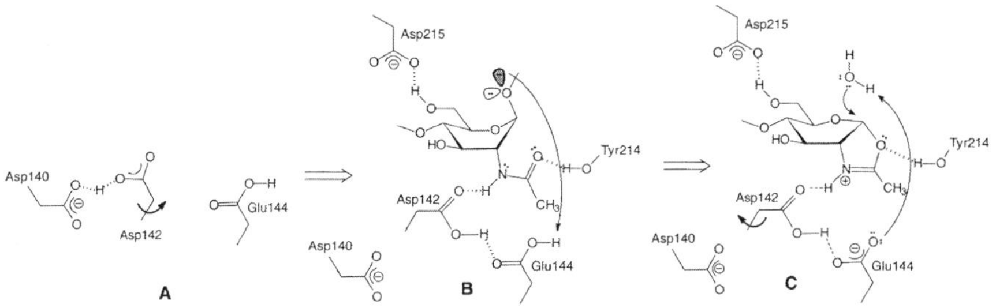

Figure 2. Substrate-assisted double displacement mechanism in ChiB, a family 18 GH from *Serratia marcescens*. The first step of chitinolysis results in cleavage of the glycosidic bond and formation of an oxazolinium ion intermediate, and hydrolysis of this ion completes the reaction. The figure shows that several residues are involved in catalysis. The interaction networks contribute to steering the acidity of Glu144, which acts both as the proton donor and acceptor during the catalytic cycle (picture taken from Van Aalten *et al.* 2001).

interactions (Maenaka *et al.*, 1994; Watanabe *et al.*, 2003). Recently, it has been shown that such aromatic residues are particularly important in processive enzymes, where they are thought to provide "fluid binding" that is needed to enable the substrate to slide through the substrate-binding cleft in between hydrolytic steps (Varrot *et al.*, 2003; Horn *et al.*, 2006a; Zakariassen *et al.*, 2009).

True chitinases (EC 3.2.1.14) occur in glycoside hydrolase families GH18 and GH19 which almost exclusively consist of chitinases. Family GH19 contains only chitinases, whereas family GH18 contains some endo-β-N-acetylglucosaminidases (EC 3.2.1.96), a few lectins (Hennig *et al.*, 1992, 1995) as well as a xylanase inhibitor (Juge *et al.*, 2004), in addition to about 3000 chitinases (January 2010). Family 18 enzymes employ the substrate-assisted retaining double displacement mechanism, whereas family 19 enzymes employ the inverting direct displacement mechanism. Both types of enzymes have been intensely studied, as discussed further below.

Enzymes with chitosanase (EC 3.2.1.132) activity hydrolyze β-1,4-glycosidic linkages in chitosan and have been found in GH families 5, 7, 8, 46, 75 and 80. GH7 contains retaining cellulases, including the well known cellobiohydrolase I (Divne *et al.*, 1994). In a very few cases chitosanase activity has been detected as a side activity of cellobiohydrolase I (Ike *et al.*, 2007). GH5 and GH8 contain a variety of enzymatic activities, including chitosanases, cellulases, licheninases, mannanase and xylanases. In GH5, chitosanase activity has been detected for only three of the over 1800 retaining enzymes listed in this family, and the activity seems to be a side activity of what probably is a cellulase (Xia *et al.*, 2008). Crystal structures have been solved for several of the enzymes belonging to GH5, but none of these structures concern enzymes with known chitosanase activity. In GH8, enzymes annotated as chitosanases occur more frequently, and this family seems to contain true inverting chitosanases (Adachi *et al.*, 2004) which are discussed further below. Families GH46, GH75 and GH80 exclusively contain chitosanases and all enzymes are thought to act by an inverting mechanism (Fukamizo *et al.*, 1995; Cheng *et al.*, 2006; www.cazy.org). Three-dimensional structures have been solved for members of GH family 46 (Marcotte *et al.*, 1996; Saito *et al.*, 1999a), but not for members of GH families 75 and 80. Generally, family 46 enzymes are the best-studied of these three classes of chitosanases. Current knowledge for these three enzyme classes is discussed below.

Chitinases and chitosanases differ in terms of their preferences for acetylated / deacetylated residues near the scissile bond (Koga *et al.*, 1999; Tremblay *et al.*, 2000). They show overlapping substrate specificities, and chitinases as well as chitosanases can hydrolyze chitosans with moderate degrees of acetylation (Sorbotten *et al.*, 2005; Heggset *et al.*, 2009). Based on their substrate specificity towards chitosan, chitosanases have been classified into subclasses I, II and III (Fukamizo *et al.*, 1994). Chitosanases in subclass I can hydrolyze GlcNAc-GlcN and GlcN-GlcN linkages (Fukamizo *et al.*, 1994), subclass II enzymes can hydrolyze GlcN-GlcN linkages only (Izume *et al.*, 1992), whereas subclass III enzymes can hydrolyze GlcN-GlcNAc and GlcN-GlcN linkages (Mitsutomi *et al.*, 1995). The ability to hydrolyze GlcNAc-GlcNAc linkages separates chitinases from chitosanases.

GH family 20 includes β-*N*-acetylhexosaminidases (EC 3.2.1.52), which can hydrolyze non-reducing terminal *N*-acetyl-D-hexosamine residues in *N*-acetyl-β-D-hexosaminides, including chitobiose [$(GlcNAc)_2$] the major product of chitinases. Some of these enzymes are also known as *N*-acetylglucosaminidases or chitobiases. Crystal structures are known for several members of this family, including β-*N*-acetylhexosaminidase from *S. marcescens* (1QBB), and show that the catalytic domain of GH20 enzymes has a TIM-barrel structure (Tews *et al.*, 1996). Family 20 glycoside hydrolases act via a retaining mechanism, and, as for family 18 chitinases, hydrolysis involves assistance by the acetamido group of the substrate (Tews *et al.*, 1997; Mark *et al.*, 2001; Vocadlo and Withers, 2005). In nature chitobiases catalyze the final step of converting chitin to monosaccharides. In an industrial setting, chitobiase activity may be needed to alleviate chitinase inhibition by chitobiose. Such product inhibition, and the resulting need for β-glucosidase activity is well known from the cellulose field (Merino and Cherry, 2007). These GH20 enzymes, which do not primarily act on polymeric substrates, are not discussed any further in this review.

Carbohydrate esterases (deacetylases) are divided into 16 different families (by January 2010; www.cazy.org). These enzymes catalyze de-O or de-N-acylation of substituted carbohydrates. Because of structural similarities between various substrates and broad substrate specificities, classification of carbohydrate esterases is less straightforward than classification of glycoside hydrolases (Davies *et al.*, 2005). Most carbohydrate esterase structures show a β/α/β-serine protease fold (Vincent *et al.*, 2003), but the largest family, carbohydrate esterase family 4 (CE4), has a different structure (Blair and van Aalten, 2004; Blair *et al.*, 2005). Some sugar deacetylases, including members of the CE4 family, have been reported to be metal ion-dependent (typically Zn^{2+} or Co^{2+}) (Blair *et al.*, 2005; Hernick and Fierke, 2005).

Microbial enzymes acting on chitin and chitosan – occurrence

Some organisms are better chitin-degraders than others. One bacterium known to be an effective chitin degrader is the Gram-negative soil bacterium *Serratia marcescens*. *S. marcescens* produces three family 18 chitinases (ChiA, ChiB & ChiC), a chitobiase (GH family 20) and a chitin-binding protein (CBM family 33) that acts synergistically in chitin degradation (Brurberg *et al.*, 1996; Suzuki *et al.*, 1998; Vaaje-Kolstad *et al.*, 2005a). Some other well known chitin-degrading microorganisms in the soil belong to the genera *Trichoderma* (fungi) and *Streptomyces* (actinomycetes). Organisms from both these genuses produce several chitinolytic enzymes (Saito *et al.*, 1999b; Seidl *et al.*, 2005). Interestingly, some *Streptomyces* species are known to produce enzymes belonging to both of the two chitinase families, GH18 and GH19.

Streptomyces are Gram-positive bacteria with a high GC-content, and are among the most numerous and ubiquitous soil bacteria (Hodgson, 2000). Compared to other bacteria, they grow in an unusual way, involving filamentous branching and formation of hyphae and spores (Hopwood, 1999). Interestingly, they are famous for their production of natural antibiotics, like neomycin, chloramphenicol and streptomycin. So far, three genomes from the genus *Streptomyces* have been published; *S. avermitilis* (Ikeda *et al.*, 2003), *S. griseus* (Ohnishi *et al.*, 2008) and *S. coelicolor* A3(2) (Bentley *et al.*, 2002). The genome sequences have revealed that these *Streptomyces* species contain a particularly large number of genes putatively coding for secreted hydrolases, including many chitosanases and chitinases.

The genome of *S. coelicolor* A3(2) contains as many as 13 chitinase genes encoding 11 family 18 and two family 19 chitinases (Kawase *et al.*, 2006). Saito *et al.* (2000) had previously described the nucleotide sequences of eigth different chitinases in *S. coelicolor* A3(2); *chiA* (SCO5003), *chiB* (SCO5673), *chiC* (SCO5376), *chiD* (SCO1429), *chiE* (SCO5954), *chiF* (SCO7263), *chiG* (SCO0482) and *chiH* (SCO6012) (Saito *et al.*, 1999b; Saito *et al.*, 2000). Chitinases ChiA-E and H belong to GH18, whereas ChiF and ChiG belong to GH19. The names of these chitinases are based on either similarities with other *Streptomyces* chitinases or the order of identification. A more systematic naming was suggested by Kawase *et al.* in 2006. This classification involves the glycoside hydrolase family number, followed by a letter indicating the subfamily (subfamily a, b or c for bacterial family 18 chitinases; Suzuki *et al.*, 1999), and a letter referring to the previous name, e.g. ChiC becomes Chi18aC. All together, *S. coelicolor* A3(2) contains four family 18 chitinases in subfamily a (Chi18aC, Chi18aD, Chi18aE and Chi18aJ), three in subfamily b (Chi18bA, Chi18bB and Chi18bI), four in subfamily c (Chi18cH, Chi18cK, Chi18cL and Chi18cM), and two family 19 chitinases (Chi19F and Chi19G).

In addition to chitinases, *S. coelicolor* A3(2) produces several chitosanases. The genome of *S. coelicolor* A3(2) contains two genes putatively encoding chitosanases belonging to family GH46 and one gene putatively encoding a chitosanase belonging to family GH75. The genome also contains genes putatively encoding family 20 glycoside hydrolases (4), and family 4 carbohydrate esterases (11). Genes encoding enzymes belonging to GH families 8 or 80 were not found in the genome of *S. coelicolor* A3(2) (www.cazy.org).

Another organism known to be a major producer of chitin-degrading enzymes in soil is the ascomycete *Trichoderma*. *Trichoderma* species are filamentous soil fungi, known for their potential as biocontrol agents against a range of phytopathogenic fungi, including *Rhizoctonia solani* and *Botrytis cinerea* (Tronsmo, 1991; Hjeljord *et al.*, 2001; Woo *et al.*, 1999). The mycoparasitic mechanism of *Trichoderma* involves secreted chitinases that hydrolyze the cell wall polymers in other fungi, causing cell lysis and death (Lorito *et al.*, 1994; Lima *et al.*, 1997). Biocontrol is of great interest because it provides a way to reduce the strong dependence of modern agriculture on fungicides, which can cause resistance problems and environmental pollution.

The genome of *Hypocrea jecorina* (previously called *Trichoderma reesei*) has recently been sequenced (Martinez *et al.*, 2008). This revealed that the chitinolytic machinery of this organism comprises as many as 18 family 18 chitinases, but no family 19 chitinases (Seidl *et al.*, 2005). Previously, *Trichoderma* chitinases were named according to their molecular weight, often combined with an abbreviation of the species from which the enzymes were cloned and/or a term alluding to the type of enzyme action (endo/exo). A more systematic nomenclature was suggested by Seidl *et al.*, resulting in a numbering from chi18-1 to chi18-18, where chi18-1 was the enzyme with the lowest theoretical pI (*Table 1*; Seidl *et al.*, 2005). Seidl *et al.* also did a phylogenetic analysis of fungal family 18 chitinases, and ended up dividing the enzymes in three phylogenetic groups (A, B and C). The subdivision of fungal GH18 into phylogenetic clusters, has recently been subject of a new, larger study also using three subgroups called A, B and C (Karlsson and Stenlid, 2008). In this latter study, *Trichoderma* enzymes classified by Seidl *et al.* (2005) as subgroup C ended up in a cluster in subgroup A. Apart from this difference, the results of the subgrouping were the same in both studies (*Table 1*).

Table 1. Overview over chitinases previously described in different *Trichoderma* species and their suggested new names and phylogenetic group (Seidl *et al.* 2005).

Previously cloned chitinases:	*Trichoderma* spp. (GenBank Acc. No.):	Nomenclature according to Seidl *et al.*, 2005:	Grouping according to Karlsson & Stenlid, 2008	References:
Ech42 / Chit42 / Tv-ech1	Various *Trichoderma* species	Chi18-5, Group A	A	(Carsolio *et al.* 1994; Garcia *et al.* 1994; Kim *et al.* 2002)
Tv-ech3	*T. virens* (AF397019)	Chi18-6, Group A	A	(Kim *et al.* 2002)
Tv-ech2	*T. virens* (AF397021)	Chi18-7, Group A	A	(Kim *et al.* 2002)
Chit33 / Tv-Cht1	*T. harzianum* (X80006) / *T. virens* (AF397017)	Chi18-12, Group B	B	(De la Cruz *et al.* 1992; Kim *et al.* 2002)
Ech30	*T. atroviride* (AY258147)	Chi18-13, Group B	B	(Hoell *et al.* 2005)
Chit36 / Chit36y	*T. harzianum* (AY028421) / *T. asperellum* (AF406791)	Chi18-15,Group not assigned	B	(Viterbo *et al.* 2001; Viterbo *et al.* 2002)
Tv-Cht2	*T. virens* (AF397018)	Chi18-17, Group B	B	(Kim *et al.* 2002)

According to the CAZy database, so far (January 2010) no genes putatively coding for chitosanases from family 8, 46 and 80 or carbohydrate esterases from family 4 have been discovered in *Trichoderma* species. One gene putatively encoding a family 75 chitosanase has been discovered in *Hypocrea lixii* (*T. harzianum*). The *Trichoderma* species contain several genes encoding enzymes belonging to GH20 (www.cazy.org).

While the CAZy database provides an excellent system for the classification of carbohydrate-active enzymes that now has been widely adopted by the glycosides hydrolase community (see above), the nomenclature of these enzymes in literature has shown little consistency. This is illustrated by the description of *Streptomyces* and *Trichoderma* machineries above. To illustrate this further, *Table 2* gives an overview over the nomenclature for a selection of enzymes studied in our laboratory in the past decade. This table includes an older classification used for plant chitinases, where the enzymes are divided into 5 subclasses labeled I - V (Cohen-Kupiec and Chet, 1998).

Family 18 Chitinases

Family 18 chitinases are retaining enzymes (Tews *et al.*, 1997), and comprise the most spread and best-studied class of chitinolytic enzymes. In January 2010, the GH18 family contained over 3000 entries in CAZY. These enzymes occur in many organisms, including archaea, bacteria, eukaryota and viruses. There is a lot of structural information available for family 18 chitinase, e.g. for ChiA and ChiB from *S. marcescens* (Perrakis *et al.*, 1994; Van Aalten *et al.*, 2000; discussed in more detail below), hevamine from the rubber tree (*Hevea brasiliensis*; Van Scheltinga *et al.*, 1994), chitinase A1 from *Bacillus circulans* (Matsumoto *et al.*, 1999) and *Af*ChiB1 from *Aspergillus fumigatus* (Rao *et al.*, 2005). The catalytic domain of family 18 chitinases has a $(\beta/\alpha)_8$ ("TIM-barrel") fold.

The bacterial family 18 chitinases have been further classified into subclass A, B, and C based on amino acid sequence similarities in parts of their catalytic domains (Suzuki *et al.*, 1999). The *S. marcescens* chitinases ChiA and ChiB both belong to subclass A, whereas ChiC belongs to subclass B (*Table 2*). All enzymes belonging to subclass A have an extra $\alpha + \beta$-fold sub-domain of approximately 75 residues, which is inserted between the seventh and eighth β-strand of the $(\beta/\alpha)_8$-barrel and which provides one "wall" of the substrate-binding cleft.

Some family 18 chitinases consist of a catalytic domain only, while others contain one or more carbohydrate binding modules. For example, in ChiB, a chitin-binding domain extends the substrate binding cleft on the side where the reducing end of the substrate binds ("+" subsites; Van Aalten *et al.*, 2000; *Figure 3*). In ChiA another type of chitin-binding domain extends the substrate-binding cleft towards the "-" subsites (*Figure 3*; Perrakis *et al.*, 1994). The structures shown in *Figure 3* suggest that ChiB degrades chitin chains from their non-reducing ends and that ChiA cleaves chitin chains from their reducing ends (Van Aalten *et al.*, 2000; Papanikolau *et al.*, 2001; Suzuki *et al.*, 2002). In other words, ChiA and ChiB are thought to have different directionalities. Studies in which the degradation of end-labeled substrates was followed by the tilt microdiffraction method (Hult *et al.*, 2005) as well as recent mutagenesis-based studies of processivity (Zakariassen *et al.*, 2009) confirm the predicted opposite directionalities of ChiA and ChiB.

Structures of enzyme-substrate complexes have been solved for both ChiA and ChiB (Papanikolau *et al.* 2001; Van Aalten *et al.* 2001). In these processive enzymes (Sikorski *et al.*, 2006) the substrate binding grooves are lined with aromatic amino acids, and so are the surfaces of the chitin-binding domains (*Figure 3*). Family 18 chitinases share

Table 2. Overview of the nomenclature of chitinases studied in the authors' laboratory.

Working name	Source organism	GH family	Catalytic domain structure	Group[1]	Plant sub-group[5]	Name according to Kawase *et al.* (2006)	Name according to Seidl *et al.* (2005) proposal[1]	Name according to CAZY
Ech30	*T. atroviride* P1	GH18	$(\beta/\alpha)_8$	B	Class III	N/A[3]	Chi18-13	TaChi18A
ChiF	*S. coelicolor* A3(2)	GH19	lysozyme like	N/A[2]	Class IV	Chi19F	N/A[4]	ScChi19F
ChiG	*S. coelicolor* A3(2)	GH19	lysozyme like	N/A[2]	Class IV	Chi19G	N/A[4]	ScChi19G
ChiA	*S. marcescens*	GH18	$(\beta/\alpha)_8$	A	Class V	Chi18aA	N/A[4]	SmChi18A
ChiB	*S. marcescens*	GH18	$(\beta/\alpha)_8$	A	Class V	Chi18aB	N/A[4]	SmChi18B
ChiC	*S. marcescens*	GH18	$(\beta/\alpha)_8$	B	Class V	Chi18bC	N/A[4]	SmChi18C

[1] According to Suzuki *et al.* (1999), Seidl *et al.* (2005) and/or Karlsson & Stenlid (2008) (all three studies result in similar grouping; see text for explanation).

[2] Not applicable; the subgrouping in this column only includes family 18 chitinases.

[3] Not applicable; the subgrouping in this column only includes bacterial chitinases.

[4] Not applicable; Seidl's nomenclature only includes family 18 chitinases from *Trichoderma*.

[5] Old classification system that has been and to some extent still is used for chitinases from plants (Neuhaus *et al.*, 1996; Cohen-Kupiec and Chet 1998).

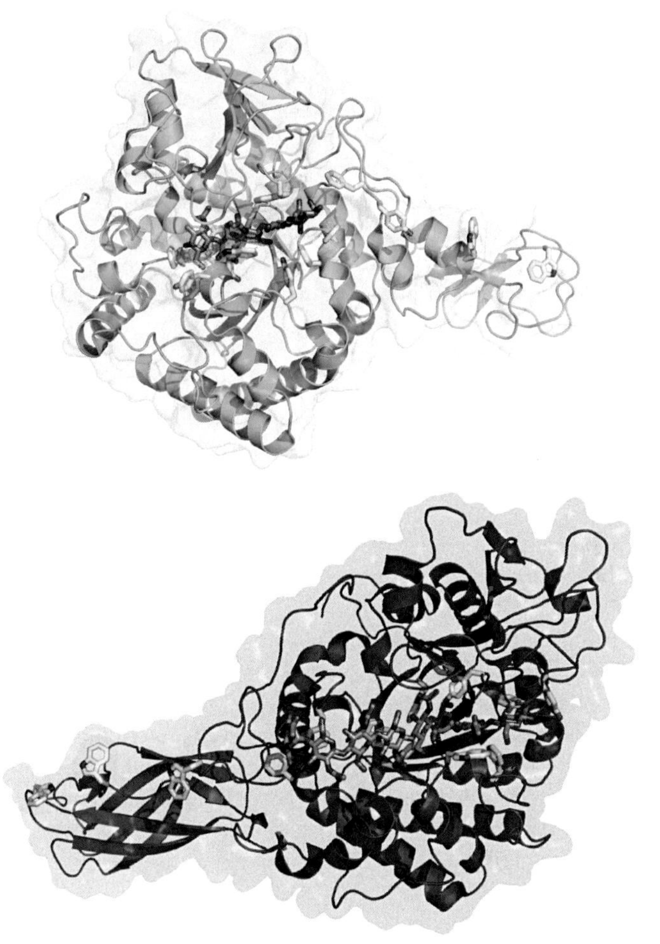

Figure 3. Structures of ChiA and ChiB from *S. marcescens*. The figure shows the structure of ChiB (upper panel, cyan, PDB: 1e6n) and ChiA (lower panel, magenta, PDB: 1ehn) in complex with $(GlcNAc)_5$ and $(GlcNAc)_8$, respectively (orange). In ChiB, the chitin-binding domain (to the right) extends the substrate binding cleft in the reducing end direction, whereas in ChiA the chitin-binding domain (to the left) extends towards the non-reducing end. The substrate-binding groove and the surfaces of the chitin binding domains are lined with aromatic side chains (yellow). Note that the catalytic domains have approximately similar orientations in both pictures; the substrates bind to subsites -2 to +3 in ChiB and subsites -6 to +2 in ChiA. The pictures were made with PYMOL (DeLano 2004). The structures are described in Papanikolau *et al.*, 2001 (ChiA) and Van Aalten *et al.*, 2001 (ChiB).

several conserved residues, including the characteristic DXXDXDXE motif, where the last Glu is the catalytic acid (Watanabe *et al.*, 1993; e.g. Glu144 in *Figure 2*). As described above, catalysis in family 18 chitinases has some modifications compared to the typical double retaining mechanism. Instead of using a carboxylate side chain of the enzyme as the catalytic nucleophile, family 18 chitinases use the acetamido group of the -1 sugar (Van Scheltinga *et al.*, 1995; Knapp *et al.*, 1996; Kobayashi *et al.*, 1996; Tews *et al.*, 1997; *Figure 2*). Other residues in the DXXDXDXE motif are thought to play important roles such as positioning and activating the acetamido group for nucleophilic attack and cycling the pK_a of the catalytic acid through the catalytic cycle (Van Aalten *et al.*, 2001; Synstad *et al.*, 2004) (see *Figure 2*; Asp140, Asp142 and Glu144 in this figure correspond to DXDXE). This special catalytic mechanism implies that for catalysis to occur, family 18 chitinases require an acetylated sugar in the -1 subsite (Sorbotten *et al.*, 2005). Other subsites show less stringent preferences for acetylated versus deacetylated sugars and most are in fact rather promiscuous in this respect (Sorbotten *et al.*, 2005; Horn *et al.*, 2006b).

The properties of the three *Serratia* chitinases have been studied in much detail, including in-depth studies of processivity using chitosan as substrate (Horn *et al.*, 2006a,b; Zakariassen *et al.*, 2009). In-depth studies on the roles of aromatic residues showed that the presence/absence of specific aromatic residues could explain the opposite directionalities of ChiA and ChiB. These studies also confirmed that aromatic residues are crucial for enzyme processivity. Most remarkably, it was shown that processivity slows down the enzymes: while non-processive mutants were less efficient towards chitin, they were much more effective towards the non-crystalline polymeric substrate chitosan (Horn *et al.*, 2006a; Zakariassen *et al.*, 2009; see Eijsink *et al.*, 2008 for further discussion). Together these studies provide a picture of how an efficient catalytic machinery may look: ChiA and ChiB are processive enzymes that degrade chitin chains in opposite directions after initial endo- or exo-binding, whereas ChiC, the third family 18 chitinase of *S. marcescens* is a non-processive endo-chitinase (Horn *et al.*, 2006b; Sikorski *et al.*, 2006).

It should be noted that, in addition to chitinases, *S. marcescens* and most other chitin-degrading microorganisms produce proteins belong to the CBM33 family that bind strongly to crystalline substrates. CBM33 binding makes the substrate more accessible for hydrolytic attack by chitinases and thus increases the efficiency of enzymatic hydrolysis (Vaaje-Kolstad *et al.*, 2005a,b; Eijsink *et al.*, 2008; Vaaje-Kolstad *et al.*, 2009). Clearly, these CBM33 proteins should be included in enzymatic processes for the conversion of crystalline chitin. The effect of these CBM33 proteins is general in the sense that they potentiate any type of chitinases, i.e. also GH19 enzymes, which do not occur in for example *S. marcescens*.

Family 19 Chitinases

In contrast to family 18 chitinases, family 19 chitinases use an inverting mechanism which leads to α-anomeric hydrolysis products (Davies and Henrissat, 1995; Iseli *et al.*, 1996). The first family 19 chitinase with known structure was from barley (Hart *et al.*, 1993; Hart *et al.*, 1995), and for many years family 19 chitinases were believed to occur in plants only. However, it is now clear that family 19 chitinases also occur in bacteria (Ohno *et al.*, 1996) and the number of family 19 chitinases genes from bacteria is increasing as massive genome sequencing proceeds. Today, the general notion is that family 19 chitinases are commonly found in plants, bacteria and viruses. By January 2010, the GH19 family in the CAZy database contained almost 1000 entries (www.cazy.org).

For a long time structural information for these chitinases was limited to the structures of two plant enzymes (Hart *et al.*, 1993; Hahn *et al.*, 2000). Recently, the structures of bacterial family 19 chitinases have become available (see below). The catalytic domains of family 19 chitinases have a lysozyme-like fold with rather shallow substrate-binding grooves that are not particularly rich in aromatic residues.

The catalytic domains of family 19 chitinases share structural similarity with family 46 chitosanases (see below) and with lysozymes in families 22, 23 and 24 of glycoside hydrolases (Hart *et al.*, 1995; Monzingo *et al.*, 1996).

The two catalytic residues in family 19 chitinases have been identified by site-directed mutagenesis experiments (Andersen *et al.*, 1997; Tang *et al.*, 2004; Hoell *et al.*, 2006; Kezuka *et al.*, 2006), but the catalytic mechanism for these enzymes is still poorly understood. One reason is that there is little structural information for enzyme-substrate complexes. It has been shown that at least two more conserved charged residues are crucial for catalysis. These residues, Glu203 and Arg215 in barley chitinase, form a triad together with the catalytic acid Glu67 (Ohnishi *et al.*, 2005) (*Figure 4*). A similar interaction network has previously been found in a family 46 chitosanase from *Streptomyces* sp. N174 (see below and Fukamizo *et al.*, 2000). While there is no doubt that electrostatic interactions between the catalytic acid and these other conserved residues are important, the precise roles of these interactions in the catalytic mechanisms are unknown.

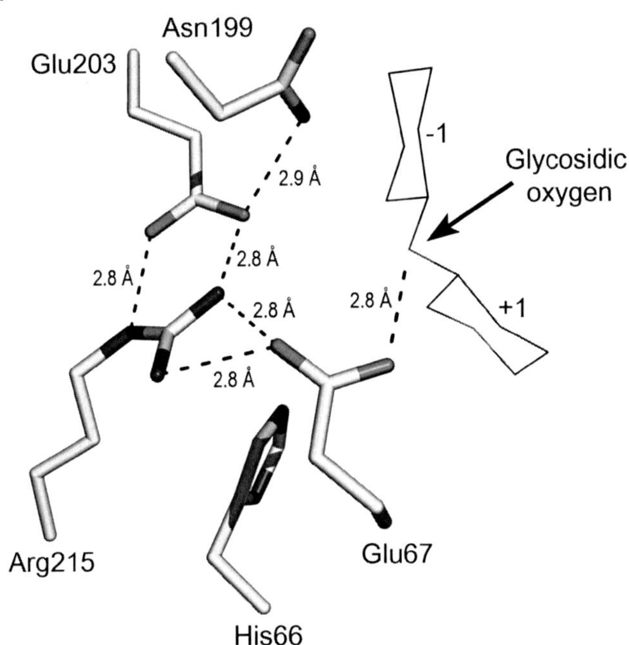

Figure 4. Conserved residues surrounding the catalytic acid, Glu67, in the family 19 papaya chitinase (PDB ID 3CQL). Amino acid side chains are shown in stick representation. Hydrogen bonds are shown as dashed lines with distances indicated in Å. A $GlcNAc_2$ unit binding in the -1 and +1 subsites is shown in narrow stick representation and an arrow indicates the position of the glycosidic oxygen. The sugar ring substituents are not shown in order to make the figure clearer. The structure of this complex comes from a model described by Huet *et al.*, 2008 (see text). Nb. Corresponding residues in the barley chitinase have the same residue numbers.

Because of the lack of structural information, little is known about how family 19 chitinases interact with their substrates. The binding of chito-oligosaccharides to family 19 chitinases has been studied in some detail by studying the anomeric distributions of products obtained after hydrolysis of chito-oligosaccharides (Hoell *et al.*, 2006; Mizuno *et al.*, 2008; Sasaki *et al.*, 2003), but the structural basis of observed substrate-binding preferences is largely unknown. Recently, a complex of papaya family 19 chitinase with GlcNAc units bound in the -2 and +1 subsites was published, together with a model of a complex with $(GlcNAc)_4$ that was build using the two crystallographically observed sugars (Huet *et al.* 2008). This is the first structure (half experimental, half modeled) of an enzyme-substrate complex in this family of glycoside hydrolases, giving a better basis for understanding the sugar binding in these enzymes. The results from Huet *et al.* (2008) may be used for further docking studies and provide a good basis for site-directed mutagenesis studies. So far, mutational data are limited and they are scattered among various family 19 enzymes (Andersen *et al.*, 1997; Tang *et al.*, 2004; Garcia-Casado *et al.*, 1998; Verburg *et al.*, 1993). Moreover, available data primarily concern plant enzymes and may not necessarily be extrapolatable to bacterial enzymes (see below).

The structures of family 19 chitinases from bacteria revealed several differences from the previously reported plant structures. Compared to plant enzymes, the bacterial enzymes lack a C-terminal extension and three loops, some of which are thought to be flexible (Ubhayasekera *et al.*, 2007; Fukamizo *et al.*, 2009) (*Figure 5*). In addition, a region referred to as 161-166 loop in *Figure 5* has a markedly different conformation. This region, also referred to as loop IV in the literature, is lacking in a subset of the plant family 19 chitinases. An overview over the regions/loops and their nomenclature (which is not fully consistent in the literature) is provided in *Table 3*.

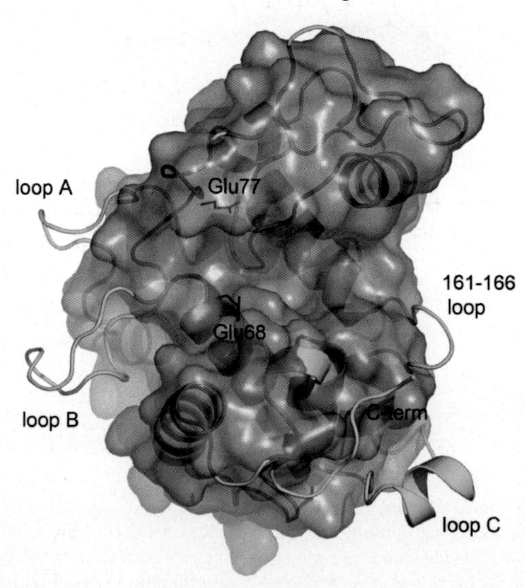

Figure 5. Structural superposition of ChiG, a bacterial family 19 chitinase from *S. coelicolor* A3(2), and the barley family 19 chitinase. The picture shows a cartoon of the barley enzyme (PDB; 2BBA; cyan) and the surface of ChiG (PDB; 2CJL; blue). Glu68 and Glu77 are the catalytic acid and the catalytic base of ChiG, respectively. As shown in the picture, bacterial enzymes lack a C-terminal extension and three loops (loop A, B and C) compared to plant enzymes. The figure is taken from Hoell *et al.* 2006. For further information on loop nomenclature, see text and *Table 3*.

Structural inspections suggest that the lack of these loops in bacterial enzymes reduces the length of the substrate-binding cleft to only four sugar binding subsites, compared to six for plant family 19 chitinases. Studies of product profiles for the (bacterial) ChiG from *S. coelicolor* A3(2) support this notion (Hoell *et al.*, 2006). The observed flexibility of the loops present in plant family 19 chitinases, as well as their location in the structure (i.e. close to what could be subsites -3 and +3) suggest that they are involved in substrate binding (Ubhayasekera *et al.*, 2007; Fukamizo *et al.*, 2009). One flexible loop referred to as loop "B" or "II" (*Table 3*) is located between the two catalytic residues (Hoell *et al.*, 2006; Kezuka *et al.*, 2006) and experimental results have shown that this loop is involved in substrate binding, primarily affecting the +3 subsite (Mizuno *et al.*, 2008; Fukamizo *et al.*, 2009). Furthermore, the presence or absence of loop III (or C) and the C-terminal extension affect the structure of the flexible 161-166 loop (*Figure 5*), which is highly likely to affect putative -3 and -4 subsites.

Table 3. Overview of major insertions and deletions in family 19 chitinases.

Name in *Figure 5*	Sequence number in barley chitinase	Alternative name	Presence / notes:
Loop A	20-26	Loop I	Extra in plant enzymes
Loop B	70-82	Loop II	Extra in plant enzymes
Loop C	176-188	Loop III	Extra in plant enzymes
161-166 loop	161-166	Loop IV	Lacking in some plant enzymes; conformational variation
C-term	237-243	C-term	Extra in plant enzymes

The roles of CBMs

Many chitinases in both families GH18 and GH19 contain one or more CBMs (e.g. *Figure 3*) and it is well known from e.g. the work by T. Watanabe and co-workers that these domains are important for binding to the substrate and hydrolytic efficiency (Watanabe *et al.*, 1994; Itoh *et al.*, 2006). While the role of these domains in increasing substrate-affinity is well established (Boraston *et al.*, 2004; Shoseyov *et al.*, 2006), it is less clear whether these domains also may play a role is disturbing the crystalline packing of the substrate, thus making the substrate more accessible to enzymatic hydrolysis (see Eijsink *et al.*, 2008 and references therein). Very little is known about the occurrence of CBMs in chitosanases, but, so far, it seems as if this is not as common as for chitinases. Only two of the known GH46 chitosanases contain a CBM, classified as "chitin-binding domain" belonging to the closely related CBM families 5 and 12. While a further discussion of the roles of these CBMs is beyond the scope of this review, it is clear that their presence and possible functions should be taken into account when studying enzyme performances. For a recent study discussing the possible roles of CBMs in chitin degradation, see Kikkawa *et al.*, 2008.

Family 8 Chitosanases

Family 8 glycoside hydrolases perform catalysis by an inverting mechanism, and contain enzymes with diverse substrate specificities. Besides chitosanases, this family includes cellulases, licheninases and endo-xylanases. Enzymes in GH8 have so far only been found in bacteria and especially *Bacillus* sp. are well known for the production of GH8 enzymes with chitosanase activity. Most of the chitosanases described in this family are in fact bifunctional enzymes combining cellulase and chitosanase activity (Pedraza-Reyes and Gutierrez-Corona, 1997; Kimoto *et al.*, 2002; Choi *et al.*, 2004; Isogawa *et al.*, 2009). These bifunctional cellulase-chitosanases are often named "chitosanases" since they often have higher chitosanolytic activity than cellulolytic activity (Xia *et al.*, 2008). The bifunctionality is most likely related to the structural similarity of chitin, chitosan and cellulose (*Figure 1*).

So far, six structures have been solved in this family of glycoside hydrolases, including one with chitosanolytic activity, namely ChoK from *Bacillus* sp. K17 with subclass II specificity (PDB: 1V5C) (Adachi *et al.*, 2004). The catalytic site is located on the scaffold of a double-α_6/α_6-barrel, which is formed by six repeating helix-loop-helix motifs. This structure is similar to the structures of a cellulase from *Clostridium thermocellum* (CelA) (Alzari *et al.*, 1996) and a xylanase from *Pseudoalteromonas haloplanktis* (Van Petegem *et al.*, 2003), both members of GH8. For all the three enzymes the catalytic acid is the same, namely a glutamate. In both CelA and *P. haloplanktis* xylanase, an aspartic acid residue acts as catalytic base (GH8 enzymes are inverting enzymes). In ChoK, this residue is however replaced by an asparagine and can thus not act as a base. Instead, a glutamate, located at a slightly different position on the α_6/α_6-barrel, was proposed to act as the catalytic base. It was further suggested that this residue is important for substrate specificity and recognition of a glucosamine (Adachi *et al.*, 2004).

One of the enzymolgically best characterized chitosanases in the GH8 family is the enzyme from *Bacillus* sp. No.7M (Izume *et al.*, 1992, Vårum *et al.*, 1996), with subclass II specificity. Other studied chitosanases in the GH8 family are produced by *Bacillus circulans* WL-12 (Mitsutomi *et al.*, 1998) and *Paenibacillus fukuinensis* strain D2 (Kimoto *et al.*, 2002).

Family 46 chitosanases

Chitosanases belonging to family 46 are the best studied of all chitosanases. So far, GH46 enzymes have been discovered in bacteria and viruses only. Among family 46 chitosanases, *Streptomyces* sp. N174 chitosanase (CsnN174) and *Bacillus circulans* MH-K1 chitosanase (MH-K1) have been studied most intensively (Marcotte *et al.*, 1996; Saito *et al.*, 1999a). Results from experiments with partially acetylated chitosan showed that CsnN174 was able to hydrolyze the GlcNAc-GlcN and GlcN-GlcN linkages (Fukamizo and Brzezinski, 1997), which classifies this enzyme into subclass I. The MH-K1 chitosanase, on the other hand, was classified into subclass III, hydrolysing GlcN-GlcNAc and GlcN-GlcN linkages (Saito *et al.*, 1999a).

The structure of CsnN174 revealed that family 46 chitosanases have a structural core shared with lysozymes from GH families 22, 23 and 24, and chitinases from family GH19. As shown in *Figure 6*, this structural core consists of two α-helixes and a β-sheet formed by three anti-parallel β-strands separated by loops of varying lengths, which together form the substrate-binding and catalytic cleft (Holm and Sander, 1994; Monzingo *et al.*, 1996; Robertus *et al.*, 1998). Even though enzymes in this "lysozyme superfamily" share this conserved structural core, they do not share significant sequence similarities (Marcotte *et al.*, 1996; Monzingo *et al.*, 1996). In addition, there are major structural differences outside the common core. Interestingly, despite the common core, catalytic mechanisms vary too: the enzymes of the lysozyme superfamily perform catalysis with an inverting mechanism, except for enzymes belonging to GH 22 (hen egg-white lysozyme, HEWL), which use a retaining mechanism.

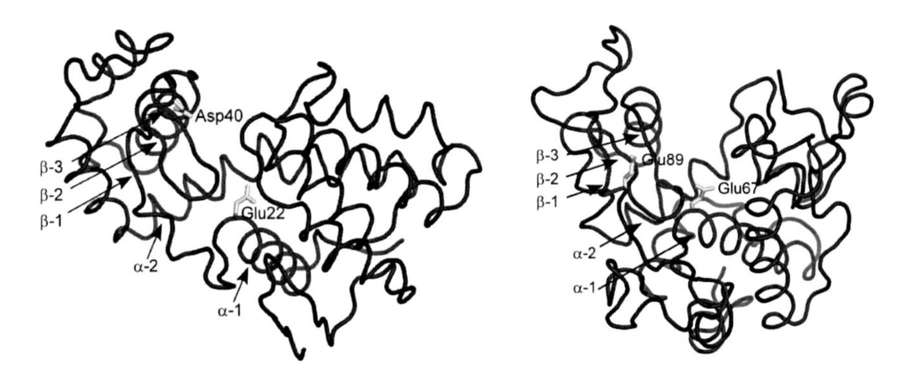

Figure 6. Structures of GH46 chitosanase CsnN174 from *Streptomyces* sp. N174 and GH19 chitinase from barley. The figure shows chitosanase CsnN174 (left; PDB ID 1CHK) and the barley chitinase (right; PDB ID 1CNS) in a ribbon view in blue color. The common structural core, consisting of two helixes (α-1 and α-2) and a three-stranded β-sheet (β-1, β2 and β-3), is colored pink. In the protein sequence, α-helix 1 is followed by the β-sheet, consisting of antiparallel β-strands 1, 2 and 3 connected by loops of different length, which is followed by α-helix 2. The side chains of the catalytic residues are shown as yellow sticks. Note that the catalytic acids (Glu67 in barley chitinase and Glu22 in CsnN174) have almost identical positions in the two structures, at the C-terminal end of α-helix 1. This picture was made with Pymol (DeLano 2004).

For CsnN174, Glu22 was identified as the catalytic acid by site-directed mutagenesis (Boucher *et al.*, 1995; Fukamizo *et al.*, 1995; Fukamizo and Brzezinski, 1997). The structure of CsnN174 has shown that the catalytic acid is in close proximity to Arg205, and the sidechain of Arg205 is in close proximity to Asp145, which in turn is in close distance to Arg190 (see *Figure 7*). Site-directed mutagenesis of residues Arg205, Asp145, and Arg190 followed by activity measurements, showed that these residues are important for catalytic activity. Fukamizo *et al.* (2000) concluded that these additional residues form an important interaction network in the catalytic cleft, but also suggested that these residues, located on three different α-helixes (*Figure 7*), can have important roles for stability of the chitosanase (Fukamizo *et al.*, 2000). It is interesting to note that a similarly complicated interaction network, involving several charged residues, is observed near the catalytic acid in family 19 chitinases (see *Figure 4*, above).

When comparing with other enzymes in the lysozyme superfamily, one observes that the catalytic acid is located at the same position in all enzymes (Lacombe-Harvey *et al.*, 2009), namely in the C-terminal end of one of the α-helixes (α-helix I) belonging to the conserved core (*Figure 6*). In contrast, the general base is not located in equivalent positions. In CsnN174, Asp40 has been proposed as the best candidate for the general base in the inverting mechanism. This was suggested by its position in the 3D-structure (Marcotte *et al.*, 1996; see *Figure 6*), and supported by observed considerable losses of activity upon mutating this residue to Glu or Asn (Boucher *et al.*, 1995). Furthermore, it has been shown that the activity of (inactive) mutant D40G could be rescued by adding sodium azide, indicating that Asp40 acts as a catalytic base and activates a water molecule (Lacombe-Harvey *et al.*, 2009).

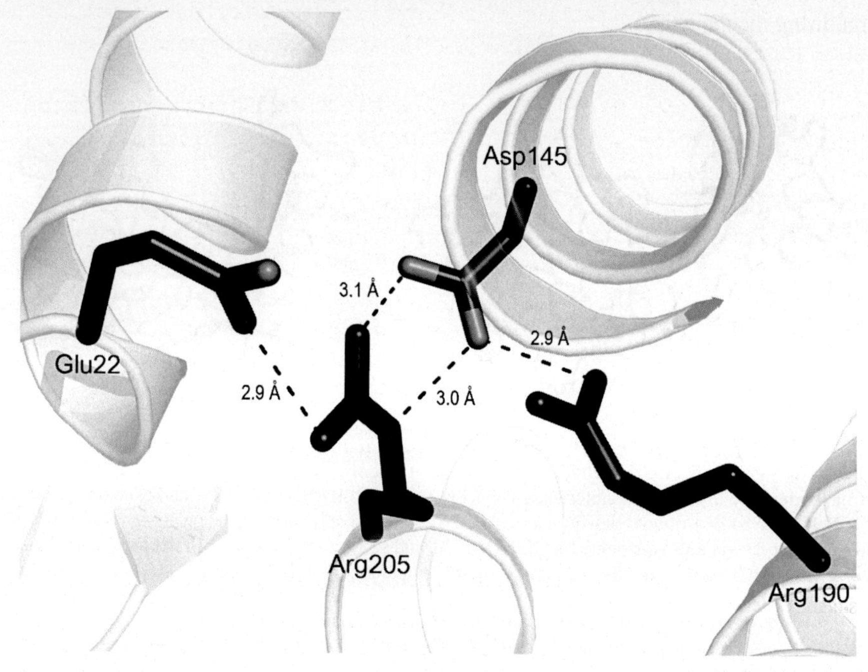

Figure 7. Detailed view of the residues surrounding the catalytic acid, Glu22 in CsnN174 from *Streptomyces sp.* N174 (PDB ID 1CHK). Residues Arg205, Asp145 and Arg190 are located on three different α-helixes, and form an interaction network including the catalytic acid, Glu22. Hydrogen bonds are indicated by dashed lines and distances are given in Å. This picture was made with Pymol (DeLano 2004).

It has been shown that Glu36 and Thr45 also are important for catalytic activity in CsnN174 (Lacombe-Harvey *et al.*, 2009). Remarkably, Glu36 and Thr45 are located in positions that are structurally equivalent to the positions of the catalytic base residue in barley family 19 chitinase (Glu89) and of the nucleophile in (retaining) HEWL (Asp52), respectively (*Table 4*). It was proposed that Glu36 acts as an alternative general base, whereas Thr45 is thought to play a role in orienting the water molecule in a position optimal for catalysis. The same role is played by Ser120 in papaya chitinase (equivalent with Ser120 in barley chitinase) (Song and Suh, 1996; Huet *et al.*, 2008). Judged from the current literature it would seem that Glu22 and Asp40 are the catalytic acid and base in CsnN174, respectively. However, more research (e.g. the structure of an enzyme-substrate complex) is needed to confirm this and to gain insight into the roles of the complex charged interaction networks that surround the catalytic residues.

All in all, these data show that the catalytic mechanisms of these enzymes, belonging to GH families 19, 22, 23, 24 and 46, are "variations on a common theme in the lysozyme superfamily" (Lacombe-Harvey *et al.*, 2009). To further illustrate this, *Table 4* gives an overview over variation and analogies in the positioning of active site residues on α-helix 1 and the β-strands in the structural cores of these enzymes. This overview strengthens the idea that there are evolutionary links between the GH families in the lysozyme superfamily. It should be noted that not all enzyme catalytic mechanisms have been investigated/described to the same level of detail. For example, the fact that there is no residue for "water positioning" listed for some of the enzymes in *Table 4* does not necessarily imply that such a residue is not present.

Table 4. Location of active site residues in equivalent positions in the conserved structural cores of enzymes in the lysozyme superfamily. GH19, 23, 24 and 46 enzymes act by an inverting mechanism, whereas GH22 acts by a retaining mechanism. This table is adapted from (Lacombe-Harvey *et al.* 2009).

Structural Motif	CsnN174 (GH46)	T4 lysozyme (GH24)	HEWL (GH22)	T. japonica lysozyme (GH22)	GEWL (GH23)	Barley chitinase (GH19)
C-terminal end of α-helix 1	E22 (acid)	E11 (acid)	E35 (acid-base)	E18 (acid-base)	E73 (acid)	E67 (acid)
β-1 strand	E36 (alternative base)			D30 (nucleophile)		E89 (base)
Loop between β-1 and β-2 strands	D40 (base)	D20 (base)				
β-2 strand	T45 (water positioning	T26 (water positioning)	D52 (nucleophile)			
β-3 strand					D97 (putative base)	S120 (water positioning)

Family 75 Chitosanases

Family 75 chitosanases have not been studied to the same extent as family 46 chitosanases and there are no reported structures of enzymes belonging to this family. Family 75 chitosanases are found in fungi such as *Aspergillus* and *Fusarium* as well as in members of the genus *Streptomyces*. One of the first genes for a family 75 enzyme was described in 1996 (Shimosaka *et al.*, 1996). In January 2010, the CAZY database contained only 36 entries in this family. One of the most studied family 75 chitosanases, is the endo-chitosanase from *Aspergillus fumigatus* (Cheng and Li, 2000; Cheng *et al.*, 2006). This fungal chitosanase was produced as inclusion bodies in *E.coli* and could be obtained in active form through refolding. Cheng *et al.* (2006) demonstrated that family 75 chitosanases use an inverting mechanism. In addition, using site-directed mutagenesis, two out of 10 conserved carboxylic amino acids, Asp160 and Glu169, were found to be important for catalytic activity (Cheng *et al.*, 2006). The same two residues (Asp175 and Glu188, respectively) were found to be essential for catalysis in a family 75 chitosanase from *Fusarium solani* produced in *Saccharomyces cerevisiae* (Shimosaka *et al.*, 2005). The *Aspergillus* chitosanase was shown to cleave GlcNAc-GlcN and GlcN-GlcN linkages, meaning that this enzyme belongs to subclass I (Cheng *et al.*, 2006).

Family 80 chitosanases

Family 80 chitosanases, first described by Park *et al.* (1999) have so far only been found in bacteria. In January 2010, the CAZY database contained only 14 entries in this family. Very little is known about these enzymes. By detailed sequence analysis Tremblay *et al.* 2000 have shown that family 80 chitosanases may resemble family 46 chitosanes, from which one may infer that the GH80 enzymes are inverting. Mutagenesis studies suggest that the catalytic acid and base both are a glutamate (Shimono *et al.*, 2002).

Chitin deacetylases

Carbohydrate esterase family 4 (CE4) includes chitin deacetylases (EC 3.5.1.41), rhizobial NodB chitooligosaccharide deacetylases (EC 3.5.1.-), peptidoglycan *N*-acetylglucosamine deacetylase (EC 3.1.1.-), peptidoglycan *N*-acetylmuramic acid deacetylase (EC 3.5.1.-) and acetyl xylan esterase (EC 3.1.1.72) (www.cazy.org). The substrates of CE4 esterases include acetyl xylan, chitin, NodB factors and peptidoglycan, which are all closely related due to structural similarities in their carbohydrate parts (Caufrier *et al.*, 2003). All enzymes in carbohydrate esterase family 4 have a conserved region called the NodB homology domain (Kafetzopoulos *et al.*, 1993).

Rhizobial NodB was the first enzyme of the CE4 family to be described in detail (Long, 1989), and is involved in de-*N*-acetylating GlcNAc residues during the synthesis of Nod factors, the morphogenic signal molecule in symbiotic leguminous plants. Peptidoglycan deacetylases modify bacterial cell walls by de-*N*-acetylation of either the GlcNAc residues or the *N*-acetylmuramic acid in peptidoglycan. The first structure of a family 4 carbohydrate esterase to be published, was that of the *Bacillus subtilis* peptidoglycan deacetylase (*Bs*PdaA) (Blair and van Aalten, 2004), which de-*N*-acetylates *N*-acetylmuramic acid. Subsequently, the structure of *Streptococcus pneumoniae* peptidoglycan deacetylase (*Sp*PgdA), and extensive mutagenesis and enzymological analyses of the streptococcal enzyme have given more insight into the catalytic mechanism employed by CE4 enzymes (Blair *et al.*, 2005).

*Sp*PgaD de-*N*-acetylates the GlcNAc residues of peptidoglycan. The catalytic domain of *Sp*PgdA contains the NodB homology domain, and comprises a $(\beta/\alpha)_8$ ("TIM-barrel") fold (Blair and van Aalten, 2004; Blair *et al.*, 2005). Family 4 carbohydrate esterases share a number of sequence motifs, including several conserved aspartic acids and histidine residues. Several reports also show that the activity of these enzymes increases when divalent cations are added (Martinou *et al.*, 2002; Caufrier *et al.*, 2003). This is consistent with the results from Blair *et al.*, which show that *Sp*PgdA is a metalloenzyme using a His-His-Asp zinc-binding triad for catalysis, with a nearby Asp and His acting as the catalytic acid and base, respectively (Blair *et al.*, 2005). Interestingly, *S. pneumonia* lacking the *Sp*PgdA is hypersensitive for exogenous lysozyme, which results in cell lysis when exposed to lysozyme (Vollmer and Tomasz, 2000). This suggests that deacetylation is used as a protection towards attack by peptidoglycan-degrading enzymes. It has been hypothesized that similar mechanisms could be employed by chitin-containing organisms in their defence against chitinases (Tsigos *et al.*, 2000).

Recently, the first structure of a chitin deacetylase, *Cl*CDA, originating from the fungal pathogen *Colletotrichum lindemuthianum*, was solved (Blair *et al.*, 2006). The overall structure of *Cl*CDA is similar to the structures of *Bs*PdaA and *Sp*PgdA, and the enzyme also contains the zinc-binding motif (His-His-Asp). Chitin deacetylases have so far not been observed in bacteria, but have been found in several fungi (Araki and Ito, 1975; Kafetzopoulos *et al.*, 1993; Tokuyasu *et al.*, 1999; Martinou *et al.*, 1993; Gao *et al.*, 1995). Chitin deacetylases are thought to be involved in cell-wall formation and remodeling and may play a role in plant-pathogen interactions (Tsigos *et al.*, 2000).

Taylor *et al.* recently presented the first two structures of distinct de-*O*-acetylases from the CE4 family, the *Clostridium thermocellum* (*Ct*CE4) and *Streptomyces lividans* (*Sl*CE4) acetyl xylan esterases (Taylor *et al.*, 2006). In *Ct*CE4 metal ion binding involves an aspartate, a histidine, and four water molecules, whereas *Sl*CE4 employs the usual His-His-Asp motif. Both enzymes are metal ion dependent with a chemical preference for cobalt (Taylor *et al.*, 2006). Generally the two structures are similar to other CE4 structures, and the catalytic mechanisms are the same.

All in all, currently available data show large variation within the CE4 family. The enzymes have similar structures and catalytic mechanisms, but they display diversity in metal coordination and binding specificity and, most importantly, in substrate preferences. It is possible that members of this family may act on several subtrates, e.g. peptidoglycan and/or xylan and/or chitin (Psylinakis *et al.*, 2005), but this phenomenon, as well as its possible biological relevance needs further investigation. There is no doubt that chitin deacetylases have a significant biotechnological potential (Tsigos *et al.*, 2000), but, so far, their application has been limited.

Concluding remark

The enzymatic modification and hydrolysis of polysaccahrides is complex and the enzymology of chitin conversion is no exception. Although much is known about some of the enzyme classes related to chitin conversion, it remains to be seen how the enzymes in the complex mixtures produced by organisms such as *Trichoderma* and *Streptomyces* work together and how they may be exploited biotechnologically. Biologically, the potential interplay between chitin synthesis, chitin modification (by deacetylases), and chitin depolymerization (for feeding, defence or morphogenesis) is far from unraveled. Biotechnologically, the enzymatic conversion of chitin e.g. to well-defined bioactive chito-oligosaccharides remains a challenge. The past decade has brought a large increase in knowledge with provides an excellent basis for further work that is both scientifically exciting and promising from an application point of view. Such further work may vary from studying the battle between chitin-containing plant fungi and their hosts, to enzyme-based process development for better exploitation of chitinous biomass. In addition, it has recently been shown that enzymatic conversion of chitin is an excellent model system for creating better understanding of the enzymatic conversion of recalcitrant polysaccharides, in particular cellulose (Eijsink *et al.*, 2008). This adds another exciting dimension to the chitin field.

References

ADACHI, W., SAKIHAMA, Y., SHIMIZU, S., SUNAMI, T., FUKAZAWA, T. SUZUKI, M., YATSUNAMI, R., NAKAMURA, S. AND TAKENAKA, A. (2004). Crystal structure of family GH-8 chitosanase with subclass II specificity from *Bacillus* sp K17. *Journal of Molecular Biology* **343,** 785-95.

ADAMS, D.J. (2004). Fungal cell wall chitinases and glucanases. *Microbiology* **150**, 2029-2035.

AIBA, S. (1991). Studies on chitosan: Evidence for the presence of random and block copolymer structures in partially *N*-acetylated chitosans. *International Journal of Biological Macromolecules* **13,** 40-44.

ALZARI, P.M., SOUCHON, H. and DOMINGUEZ, R. (1996). The crystal structure of endoglucanase CelA, a family 8 glycosyl hydrolase from *Clostridium thermocellum. Structure* **4,** 265-75.

ANDERSEN, M.D., JENSEN, A., ROBERTUS, J.D., LEAH, R. AND SKRIVER, K. (1997). Heterologous expression and characterization of wild-type and mutant forms of a 26 kDa endochitinase from barley (*Hordeum vulgare* L.). *Biochemical Journal* **322,** 815-22.

ANTHONSEN, M.W., VARUM, K.M. and SMIDSROD, O. (1993). Solution properties of chitosans - conformation and chain stiffness of chitosans with different degrees of *N*-acetylation. *Carbohydrate Polymers* **22,** 193-201.

ARAKI, Y. AND ITO, E. (1975). Pathway of chitosan formation in *Mucor rouxii* - enzymatic deacetylation of chitin. *European Journal of Biochemistry* **55**, 71-78.

BENTLEY, S.D., CHATER, K.F., CERDEÑO-TÁRRAGA, A.M., CHALLIS, G.L., *ET AL.* (2002). Complete genome sequence of the model actinomycete *Streptomyces coelicolor* A3(2). *Nature* **417**, 141-47.

BIELY, P., KRATKY, Z. AND VRSANSKA, M. (1981) Substrate-binding site of endo-b-1,4-xylanase of the yeast *Cryptococcus albidus. European Journal of Biochemistry* **119**, 559-64.

BLACKWELL, J. (1988). Physical methods for determinination of chitin structure and conformation. *Methods in enzymology*. J. N. Abelson and M. I. Simon. San Diego, Academic Press, Inc. **161,** 435-42.

BLAIR, D. E., HEKMAT, O., SCHÜTTELKOPF, A.W., SHRESTHA, B., TOKUYASU, K., WITHERS, S.G. AND VAN AALTEN, D.M. (2006). Structure and mechanism of chitin deacetylase from the fungal pathogen *Colletotrichum lindemuthianum. Biochemistry* **45**, 9416-26.

BLAIR, D.E., SCHÜTTELKOPF, A.W., MACRAE, J.I. AND VAN AALTEN, D.M.F. (2005). Structure and metal-dependent mechanism of peptidoglycan deacetylase, a streptococcal virulence factor. *Proceedings of the National Academy of Sciences of the United States of America* **102**, 15429-34.

BLAIR, D. E. AND VAN AALTEN, D.M.F. (2004). Structures of *Bacillus subtilis* PdaA, a family 4 carbohydrate esterase, and a complex with *N*-acetyl-glucosamine. *FEBS Letters* **570**, 13-19.

BOOT, R.G., BLOMMAART, E.F., SWART, E., GHAUHARALI-VAN DER VLUGT, K., BIJL, N., MOE, C., PLACE, A. AND AERTS, J.M. (2001). Identification of a novel acidic mammalian chitinase distinct from chitotriosidase. *Journal of Biological Chemistry* **276**, 6770-8.

BOOT, R.G., RENKEMA, G.H., STRIJLAND, A., VAN ZONNEVELD, A.J. AND AERTS, J.M. (1995). Cloning of a cDNA encoding chitotriosidase, a human chitinase produced by macrophages. *Journal of Biological Chemistry* **270**, 26252-6.

BORASTON, A.B., BOLAM, D.N., GILBERT, H.J. AND DAVIES, G.J. (2004). Carbohydrate-binding modules: fine-tuning polysaccharide recognition. *Biochemical Journal* **382,** 769-81.

BOUCHER, I., FUKAMIZO, T., HONDA, Y., WILLICK, G.E., NEUGEBAUER, W.A. AND BRZEZINSKI, R. (1995). Site-directed mutagenesis of evolutionary conserved carboxylic amino acids in the chitosanase from *Streptomyces* sp. N174 reveals two residues essential for catalysis. *Journal of Biological Chemistry* **270**, 31077-82.

BRURBERG, M. B., NES, I.F. AND EIJSINK, V.G.H. (1996). Comparative studies of chitinases A and B from *Serratia marcescens*. *Microbiology-Uk* **142,** 1581-9.

BUSSINK, A. P., SPEIJER, D., AERTS, J.M. AND BOOT, R.G. (2007). Evolution of mammalian chitinase (-like) members of family 18 glycosyl hydrolases. *Genetics* **177**, 959-70.

CANTAREL, B.L., COUTINHO, P.M., RANCUREL, C., BERNARD, T., LOMBARD, V. and HENRISSAT, B. (2009). The Carbohydrate-Active EnZymes database (CAZy): an expert resource for Glycogenomics. *Nucleic Acids Research* **37,** D233-D238.

CARSOLIO, C., GUTIÉRREZ, A., JIMÉNEZ, B., VAN MONTAGU, M. and HERRERA-ESTRELLA, A. (1994). Characterization of *ech-42*, a *Trichoderma harzianum* endochitinase gene expressed during mycoparasitism. *Proceedings of the National Academy of Sciences of the United States of America* **91**, 10903-7.

CAUFRIER, F., MARTINOU, A., DUPONT, C. and BOURITIS, V. (2003). Carbohydrate esterase family 4 enzymes: substrate specificity. *Carbohydrate Research* **338**, 687-92.

CHENG, C. Y., CHANG, C.H., WU, Y.J. and LI, Y.K. (2006). Exploration of glycosyl hydrolase family 75, a chitosanase from *Aspergillus fumigatus*. *Journal of Biological Chemistry* **281**, 3137-44.

CHENG, C.Y. AND LI, Y.K. (2000). An *Aspergillus* chitosanase with potential for large-scale preparation of chitosan oligosaccharides. *Biotechnologi and Applied Biochemistry* **32**, 197-203.

CHOI, Y.J., KIM, E.J., PIAO, Z., YUN, Y.C. and SHIN, Y.C. (2004). Purification and characterization of chitosanase from *Bacillus sp* strain KCTC 0377BP and Its application for the production of chitosan oligosaccharides. *Applied and environmental microbiology* **70**, 4522-31.

COHEN-KUPIEC, R. AND CHET, I. (1998). The molecular biology of chitin digestion. *Current Opinion in Biotechnology* **9**, 270-7.

DAVIES, G. AND HENRISSAT, B. (1995). Structures and mechanisms of glycosyl hydrolases. *Structure* **3**, 853-59.

DAVIES, G. J., GLOSTER, T.M. and HENRISSAT, B. (2005). Recent structural insights into the expanding world of carbohydrate-active enzymes. *Current Opinion in Structural Biology* **15**, 637-45.

DAVIES, G. J., WILSON, K.S. and HENRISSAT, B. (1997). Nomenclature for sugar-binding subsites in glycosyl hydrolases. *Biochemical Journal* **321**, 557-9.

DE LA CRUZ, J., HIDALGO-GALLEGO, A., LORA, J.M., BENITEZ, T., PINTOR-TORO, J.A. and LLOBELL, A. (1992). Isolation and characterization of three chitinases from *Trichoderma harzianum. European Journal of Biochemistry* **206**, 859-67.

DIVNE, C., STÅHLBERG, J., REINIKAINEN, T., RUOHONEN, L., PETTERSSON, G., KNOWLES, J.K., TEERI, T.T. and JONES, T.A. (1994) The three-dimensional crystal structure of the catalytic core of cellobiohydrolase I from *Trichoderma reesei. Science* **265**, 524-528.

DUTTA, P. K., RAVIKUMAR, M.N. and DUTTA, J. (2002). Chitin and chitosan for versatile applications. *Journal of Macromolecular Science* **C42**, 307-54.

EIJSINK, V. G. H., VAAJE-KOLSTAD, G., VÅRUM, K.M. and HORN, S.J. (2008). Towards new enzymes for biofuels: lessons from chitinase research. *Trends in Biotechnology* **26**, 228-35.

FELSE, P. A. AND PANDA, T. (1999). Studies on applications of chitin and its derivatives. *Bioprocess Engineering* **20**, 505-12.

FELT, O., BURI, P. and GURNY, R. (1998). Chitosan: A unique polysaccharide for drug delivery. *Drug Development and Industrial Pharmacy* **24**, 979-93.

FERNANDES, J. C., TAVARIA, F.K., SOARES, J., RAMOS, O.S., MONTEIRO, M.J, PINTADO, M.E. and MALCATA, F.X. (2008). Antimicrobial effects of chitosans and chitooligosaccharides, upon *Staphylococcus aureus* and *Escherichia coli*, in food model systems. *Food Microbiology* **25**, 922-28.

FUKAMIZO, T. AND BRZEZINSKI, R. (1997). Chitosanase from *Streptomyces* sp. strain N174: a comparative review of its structure and function. *Biochemical Cell Biology* **75**, 687-96.

FUKAMIZO, T., HONDA, Y., GOTO, S., BOUCHER, I. and BRZEZINSKI, R. (1995). Reaction mechanism of chitosanase from *Streptomyces* sp. N174. *Biochemical Journal* **311**, 377-83.

FUKAMIZO, T, JUFFER, A.H, VOGEL, H.J., HONDA, Y., TREMBLAY, H., BOUCHER, I., NEUGEBAUER, W.A. and BRZEZINSKI, R. (2000). Theoretical calculation of p*K*a reveals an important role of Arg205 in the activity and stability of *Streptomyces* sp N174 chitosanase. *Journal of Biological Chemistry* **275**, 25633-40.

FUKAMIZO, T, MIYAKE, R., TAMURA, A., OHNUMA, T., SKRIVER, K., PURSIAINEN, N.V. and JUFFER, A.H. (2009). A flexible loop controlling the enzymatic activity and specificity in a glycosyl hydrolase family 19 endochitinase from barley seeds (*Hordeum vulgare* L.). *Biochimica et Biophysica Acta.* **1794,** 1159-67.

FUKAMIZO, T, OHKAWA, T., IKEDA, Y. and GOTO, S. (1994). Specificity of chitosanase from *Bacillus pumilus. Biochimica et Biophysica Acta* **1205**, 183-8.

GAO, X.D., KATSUMOTO, T. and ONODERA, K. (1995). Purification and characterization of chitin deacetylase from *Absidia coerulea. Journal of Biochemistry* **117,** 257-63.

GARCÍA, I., LORA, J.M., DE LA CRUZ, J., BENÍTEZ, T., LLOBELL, A. and PINTOR-TORO, J.A. (1994) Cloning and characterization of a chitinase (chit42) cDNA from the mycoparasitic fungus *Trichoderma harzianum. Current Genetics* **27**, 83-9.

GARCIA-CASADO, G., COLLADA, C. ALLONA, I., CASADO, R., PACIOS, L.F., ARAGONCILLO, C. and GOMEZ, L (1998). Site-directed mutagenesis of active site residues in a class I endochitinase from chestnut seeds. *Glycobiology* **8**, 1021-28.

GARDNER, K. H. AND BLACKWELL, J. (1975). Refinement of the structure of beta-chitin.

Biopolymers **14**, 1581-95.

GIRAUD-GUILLE, M. M. (1986). Chitin-protein molecular organization in arthropods. Chitin in nature and technology. R. Muzarelli, Jeuniaux C. and Gooday G. W. New York, Plenum Press, 29-35.

GOODAY, G. W. (1990). The ecology of chitin degradation. In *Advances in Microbial Ecology* ed. K. C. Marshall, pp 387–430. New York: Plenum Press.

GUAN, B.H., NI, W.M., WU, Z.B. AND LAI, Y. (2009). Removal of Mn(II) and Zn(II) ions from flue gas desulfurization wastewater with water-soluble chitosan. *Separation and Purification Technology* **65,** 269-74.

GUAN, S.P., MOK, Y.K., KOO, K.N., CHU, K.L. AND WONG, W.S. (2009) Chitinases: biomarkers for human diseases. *Protein and Peptide Letters* **16**, 490-98.

HAHN, M., HENNIG, M., SCHLEISER, B. AND HOHNE, W. (2000). Structure of jack bean chitinase. *Acta Crystallographica. Section D, Biological Crystallography* **56,** 1096-9.

HART, P. J., MONZINGO, A.F., READY, M.P., ERNST, S.R. AND ROBERTUS, J.D. (1993). Crystal structure of an endochitinase from *Hordeum vulgare* L seeds. *Journal of Molecular Biology* **229,** 189-93.

HART, P. J., PFLUGER, H.D., MONZINGO, A.F., HOLLIS, T. AND ROBERTUS, J.D. (1995). The refined crystal-structure of an endochitinase from *Hordeum vulgare* L seeds at 1.8 Angstrom resolution. *Journal of Molecular Biology* **248,** 402-13.

HEGGSET, E. B., HOELL, I.A., KRISTOFFERSEN, M., EIJSINK, V.G.H. AND VÅRUM, K.M. (2009). Degradation of chitosans with chitinase G from *Streptomyces coelicolor* A3(2): production of chito-oligosaccharides and insight into subsite specificities. *Biomacromolecules* **10**, 892-9.

HENNIG, M., JANSONIUS, J.N., TERWISSCHA VAN SCHELTINGA, A.C., DIJKSTRA, B.W. and SCHLEISIER, B. (1995). Crystal structure of concanavalin B at 1.65 A resolution, An "inactivated" chitinase from seeds of Canavalia ensiformis. *Journal of Molecular Biology*. **254**, 237-46.

HENNIG, M., SCHLESIER, B., DAUTER, Z., PFEFFER, S., BETZEL, C., HÖHNE, W.E. and WILSON, K.S. (1992) A TIM barrel protein without enzymatic activity? Crystal-structure of narbonin at 1.8 A resolution. *FEBS Letters* **306**, 80-4.

HENRISSAT, B. and DAVIES, G. (1997). Structural and sequence-based classification of glycoside hydrolases. *Current Opinion in Structural Biology* **7**, 637-44.

HENRISSAT, B., TEERI, T.T. and WARREN, R.A.J. (1998) A scheme for designating enzymes that hydrolyse the polysaccharides in the cell walls of plants. *FEBS Letters* **425**, 352-4.

HERNICK, M. and FIERKE, C.A. (2005). Zinc hydrolases: the mechanisms of zinc-dependent deacetylases. *Archives of Biochemistry and Biophysics* **433**, 71-84.

HJELJORD, L.G., STENSVAND, A. and TRONSMO, A. (2001). Antagonism of nutrient-activated conidia of *Trichoderma harzianum* (*atroviride*) P1 against *Botrytis cinerea. Phytopathology* **91**, 1172-80.

HODGSON, D. A. (2000). Primary metabolism and its control in *Streptomycetes*: a most unusual group of bacteria. *Advances in Microbial Physiology* **42,** 47-238.

HOELL, I. A., DALHUS, B., HEGGSET, E.B., ASPMO, S.I. AND EIJSINK, V.G.H. (2006). Crystal structure and enzymatic properties of a bacterial family 19 chitinase reveal differences from plant enzymes. *FEBS Journal* **273,** 4889-900.

HOELL, I.A., KLEMSDAL, S.S., VAAJE-KOLSTAD, G., HORN, S.J. and EIJSINK, V.G.H.

(2005). Overexpression and characterization of a novel chitinase from *Trichoderma atroviride* strain P1. *Biochimica et Biophysica Acta* **1748**, 180-90.

Holm, L. and Sander, C. (1994). Structural similarity of plant chitinase and lysozymes from animals and phage - an evolutionary connection. *FEBS Letters* **340,** 129-32.

Hopwood, D. A. (1999). Forty years of genetics with *Streptomyces*: from in vivo through in vitro to in silico. *Microbiology* **145,** 2183-2202.

Horn, S. J., Sikorski, P., Cederkvist, J.B., Vaaje-Kolstad, G., Sorlie, M., Synstad, B., Vriend, G., Vårum, K.M. and Eijsink, V.G.H. (2006a). Costs and benefits of processivity in enzymatic degradation of recalcitrant polysaccharides. *Proceedings of the National Academy of Sciences of the United States of America* **103,** 18089-94.

Horn, S. J., Sorbotten, A., Synstad, B., Sikorski, P., Sorlie, M., Vårum, K.M. and Eijsink, V.G.H. (2006b). Endo/exo mechanism and processivity of family 18 chitinases produced by *Serratia marcescens*. *FEBS Journal* **273,** 491-503.

Huet, J., Rucktooa, P., Clantin, B., Azarkan, M., Looze, Y., Villeret, V. and Wintjens, R. (2008). X-ray structure of papaya chitinase reveals the substrate binding mode of glycosyl hydrolase family 19 chitinases. *Biochemistry* **47,** 8283-91.

Hult, E. L., Katouno, F., Clantin, B., Azarkan, M., Looze, Y., Villeret, V. and Wintjens, R. (2005). Molecular directionality in crystalline beta-chitin: hydrolysis by chitinases A and B from *Serratia marcescens* 2170. *Biochemical Journal* **388,**: 851-56.

Ike, M., Ko, Y., Yokoyama, K., Sumitani, J.I., Kawaquchi, T., Ogasawara, W., Okada, H. and Morikawa, Y. (2007). Cellobiohydrolase I (Cel7A) from *Trichoderma reesei* has chitosanase activity. *Journal of Molecular Catalysis B-Enzymatic* **47,** 159–63.

Ikeda, H., Ishikawa, J., Hanamoto, A., Shinose, M., Kikuchi, H., Shiba, T., Sakaki, Y., Hattori, M. and Omura, S. (2003). Complete genome sequence and comparative analysis of the industrial microorganism *Streptomyces avermitilis*. *Nature Biotechnology* **21,** 526-31.

Iseli, B., Armand, S., Boller, T., Neuhaus, J.M. and Henrissat, B. (1996). Plant chitinases use two different hydrolytic mechanisms. *FEBS Letters* **382,** 186-8.

Isogawa, D., Fukuda, T., Kuroda, K., Kusaoke, H., Kimoto, H., Suye, S. and Ueda, M. (2009). Demonstration of catalytic proton acceptor of chitosanase from *Paenibacillus fukuinensis* by comprehensive analysis of mutant library. *Applied Microbiology and Biotechnology* **85,** 95-104.

Itoh, Y., Watanabe, J., Fukada, H., Mizuno, R., Kezuka, Y., Nonaka, T. and Watanabe, T. (2006). Importance of Trp59 and Trp60 in chitin-binding, hydrolytic, and antifungal activities of *Streptomyces griseus* chitinase C. *Applied Microbiology and Biotechnology* **72,** 1176-84.

Izume, M., Nagae, S., Kawagishi H., Mitsutomi M. and Ohtakara A. (1992). Action pattern of *Bacillus* sp no7 M chitosanase on partially *N*-acetylated chitosan. *Bioscience, Biotechnology, and Biochemistry* **56,** 448-53.

Jeon, Y. J., Park, P. J. and Kim, S.K. (2001). Antimicrobial effect of chitooligosaccharides produced by bioreactor. *Carbohydrate Polymers* **44,** 71-6.

Kafetzopoulos, D., Thireos, G. , Vournakis, J.N. and Bouriotis, V. (1993). The primary structure of a fungal chitin deacetylase reveals the function for 2 bacterial gene products. *Proceedings of the National Academy of Sciences of the United States of America* **90,** 8005-8.

Karlsson, M. and Stenlid, J. (2008). Comparative evolutionary histories of the fungal chitinase gene family reveal non-random size expansions and contractions due to adaptive natural selection. *Evolutionary Bioinformatics Online* **4,** 47-60.

KARLSON, M. and STENLID, J. (2009). Evolution of family 18 glycoside hydrolases: diversity, domain structures and phylogenetic relationships. *Journal of Molecular Microbiology and Biotechnology* **16,** 208-23.

KAWASE, T., YOKOKAWA, S., SAITO, A., FUJII, T., NIKAIDOU, N., MIYASHITA, K. and WATANABE, T. (2006). Comparison of enzymatic and antifungal properties between family 18 and 19 chitinases from *S. coelicolor* A3(2). *Bioscience, Biotechnology, and Biochemistry* **70,** 988-98.

KEZUKA, Y., OHISHI, M., ITOH, Y., WATANABE, J., MITSUTOMI, M., WATANABE, T. and NONAKA, T. (2006). Structural studies of a two-domain chitinase from *Streptomyces griseus* HUT6037. *Journal of Molecular Biology* **358,** 472-84.

KHOR, E. (2002). Chitin: a biomaterial in waiting. *Current Opinion in Solid State & Materials Science* **6,** 313-17.

KIKKAWA, Y., TOKUHISA, H., SHINGAI, H., HIRAISHI, T., HOUJOU, H., KANESATO, M., IMANAKA, T. and TANAKA, T. (2008). Interaction force of chitin-binding domains onto chitin surface. *Biomacromolecules* **9,** 2126-31.

KIM, D.J., BAEK, J.M., URIBE, P., KENERLEY, C.M. and COOK, D.R. (2002). Cloning and characterization of multiple glycosyl hydrolase genes from *Trichoderma virens. Current Genetics* **40**, 374-84.

KIM, S. K. and RAJAPAKSE, N. (2005). Enzymatic production and biological activities of chitosan oligosaccharides (COS): a review. *Carbohydrate Polymers* **62,** 357-68.

KIMOTO, H., KUSAOKE, H., YAMAMOTO, I., FUJII, Y., ONODERA, T. and TAKETO, A. (2002). Biochemical and genetic properties of *Paenibacillus* glycosyl hydrolase having chitosanase activity and discoidin domain. *Journal of Biological Chemistry* **277,** 14695-702.

KNAPP, S., VOCADLO, D., GAO, Z.N., KIRK, B., LOU, J.P. and WITHERS, S.G. (1996). NAG-thiazoline, an *N*-acetyl-beta-hexosaminidase inhibitor that implicates acetamido participation. *Journal of the American Chemical Society* **118,** 6804-5.

KOBAYASHI, S., KIYOSADA, T. and SHODA, S. (1996). Synthesis of artificial chitin: Irreversible catalytic behavior of a glycosyl hydrolase through a transition state analogue substrate. *Journal of the American Chemical Society* **118,** 13113-4.

KOSHLAND, D. E. (1953). Stereochemistry and the mechanism of enzymatic reactions. *Biological Reviews of the Cambridge Philosophical Society* **28,** 416-36.

KRAJEWSKA, B. (2004). Application of chitin- and chitosan-based materials for enzyme immobilizations: a review. *Enzyme and Microbial Technology* **35,** 126-39.

KUBOTA, N. and EGUCHI, Y. (1997). Facile preparation of water-soluble *N*-acetylated chitosan and molecular weight dependence of its water-solubility. *Polymer Journal* **29,** 123-7.

KURITA, K. (2001). Controlled functionalization of the polysaccharide chitin. *Progress in Polymer Science* **26,** 1921-71.

LACOMBE-HARVEY, M. E., FUKAMIZO, T., GAGNON, J., GHINET, M.G., DENNHART, N., LETZEL, T. and BRZEZINSKI, R. (2009). Accessory active site residues of *Streptomyces* sp. N174 chitosanase: variations on a common theme in the lysozyme superfamily. *FEBS Journal* **276,** 857-69.

LADET, S., DAVID, L. and DOMARD, A. (2008). Multi-membrane hydrogels. *Nature* **452,** 76-9.

LANGER, R.C. and VINETZ, J.M. (2001). Plasmodium ookinete-secreted chitinase and parasite penetration of the mosquito peritrophic matrix. *Trends in Parasitology* **17,** 269-72.

LIMA, L.H.C., ULHOA, C.J., FERNANDES, A.P. and FELIX, C.R. (1997). Purification of a chitinase from *Trichoderma* sp. and its action on *Sclerotium rolfsii* and *Rhizoctonia solani* cell walls. *Journal of Genetics and Applied Microbiology* **43,** 31-7.

LIU, H. T., LI, W.M., XU, G., LI, X.Y., BAI, X.F., WEI, P., YU, C. and DU, Y.G. (2009). Chitosan oligosaccharides attenuate hydrogen peroxide-induced stress injury in human umbilical vein endothelial cells. *Pharmacological Research* **59,** 167-75.

LONG, S. R. (1989). *Rhizobium* legume nodulation - life together in the underground. *Cell* **56,** 203-14.

LORITO, M., HAYES, C.K., DIPIETRO, A., WOO, S.L. and HARMAN, G.E. (1994). Purification, characterization, and synergistic activity of a glucan 1,3-beta-glucosidase and an *N*-acetyl-beta-glucosaminidase from *Trichoderma harzianum*. *Phytopathology* **84,** 398-405.

MAENAKA, K., KAWAI, G., WATANABE, K., SUNADA, F. and KUMAGAI, I. (1994). Functional and structural role of a tryptophan generally observed in protein-carbohydrate interaction - Trp62 of hen egg-white lysozyme. *Journal of Biologucal Chemistry* **269,** 7070-5.

MARCOTTE, E. M., MONZINGO, A.F., ERNST, S.R., BRZEZINSKI, R. and ROBERTUS, J.D. (1996). X-ray structure of an anti-fungal chitosanase from *Streptomyces* N174. *Nature Structural Biology* **3,** 155-62.

MARK, B. L., VOCADLO, D.J., KNAPP, S., TRIGGS-RAINE, B.L., WITHERS, S.G. and JAMES, M.N.G. (2001). Crystallographic evidence for substrate-assisted catalysis in a bacterial beta-hexosaminidase. *Journal of Biological Chemistry* **276,** 10330-7.

MARTINEZ, D., BERKA, R.M., HENRISSAT, B., SALOHEIMO, M., *ET AL.* (2008). Genome sequencing and analysis of the biomass-degrading fungus *Trichoderma reesei* (syn. *Hypocrea jecorina*) *Nature Biotechnology* **26,** 553-560.

MARTINOU, A., KAFETZOPOULOS, D. and BOURIOTIS, V. (1993). Isolation of chitin deacetylase from *Mucor rouxii* by immunoaffinity chromatography. *Journal of Chromatography* **644,** 35-41.

MARTINOU, A., KOUTSIOULIS, D. and BOURIOTIS, V. (2002). Expression, purification, and characterization of a cobalt-activated chitin deacetylase (Cda2p) from *Saccharomyces cerevisiae*. *Protein Expression and Purification* **24,** 111-6.

MERINO, S. T. and CHERRY, J. (2007). Progress and challenges in enzyme development for biomass utilization. *Advances in Biochemical Engineering/Biotechnology* **108,** 95-120.

MERZENDORFER, H. and ZIMOCH, L. (2003). Chitin metabolism in insects: structure, function and regulation of chitin synthases and chitinases. *Journal of Experimental Biology* **206,** 4393-4412.

MINKE, R. and BLACKWELL J. (1978). The structure of alpha-chitin. *Journal of Molecular Biology* **120,** 167-81.

MITSUTOMI, M., KIDOH, H. TOMITA, H. and WATANABE, T. (1995). The action of *Bacillus circulans* Wl-12 chitinases on partially *N*-acetylated chitosan. *Bioscience Biotechnology and Biochemistry* **59,** 529-31.

MITSUTOMI, M., ISONO, M., UCHIYAMA, A., NIKAIDOU, N., IKEGAMI, T. and WATANABE, T. (1998). Chitosanase activity of the enzyme previously reported as beta-1,3-1,4-glucanase from *Bacillus circulans* WL-12. *Bioscience Biotechnology and*

Biochemistry **62,** 2107-14.

MIZUNO, R., FUKAMIZO, T., SUGIYAMA, S. , NISHIZAWA, Y., KEZUKA, Y., NONAKA, T., SUZUKI, K. and WATANABE, T. (2008). Role of the loop structure of the catalytic domain in rice class I chitinase. *Journal of Biochemistry* **143,** 487-95.

MONZINGO, A. F., MARCOTTE, E.M., HART, P.J. and ROBERTUS, J.D. (1996). Chitinases, chitosanases, and lysozymes can be divided into procaryotic and eucaryotic families sharing a conserved core. *Nature Structural Biology* **3,** 133-40.

NEUHAUS, J.M., FRITIG, B., LINTHORST, H.J.M., MEINS, F., MIKKELSEN, J.D. AND RYALS, J. (1996). A revised nomenclature for chitinase genes. *Plant Molecular Biology Reporter* **14**, 102-4.

NGO, D. N., KIM, M.M. and KIM, S.K. (2008). Chitin oligosaccharides inhibit oxidative stress in live cells. *Carbohydrate Polymers* **74,** 228-34.

NISHIYAMA, Y., LANGAN, P. and CHANZY, H. (2002). Crystal Structure and Hydrogen-Bonding System in Cellulose Iβ from Synchrotron X-ray and Neutron Fiber Diffraction. *Journal of American Chemical Society* **124,** 9074-82.

NO, H. K. and MEYERS, S.P. (2000). Application of chitosan for treatment of wastewaters. *Reviews of Environmental Contamination and Toxicology* **163,** 1-27.

OHNISHI, T., JUFFER, A.H., TAMOI, M., SKRIVER, K. and FUKAMIZO, T. (2005). 26 kDa endochitinase from barley seeds: an interaction of the ionizable side chains essential for catalysis. *Journal of Biochemistry* **138,** 553-62.

OHNISHI, Y., ISHIKAWA, J., HARA, H., SUZUKI, H., IKENOYA, M., IKEDA, H., YAMASHITA, A., HATTORI, M. and HORINOUCHI, S. (2008). Genome sequence of the streptomycin-producing microorganism *Streptomyces griseus* IFO 13350. *Journal of Bacteriology* **190,** 4050-60.

OHNO, T., ARMAND, S., HATA, T., NIKAIDOU, N., HENRISSAT, B., MITSUTOMI, M. and WATANABE, T. (1996). A modular family 19 chitinase found in the prokaryotic organism *Streptomyces griseus* HUT 6037. *Journal of Bacteriology* **178,** 5065-70.

PAPANIKOLAU, Y., PRAG, G., TAVLAS, G., VORGIAS, C.E., OPPENHEIM, A.B. and PETRATOS, K. (2001). High resolution structural analyses of mutant chitinase A complexes with substrates provide new insight into the mechanism of catalysis. *Biochemistry* **40,** 11338-43.

PARK, J.K., SHIMONO, K., OCHIAI, N., SHIGERU, K., KURITA, M., OHTA, Y., TANAKA, K., MATSUDA, H. and KAWAMUKAI, M. (1999). Purification, characterization, and gene analysis of a chitosanase (ChoA) from *Matsuebacter chitosanotabidus* 3001. *Journal of Bacteriology* **181,** 6642-9.

PAUL, W. and SHARMA, C.P. (2000). Chitosan, a drug carrier for the 21st century: a review. *STP Pharma Sciences* **10,** 5-22.

PEDRAZA-REYES, M. and GUTIERREZ-CORONA, F. (1997). The bifunctional enzyme chitosanase-cellulase produced by the gram-negative microorganism *Myxobacter sp.* AL-1 is highly similar to *Bacillus subtilis* endoglucanases. *Archives of Microbiology* **168,** 321-7.

PERRAKIS, A., TEWS, I., DAUTER, Z., OPPENHEIM, A.B., CHET, I., WILSON, K.S. and VORGIAS, C.E. (1994). Crystal-structure of a bacterial chitinase at 2.3 angstrom resolution. *Structure* **2,** 1169-80.

PRASHANTH, K. V. H. and THARANATHAN, R.N. (2005). Depolymerized products of chitosan as potent inhibitors of tumor-induced angiogenesis. *Biochimica et Biophysica Acta* **1722,**: 22-29.

Psylinakis, E., Boneca, I.G., Mavromatis, K., Deli, A., Hayhurst, E., Foster, S.J., Vårum, K.M. and Bouriotis, V. (2005). Peptidoglycan N-acetylglucosamine deacetylases from *Bacillus cereus*, highly conserved proteins in *Bacillus anthracis*. *Journal of Biological Chemistry* **280,** 30856-63.

Raabe, D., Al-Sawalmih, A., Yi, S. B. and Fabritius, H. (2007). Preferred crystallographic texture of alpha-chitin as a microscopic and macroscopic design principle of the exoskeleton of the lobster *Homarus americanus*. *Acta Biomaterialia* **3,** 882-95.

Rao, F. V., Houston, D.R., Boot, R.G., Aerts, J.M.F.G., Hodkinson, M., Adams, D.J., Shiomi, K., Omura, S. and van Aalten, D.M.F. (2005). Specificity and affinity of natural product cyclopentapeptide inhibitors against *A. fumigatus*, human, and bacterial chitinases. *Chemistry and Biology* **12,** 65-76.

Renkema, G. H., Boot, R.G., Muijsers, A.O., Donkerkoopman, W.E. and Aerts, J.M.F.G. (1995). Purification and characterization of human chitotriosidase, a novel member of the chitinase family of proteins. *Journal of Biological Chemistry* **270,** 2198-202.

Rinaudo, M. (2006). Chitin and chitosan: Properties and applications. *Progress in Polymer Science* **31,** 603-32.

Rinaudo M. and Domard A. (1989). Solution properties of chitosan. In *Chitin and Chitosan: Sources, Chemistry, Biochemistry, Physical Properties and Applications* eds G. Skjak-Braeck, T. Anthonsen and P.A. Sandford, pp. 71-86. New York: Elsevier Applied Science.

Robertus, J. D., Monzingo, A.F., Marcotte, E.M. and Hart, P.J. (1998). Structural analysis shows five glycohydrolase families diverged from a common ancestor. *Journal of Experimental Zoology* **282,** 127-32.

Robyt, J. F. and French, D. (1967). Multiple attack hypothesis of alpha-amylase action - action of porcine pancreatic human salivary and *Aspergillus oryzae* alpha-amylases. *Archives of Biochemistry and Biophysics* **122,** 8-16.

Robyt, J. F. and French, D. (1970). Multiple attack and polarity of action of porcine pancreatic alpha-amylase. *Archives of Biochemistry and Biophysics* **138,** 662-70.

Rye, C.S. and Withers, S.G. (2000). Glucosidase mechanisms. *Current Opinion in Chemical Biology* **4,** 573-80.

Saito, J., A. Kita, Higuchi, Y., Nagata, Y., ando, A. and Miki, K. (1999a). Crystal structure of chitosanase from *Bacillus circulans* MH-K1 at 1.6-A resolution and its substrate recognition mechanism. *Journal of Biological Chemistry* **274,** 30818-25.

Saito, A., Fujii, T., Redenbach, M., Ohno, T., Watanabe, T. and Miyashita, K. (1999b). High-multiplicity of chitinase genes in *Streptomyces coelicolor* A3(2). *Bioscience, Biotechnology, and Biochemistry* **63,** 710-8.

Saito, A., M. Ishizaka, Francisco, P.B., Fujii, T. and Miyashita, K. (2000). Transcriptional co-regulation of five chitinase genes scattered on the *Streptomyces coelicolor* A3(2) chromosome. *Microbiology-UK* **146,** 2937-46.

Sasaki, C., Itoh, Y., Takehara, H., Kuhara, S. and Fukamizo, T. (2003). Family 19 chitinase from rice (*Oryza sativa* L.): substrate-binding subsites demonstarted by kinetic and molecular modeling studies. *Plant Molecular Biology* **52,** 43-52.

Seidl, V., Huemer, B., Seiboth, B. and Kubicek, C.P. (2005). A complete survey of *Trichoderma* chitinases reveals three distinct subgroups of family 18 chitinases. *FEBS Journal* **272,** 5923-39.

Shahidi, F., Arachchi, J.K.V. and Jeon, Y.J. (1999). Food applications of chitin and chitosans. *Trends in Food Science and Technology* **10,** 37-51.

SHIMONO, K., SHIGERU, K., TSUCHIYA, A., ITOU, N., OHTA, Y., TANAKA, K., NAKAGAWA, T., MATSUDA, H. and KAWAMUKAI, M. (2002). Two glutamic acids in chitosanase A from *Matsuebacter chitosanotabidus* 3001 are the catalytically important residues. *Journal of Biochemistry* **131,** 87-96.

SHIMOSAKA, M., KUMEHARA, M., ZHANG, X.Y., NOGAWA, M. and OKAZAKI, M. (1996). Cloning and characterization of a chitosanase gene from the plant pathogenic fungus *Fusarium solani*. *Journal of Fermentation and Bioengineering* **82,** 426-31.

SHIMOSAKA, M., SATO, K., NISHIWAKI, N., MIYAZAWA, T. and OKAZAKI, W. (2005). Analysis of essential carboxylic amino acid residues for catalytic activity of fungal chitosanases by site-directed mutagenesis. *Journal of Bioscience and Bioengineering* **100,** 545-50.

SHOSEYOV, O., SHANI, Z. and LEVY, I. (2006). Carbohydrate binding modules: biochemical properties and novel applications. *Microbiology and Molecular Biology Reviews* **70,** 283-95.

SIKORSKI, P., SORBOTTEN, A., HORN, S.J., EIJSINK, V.G.H. and VÅRUM, K.M. (2006). *Serratia marcescens* chitinases with tunnel-shaped substrate-binding grooves show endo activity and different degrees of processivity during enzymatic hydrolysis of chitosan. *Biochemistry* **45,** 9566-74.

SINNOTT, M. L. (1990). Catalytic mechanisms of enzymatic glycosyl transfer. *Chemical Reviews* **90,** 1171-1202.

SONG, H. K. and SUH, S.W. (1996). Refined structure of the chitinase from barley seeds at 2.0 angstrom resolution. *Acta Crystallographica. Section D, Biological Crystallography* **52,** 289-98.

SORBOTTEN, A., HORN, S.J., EIJSINK, V.G.H. and VÅRUM, K.M. (2005). Degradation of chitosans with chitinase B from *Serratia marcescens* - production of chito-oligosaccharides and insight into enzyme processivity. *FEBS Journal* **272,** 538-49.

STRAND, S. P., ISSA, M.M., CHRISTENSEN, B.E., VARUM, K.M. and ARTURSSON, P. (2008). Tailoring of chitosans for gene delivery: novel self-branched glycosylated chitosan oligomers with improved functional properties. *Biomacromolecules* **9,** 3268-76.

SUZUKI, K., SUGAWARA, N., SUZUKI, M., UCHIYAMA, T., KATOUNO, F., NIKAIDOU, N. and WATANABE, T. (2002). Chitinases A, B, and C1 of *Serratia marcescens* 2170 produced by recombinant *Escherichia coli:* Enzymatic properties and synergism on chitin degradation. *Bioscience, Biotechnology, and Biochemistry* **66,** 1075-83.

SUZUKI, K., SUZUKI, M., TAIYOJI M., NIKAIDOU, N. and WATANABE, T. (1998). Chitin binding protein (CBP21) in the culture supernatant of *Serratia marcescens* 2170. *Bioscience, Biotechnology, and Biochemistry* **62,** 128-35.

SUZUKI, K., TAIYOJI, M., SUGAWARA, N., NIKAIDOU, N., HENRISSAT, B. and WATANABE, T. (1999). The third chitinase gene (chiC) of *Serratia marcescens* 2170 and the relationship of its product to other bacterial chitinases. *Biochemical Journal* **343,** 587-96.

SYNSTAD, B., GASEIDNES, S., VAN AALTEN, D.M.F., VRIEND, G., NIELSEN, J.E. and EIJSINK, V.G.H. (2004). Mutational and computational analysis of the role of conserved residues in the active site of a family 18 chitinase. *European Journal of Biochemistry* **271,** 253-62.

TANG, C. M., CHYE, M.L., RAMALINGAM, S., OUYANG, S.W., ZHAO, K.J., UBHAYASEKERA, W. and MOWBRAY, S.L. (2004). Functional analyses of the chitin-binding domains and the catalytic domain of *Brassica juncea* chitinase BjCHI1. *Plant Molecular Biology* **56,** 285-98.

Taylor, E. J., Gloster, T.M., Turkenburg, J.P., Vincent, F., Brzozowski, A.M., Dupont, C., Shareck, F., Centeno, M.S.J., Prates, J.A.M., Puchart, V., Ferreira, L.M.A., Fontes, C.M.G.A., Biely, P. and Davies, G.J. (2006). Structure and activity of two metal ion-dependent acetylxylan esterases involved in plant cell wall degradation reveals a close similarity to peptidoglycan deacetylases. *Journal of Biological Chemistry* **281,** 10968-75.

Tews, I., Perrakis, A., Oppenheim, A., Dauter, Z., Wilson, K.S. and Vorgias, C.E. (1996). Bacterial chitobiase structure provides insight into catalytic mechanism and the basis of Tay-Sachs disease. *Nature Structural Biology* **3,** 638-48.

Tews, I., Van Scheltinga, A.C.T., Perrakis, A., Wilson, K.S. and Dijkstra, B.W. (1997). Substrate-assisted catalysis unifies two families of chitinolytic enzymes. *Journal of American Chemical Society* **119,** 7954-9.

Tharanathan, R. N. and Kittur, F.S. (2003). Chitin - The undisputed biomolecule of great potential. *Critical Reviews in Food Science and Nutrition* **43,** 61-87.

Tokuyasu, K., Ohnishi-Kameyama, M., Hayashi, K. and Mori, Y. (1999). Cloning and expression of chitin deacetylase gene from a Deuteromycete, *Colletotrichum lindemuthianum. Journal of Bioscience and Bioengineering* **87**, 418-23.

Tronsmo, A. (1991). Biological and Integrated Controls of Botrytis cinerea on Apple with Trichoderma harzianum. *Biological Control* **1,** 59-62 .

Tsigos, I., Martinou, A., Kafetzopoulos, D. and Bouriotis, V. (2000). Chitin deacetylases: new, versatile tools in biotechnology. *Trends in Biotechnology* **18,** 305-12.

Ubhayasekera, W., Tang, C.M., Ho, S.W.T., Berglund, G., Bergfors, T., Chye, M.L. and Mowbray, S.L. (2007). Crystal structures of a family 19 chitinase from *Brassica juncea* show flexibility of binding cleft loops. *FEBS Journal* **274,** 3695-703.

Vaaje-Kolstad, G., Horn, S.J., van Aalten, D.M.F., Synstad, B. and Eijsink, V.G.H. (2005a). The non-catalytic chitin-binding protein CBP21 from *Serratia marcescens* is essential for chitin degradation. *Journal of Biological Chemistry* **280,** 28492-7.

Vaaje-Kolstad, G., Houston, D.R., Riemen, A.H.K., Eijsink. V.G.H. and van Aalten, D.M.F. (2005b). Crystal structure and binding properties of the *Serratia marcescens* chitin-binding protein CBP21. *Journal of Biological Chemistry* 280, 11313-9.

Vaaje-Kolstad, G., Bunaes, A.C., Mathiesen, G. and Eijsink, V.G.H. (2009). The chitinolytic system of *Lactococcus lactis ssp lactis* comprises a nonprocessive chitinase and a chitin-binding protein that promotes the degradation of alpha- and beta-chitin. *FEBS Journal* **276**, 2402-15.

Van Aalten, D.M.F., Komander, D., Synstad, B., Gaseidnes, S., Peter, M.G. and Eijsink, V.G.H. (2001). Structural insights into the catalytic mechanism of a family 18 exo-chitinase. *Proceedings of the National Academy of Sciences of the United States of America* **98,** 8979-84.

Van Aalten, D.M.F., Synstad, B., Brurberg, M.B., Hough, E., Riise, B.W., Eijsink, V.G.H. and Wierenga, R.K. (2000). Structure of a two-domain chitotriosidase from *Serratia marcescens* at 1.9 angstrom resolution. *Proceedings of the National Academy of Sciences of the United States of America* **97,** 5842-7.

Van Petegem, F., Collins, T., Meuwis, M.A., Gerday, C., Feller, G. and Van Beeumen, J. (2003). The structure of a cold-adapted family 8 xylanase at 1.3 angstrom resolution - Structural adaptations to cold and investigation of the active site. *Journal of Biological Chemistry* **278,** 7531-9.

Van Scheltinga, A. C. T., Armand, S., Kalk, K.H., Isogai, A., Henrissat, B. and Dijkstra, B.W. (1995). Stereochemistry of chitin hydrolysis by a plant chitinase lysozyme and X-ray structure of a complex with allosamidin - evidence for substrate assisted catalysis. *Biochemistry* **34,** 15619-23.

Van Scheltinga, A. C. T., Kalk, K.H., Beintema, J.J. and Dijkstra, B.W. (1994). Crystal-structures of hevamine, a plant defense protein with chitinase and lysozyme activity, and its complex with an inhibitor. *Structure* **2,** 1181-9.

Varrot, A., Frandsen, T.P., von Ossowski, I., Boyer, V., Cottaz, S., Driguez, H., Schulein, M. and Davies, G.J. (2003). Structural basis for ligand binding and processivity in cellobiohydrolase Cel6A from *Humicola insolens*. *Structure* **11,** 855-64.

Vårum, K. M., Ottoy, M.H. and Smidsrød, O. (1994). Water-solubility of partially *N*-acetylated chitosans as a function of pH - effect of chemical-composition and depolymerization. *Carbohydrate polymers* **25,** 65-70.

Vårum, K.M., Holme, H.K., Izume, M., Stokke, B.T. and Smidsrød, O. (1996). Determination of enzymatic hydrolysis specificity of partially N-acetylated chitosans. *Biochimica et Biophysica Acta* **1291,** 5-15.

Verburg, J.G., Rangwala, S.H., Samac, D.A., Luckow, V.A. and Huynh, Q.K. (1993). Examination of the role of tyrosine-174 in the catalytic mechanism of the *Arabidopsis thaliana* chitinase: comparison of variant chitinases generated by site-directed mutagenesis and expressed in insect cells using baculovirus vectors. *Archives of Biochemistry and Biophysics* **300,** 223-30.

Vincent, F., Charnock, S.J., Verschueren, K.H.G., Turkenburg, J.P., Scott, D.J., Offen, W.A., Roberts, S., Pell, G., Gilbert, H.J., Davies, G.J. and Brannigan, J.A. (2003). Multifunctional xylooligosaccharide/cephalosporin C deacetylase revealed by the hexameric structure of the *Bacillus subtilis* enzyme at 1.9 angstrom resolution. *Journal of Molecular Biology* **330,** 593-606.

Viterbo, A., Haran, S., Friesem, D., Ramot, O. and Chet, I. (2001). Antifungal activity of a novel endochitinase gene (chit36) from *Trichoderma harzianum* Rifai TM. *FEMS Microbiology Letters* **200**, 169-74.

Viterbo, A., Montero, M., Ramot, O., Friesem, D., Monte, E., Llobell, A. and Chet, I. (2002). Expression regulation of the endochitinase chit36 from *Trichoderma asperellum* (*T. harzianum* T-203). *Current Genetics* **42**, 114-22.

Vocadlo, D. J., Davies, G.J., Laine, R. and Withers, S.G. (2001). Catalysis by hen egg-white lysozyme proceeds via a covalent intermediate. *Nature* **412,** 835-8.

Vocadlo, D. J. and Withers, S.G. (2005). Detailed comparative analysis of the catalytic mechanisms of beta-*N*-acetylglucosaminidases from families 3 and 20 of glycoside hydrolases. *Biochemistry* **44,** 12809-18.

Vollmer, W. and Tomasz, A. (2000). The pgdA gene encodes for a peptidoglycan *N*-acetylglucosamine deacetylase in *Streptococcus pneumoniae*. *Journal of Biological Chemistry* **275,** 20496-501.

Watanabe, T., Ito, Y., Yamada, T., Hashimoto, M., Sekine, S. and Tanaka, H. (1994). The roles of the C-terminal domain and type III domains of chitinase A1 from *Bacillus circulans* WL-12 in chitin degradation. *Journal of Bacteriology* **176**, 4465-72.

Watanabe, T., Ariga, Y., Sato, U., Toratani, T., Hashimoto, M., Nikaidou, N., Kezuka, Y., Nonaka, T. and Sugiyama, J. (2003). Aromatic residues within the substrate-binding cleft of *Bacillus circulans* chitinase A1 are essential for hydrolysis of crystalline chitin. *Biochemical Journal* **376,** 237-44.

Watanabe, T., Kobori, K., Miyashita, K., Fujii, T., Sakai, H., Uchida, M. and Tanaka, H. (1993). Identification of glutamic acid 204 and aspartic acid 200 in chitinase A1 of *Bacillus circulans* Wl12 as essential residues for chitinase activity. *Journal of Biological Chemistry* **268,** 18567-72.

Woo, S. L., Donzelli, B., Scala, F., Mach, R., Harman, G.E., Kubicek, C.P., Del Sorbo, G. and Lorito, M. (1999). Disruption of the ech42 (endochitinase-encoding) gene affects biocontrol activity in *Trichoderma harzianum* P1. *Molecular Plant Microbe Interaction* **12,** 419-29.

Wu, H. G., Yao, Z., Bal, X.F., Du, Y.G. and Lin, B.C. (2008). Anti-angiogenic activities of chitooligosaccharides. *Carbohydrate Polymers* **73,** 105-10.

Xia, W., Liu, P., and Liu, J. (2008). Advance in chitosan hydrolysis by non-specific cellulases. *Bioresource Technology* **99,** 6751–62.

Xu, J. G., Zhao, X.M., Han, X.W. and Du, Y.G. (2007). Antifungal activity of oligochitosan against *Phytophthora capsici* and other plant pathogenic fungi in vitro. *Pesticide Biochemistry and Physiology* **87,** 220-8.

Zakariassen, H., Aam, B.B., Horn, S.J. , Vårum, K.M., Sorlie, M. and Eijsink, V.G.H. (2009). Aromatic residues in the catalytic center of chitinase A from *Serratia marcescens* affect processivity, enzyme activity, and biomass converting efficiency. *Journal of Biological Chemistry* **284,** 10610-7.

Biotechnology and Genetic Engineering Reviews - Vol. 27, 367-382 (2010)

Usefulness of Kinetic Enzyme Parameters in Biotechnological Practice

NÉSTOR CARRILLO[1], EDUARDO A. CECCARELLI[1] AND OSCAR A. ROVERI[2*]

[1]Instituto de Biología Molecular y Celular de Rosario (IBR) – CONICET and [2]Área Biofísica, Facultad de Ciencias Bioquímicas y Farmacéuticas, Universidad Nacional de Rosario, Suipacha 531, (S2002LRK) Rosario, Argentina

Abstract

The k_{cat}/K_M ratio, where k_{cat} is the catalytic constant for the conversion of substrate into product, and K_M is the Michaelis constant, has been widely used as a measure of enzyme performance, but recent analyses have underscored the inadequacy of this ratio to describe the efficiency of a biocatalyst, particularly when employed as a criterion for selecting between enzyme variants for industrial purposes. The main problem with this kinetic relationship is that it neglects the contribution of important factors operating in actual bioprocess conditions, such as substrate concentrations and product inhibition, leading to unreal expectations on enzyme performance and erroneous selection of the most adequate biocatalyst. Two complementary formalisms, the efficiency function and the catalytic effectiveness, have been introduced to incorporate important features of any bio-industrial system. We review herein the rationales underlying each derivation, and the strengths and fields of application of both strategies. Examples of different situations, including continuous and batch-type reactors, as well as reversible and irreversible processes, are provided, together with recommendations on the use of both approaches.

Introduction

The use of enzymes as biocatalysts has gained momentum in recent years, playing increasingly important roles in chemical, pharmaceutical and bio-industrial

*To whom correspondence may be addressed (oroveri@fbioyf.unr.edu.ar)

Abbreviations: FNR, ferredoxin-$NADP^+$ reductase; M1, M2, and M3, mutant enzymes; WT, wild-type enzyme; Y308S, ferredoxin-$NADP^+$ reductase site-directed mutant.

manufacturing and in the development of green chemistries friendly to the environment (Pollard and Woodley, 2007; Schoemaker *et al.*, 2003; Thayer, 2006). This success has encouraged research aimed at the improvement of enzymatic properties and activities to suit the needs of industrial use. Until the 1970's, only those enzymes found in nature had been used as commercial biocatalysts (Fox and Huisman, 2008). Unfortunately, natural enzymes are shaped by evolution to meet physiological demands in their original organisms, and are generally ill-suited for the conditions employed in industrial processes. Accordingly, biocatalysts often need to be optimised or even developed *de novo* to function in non-natural environments (including high substrate and/or product concentrations) and, to withstand wide fluctuations in process conditions, the exposure to solvents and non-physiological temperatures or pH values (Rubin-Pitel and Zhao, 2006).

With the advent of site-directed mutagenesis, the possibility of replacing amino acids in protein structures almost at will, became a reality. However, decisions concerning which particular residues are worth being mutagenised for a particular purpose, require an intimate knowledge of the relationship between enzyme structure and function. Despite the great advances made in recent years on the development of computational methods and strategies for the rational design of proteins, this goal has not yet been achieved (Park *et al.*, 2006; Robertson and Scott, 2007).

Given the promising but presently limited possibilities of purely rational approaches to enzyme optimisation (Fox and Huisman, 2008), non-rational or semi-rational strategies have been pursued with vigour. Using non-specific chemical, physical or biological mutagens, any position of the sequence could, in principle, be altered at random without previous knowledge of the role played by individual residues. In this approach, enzyme variants are generated by iterative rounds of random mutagenesis in which the best sequences are identified at each step and used as templates for additional mutagenesis. However, due to the fact that most variants are produced through point mutations, generation of diversity by this strategy is rather limited (see Giver and Arnold, 1998 for a review on this).

In 1994, Stemmer took advantage of the superior efficiency of sexual recombination to increase diversity by introducing the method of gene shuffling, virtually creating the new field of directed evolution (Stemmer, 1994). Since then, significant efforts have been devoted to develop new ways of creating combinatorial libraries (Wong *et al.*, 2007; Yuan *et al.*, 2005). Although these *in vitro* recombination methods are extraordinarily efficient compared with purely random strategies, they are not based on previous knowledge of the protein structure-function relationships and are essentially non-rational in nature. In these approaches, the challenge is to obtain a highly improbable outcome through a blind, evolutionary process, as reflected by the known aphorism of R. A. Fisher, co-founder of neo-Darwinian synthesis: "... selection is a mechanism for generating an exceedingly high degree of improbability" (see Edwards, 2000). In recent years, these tasks have been made less "blind" by the introduction of additional methods and computational aids. For instance, semi-synthetic shuffling (Stutzman-Engwall *et al.*, 2005) and machine-learning guided evolution (Fox and Huisman, 2008) have improved the process of artificial selection by providing more efficient strategies to identify and recombine beneficial diversity.

Despite these advances, the key factor for success in a directed evolution programme still relies on the design of the screen which should resemble the required application.

Subsequently, the strategy serves as fitness functions for search algorithms used to sift through the myriad of possible variants generated. Screen strategies need to consider several specific properties depending on the entail enzyme operating conditions such as catalytic performance, thermal stability, temperature and pH optima, and tolerance to solvents or inactivating by-products. The question then arises as to whether it is possible to describe these parameters in quantitative terms, in a way that is meaningful for the manufacturing process.

The specificity constant k_{cat}/K_M has been frequently, and sometimes misguidedly, used as a measure of catalytic performance (Koshland, 2002), although its application to industrial biocatalysts has been recently criticised (Ceccarelli *et al.*, 2008; Eisenthal *et al.*, 2007; Fox and Clay, 2009). In this article, we review recent contributions to solve these questions which provide general formalisms to calculate enzymatic parameters that could be applied to real industrial conditions.

Use and misuse of the k_{cat}/K_M ratio

Enzyme catalysis is generally described by the relevant parameters k_{cat}, the catalytic constant for the conversion of substrate into product, and K_M, the Michaelis constant, which provides an estimation of the affinity of the enzyme for a substrate and is operationally defined as the substrate concentration at which initial rates are half of the maximum velocity. The ratio between these two parameters, k_{cat}/K_M, usually referred to as the specificity constant, is in turn a useful indicator of the relative efficiency of an enzyme acting simultaneously on two competing substrates, A and B, whose K_M values equal the K_i for the other reaction. Assuming the steady-state condition and using different approaches, Fersht (1999) and Cornish-Bowden (2002) derived a formalism describing the kinetics of an enzyme able to catalyse the conversion of two substrates, A and B, at rates v_A and v_B

$$\frac{v_A}{v_B} = \frac{\left(k_{cat}^A / K_M^A\right)[A]}{\left(k_{cat}^B / K_M^B\right)[B]}. \tag{1}$$

Kinetic parameters can be determined from measurements carried out with each substrate separately, and Equation 1 thus used to compare enzyme velocities at any concentrations of the competing substrates. It is worth noting, however, that when the performance of an enzyme has to be predicted under a real situation, namely, in a reactor or transgenic organism where alternative substrates could be present at widely different concentrations, comparison of the k_{cat}/K_M parameters will not provide, *per se*, an indication of which of the possible reactions will be catalysed at the higher rate, because substrate concentrations also contribute to the v_A/v_B relationship (Equation 1). We will elaborate further on this issue at the end of this section.

A far more serious misuse of the specificity constant has been commonplace in the scientific literature for at least the last two decades. The k_{cat}/K_M ratio has been regarded as a proxy for "catalytic efficiency" and employed to compare *different* enzymes acting on the same substrate. This trend is especially problematic in the description of the properties of mutated and/or engineered proteins. Although the limitations and pitfalls of this approach had been addressed before (see, for instance, Carrillo and Ceccarelli, 2003), Eisenthal *et al.* (2007) were the first to analyse in detail the inadequacy of using k_{cat}/K_M to compare the efficiencies of two enzymes catalysing the same reaction. Those authors made their case compelling by showing that the ratio of velocities displayed by two different enzyme forms having identical k_{cat}/K_M values for the same substrate will depend, in most cases, on the substrate concentration $[S]$ at which the biocatalysts are assayed (Eisenthal *et al.*, 2007). Consider, for instance, a wild-type (WT) enzyme that catalyses a single-substrate reaction with $k_{cat} = 50\ s^{-1}$ and $K_M = 10\ \mu M$ ($k_{cat}/K_M = 5\ s^{-1}\ \mu M^{-1}$), and three mutants with altered kinetic parameters (*Table 1*). M1 and M2 have the same or moderately increased k_{cat}/K_M values, but the maximum velocities are increased 10- and 20-fold, respectively (*Table 1*). The M3 mutant appears to be, at first inspection, the most improved enzyme version, since it displays a $k_{cat}/K_M = 50\ s^{-1}\ \mu M^{-1}$ which stems entirely from a 10-fold decrease of K_M relative to the WT (*Table 1*).

Table 1. Kinetic parameters of a chosen model wild-type (WT) enzyme and three catalytically-improved mutants (M1 to M3).

Enzyme variants	k_{cat} (s^{-1})	K_M (μM)	k_{cat}/K_M ($\mu M^{-1}\ s^{-1}$)
WT	50	10	5
M1	500	100	5
M2	1000	100	10
M3	50	1	50

Figure 1 shows rate *vs.* $[S]$ curves for the four enzyme variants, illustrating that M3 behaves as the best catalyst only at low $[S]$ (see *Figure 1B*). As the substrate concentration is raised, the velocity ratio (v_{M3} / v_{WT}) approaches unity (*Figure 2*), indicating that this enzyme form is not more active than the WT under these conditions. Despite their lower k_{cat}/K_M values, the M1 and M2 mutants are better catalysts than M3 at high $[S]$ (*Figure 1*), where the v_{Mx} / v_{M3} ratios tend to close up with $k_{cat}^{Mx} / k_{cat}^{M3}$. The rate-substrate curves of M1 and M2 cross that of M3 at a defined substrate concentration $[S]_c$ (*Figure 1B*), at which the v_{Mx} / v_{M3} ratios become unity. The value of $[S]_c$ is a function of the kinetic parameters of the two enzyme considered. For the case of the M2 and M3 mutants, for instance, $[S]_c$ is given by the following expression (Eisenthal *et al.*, 2007):

$$[S]_c = \frac{k_{cat}^{M2} K_M^{M3} - k_{cat}^{M3} K_M^{M2}}{k_{cat}^{M3} - k_{cat}^{M2}} \tag{2}$$

The ratio between the rates of M2 and M3 (v_{M2} / v_{M3}) will vary from $\frac{k_{cat}^{M2} / K_M^{M2}}{k_{cat}^{M3} / K_M^{M3}}$ at $[S] / K_M$ close to zero to $k_{cat}^{M2} / k_{cat}^{M3}$ as $[S] / K_M$ approaches infinity. Only when the K_M values of the two enzyme variants are identical, as in the case of the M1 and M2 mutants, the velocity ratio is constant and equal to $k_{cat}^{M2} / k_{cat}^{M1}$ (*Figure 2B*).

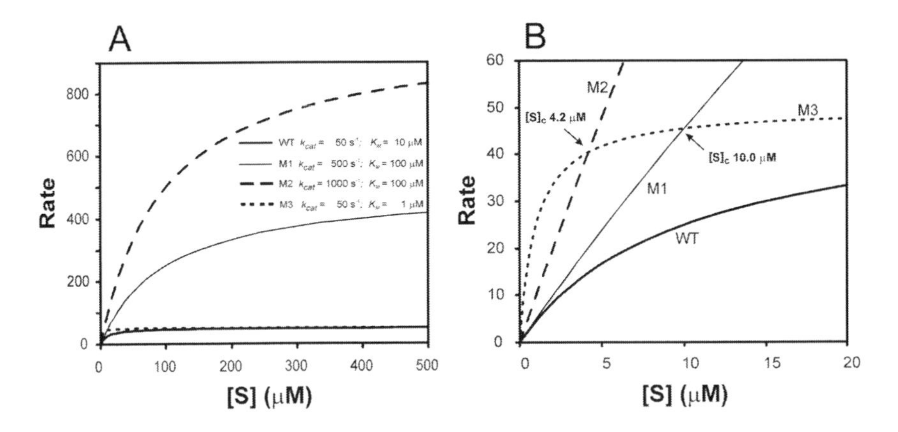

Figure 1. Rate-substrate plots for a chosen wild-type (WT) enzyme and three mutants M1-M3. Curves were generated using the Michaelis-Menten equation and the kinetic parameters of *Table 1*. Panel *B* shows an amplification of the initial part of the profiles. The arrows in panel *B* indicate the $[S]_c$ values for the corresponding pair of enzymes (M1/M3 and M2/M3).

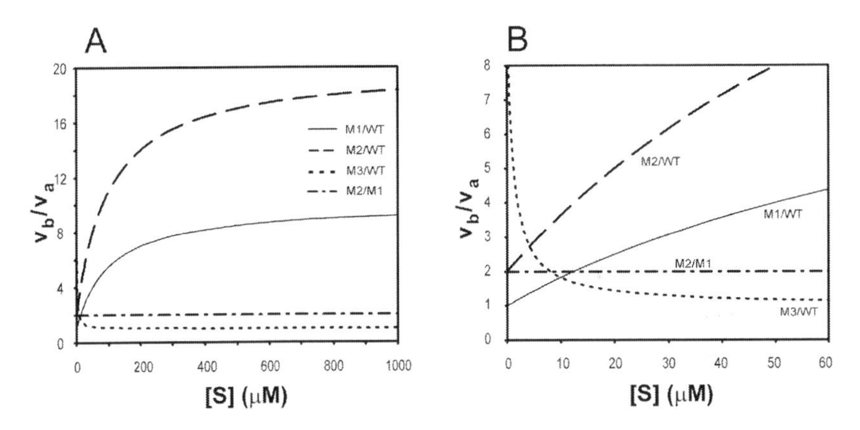

Figure 2. Velocity ratios of selected reactions as a function of the substrate concentration. Activities and conditions for each enzyme species are those of *Figure 1*. Panel *B* shows an amplification of the initial part of the curves.

Figure 3 shows a different example taken from the literature and intended to illustrate that, even when the k_{cat}/K_M ratio is used as a true specificity constant to discriminate between alternative substrates, it must be handled with caution in a real situation. Ferredoxin-NADP$^+$ reductase (FNR) is an ubiquitous flavoenzyme found in animals, plants and bacteria that catalyses NAD(P)H oxidation by a number of electron acceptors (Carrillo and Ceccarelli, 2003). WT FNR displays a strong preference for NADPH against NADH as substrate, but a site-directed mutant (Y308S) in which the C-terminal tyrosine 308 was replaced by a serine residue, exhibited major changes in pyridine nucleotide discrimination (Piubelli *et al.*, 2000). The k_{cat}/K_M value for NADH increased from nearly zero in the WT enzyme to about 1 μM^{-1} s^{-1} in the Y308S FNR, although it was ≅ 75-fold lower than that of NADPH, suggesting that NADPH was still the preferred substrate for the Y308S mutant. Although comparison of k_{cat}/K_M parameters has generally been regarded as a valid estimation for enzyme specificity (Cornish-Bowden, 2002; Fersht, 1999), rate *vs.* [*S*] plots indicate that only at [*S*] below ≅ 36 μM the v_{NADPH}/v_{NADH} ratio is actually higher than unity (*Figure 3B*). Above this threshold concentration the enzyme-catalysed reaction is faster with NADH.

These examples illustrate that the k_{cat}/K_M ratio alone is meaningless for estimation of the efficiencies of different enzymes or of a single enzyme acting on different substrates, and that the actual [*S*] at which an enzyme will operate in an organism or reactor has to be taken into consideration for any useful comparative analysis (Eisenthal *et al.*, 2007).

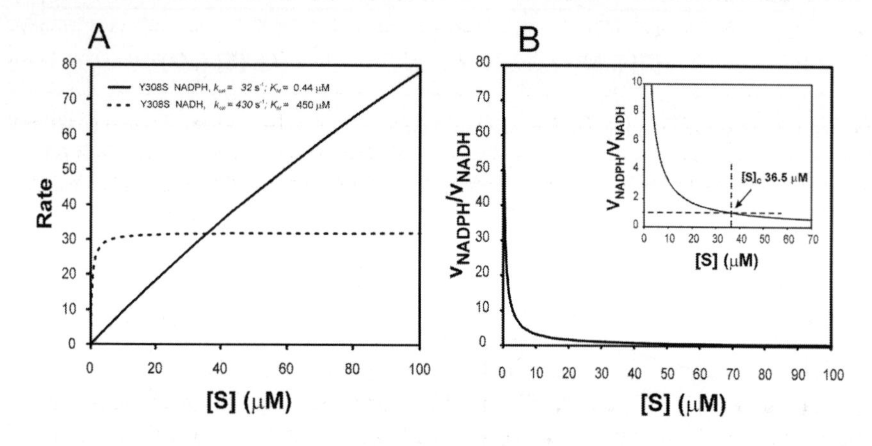

Figure 3. Rate-substrate plots for an enzyme acting on two different substrates. Kinetic parameters for NADPH or NADH oxidation catalysed by pea Y308S ferredoxin-NADP+ reductase are indicated in panel *A* and were taken from Piubelli *et al.* (2000). Panel *B* shows the ratio of rates for the two pyridine nucleotides as a function of substrate concentration. The initial part of the curve is amplified in the inset, with the arrow indicating the $[S]_c$ value at which the two velocities become equal.

Catalytic effectiveness

Fox and Clay (2009) followed a different approach to address the same issues. Instead of using the steady-state approximation to compare initial velocities, they integrated the rate equation with respect to [*S*] over the time course, on the basis that the time required to complete the reaction is the most relevant feature of any enzyme-mediated biotechnological process. For an irreversible reaction involving the conversion of a single substrate *S* into a single product *P* (uni-uni reaction), assuming that: (*i*) $[S] = [S]_0$ at $t = 0$; (*ii*) $[P] = [S]_0 - [S]$ at all times; and (*iii*) the product is a competitive inhibitor, integration of the general velocity equation with respect to [*S*] yields the following expression (Fox and Clay, 2009):

$$t = \frac{[P] + K_M^{eff}}{V_{\max}} \tag{3}$$

where v_{max} is the maximum velocity and K_M^{eff} the effective K_M, defined by

$$K_M^{eff} = K_M \left(1 + \frac{[S]_0 - \dfrac{[P]}{\ln \dfrac{1}{1-C}}}{K_{ic}} \right) \ln \frac{1}{1-C} \tag{4}$$

where K_{ic} is the competitive product inhibition constant, and C represents the fraction of substrate conversion into product, varying between 0 and 1 (Fox and Clay, 2009). Equations 3 and 4 allow the calculation of $[P]$ at any time point of the process, and the timeframe required to reach a given value of C. Based on the time course of $[P]$, these relations are particularly suited for batch-type reactions, although alternative metrics could be obtained for substrate-fed reactors by assuming that $[S] = [S]_0$ at all times (Fox and Clay, 2009). Despite $[P]$ remaining constant in continuous reactors that operate under steady-state conditions ($d[S]/dt = d[P]/dt = 0$), the formalism can be adapted to these reactor types by integrating $[P]$ with respect to reaction time.

Those authors also defined the average velocity $\bar{v}$, and termed it the catalytic effectiveness of the enzyme which can be expressed as follows:

$$\bar{v} = \frac{[P]}{t} = \frac{V_{max}[P]}{[P]\left(1 - \dfrac{K_M}{K_{ic}}\right) + K_M\left(1 + \dfrac{[S]_0}{K_{ic}}\right) \ln \dfrac{1}{1-C}} \tag{5}$$

Then, the catalytic effectiveness is a function of the kinetic parameters, the initial substrate concentration and the fraction of conversion C. Plots of $[P]$ or C *vs.* time are useful to compare enzyme variants with different k_{cat}/K_M values (see *Figure 4*). For obvious economic reasons, in most instances of industrial application the substrate is fed into the reactor at high concentrations, well above the K_M (Fox and Clay, 2009). Using the same examples provided in the previous section (see also *Table 1*), *Figure 4A* shows that although the k_{cat}/K_M of the M3 mutant is 5-fold higher than that of the M2 variant and could be mistakenly classed on this basis as a superior biocatalyst, it actually takes almost 14-times as long to reach 99% completion. Moreover, the average velocity of M3 ($\bar{v} = 50$ μM s^{-1}), as determined by the slope of the $[P]$ *vs.* time plot (*Figure 4A*), is much lower than those of M1 ($\bar{v} = 435$ μM s^{-1}) or M2 ($\bar{v} = 893$ μM s^{-1}). In fact, the M3 mutant does not significantly differ from the WT (49 μM s^{-1}) in either catalytic effectiveness or time to completion (*Figure 4A*), despite having a 10-fold higher k_{cat}/K_M ratio (*Table 1*).

One of the most important factors to be considered when screening for the most efficient enzyme, with an even less intuitive prognosis, is product inhibition. This is particularly relevant for batch-type reactions where $[P]$ increases continuously during the course of the reaction. *Figure 4B* shows the same plots of *Figure 4A* but assuming product inhibition with $K_{ic} = 10$ μM. In this case, the enzyme variants that start faster (M2 and M1) find difficulties in completing the reaction because they spend a large fraction of time in the product-inhibited region. Surprisingly, M3 now becomes the most efficient enzyme in terms of time to completion, outrunning the other mutants in the final part of the reaction (*Figure 4B*).

The examples of *Figure 4* indicate that screens designed to select the most fit enzyme need to consider the substrate concentration at which it is expected to operate, the potential inhibitory effects of rising product contents and the ability of the biocatalyst to finish off the reaction (Fox and Clay, 2009).

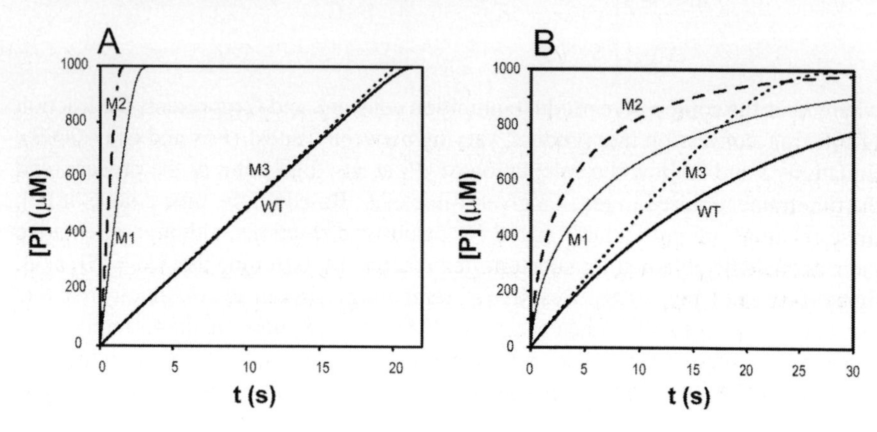

Figure 4. Time course of product formation for a chosen wild-type (WT) enzyme and three mutants M1-M3. Examples are those of *Figure 1* and *Table 1*, at $[S]$ = 1000 μM. (*A*) Neither enzyme is product-inhibited, $K_{ic} \to \infty$. (*B*) All enzymes undergo competitive product inhibition with K_{ic} = 10 μM.

The efficiency function

An efficient enzyme is the one that catalyses a high flux of substrate to product. For a uni-uni reaction, this can be expressed by the following equation (Albery and Knowles, 1976):

$$y = [E]_0 \frac{1 - \frac{[P]}{K_{eq}[S]}}{v_{net}} \tag{6}$$

where K_{eq} is the overall equilibrium constant, v_{net} is the net flux from S to P ($v_{net} = \vec{v} - \overleftarrow{v}$), where $\to$ and $\leftarrow$ indicate the forward and reverse directions respectively and *y* is the reciprocal of the specific velocity of the reaction in the forward direction ($\vec{v}$). The rate equation, derived using the steady-state approximation, is:

$$v_{net} = [E]_0 \frac{\vec{k}_{cat}\left([S] - \frac{[P]}{K_{eq}}\right)}{\vec{K}_M\left(1 + \frac{[P]}{\overleftarrow{K}_M}\right) + [S]} \tag{7}$$

where $\vec{k}_{cat}$ is the turnover number for the forward ($S \to P$) reaction, $\vec{K}_M$ and $\overleftarrow{K}_M$ the Michaelis constants for *S* (forward reaction) and *P* (reverse reaction), respectively.

Equation 6 can be written in terms of kinetic constants (Albery and Knowles, 1976):

$$y = \frac{\overrightarrow{K}_M}{\overrightarrow{k}_{cat}} \frac{1}{[S]} + \frac{1}{\overrightarrow{k}_{cat}} + \theta \frac{1}{\overleftarrow{k}_{cat}} \tag{8}$$

$\overleftarrow{k}_{cat}$ is the turnover number for the reverse ($P \rightarrow S$) reaction and θ a factor that accounts for the "reversibility" of the process, being zero for the "irreversible" and unity for the "fully reversible" cases (Albery and Knowles, 1976), respectively (see below).

The maximal theoretical catalytic efficiency achievable will be limited by the rate of diffusion of substrates to the active site, hence -as long as a catalyst evolves to perfection- the value of y will decrease till it reaches a minimum:

$$y_{\min} = \frac{1}{k_{dif}[S]} \left(1 + \frac{1}{K_{eq}} \right) \tag{9}$$

where k_{dif} is the rate constant for a diffusion-controlled process (~ 10^9 M^{-1} s^{-1}). Accordingly, Albery and Knowles (1976) defined the Efficiency Function E_f as:

$$E_f = \frac{y_{\min}}{y} \tag{10}$$

E_f equals unity for a "perfect catalyst", whereas values smaller than one are obtained for catalysts that have not reached perfection. When E_f is applied to a uni-uni reaction (Carrillo and Ceccarelli, 2003) in the absence of product inhibition ($[P] = 0$ or $\overleftarrow{K}_M \rightarrow \infty$, see Equation 7):

$$E_f = \frac{\overrightarrow{k}_{cat} \left(1 + \frac{1}{K_{eq}} \right)}{k_{dif} \left(\overrightarrow{K}_M + [S] + \frac{\theta \overrightarrow{k}_{cat} [S]}{\overleftarrow{k}_{cat}} \right)} \tag{11}$$

IRREVERSIBLE CONDITIONS

The factor θ accounts for the "reversibility" of the reaction and equals to:

$$\theta = \frac{[P]}{K_{eq}[S]} \tag{12}$$

The value of θ approaches zero when the system is far from equilibrium, *viz* $K_{eq} \rightarrow \infty$ or $[P] \rightarrow 0$. The so-called "irreversible" condition defined by Albery and Knowles

(1976) stands for an enzyme that is rate-determining in a reaction pathway ($[P] \rightarrow 0$), hence no product inhibition is expected. If in addition $K_{eq} \rightarrow \infty$, Equation 11 is simplified as follows:

$$E_f = \frac{\vec{k}_{cat}}{k_{dif}\left(\vec{K}_M + [S]\right)} \tag{13}$$

Unlike the k_{cat}/K_M ratio, E_f is a practical catalytic efficiency that can be used to compare different enzyme forms mediating the same reaction (Ceccarelli *et al.*, 2008). *Figure 5* shows E_f *vs.* [S] plots for the WT and mutant enzymes of *Table 1*. Once again, the M3 variant with the highest k_{cat}/K_M relationship is the best catalyst only at low [S], becomes less efficient than the M2 and M3 mutants after defined substrate concentrations and finally equals the efficiency of the WT (*Figure 5B*). The value of E_f approaches $\vec{k}_{cat} / k_{dif}\,\vec{K}_M$ at $[S] << \vec{K}_M$. Half of that maximal E_f is obtained when $[S] = \vec{K}_M$.

If the enzyme catalyses an irreversible reaction ($K_{eq} \rightarrow \infty$), but competitive product inhibition is observed, the equations for v_{net}, y, and E_f are:

$$v_{net} = [E]_0 \frac{\vec{k}_{cat}[S]}{\vec{K}_M\left(1 + \frac{[P]}{K_{ic}}\right) + [S]} \tag{14}$$

$$y = \frac{\vec{K}_M}{\vec{k}_{cat}[S]}\left(1 + \frac{[P]}{K_{ic}}\right) + \frac{1}{\vec{k}_{cat}} \tag{15}$$

$$E_f = \frac{\vec{k}_{cat}}{k_{dif}\left(\vec{K}_M\left(1 + \frac{[P]}{K_{ic}}\right) + [S]\right)} \tag{16}$$

Figure 6 shows E_f *vs.* [S] plots for a mutant with a k_{cat}/K_M ratio as that of M2 but with different K_{ic} values for product inhibition. In *Figure 6A*, a fixed [P] was used for the simulation, revealing that product inhibition decreases E_f at all [S]. When [S] / K_M approaches zero, smaller values of E_f are obtained as K_{ic} decreases (stronger product inhibition):

$$\lim_{[S] \rightarrow 0} E_f = \frac{\vec{k}_{cat}}{k_{dif}\left(\vec{K}_M\left(1 + \frac{[P]}{K_{ic}}\right)\right)} \tag{17}$$

Similarly, $[S]_{0.5}$, the substrate concentration at which half-maximal E_f is obtained, is affected by product inhibition as follows:

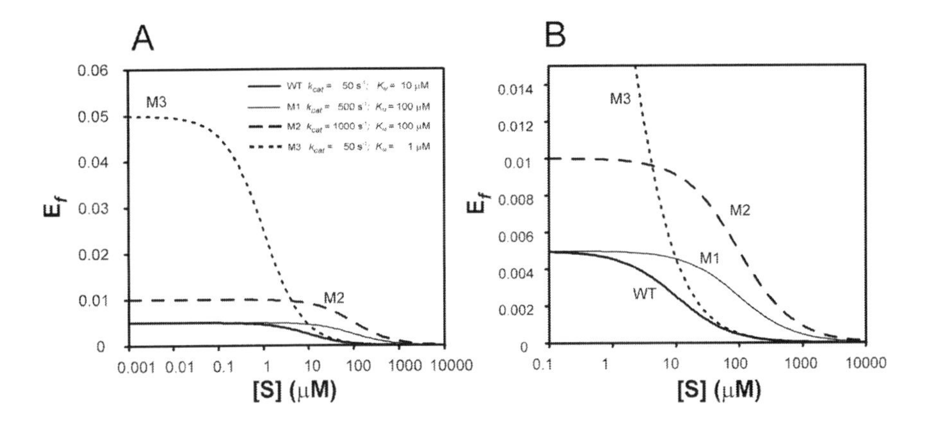

Figure 5. Variation of the efficiency function (E_f) with substrate concentration. The WT and mutant enzymes of *Figure 1* and *Table 1* are used for illustration. Panel *B* shows an amplification of the final part of the profiles.

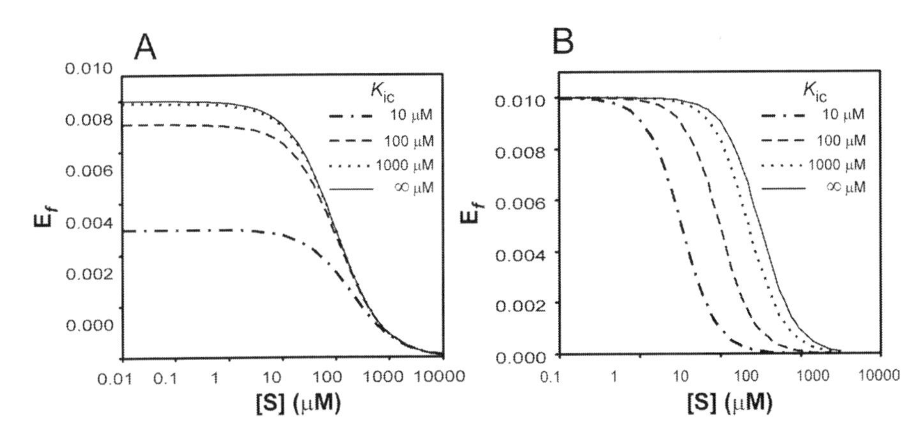

Figure 6. Variation of the efficiency function (E_f) with [*S*] in the presence of product inhibition. The M2 mutant (see *Figure 1* and *Table 1*) has been chosen as the example. (*A*) Product concentration was fixed at 10 μM, and K_{ic} varied as indicated. (*B*) The condition $[P]=\frac{C}{1-C}[S]$ (see text, *C*=0.9) was simulated for the same K_{ic} values as indicated in Panel *A*.

$$[S]_{0.5}=\vec{K}_M\left(1+\frac{[P]}{K_{ic}}\right) \tag{18}$$

An interesting situation from a biotechnological point of view is shown in *Figure 6B*, corresponding to the condition, in which there is a constant relationship between [*P*] and [*S*] ($[P]=\frac{C}{1-C}[S]$).

$$E_f=\frac{\vec{k}_{cat}}{k_{dif}\left(\vec{K}_M+\left(\frac{\vec{K}_M}{K_{ic}}\frac{C}{1-C}+1\right)[S]\right)} \tag{19}$$

This is the case a biotechnologist must deal with when analysing the behaviour of a continuous reactor at a fixed conversion fraction C. At low $[S]$, E_f is not affected by product inhibition. However, as product inhibition becomes stronger (lower K_{ic}), the decrease of E_f will occur at lower $[S]$. For this case $[S]_{0.5}$ increases with K_{ic} and $\overrightarrow{K}_M$ whereas it decreases with C:

$$[S]_{0.5} = \frac{1}{\frac{C}{(1-C)}\frac{1}{K_{ic}} + \frac{1}{\overrightarrow{K}_M}} \tag{20}$$

REVERSIBLE CONDITIONS

When factor θ, first introduced in Equation 8, is different from zero, $[P] \neq 0$ and the reaction catalysed is reversible and much faster than the uphill reaction pathway (see Albery and Knowles, 1976). This situation is representative of a continuous stirred reactor that operates at constant $[S]$ and $[P]$. Its kinetic performance can be satisfactorily characterised by E_f since the Efficiency Function compares the flux from S to P with the flux of a diffusion-controlled process at the same $[S]$. Even when $\theta = 1$, which means that $[P]/[S] = K_{eq}$ (see Equation 12), and hence $v_{net} = 0$ (see Equation 7), such flux determines the rate of product removal from the reactor. This particular case is the so-called "fully-reversible" case by Albery and Knowles (1976), and Equation 11 becomes:

$$E_f = \frac{\overrightarrow{k}_{cat}\left(1 + \frac{1}{K_{eq}}\right)}{k_{dif}\left(\overrightarrow{K}_M + [S] + \frac{\overrightarrow{k}_{cat}[S]}{\overleftarrow{k}_{cat}}\right)} \tag{21}$$

Since the Haldane relationship for a reversible uni-uni reaction is:

$$K_{eq} = \frac{\overrightarrow{k}_{cat}\overleftarrow{K}_M}{\overleftarrow{k}_{cat}\overrightarrow{K}_M} \tag{22}$$

Replacing K_{eq} in Equation 21:

$$E_f = \frac{\overrightarrow{k}_{cat}\left(1 + \frac{\overleftarrow{k}_{cat}\overrightarrow{K}_M}{\overrightarrow{k}_{cat}\overleftarrow{K}_M}\right)}{k_{dif}\left(\overrightarrow{K}_M + [S] + \frac{\overrightarrow{k}_{cat}[S]}{\overleftarrow{k}_{cat}}\right)} = \frac{\overleftarrow{k}_{cat}}{k_{dif}\overleftarrow{K}_M}\frac{\frac{\overrightarrow{k}_{cat}\overleftarrow{K}_M + \overleftarrow{k}_{cat}\overrightarrow{K}_M}{\overleftarrow{k}_{cat} + \overrightarrow{k}_{cat}}}{\left(\frac{\overrightarrow{K}_M\overleftarrow{k}_{cat}}{\overleftarrow{k}_{cat} + \overrightarrow{k}_{cat}} + [S]\right)} \tag{23}$$

As $[S]$ approaches zero, E_f equals $\left(\overrightarrow{k}_{cat}/\overrightarrow{K}_M + \overleftarrow{k}_{cat}/\overleftarrow{K}_M\right)/k_{dif}$, while $[S]_{0.5}$ is given by the following equation:

$$[S]_{0.5} = \frac{\overleftarrow{k}_{cat}\overrightarrow{K}_M}{\overleftarrow{k}_{cat} + \overrightarrow{k}_{cat}} = \frac{\overrightarrow{K}_M}{1 + \dfrac{\overrightarrow{k}_{cat}}{\overleftarrow{k}_{cat}}} \tag{24}$$

Replacing $\overrightarrow{k}_{cat}/\overleftarrow{k}_{cat}$ from Haldane relationship (Equation 22):

$$[S]_{0.5} = \frac{\overrightarrow{K}_M}{1 + K_{eq}\dfrac{\overrightarrow{K}_M}{\overleftarrow{K}_M}} = \frac{1}{\dfrac{1}{\overrightarrow{K}_M} + \dfrac{K_{eq}}{\overleftarrow{K}_M}} \tag{25}$$

Hence, $[S]_{0.5}$ increases as long as $\overrightarrow{K}_M$ and $\overleftarrow{K}_M$ decrease (*Figure 7*). However, the relative influence of both kinetic parameters depends on the equilibrium constant (K_{eq}); when $K_{eq} >> 1$, $[S]_{0.5}$ shows a stronger dependency on $\overleftarrow{K}_M$ than on $\overrightarrow{K}_M$ (*Figure 7A*), whereas the opposite occurs when $K_{eq} << 1$ (*Figure 7B*).

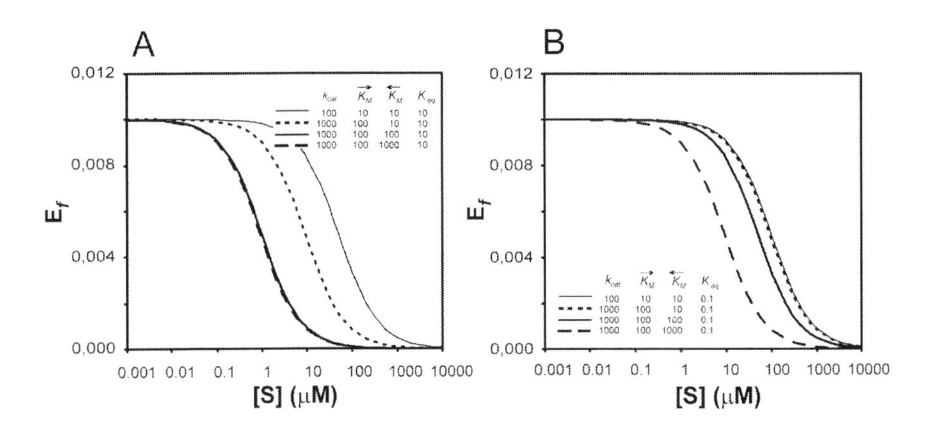

Figure 7. Variation of the efficiency function (E_f) with $[S]$ for a system under "fully reversible" (equilibrium) conditions . Four mutants have been chosen as example with kinetic parameters as indicated.

Concluding Remarks

The various analyses described in this review clearly indicate that an enzyme variant displaying a higher k_{cat}/K_M value can, at certain substrate concentrations, catalyse a reaction at lower rates than one having a lower k_{cat}/K_M relationship (*Figures 1* and *2*).

Even when comparing a single enzyme acting on two different substrates, comparison of k_{cat}/K_M parameters does not provide an unequivocal diagnosis of efficiency (*Figure 3*). In both cases, the ratio between the two velocities will strictly depend on the $[S] / K_M$ value. These examples also apply to natural systems, illustrating that enzymes evolve by maximising k_{cat}/K_M while keeping K_M at or above physiological substrate concentrations (Ceccarelli *et al.*, 2008). However, the use of k_{cat}/K_M as an efficiency aid can be particularly misleading in driving conclusions about the performance of industrial biocatalysts, which normally operate under extreme, non-natural conditions including high substrate and/or product concentrations (Fox and Clay, 2009).

Two alternative formalisms, the efficiency function (Ceccarelli *et al.*, 2008) and the catalytic effectiveness (Fox and Clay, 2009) have been introduced which overcome these limitations. Both approaches capture important features of the system that have been so far largely ignored, integrating the various colliding effects of enzyme kinetic parameters and variable substrate and product concentrations. They predict the performance of an enzyme under the real conditions of industrial manufacturing, thus providing rational clues for decision making at the level of the bioconversion process.

The two functions made use of a similar underlying principle, namely that the actual reaction rate depends not only on the value of k_{cat}/K_M but also on $[S]$, $[P]$ and their apparent dissociation constants. Although expressions can be derived for different types of reactor conditions, they are to a certain extent complementary in their potential applications. Catalytic effectiveness, as defined in Equation 5, is best suited for batch reactors because it provides a rigorous way to calculate the most important feature in this type of bioconversion processes, namely, time to completion. The E_f function, on the other hand, allows straightforward interpretation of enzyme behaviour under steady-state conditions in terms of efficiency, which has a longstanding tradition in biotechnological practice. More important, it offers the possibility of comparing the relative rates of different enzyme variants between each other and with the highest achievable one, that of a diffusion-controlled reaction (Equation 11). This information is critical during the design stage of an improvement programme to estimate how far a mutant or engineered enzyme is from "catalytic perfection" (Albery and Knowles, 1976), helping to decide whether further directed evolution or mutagenesis of a biocatalyst deserve the effort and investment involved.

Finally, it is worth noting that although all examples provided herein, for the sake of simplicity, were based on uni-uni reactions, it is possible to generalise the formalisms to include higher order processes. The resulting solutions will be more complex, but still closed-formed and amenable to mathematical analysis, providing numerical metrics to adequately compare the kinetic properties of enzyme variants (Fox and Clay, 2009).

Acknowledgements

Supported by the National Agency for the Promotion of Science and Technology (ANPCyT), the National Research Council (CONICET) and the National University of Rosario (UNR), Argentina. NC, EAC and OAR are Members of the Carrera del Investigador from CONICET.

References

ALBERY W.J., KNOWLES J.R. (1976). Evolution of enzyme function and the development of catalytic efficiency. *Biochemistry*, **15:** 5631-5640.

CARRILLO N., CECCARELLI E.A. (2003). Open questions in ferredoxin-NADP+ reductase catalytic mechanism. *European Journal of Biochemistry*, **270:** 1900-1915.

CECCARELLI E.A., CARRILLO N., ROVERI O.A. (2008). Efficiency function for comparing catalytic competence. *Trends in Biotechnology*, **26:** 117-118.

CORNISH-BOWDEN A. (2002). *Fundamentals of Enzyme Kinetics*. Portland Press: London.

EDWARDS A.W. (2000). The genetical theory of natural selection. *Genetics*, **154:** 1419-1426.

EISENTHAL R., DANSON M.J., HOUGH D.W. (2007). Catalytic efficiency and k_{cat}/K_M: a useful comparator? *Trends in Biotechnology*, **25:** 247-249.

FERSHT A. (1999). *Structure and Mechanism in Protein Science. A Guide to Enzyme Catalysis and Protein Folding*. W.H. Freeman and Co.

FOX R.J., CLAY M.D. (2009). Catalytic effectiveness, a measure of enzyme proficiency for industrial applications. *Trends in Biotechnology*, **27:** 137-140.

FOX R.J., HUISMAN G.W. (2008). Enzyme optimization: moving from blind evolution to statistical exploration of sequence-function space. *Trends in Biotechnology*, **26:** 132-138.

GIVER L., ARNOLD F.H. (1998). Combinatorial protein design by *in vitro* recombination. *Current Opinion in Chemical Biology*, **2:** 335-338.

KOSHLAND D.E. (2002). The application and usefulness of the ratio k_{cat}/K_M. *Bioorganic Chemistry*, **30:** 211-213.

PARK H.S., NAM S.H., LEE J.K., YOON C.N., MANNERVIK B., BENKOVIC S.J., *et al.* (2006). Design and evolution of new catalytic activity with an existing protein scaffold. *Science*, **311:** 535-538.

PIUBELLI L., ALIVERTI A., ARAKAKI A.K., CARRILLO N., CECCARELLI E.A., KARPLUS P.A., *et al.* (2000). Competition between C-terminal tyrosine and nicotinamide modulates pyridine nucleotide affinity and specificity in plant ferredoxin-NADP(+) reductase. *Journal of Biological Chemistry*, **275:** 10472-10476.

POLLARD D.J., WOODLEY J.M. (2007). Biocatalysis for pharmaceutical intermediates: the future is now. *Trends in Biotechnology*, **25:** 66-73.

ROBERTSON M.P., SCOTT W.G. (2007). Biochemistry: designer enzymes. *Nature*, **448:** 757-758.

RUBIN-PITEL S.B., ZHAO H. (2006). Recent advances in biocatalysis by directed enzyme evolution. *Combinatorial Chemistry & High Throughput Screening*, **9:** 247-257.

SCHOEMAKER H.E., MINK D., WUBBOLTS M.G. (2003). Dispelling the myths-biocatalysis in industrial synthesis. *Science*, **299:** 1694-1697.

STEMMER W.P. (1994). Rapid evolution of a protein *in vitro* by DNA shuffling. *Nature*, **370:** 389-391.

STUTZMAN-ENGWALL K., CONLON S., FEDECHKO R., MCARTHUR H., PEKRUN K., JENNE S., *et al.* (2005). Semi-synthetic DNA shuffling of *aveC* leads to improved industrial scale production of doramectin by *Streptomyces avermitilis*. *Metabolic Engineering*, **7:** 27-37.

THAYER A.M. (2006). Custom Chemicals. *Chemical and Engineering News*, **84:** 17-24.

WONG T.S., ROCCATANO D., SCHWANEBERG U. (2007). Steering directed protein evolution: strategies to manage combinatorial complexity of mutant libraries. *Environmental Microbiology*, **9:** 2645-2659.

YUAN L., KUREK I., ENGLISH J., KEENAN R. (2005). Laboratory-directed protein evolution. *Microbiology and Molecular Biology Reviews*, **69:** 373-392.

Index